Eukaryotic mRNA Processing

Frontiers in Molecular Biology

SERIES EDITORS

B. D. Hames

Department of Biochemistry and Molecular Biology
University of Leeds, Leeds LS2 9JT, UK

AND

D. M. Glover

Cancer Research Campaign Laboratories
Department of Anatomy and Physiology
University of Dundee, Dundee DD1 4HN, UK

TITLES IN THE SERIES

---

1. Human Retroviruses
Bryan R. Cullen

2. Steroid Hormone Action
Malcolm G. Parker

3. Mechanisms of Protein Folding
Roger H. Pain

4. Molecular Glycobiology
Minoru Fukuda and Ole Hindsgaul

5. Protein Kinases
Jim Woodgett

6. RNA–Protein Interactions
Kyoshi Nagai and Iain W. Mattaj

7. DNA–Protein: Structural Interactions
David M. J. Lilley

8. Mobile Genetic Elements
David J. Sherratt

9. Chromatin Structure and Gene Expression
Sarah C. R. Elgin

10. Cell Cycle Control
Chris Hutchinson and D. M. Glover

11. Molecular Immunology (Second Edition)
B. David Hames and David M. Glover

12. Eukaryotic Gene Transcription
Stephen Goodbourn

13. Molecular Biology of Parasitic Protozoa
Deborah F. Smith and Marilyn Parsons

14. Molecular Genetics of Photosynthesis
Bertil Andersson, A. Hugh Salter, and James Barber

15. Eukaryotic DNA Replication
J. Julian Blow

16. Protein Targeting
Stella M. Hurtley

17. Eukaryotic mRNA Processing
Adrian R. Krainer

Eukaryotic mRNA Processing

EDITED BY

Adrian R. Krainer

Cold Spring Harbor Laboratory, Cold Spring Harbor. NY 11724, USA

Oxford University Press, Great Clarendon Street, Oxford OX2 6DP

Oxford New York
Athens Auckland Bangkok Bogota Bombay Buenos Aires
Calcutta Cape Town Dar es Salaam Delhi Florence Hong Kong
Istanbul Karachi Kuala Lumpur Madras Madrid Melbourne
Mexico City Nairobi Paris Singapore Taipei Tokyo Toronto

and associated companies in
Berlin Ibadan

Published in the United States
by Oxford University Press Inc., New York

A catalogue record for this book is available from the British Library

Library of Congress Cataloging in Publication Data
(Data available)

ISBN 0 19 963418 1 (Hbk)
ISBN 0 19 963417 3 (Pbk)

Typeset by Footnote Graphics, Warminster, Wilts
Printed in Great Britain by The Bath Press, Bath

Preface

The control of gene expression occurs at many levels, including transcription, RNA maturation, and translation. Eukaryotic messenger RNAs undergo a variety of complex processing reactions, which begin on nascent transcripts as soon as a few nucleotides have been synthesized by RNA polymerase II, and continue with the export of mature mRNA to the cytoplasm for translation. Further processing can take place in the cytoplasm, culminating with mRNA turnover. Each of these steps is essential for gene expression or for its proper regulation. This volume focuses on the major mRNA maturation reactions, primarily those that take place in the nucleus of eukaryotic cells. Major advances in these areas have been made in the last two decades, including the discovery of split genes and RNA splicing, of ribozymes, and of RNA editing. These discoveries were paralleled by technical advances that have allowed the biochemical and genetic dissection of the corresponding mechanisms. The characterization of these reactions is, of course, still ongoing, and much remains to be done. The aim of this volume is to provide an in-depth analysis of several of the major nuclear mRNA processing events, of the underlying enzymatic reactions, and of the structure and function of the RNA and protein machines responsible for these processing reactions. Leading researchers have provided their unique perspectives about current knowledge in these areas, and contributed comprehensive, authoritative accounts of recent and exciting developments.

The first chapter, by Alain Jacquier, reviews in detail the structure and function of the known ribozmes, their reaction mechanisms and kinetics, and the implications of RNA catalysis for the evolution and processing of nuclear mRNA introns. In the second chapter, Sui Huang and David Spector discuss mRNA processing in relation to nuclear architecture, reviewing current knowledge and ideas about the spatial localization and dynamics of the processing factors and assemblies in the context of transcription, mRNA maturation, and mRNA export across the nuclear pores. In the third chapter, Wallace LeStourgeon and co-workers discuss how nascent transcripts are packaged with hnRNP proteins, and provide an overview of the assembly process and of the RNA-binding properties of the abundant hnRNP proteins. In the fourth chapter, Robin Reed and Leon Palandjian describe the sequential *in vitro* assembly of non-specific and specific complexes on the pre-mRNA, which culminates with the formation of a catalytically active spliceosome, as well as the composition of the various complexes. In the fifth chapter, Cindy Will and Reinhard Lührmann review the composition, higher order structure, and function of each of the spliceosomal small nuclear ribonucleoprotein particles, which are essential for splice-site recognition and for splicing catalysis. In the sixth chapter, Javier Cáceres and I present an overview of mammalian proteins involved in pre-mRNA splicing, with an emphasis on the SR proteins, which are involved in diverse aspects of spliceosome assembly, splice-site selection, and splicing regulation. In the seventh chapter, Peter Hodges, Mary Plumpton, and Jean Beggs discuss the significant advances made in the dissection of the splicing apparatus of the budding yeast *S. cerevisiae,* in which many spliceosomal constituents and the underlying interactions have been identified through powerful genetic methods. In the eighth chapter, Jung-Chih Wang, Meena Selvakumar, and David Helfman describe the generation of protein isoform diversity by alternative splicing and the developmental or tissue-specific regulation of this process, and review current knowledge about *cis*-acting elements and *trans*-acting factors in several of the best characterized systems. In the ninth chapter, Elmar Wahle and Walter Keller discuss several pre-mRNA modifications, including 5′ end capping and internal methylation, but focusing specifically on the maturation of mRNA 3′ ends, especially by cleavage and polyadenylation. In the tenth chapter, Timothy Nilsen reviews the mechanisms of intermolecular *trans*-splicing in several systems, particularly the addition of spliced leaders to nematode and trypanosome mRNAs. In the eleventh and final chapter, Larry Simpson and Otavio Thiemann present an overview of the mechanisms of mRNA editing, focusing on an extreme example of this remarkable process, the extensive insertion/deletion type of editing found in kinetoplastid protozoa.

I wish to thank all the authors for their enthusiasm and dedication, their excellent contributions, and their patience. The best reward for their investment in this joint project will be if the scientific community and especially new students in the mRNA processing field, find this volume useful, and if it inspires new ideas and continued progress. Thanks to the quality of the individual contributions, I am confident that it will.

Cold Spring Harbor
April 1997

A. R. K.

Contents

Contributors

JEAN D. BEGGS
Institute of Cell and Molecular Biology, University of Edinburgh, Darwin Building, King's Buildings, Mayfield Road, Edinburgh EH9 3JR, UK.

JAVIER F. CÁCERES
Cold Spring Harbor Laboratory, Cold Spring Harbor, NY 11724, USA.

DAVID M. HELFMAN
Cold Spring Harbor Laboratory, Cold Spring Harbor, NY 11724, USA.

PETER E. HODGES
Institute of Cell and Molecular Biology, University of Edinburgh, Darwin Building, King's Buildings, Mayfield Road, Edinburgh EH9 3JR, UK.

MAE HUANG
Department of Biochemistry, University of Tennessee College of Medicine, Memphis, TN 38163, USA.

SUI HUANG
Cold Spring Harbor Laboratory, Cold Spring Harbor, NY 11724, USA.

SUNITA IYENGAR
Department of Molecular Biology, Vanderbilt University, Nashville, TN 37235, USA.

ALAIN JACQUIER
Laboratoíre du Métabolisme des ARN, Département des Biotechnologies, Institut Pasteur, 28 Rue du Dr. Roux, 75724 Paris Cedex 15, France.

WALTER KELLER
Biozentrum der Universität Basel, Abteilung Zellbiologie, Klingelbergstrasse 70, CH-4056 Basel, Switzerland.

ADRIAN R. KRAINER
Cold Spring Harbor Laboratory, Cold Spring Harbor, NY 11724, USA.

WALLACE M. LeSTOURGEON
Department of Molecular Biology, Vanderbilt University, Nashville, TN 37235, USA.

REINHARD LÜHRMANN
Institut für Molekularbiologie und Tumorforschung, Philipps-Universität Marburg, Emil Mannkopff Straße 2, D-35037 Marburg, Germany.

JAMES G. McAFFEE
Department of Chemistry, Pittsburgh State University, Pittsburgh, Kansas 66772, USA.

TIMOTHY W. NILSEN
Department of Molecular Biology and Microbiology, Case Western Reserve University School of Medicine, 10900 Euclid Avenue, Cleveland, Ohio 44106, USA.

MARY PLUMPTON
Wellcome Research Laboratories, Department of Cell Biology, Langley Court, South Eden Park Road, Beckenham, Kent BR3 3BS, UK.

LEON PALANDJIAN
Department of Cell Biology, Harvard Medical School, 45 Shattuck Street, Boston, MA 02115, USA.

JANE E. RECH
Department of Molecular Biology, Vanderbilt University, Nashville, TN 37235, USA.

ROBIN REED
Department of Cell Biology, Harvard Medical School, 45 Shattuck Street, Boston, MA 02115, USA.

MEENAKSHI SELVAKUMAR
Cold Spring Harbor Laboratory, Cold Spring Harbor, NY 11724, USA.

L. SIMPSON
Howard Hughes Medical Institute, Department of Molecular, Cellular, and Developmental Biology, MacDonald Research Laboratory, University of California at Los Angeles, Los Angeles, CA 90095, USA.

SYRUS SOLTANINASSAB
Department of Molecular Biology, Vanderbilt University, Nashville, TN 37235, USA.

DAVID L. SPECTOR
Cold Spring Harbor Laboratory, Cold Spring Harbor, NY 11724, USA.

OTAVIO H. THIEMANN
Howard Hughes Medical Institute, Department of Molecular, Cellular, and Developmental Biology, MacDonald Research Laboratory, University of California at Los Angeles, Los Angeles, CA 90095, USA.

ELMAR WAHLE
Institut für Biochemie, Justus-Liebig-Universität Giessen, Heinrich-Buff-Ring 58, 35392 Giessen, Germany.

YUNG-CHIH WANG
Cold Spring Harbor Laboratory, Cold Spring Harbor, NY 11724, USA.

CINDY L. WILL
Institut für Molekularbiologie und Tumorforschung, Philipps-Universität Marburg, Emil Mannkopff Straße 2, D-35037 Marburg, Germany.

Abbreviations

AMT	aminomethyltrioxsalen
ASF	alternative splicing factor
ASLV	avian sarcoma-leukosis virus
bGH	bovine growth hormone
BPS	branchpoint sequence
BR	Balbiani ring
CF1	cleavage factor 1
CGRP	calcitonin gene-related peptide
COII	cytochrome oxidase subunit II
CPE	cytoplasmic polyadenylation element
CPSF	cleavage and polyadenylation specificity factor
CStF	cleavage stimulation factor
cTNT	cardiac troponin T
CYb	cytochrome b
DCS	domain-connection sequence
DRB	5,6-dichloro-1-β-D-ribofuranosylbenzimidazole
dsx	*doublesex* gene
EBS	exon binding site
EBV	Epstein–Barr virus
ERS	exon-recognition sequence
ESE	exon-splicing enhancer
ESS	exon-splicing silencer
FISH	fluorescence *in situ* hybridization
G_β	G protein β subunit
HCMV-IE	human cytomegalovirus immediate-early transcripts
hGH-N	normal human growth hormone
hGH-V	human growth hormone variant
HIV	human immunodeficiency virus
hnRNA	heterogeneous nuclear RNA
hnRNP	heterogeneous nuclear ribonucleoprotein
HRH1	human RNA-helicase 1
IBS	intron-binding site
Ig	immunoglobulin
IVS	intervening sequence
kDNA	kinetoplast DNA
KH	hnRNP K homology
K_M	Michaelis constant
mAb	monoclonal antibody
m_3G	2,2,7-trimethylguanosine
m^7G	7-methylguanosine

mud	*mutant-u-die* gene
MURF	maxicircle unidentified reading frames
N-CAM	neural cell adhesion molecule
ND1–5	NADH dehydrogenase subunits 1 to 5
NLS	nuclear localization signal
NMR	nuclear magnetic resonance
NOR	nucleolar-organizing region
NRS	negative regulator of splicing
ORF	open reading frame
PAB	poly(A)-binding protein
PCR	polymerase chain reaction
PRP	precursor RNA processing
PSF	PTB-associated splicing factor
PSI	P-element somatic inhibitor
PTB	polypyrimidine tract binding protein
RACE-PCR	rapid amplification of cDNA ends by PCR
RNase P	ribonuclease P
RNP	ribonucleoprotein
RPS12	ribosomal protein S12
RRM	RNA-recognition motif
RS domain	arginine/serine-rich domain
RSV	Rous sarcoma virus
RT-PCR	reverse transcriptase PCR
SAP	spliceosome-associated protein
SC35	spliceosome complex protein 35 (splicing factor)
SDS-PAGE	sodium dodecyl sulphate polyacrylamide gel electrophoresis
SELEX	systematic evolution of ligands by exponential enrichment
SF2/ASF	splicing factor 2/alternative splicing factor
SL RNA	*trans*-spliced leader RNA
SLA RNA	spliced-leader-associated RNA
slu	*synergistic lethal with U5* gene
Sm	Smith autoantigen
snoRNA	small nucleolar RNA
snRNP	small nuclear RNP
SR protein	serine/arginine-rich protein
SSS	*src* suppressor of splicing
sTobRV	satellite RNA of tobacco ringspot virus
Sxl	*Sex lethal* gene
SWAP	suppressor of white apricot
TM	tropomyosin
TPR	tetratricopeptide repeat
tra	*transformer* gene
TUTase	3′ terminal uridylyl transferase
U2AF	U2 auxiliary factor
VSG	variant surface glycoprotein

1 | RNA-catalysed RNA processing

ALAIN JACQUIER

1. Introduction

RNA processing reactions, as most chemical reactions occurring in living cells, are greatly accelerated by enzymatic catalysis. Proteins were, for a long time, the only class of molecules known to support enzymatic activity. The discovery, a decade ago, that several RNA processing events could be catalysed by RNA molecules (1, 2) thus revolutionized our views in several areas of biology and, most particularly of course, in the RNA processing field. These catalytic RNA molecules have been called 'ribozymes'.

Enzymes are biological catalysts which exhibit the following criteria:

(1) they show a very high specificity toward their substrates;

(2) they greatly accelerate the rate of the reactions they act upon;

(3) they are not consumed by the reaction and are thus recycled.

Although only ribonuclease P (RNase P) among known RNA catalysts (ribozymes) meets the third criterion, the other ribozymes can be artificially engineered to sequentially process several substrate molecules *in trans* (see Section 6.2). Moreover, several protein enzymes also act stoichiometrically rather than catalytically (3). It thus appears that the recycling ability does not fundamentally distinguish enzymatic from non-enzymatic macromolecules.

This chapter will give an overview of some approaches used to study RNA catalysis as well as the current state of understanding of how ribozymes achieve high substrate specificity and rate enhancement. Finally, the possibility that RNA catalysis is also involved in the processing of nuclear pre-mRNAs by the spliceosome machinery will be discussed.

2. Known catalytic RNAs

This section will present a short overview of the proven natural ribozymes to date. Interestingly, all of them are involved in RNA processing events. As will be discussed in Section 8, however, it is likely that RNA catalysis will be demonstrated in the future not to be restricted to reactions involving RNA substrates.

2.1 Intronic catalytic RNAs

Introns are sequences that interrupt coding sequences of genes. They are transcribed along with the exons, the sequences found in the mature RNA, but are excised from the primary transcripts by a process called RNA splicing. On the basis of structural features such as nucleotide sequence conservation or potential RNA folding, introns have been classified into at least four distinct groups: group I, group II, nuclear pre-mRNA, and nuclear pre-tRNA introns. Group I and group II introns are characterized by short conserved sequences and the potential ability of their RNA to fold into complex secondary structures characteristic of each group (see Fig. 1a and b). Distinct reaction pathways distinguish the two types of introns (see Section 3.1.2 and Fig. 2). Remarkably, the splicing pathway of group II introns is identical to the splicing pathway of nuclear pre-mRNA introns, suggesting that the latter might originate from the former (see Section 7). Correlated with the presence of complex RNA structures, it was found that members of both group I and group II intron families carry the active site responsible for their own excision from the RNA (1, 4, 5). Splicing of many of these introns *in vivo*, however, has been shown to rely on the presence of either maturases encoded by intronic open reading frames (ORFs) or nuclear-encoded proteins (reviewed in 6).

Group I introns are extremely widespread. They have been found in mRNA, rRNA, and tRNA genes in the mitochondrial, chloroplast, and nuclear genomes of eukaryotes (although not in vertebrates); in bacteriophages; and in eubacterial genomes (see 7, 8). Although some phylogenetic evidence indicates that horizontal transmission might have contributed to the spreading of these elements (9), their very wide distribution is generally interpreted as a clue of their ancient origin (10). The distribution of group II introns was long thought to be restricted to eukaryotic organelles (11) but recently they have been found in the genomes of cyanobacteria and proteobacteria (believed to be respectively the ancestors of chloroplasts and mitochondria), suggesting that these introns are also of ancient origin (12).

In addition to their fascinating self-splicing abilities, group I and group II introns are associated with remarkable biological phenomena. Indeed, a number of them can achieve site-specific insertion, known as 'homing': the introns of a given gene can invade, at their insertion sites, the intronless copies of the same gene. For group I introns, this process is promoted by a highly specific DNA endonuclease encoded by the intronic ORF, which induces a gene conversion-like process. For group II introns, the duplicative site-specific transposition is mediated by a reverse transcriptase encoded by the intronic ORF (reviewed in 13).

2.2 Ribonuclease P

Ribonuclease P (RNase P) is a ribonucleoprotein (RNP) complex responsible for the endonucleolytic cleavage that removes the 5′-terminal leader sequence of all the different precursor tRNAs in the cell (see Fig. 1e). It cleaves by hydrolysis to produce 5′-phosphate and 3′-hydroxyl termini, a rare cleavage specificity among ribonucleases. RNase P activity seems universal since it is found in every cell and every eukaryotic sub-cellular compartment that synthesizes tRNAs. Moreover, the enzyme has generally been shown to contain RNA (reviewed in 14, 15). In eubacteria, the RNP consists of an RNA subunit of 350–410 nucleotides and a basic protein of about 14 kDa. In *Escherichia coli* both the protein subunit (the C5 protein) and the RNA subunit (the M1 RNA) of RNase P are essential for cell viability. However, the RNA subunit of RNase P from *Escherichia coli* or *Bacillus subtilis* was shown to be capable of catalysing the cleavage of appropriate substrates in the absence of the protein component *in vitro*, at high salt concentrations (2). This represents the only known natural example of an RNA behaving as a classical enzyme, in the sense that it naturally catalyses the cleavage reaction with a multiple turnover.

2.3 Small self-cleaving RNA domains

2.3.1 Hammerhead ribozymes

Viroids and virusoids are very small plant infectious agents or plant virus satellites composed of single-stranded circular RNA. Their replication cycle involves a rolling circle mechanism that initially generates greater-than-unit-length plus and minus (complementary) RNAs. These RNAs are specifically cleaved to produce plus or minus linear monomers, which are subsequently ligated to form circular monomers. Cleavage of multimeric forms of the plus strand of the satellite RNA of tobacco ringspot virus (sTobRV) (16) and of the plus and minus strands of the avocado sunblotch viroid (17) is an autolytic reaction, occurring *in vitro* in the absence of protein. The cleavage reaction produces 5′-hydroxyl and 2′,3′-cyclic phosphate termini, as those found *in vivo*. Comparison of the sequences of several other related plant pathogenic RNAs has revealed a common structural motif, containing the cleavage site, called the hammerhead structure (see Fig. 1c) (18). Isolation of this structural motif showed that it was sufficient to promote self-cleavage in the presence of magnesium (19). Hammerhead ribozymes have been described in a number of virus or virus satellites of plants (see, for example, 18). They are not, however, restricted to plants, since a hammerhead ribozyme has been found in the transcript of a satellite DNA from newt (20).

2.3.2 Other types of autolytic RNA structures

Self-cleavage reactions that give rise to 5′-hydroxyl and 2′,3′-cyclic phosphate termini, have been described for several RNA structures different from the hammerhead motif. They are briefly described below. The fact that only one or two examples of each type of structure are known suggests that there should be a number of other, yet undiscovered, natural RNA structures capable of promoting this kind of reaction.

The hairpin ribozyme. The minus strand of sTobRV, like the plus strand (see above), is able to undergo autolytic cleavage *in vitro* (21). No hammerhead structure, how-

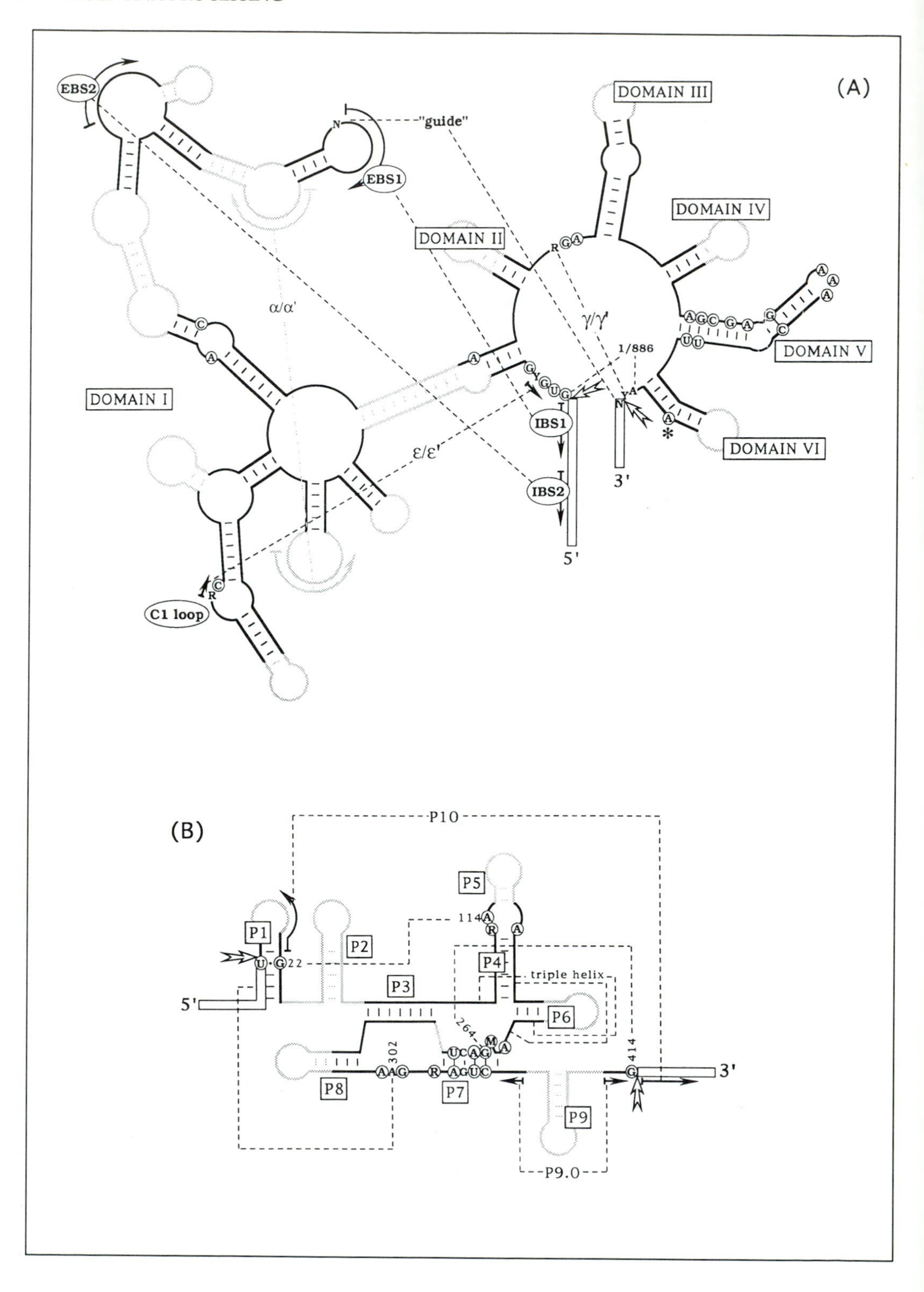
(A)
EBS2
"guide"
N
EBS1
DOMAIN III
DOMAIN IV
DOMAIN II
DOMAIN V
DOMAIN VI
DOMAIN I
α/α'
γ/γ'
ε/ε'
1/886
IBS1
IBS2
5'
3'
C1 loop
(B)
P10
P5
P1
P2
P3
P4
P6
P7
P8
P9
P9.0
triple helix
5'
3'
114
264
302
414

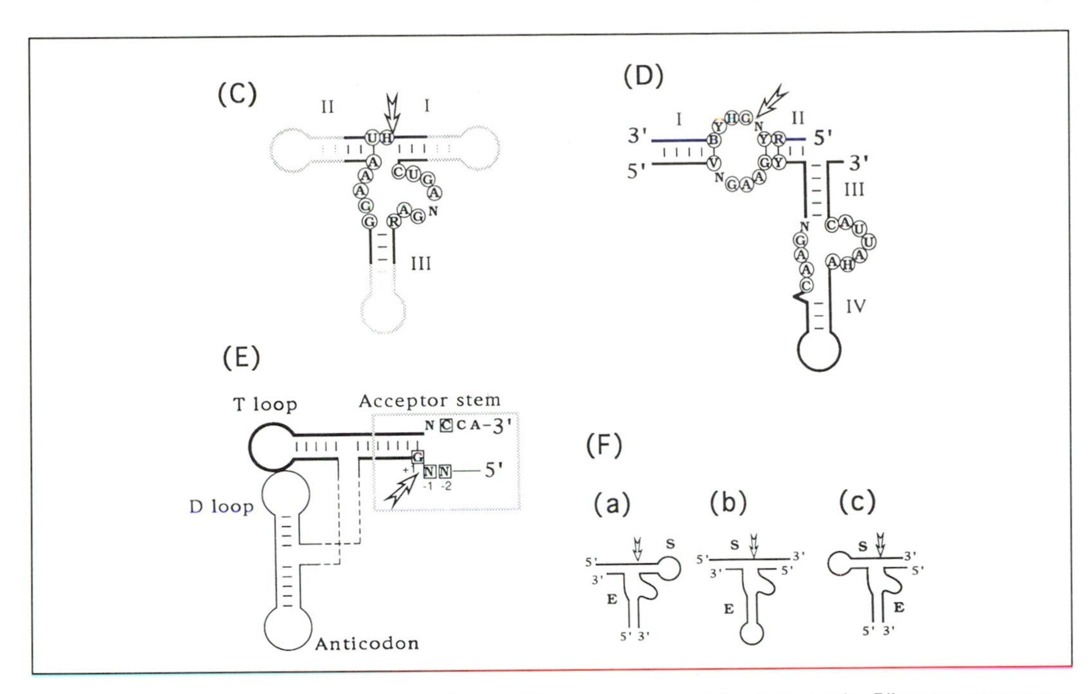

Fig. 1 (a–f) The conserved core structures of some ribozymes, shown at the same scale. Ribozyme sequences are shown as thick lines. The most conserved structures are in black, the divergent regions in grey. The most conserved nucleotides are shown circled. Y: U or C; R: A or G; M: A or C; V: A, C, or G; B: U, C, or G; H: A, C, or U. The dashed lines indicate tertiary interactions. Their regions of complementarity are shown by thin arrows. The white arrows indicate the cleavage sites. (a) Conserved structures of group II introns (subgroup IIb) as defined in (11). The guide, 1/886, and ε/ε' interactions are from (40, 51, 105), respectively (numbering is by reference to *Saccharomyces cerevisiae* ai5γ). Exon sequences are shown as thin boxes. The branch site adenosine is indicated by the asterisk. (b) Conserved structures in group I introns (as deduced from the appendix in 7). Numbering is by reference to the *Tetrahymena* rRNA intron. Not all tertiary interactions are shown (see also 7, 65). (c) Minimal structure requirement for a hammerhead ribozyme (152). (d) Minimal structure requirement for the hairpin ribozyme (24). (e) Summary of the structural features recognized by RNase P. The coaxial T and acceptor stems, the main structures recognized by RNase P in pre-tRNA, are shown as thick lines. The grey box defines the minimal structural requirement for RNase P recognition. The boxed nucleotides are those for which the 2′ hydroxyl has been shown to be important for cleavage (see text). (f) Schematic representation of the different enzyme–substrate combinations studied in hammerhead ribozymes. E: enzyme subunit; S: substrate subunit. (a) is from (153), (b) from (85), and (c) from (154).

ever, could be recognized around the cleavage site; instead, another type of structure, called the 'hairpin' ribozyme, was proposed (22). The absence of extensive phylogenetic data from natural sequences to confirm the model was overcome by the construction of an artificial library of sequence variants generated with the use of a powerful *in vitro* selection technique (see Section 5.2) (23). This approach allowed the precise definition of the minimal structure and sequence requirements of the hairpin ribozyme (see Fig. 1d) (24). One peculiarity of this catalytic RNA domain is that it efficiently catalyses the reverse reaction—the ligation of the termini generated by the cleavage reaction (see Section 3.1.1). This activity is thought to be involved in the replication cycle as well.

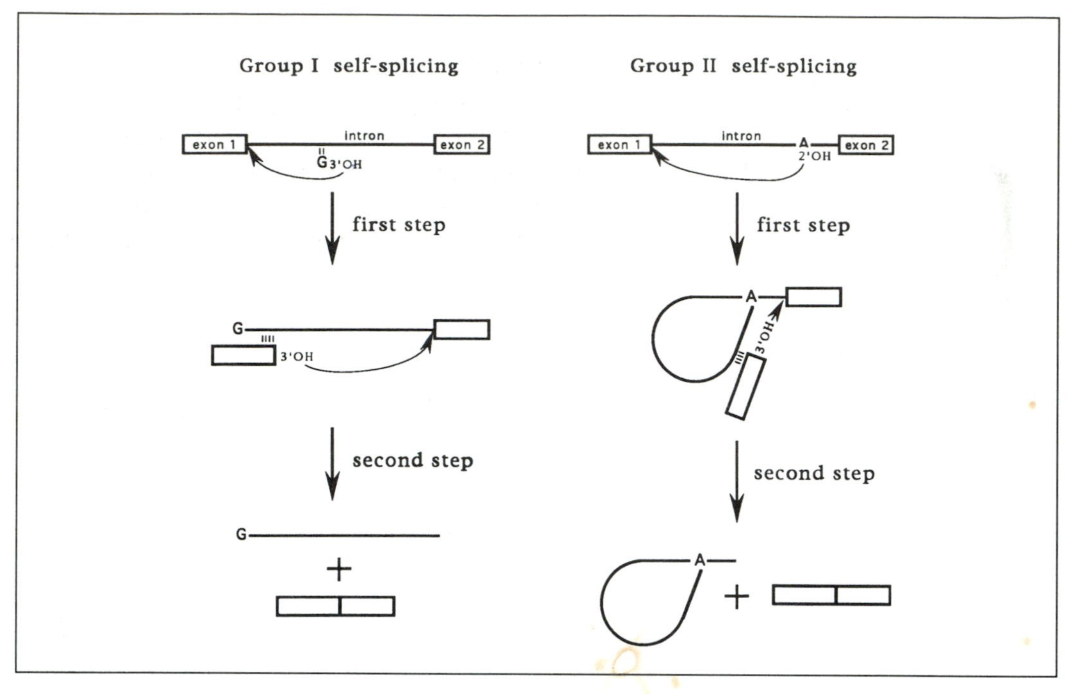

Fig. 2 Splicing pathways of group I and group II introns. Thick lines indicate introns; boxes indicate exons.

The hepatitis delta virus ribozyme. This is a satellite RNA of human hepatitis B virus. It contains an autolytic motif in both the genomic and antigenomic strands that is required for the replication cycle (25, 26). The folding of this RNA domain appears different from that of the hammerhead or the hairpin ribozymes.

The Neurospora ribozyme. The *Neurospora* VS RNA is a mitochondrial single-stranded RNA capable of self-cleavage of multimeric forms and religation. It lacks any of the previously described RNA catalytic domains (27, 28).

3. The chemistry of self-splicing or self-cleaving RNA

3.1 The reaction pathways and nature of the attacking hydroxyl groups

All naturally occurring RNA-catalysed reactions described to date can be summarized as the nucleophilic attack of a phosphodiester bond by a hydroxyl group. Two classes of nucleophilic attack giving rise to different types of products can be distinguished, depending on whether the reactive hydroxyl is adjacent to the cleaved phosphate or remote from it in the primary structure (reviewed in 29).

3.1.1 Attack by an adjacent hydroxyl group

All self-cleaving motifs, the hammerhead ribozymes (16, 17, 18, 20), the hairpin ribozyme (21), and the hepatitis delta virus RNA (25) self-cleave to give rise to 5′-hydroxyl and cyclic 2′,3′-monophosphate termini. This is considered to be the diagnostic feature of a transesterification reaction induced by the nucleophilic attack of the 3′–5′ phosphodiester by the adjacent 2′-OH (see Fig. 3a). Consistent with this model, the 2′-OH adjacent to the cleavage site was found to be absolutely

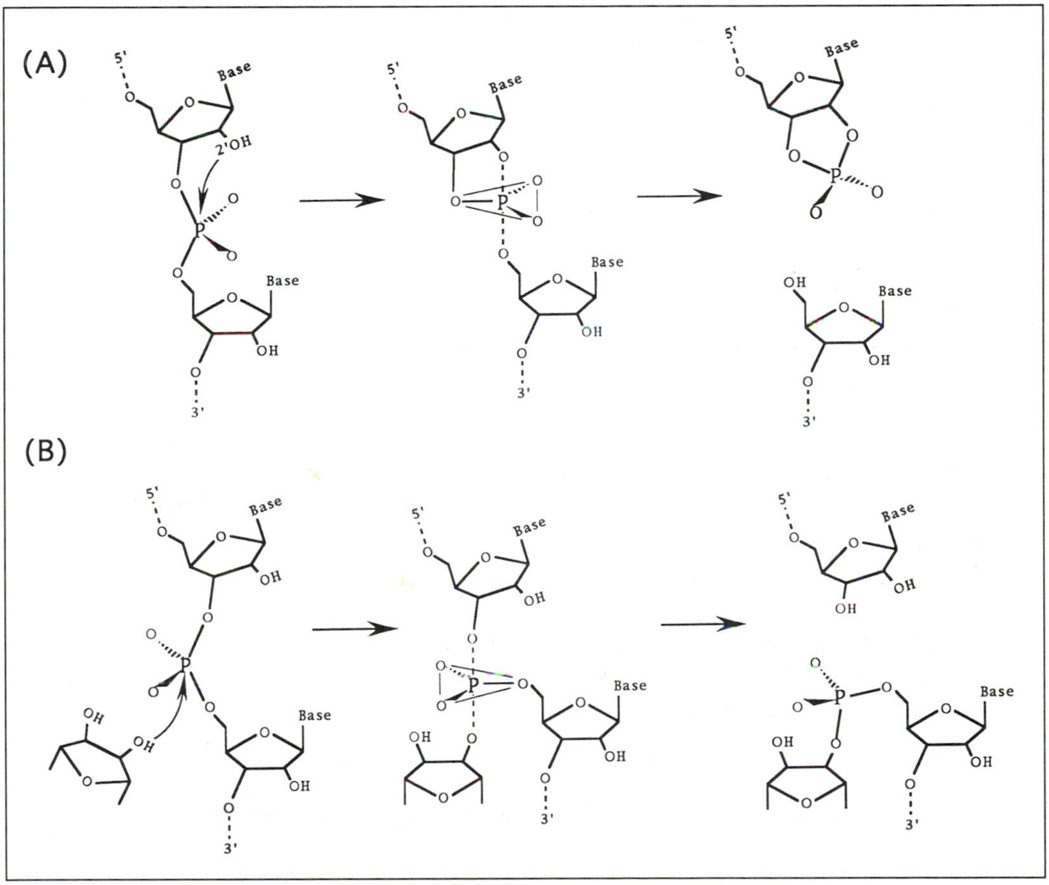

(A)

(B)

Fig. 3 Mechanisms of ribozyme phosphodiester cleavage. In each case, the reaction is induced by an in-line nucleophilic attack of the phosphate that proceeds with inversion of configuration around the phosphorus. (a) Mechanism of transesterification reaction, induced by an internal nucleophile, which gives rise to 2′,3′-cyclic phosphate and 5′-hydroxyl termini (self-cleaving RNA domains). The 2′-hydroxyl group acts as a nucleophile on the neighbouring phosphate. (b) Mechanism of transesterification induced by an external nucleophile. In the case of the first splicing step of group I introns or the second splicing step of group I and group II introns, the nucleophilic attack is induced by the 3′-OH of a free guanosine or of the liberated 5′ exon respectively, resulting in 5′,3′-phosphate linkages. In the case of the first splicing step of group II introns, the attacking nucleophile is the 2′-OH of the branch site adenosine, resulting in a 5′,2′-phosphate linkage. In the hydrolysis reaction catalysed by RNase P, the attacking hydroxyl is water, resulting in 5′-phosphate and 3′-hydroxyl termini.

required for self-cleavage of both hammerhead (30) and hairpin ribozymes (31). Also consistent with a transesterification mechanism, the reaction was found to be fully reversible for the hairpin ribozyme (21) and, with reduced efficiency, for the hammerhead (16).

3.1.2 Attack by a remote hydroxyl group

RNase P. The RNase P-catalysed reaction generates a 5′-phosphate and a 3′-hydroxyl group. Since no covalent intermediate in the reaction has ever been found, it is very likely that the nucleophilic attack is simply performed by water, possibly complexed with a magnesium ion (32).

Group I introns. Splicing reactions, which consist of the excision of an intron and the concomitant ligation of the flanking exons, require two successive transesterification reactions. In contrast with what is observed for the self-cleavage reactions, the attacking hydroxyls are provided by nucleotides located away from the cleavage sites. In group I introns, the first transesterification is initiated by the nucleophilic attack of the 3′-OH of a free guanosine on the 5′ splice site. This results in the guanosine being joined to the first nucleotide of the intron by a normal 3′,5′-phosphodiester bond and the upstream exon terminating with a free 3′-OH. This free 3′-OH then attacks the 3′ splice site, initiating the second transesterification, which is responsible for the release of the intron in a linear form with the simultaneous ligation of the exons (see Fig. 2a). As expected for transesterification reactions, which conserve the number of phosphodiester bonds, both steps have been shown to be fully reversible (33). In the absence of exogenous guanosine providing the nucleophile involved in the first transesterification, the 5′ splice site can be hydrolysed by a direct nucleophilic attack of a water molecule or a hydroxide ion (34), leaving 5′-phosphate and 3′-hydroxyl termini identical to those generated by the RNase P-induced cleavage. Likewise, in the absence of a 5′ exon providing its 3′-OH as a nucleophile, the 3′ splice site can also undergo site-specific hydrolysis (34). Kinetic analysis showed that the rate enhancement of the ribozyme-catalysed 5′ splice site hydrolysis over non-enzymatic random hydrolysis is only about 500-fold lower than the rate enhancement observed for the cleavage with guanosine (which is about 10^{11}) (35). Thus, the guanosine contributes little to the transition-state stabilization and rate enhancement provided by the ribozyme.

Group II introns. In group II introns, the two transesterification steps resemble those of group I introns, except that the hydroxyl responsible for the initiation of the first transesterification reaction is now the 2′-OH of an internal intronic adenosine, located a few nucleotides upstream from the 3′ splice site (see Figs 1a and 2). This transesterification thus results in the ligation of this adenosine to the first nucleotide of the intron by a characteristic 2′–5′-phosphodiester bond responsible for the lariat form of the released intron. The second step is analogous to that of group I introns: the 3′-OH liberated at the end of the upstream exon is now able to attack the 3′ splice site, resulting in intron release and exon ligation (see Fig. 2). As for group I introns, both group II intron transesterification steps have been shown to be fully reversible under appropriate reaction conditions (36, 37). While hydrolysis of the 3′ splice site has never been reported, hydrolysis of the 5′ splice site has been described in the presence of high potassium chloride concentrations (38) or when the 5′ splice site is provided in *trans* (39) or mutated (40). A 5′ splice site, when provided in *trans*, can be hydrolysed with a half-life of about five minutes at pH 7.5 and 45°C (40). This would correspond to a rate enhancement of at least 10^8 over the rate constant for hydrolysis giving rise to the same termini (rate constant estimated in 35). Since the rate-limiting step is unknown, it is possible that the rate enhancement calculated for the catalytic step would be even greater. Thus, although more detailed kinetic analyses remain to be done for group II introns, it appears that, as for group I introns, the rate enhancement provided by the ribozyme for 5′ splice-site cleavage involves more than the correct positioning of the attacking hydroxyl (the 2′-OH of the branch site adenosine in this case).

3.2 One or two catalytic sites for introns?

An important question raised by the occurrence of two successive transesterification reactions in the splicing pathway is whether these two steps are catalysed by a single active site, or whether there are two distinct catalytic sites, one for each reaction. For group I introns, several observations argue in favour of a single active site acting for both splicing steps. The first indication in favour of this hypothesis came simply from the nature of the reactants and products in each step: a G is added during the first step at the 5′ terminus of the intron (see Fig. 2) while a G always precedes the 3′ splice site (see Fig. 1b). Therefore, in each step, a G is located upstream of the reactive bond: in the first step, the G is located upstream of the reacted bond in the product, whereas in the second step the G is located upstream of the reactive bond in the substrate. It was thus proposed that the second step, which can be described as a G-leaving reaction, corresponds to the reversal of the first step, which constitutes a G-addition reaction (34, 41). A strong prediction of this model was that a single G binding site will be used for the attacking G in the first step and for the leaving G in the second step. Identification of the G binding site (see Section 6.3.3) (42) and alteration of its binding specificity from G to A allowed the confirmation of that prediction (43). Moreover, an opposite stereochemistry specificity towards phosphorothioates introduced at the reactive sites was observed for the first and the second splicing steps, providing further evidence in favour of a single active site catalysing successively a forward and a reverse reaction (see Section 3.3) (44–46). Finally, the analysis of tertiary interactions involved in the stabilization of the intron core led Jaeger *et al.* (47) to propose that group I introns are composed of a static, stable ribozyme, which accommodates successive substrates.

In group II introns, as in nuclear pre-mRNA introns, the nature of the substrates and products of the reactions is not consistent with a single active site performing successively a forward and a reverse reaction. In the first transesterification reaction, an incoming 2′ oxygen atom forms a 2′–5′-phosphodiester bond while the second transesterification involves an attacking 3′ oxygen. Also, the nature of the nucleotides flanking the reactive bond in the substrates or products of both reactions does not suggest any type of relationship between the two steps, in contrast to group I introns. Finally, the observed identical stereochemical specificity towards phosphorothioates introduced at the 5′ or 3′ splice site in nuclear pre-mRNAs provides further evidence against the idea that the two splicing steps represent forward and reverse directions of the same reaction (48). The existence of two different, probably overlapping active sites in the spliceosome and in group II introns was thus proposed (reviewed in 49). Alternatively, the two steps could correspond to forward reactions catalysed by a single catalytic centre associated with substrate binding sites undergoing a rearrangement at the intermediate stage of splicing in order to accommodate the different substrates of the two steps (50). In both group II (51) and nuclear pre-mRNA (52) introns, an interaction between the first nucleotide of the intron (G^1) and the 3′ junction is required for an efficient second step. In group II introns at least, this interaction does not seem to occur in the absence of branch formation. Thus, part of a binding site for the second step is formed as a product of the first step, an observation which strengthens the hypothesis of a structural rearrangement of the active sites and/or the substrate binding sites at the intermediate stage of the splicing reaction.

3.3 The reactive phosphoryl groups

The two non-bridging oxygens in phosphate diesters are not stereochemically equivalent and have been designated as pro-*R* and pro-*S*. It is possible, by replacing one or the other of these two oxygens by a sulphur, to obtain two chirally distinct phosphorothioate diastereomers, the Rp and the Sp forms. These have been used to analyse the stereochemical course of a number of enzyme mechanisms that involve phosphorus in nucleic acids. Most protein-catalysed reactions that involve nucleophilic displacement of a phosphodiester bond were found to induce an inversion of configuration of the phosphorus (53). This is usually interpreted as the diagnostic feature of an in-line attack of the incoming group relative to the leaving hydroxyl group. The phosphorus in the transition state adopts a trigonal bipyramidal geometry, with incoming and leaving groups in the axial positions (S_N2 type of nucleophilic substitution reaction) (see Fig. 3). This type of analysis was performed on group I introns for a reaction analogous to the first splicing step (54) or its reversal (45), and for the reverse of the second splicing step (46). In all cases, inversion of configuration from the Rp to the Sp isomer was observed, consistent with an in-line reaction mechanism. The same observations were made for the cleavage reaction of hammerhead and hairpin ribozymes (55). In conclusion, all the RNA-catalysed transesterification reactions for which the stereochemical course has been analysed, appear to consist of a direct in-line S_N2 nucleophilic substitution reaction (Fig. 3).

Substituting a phosphate by a thio-phosphate at the cleavage site is also interesting for the analysis of the stereospecificity of catalysis. Indeed, thio substitution in RNA has no effect on its non-enzymatic intramolecular alkaline cleavage and only a modest negative effect on non-enzymatic model chemical reactions (44). Therefore, a strong 'thio effect' (inhibition of the ribozyme-catalysed reaction) induced by the substitution of sulphur for either of the non-bridging oxygens at the cleavage site indicates that the corresponding oxygen plays an important role in catalysis. This type of analysis was particularly informative in the case of group I introns to test the hypothesis that the two splicing steps represent the forward and reverse directions of a reaction catalysed by a single active site (see Section 3.2). Indeed, since the reaction proceeds by inversion of the configuration, the pro-*R* oxygen in one direction is the pro-*S* oxygen in the other direction. Thus, a given reaction should show an inverse stereospecificity for the forward and the reverse directions. This is indeed what was observed for the group I intron-catalysed transesterification reactions. The first step is not strongly inhibited by an Rp phosphorothioate at the 5′ splice site (44), while a reaction analogous to the reverse of the first splicing step is inhibited about a thousand-fold by an Rp phosphorothioate (45). As predicted, if the second splicing step is analogous to the reverse of the first splicing step, the exon ligation reaction is inhibited by an Rp phosphorothioate at the reactive site, while the reversal of this reaction is not (46). Analogous studies performed on nuclear pre-mRNA splicing show that the two splicing steps are inhibited by the Rp phosphorothioate but not by the Sp, indicating that the second step is not analogous to the reversal of the first step (see Section 3.2) (48).

3.4 Ribozymes are metalloenzymes

3.4.1 Role of metal ions in ribozyme folding

Studies of tRNAs showed that, in addition to a general role as counterions neutralizing negative charges of phosphates, divalent cations such as Mg^{2+}, Mn^{2+}, and Ca^{2+} are associated with high-affinity binding pockets, involving specific hydrogen bonding to the RNA (reviewed in 56, 57). All ribozymes studied so far require divalent cations for efficient catalysis. Studies of the metal ion requirement of RNase P (32) or group I introns (57) suggest the existence of two classes of metal ion binding sites: 'activity' sites, which must be filled by a restricted class of ions (usually Mg^{2+} and/or Mn^{2+}) in order for the ribozyme to be active, and 'structural' sites, which can accommodate a wider range of divalent cations. A dual role for Mg^{2+} has also been suggested for group II introns (38) and the hammerhead ribozymes (56), in which only part of the Mg^{2+} requirement can be alleviated by spermidine (known to play a structural role).

3.4.2 Role of metal ions in ribozyme chemistry

The direct involvement of metal ions in catalysis was revealed by several types of experiments. In hammerheads, the rate of the chemical cleavage step increases with pH and, when different metal ions are used, the pH profiles shift in correlation with the pK_a of the metal ion. This strongly suggests that a metal hydroxyde is directly implicated in catalysis (reviewed in 58). Such direct involvement of a metal ion-bound hydroxyl was described for the Pb^{2+}-induced specific cleavage of yeast $tRNA^{phe}$. Despite its lack of relevance for biological function, this reaction was extensively studied because, being amenable to X-ray crystallographic studies, it provides a useful model for the role of metal ions in RNA catalysis. It was proposed that the Pb^{2+}-bound hydroxyl group $(Pb\text{-}OH)^+$, because it is fortuitously correctly positioned, increases the hydrophilicity and probably even abstracts the proton of the 2′-OH at the cleavage site. The resulting $2'\text{-}O^-$ then attacks the phosphorus atom of the neighbouring phosphate junction, resulting in strand cleavage by a reaction analogous to alkaline hydrolysis of RNA. This mechanism is an example of general acid–base catalysis in which the base is the hydroxyl group coordinated to the metal ion (reviewed in 3, 58). In hammerheads, metal ions seem to play at least one additional role. A strong thio effect was observed when an Rp phosphorothioate was introduced at the cleavage site of a hammerhead ribozyme (59), but this effect was apparent with Mg^{2+} as the divalent cation and not with Mn^{2+}. Since Mn^{2+} coordinates sulphur much more strongly than Mg^{2+}, this result indicates that a metal ion is directly coordinated to this non-bridging oxygen at the cleavage site (56). It is not yet known precisely how this metal ion participates in the reaction but it probably contributes to the stabilization of the transition state (56). Finally, this metal ion could be the one that coordinates the hydroxyl group responsible for 2′-OH activation (29).

A differential thio-effect between Mg^{2+} and Mn^{2+} was also observed for the bridging 3′-oxygen atom at the 5′ cleavage site of the *Tetrahymena* group I intron, indicating that a metal ion is directly coordinated to the oxygen leaving group (60). It is proposed that Mg^{2+} (or Mn^{2+}) stabilizes the developing negative charge on the oxyanion leaving group. In group I introns, a strong thio effect is also observed when the non-bridging pro-Sp oxygen is replaced by a sulphur, but no suppressor effect of Mn^{2+} has been reported.

Recently, a mechanism for phosphoryl transfer, catalysed by the 3′–5′ editing exonuclease activity of *Escherichia coli* DNA polymerase I, was proposed on the basis of a refined crystal structure of this enzyme (61). Two metal ions are complexed in the pre-transition state active site, one on each side of a non-bridging oxygen of the scissile phosphate. They are proposed to facilitate the formation of the attacking hydroxyl group, the leaving of the 3′-hydroxyl and, by contacting simultaneously bridging and non-bridging oxygens, to stabilize the 90° O-P-O angle between apical and equatorial oxygen atoms in the transition state. Interestingly, none of the protein side-chains is implicated in the chemistry of catalysis. It was thus proposed that this phosphoryl-transfer mechanism could be directly applied to RNA catalysis (50). Although this model is compatible with the available data, it remains to be proven.

4. Kinetics of RNA catalysis

Despite the fact that most ribozymes (except RNase P) are not recycled *in vivo*, it is often possible to dissect the ribozymes into substrates and enzymes (see Section 6.2).

These enzymes are able to act catalytically, allowing measurements of the steady-state rate constants of the reactions as would be done for regular enzymes. For hammerhead ribozymes, K_M values of 40 nM and k_{cat} of 1.5 min^{-1} have been reported (62). Similar values were observed for the hairpin (63) and the HDV (64) ribozymes. These k_{cat} values correspond to a rate enhancement of about six orders of magnitude over a similar, noncatalysed reaction (58). The k_{cat}/K_M values of about 10^7 M^{-1} min^{-1} are similar to those of protein ribonucleases, which have faster k_{cat} but higher K_M (58).

For group I introns, the kinetic parameters of the guanosine-induced cleavage by a *Tetrahymena*-derived ribozyme of an oligonucleotide substrate carrying the 5′ splice site were studied in detail (35). The chemical step was estimated to have a rate constant $k_c \approx 350$ min^{-1}. This would correspond to a rate enhancement of about 10^{11} over an uncatalysed hydrolysis reaction, giving rise to the same 5′-phosphate and 3′-hydroxyl termini. This is comparable to the rate enhancement achieved by protein enzymes (65). The k_{cat}/K_m of about 10^8 M^{-1} min^{-1} (35) appears limited by the diffusive step, which corresponds to the rate of formation of the base-paired duplex between the substrate and the enzyme (66). As a multiple turnover enzyme, however, this ribozyme is limited by the rate of release of the cleaved 5′ exon product. This makes sense since, in the wild-type intron, this is the product of the first reaction step and the cleaved exon must remain associated with the enzyme in order for the second reaction step to proceed.

More unexpectedly, the RNA subunit of RNase P, which naturally acts as a multiple turnover enzyme, is also rate-limited by the release of the product, at least for some tRNAs (67). In fact, it is proposed that one role of the protein component of RNase P is to increase the release of the product and hence to facilitate the recycling of the enzyme *in vivo* (67, 68).

5. RNA structure characterization

As for protein enzymes, the specific folding of the molecule confers the enzymatic properties to ribozymes. A great deal of the effort to understand the basis of RNA catalysis has thus been aimed at the description of the higher order structure of these molecules. Moreover, the structures of ribozymes have not only been studied with the aim of understanding RNA catalysis; indeed, catalytic RNAs share distinctive features which make them particularly attractive models for more general studies on RNA structure and folding:

(1) they are highly structured RNA molecules;

(2) since they are functionally active *in vitro* in the absence of proteins, structural studies on naked RNA can be expected to be directly relevant to biological activity, in contrast to, for example, rRNAs or snRNAs, for which the structure of the naked RNA in solution might differ from that in the RNP particle;

(3) the catalytic activity can be exploited as a direct assay of structure integrity under various conditions or for specific variants.

Although the most direct and accurate techniques to study the structure of macromolecules are X-ray crystallography or nuclear magnetic resonance (NMR), ribozymes have so far resisted crystallization and are usually too big to enable detailed NMR studies (some NMR studies have nevertheless been performed on the smaller ribozymes: see, for example, 69). Thus, the study of ribozyme structures has been carried out using more indirect, although often classical approaches, which will be briefly reviewed below.

The determination of the higher order structures of RNAs typically requires several stages. The general approach consists of the construction, in a first step, of a secondary structure model, which is then tested experimentally through a wide range of techniques. Once the secondary structure of the molecule is firmly established, tertiary interactions can be looked for. It is only once a sufficiently large number of tertiary contacts are known, that stereochemical modelling of the tertiary structure can be envisioned. If tertiary structure modelling is successful, it allows the prediction of new tertiary contacts that can in turn be specifically tested to assess the validity of the model (7).

5.1 Structure modelling

There are classically two methods to model RNA structure. The first one, called minimum free energy calculation, makes use of computer programs that estimate, on the basis of thermodynamic data experimentally generated for specific double-stranded oligonucleotides, the most energetically favourable structures. Despite great improvements, the latest programs remain unable to predict all but some of the secondary structures of long RNA molecules (70). The second method, called phylogenetic comparative analysis, is based on the principle that functionally equivalent molecules will exhibit conserved structures, despite a certain degree of primary sequence divergence. This method (reviewed in 71) is most powerful when a relatively large number of homologous sequences is known, which is the case for most ribozymes (7, 11, 18, 72). The power of the method has been best demonstrated by the modelling of the three-dimensional architecture of group I introns (7), a model that has so far withstood experimental challenge (65).

5.2 Experimental approaches

In several instances, it is known that a given RNA molecule can adopt several alternative configurations in solution (see, for example, 73). Since the active conformation is likely to be the only one that is biologically significant, the techniques that are able to discriminate between active and inactive configurations represent the ultimate methods to assess the validity of an RNA-folding model. The most classical of these approaches is a genetic one. To test a potential interaction between two nucleotides, both partners are substituted in such a way that the novel combination will restore an isomorphic equivalent of the postulated structure. The result will be positive if the single mutants are catalytically defective but the combination of the mutations at both positions restores the activity. Another approach, *in vitro* selection, has been recently developed. It makes use of the dual role, informational and enzymatic, of these RNA molecules. It is a very powerful method based on the use of very large pools of mutant transcripts (up to 10^{15} different sequences) (74), which are screened in an iterative way by using several selection–amplification cycles, thus allowing the extremely sensitive identification of active revertants or variants (23, 75; reviewed in 76).

Other methods that do not rely on the functionality of the molecules have nevertheless been useful for RNA structure characterization. Enzymatic or chemical structure probing methods are among the most widely used (77). For example, Fe(II)-EDTA induces oxidative cleavage of the nucleic acid backbone in the presence of O_2 and reducing agents. Fe(II)-EDTA is particularly useful because, as a large molecule, it is sterically excluded from the interior of folded structures. Thus, it allows determination of the exterior, solvent-exposed backbone regions, which constitutes invaluable information to probe the tertiary folding of a transcript (78). Also, chemical accessibility has proved most powerful when combined with mutational analysis to investigate detailed specific interactions (see, for example, 79). Cross-linking has also been in extensive use since techniques became available that allow the incorporation of photo-reactive chemical groups at specific positions along the RNA molecule (80, 81). This allows the probing of structures proximal to essential nucleotides (see, for example, 82).

6. Functional dissection of ribozymes

Understanding RNA catalysis requires the analysis of the specific roles played by different regions of a ribozyme. Analysis of the RNA structure and of mutations affecting it has allowed the assignment of certain sequences to specific functions.

6.1 The catalytic core

The structural elements that are phylogenetically conserved in all members of a given type of ribozyme generally coincide with the set of sequences that were experimentally defined as essential for catalysis. This body of sequences constitutes what is commonly called the core of the ribozyme. Although primarily based on a practical definition, this term actually carries a more fundamental notion: some residues are directly involved in the catalytic site, for example because they carry the functional groups that interact with the substrate(s) to stabilize the transition state, or because they are directly involved in the specific geometry of this catalytic site. In contrast, some interactions, usually located at the periphery of the ribozyme, serve to stabilize the overall structure and affect the catalytic site only indirectly. The latter type of structures are generally variable, usually leading to the definition of subgroups. Experimentally, the assignment of particular structures to an RNA folding function versus a direct catalytic role is often based on the fact that the lack of one or a few interactions of the former type can be compensated by elevated concentrations of ions that are known to stabilize RNA structures. However, a more rigorous approach requires the assessment that mutations that disrupt these types of interactions will affect the folding of the molecule but that, once properly folded, the mutant transcripts will exhibit kinetic characteristics similar to the wild type (47, 83). Nevertheless, the different experimental results usually coincide fairly well with the phylogenetic approach. For example, all group I introns have a conserved core structure constituted by the P3 to P8 helices and their connecting segments (see Fig. 1b and 7 for review). This core structure also corresponds to the minimal sequences experimentally required for activity (see 84). The conserved core structures of different ribozymes are schematically represented in Fig. 1a–d.

6.2 Dissection of the ribozyme molecule into enzyme and substrate portions

For all demonstrated ribozymes to date, except the M1 RNA of RNase P, the molecule acts on itself and so cannot be considered as a true enzyme. However, in most systems it has been possible artificially to divide the molecule in two parts: a substrate, on which the chemical reaction occurs; and an enzyme, which remains unchanged during the course of the reaction and can be recycled. The somewhat artificial nature of this approach is illustrated in Fig. 1f, which shows that a given ribozyme, the hammerhead ribozyme in this example, can be divided into various combinations of enzyme and substrate. However, it is often possible to isolate minimal substrate sequences carrying essentially the reactive sites, with the core structures being carried by the enzyme part. For example, in the substrate–enzyme combination first described by Haseloff and Gerlach (85) for hammerheads, the substrate subunit carries only two out of the 14 conserved nucleotides (see Fig. 1f(b)). These systems have been essential for the detailed analysis of the interactions responsible for enzyme specificity (see below) and thus, they carry more biological sense than their artificial nature may suggest. They have also been essential for the kinetic analyses of ribozyme reactions (see 35 and Section 4).

6.3 Substrate binding sites and specificity

6.3.1 Substrate binding in hammerhead and hairpin ribozymes

As discussed above, in hammerheads, the enzyme–substrate combination first described by Haseloff and Gerlach (85) (see Fig. 1f(ii)) uses substrate molecules with minimal structural requirements. This allows the design of ribozymes potentially able to cleave specifically virtually any kind of substrate sequences. The binding constant (K_m) of the reaction is related to the stability of the ribozyme–substrate hybrid (62). But the Watson–Crick base pairing does not seem to play the sole role in this binding constant, as the replacement of some of the ribonucleotides of the substrate by deoxyribonucleotides has a great influence on the K_M of the reaction. However, it was not possible to ascribe the loss of binding strength to interactions with specific 2′-hydroxyls of the substrate. Rather, differences in the conformation of the catalytic core induced by the presence of a DNA–RNA heteroduplex and/or interference with binding of an essential magnesium seem to be responsible, at least in part, for the observed effect on the K_M (86, 87).

With the hairpin ribozyme, the distinction between enzyme and substrate was straightforward. Indeed, a small substrate sequence (as short as 10 nucleotides), including the normal cleavage site, can be processed by a longer catalytic RNA sequence (more than 46 nucleotides) normally localized 125 nucleotides away in the STobRV (-)RNA molecule (88). Analysis of the secondary structure of the hairpin ribozyme (22, 63, 89) showed that the substrate sequence binds the catalytic sequence via two successive stems called helix 1 and helix 2. The cleavage site is located within the four-nucleotide loop that separates the two helices (see Fig. 1d). As for hammerhead ribozymes, some 2′-hydroxyls are important for binding, but it is not known whether this stabilization is mediated by direct contacts between the ribozyme and the 2′-OH of the substrate (31).

6.3.2 Substrate binding and specificity in self-splicing introns

Binding of the 5′ exon and recognition of the upstream junction. In both group I and group II introns, the first splicing step results in the cleavage of the upstream junction. The free 5′ exon must remain bound to the intron in order for the exon ligation to take place. In both systems, separation of the enzyme and substrate for the first reaction was possible and used to analyse the interactions involved in recognition of the upstream junction and binding of the 5′ exon.

In group I introns, the early versions of the secondary structure models (90, 91) already included a Watson–Crick base pairing between the 5′ exon and intron upstream sequences, which forms the P1 helix (see Fig. 1b). The only conserved sequence of the P1 helix is the U immediately preceding the 5′ splice site, which is paired to an invariant G on the intron side of the helix. Single-base changes and second-site suppressors provide genetic evidence for the importance of the P1 helix in splicing both *in vivo* (92) and *in vitro* (93, 94).

Various experimental systems have been developed that use molecules containing a 5′ exon, or a related sequence, as substrate in a reaction catalysed in *trans* by the core structure of the ribozyme (see, for examples, 95, 96). Separation of the substrate from the enzyme allows measurement of the affinities for these various substrates. Kinetic data (35) or direct measurement of oligonucleotide substrate binding (97, 98) indicate that the affinity of the ribozyme for 5′ exon substrate homologues is greater, by several orders of magnitude, than expected for a simple Watson–Crick base-paired duplex. Studies using mixed ribo- and deoxyribonucleotide substrates showed that the 2′-OH groups at positions -2 and -3 relative to the 5′ splice site are directly involved in the enzyme–substrate complex stabilization (99, 100). Finally, a nucleotide of the core (nucleotide A^{302}; see Fig. 1b) was shown to interact directly with the 2′-OH at position -3 within the 5′ exon (79).

The fact that tertiary interactions play an important role in substrate binding is also consistent with the finding that a detached P1 helix can serve as a substrate for cleavage by the ribozyme (101). Moreover, analysis of the cleavage of mutant P1-containing substrates indicated that the structure, rather than the sequence, determines the specificity of the reaction: cleavage occurs after a wobble base pair (G·U in the wild type, but A·C is also recognized) that must be located within a three nucleotide window centred on the fifth base pair from the bottom of the P1 helix in this system (102). Finally, helices of virtually any sequence can be used as substrates in a reversal of the G-induced cleavage reaction catalysed by a group I intron, illustrating the importance of the tertiary interactions involving the sugar–phosphate backbone for substrate binding (101). Studies of mutant ribozymes affected in their ability to form accurate tertiary interactions with the P1 helix indicate that the binding of the 5′ exon substrate involves at least two steps: first, the 5′ exon is held only by base-pairing interactions and forms an 'open complex' with the enzyme core; then, the base-paired duplex (P1) can 'dock' into tertiary interactions to form a 'closed complex', which is required for activity (103).

In group II introns, early secondary structure models did not propose direct interactions between the 5′ exon and the intron. However, the simple fact that, after the first reaction step, these two sequences are no longer covalently linked, suggested the existence of a 5′ exon binding site in the intron. The observation that the 5′ exon could efficiently serve as a substrate when provided in *trans* soon confirmed this hypothesis (39). Two internal loop sequences, located in domain I, were identified as the main structural elements required for 5′ exon recognition and binding. These sequences were called EBS1 and EBS2 for exon binding sites 1 and 2. Their Watson–Crick base pairing with two complementary 5′ exon distal sequences (IBS 1 and 2 for intron binding sites 1 and 2) was shown, by *in vitro* genetic analyses, to be essential for the splicing reaction (see Figs 1a and 4a) (104). Watson–Crick interactions involving intron sequences following the 5′ splice site (nucleotides 3 and 4 of the intron) also play a role in 5′ splice site recognition in the normal *cis*-splicing reaction (interaction ε/ε': see Figs 1a and 4a) (105). The specificity of the reaction seems to be a result of both the correct positioning of the 5′ exon relative to the EBS1 sequence (40) and the recognition of the 5′ terminal G of the intron during branch formation (106). Although a DNA version of the 5′ exon can serve as a substrate for the second reaction step (3′ splice site cleavage and exon ligation) (107), the specific role of 2′-OH moieties in exon binding has not yet been analysed.

Recognition of the downstream junction. For group I introns, Davies *et al.* (91) proposed that the base pairing of the intron sequence with the 5′ exon (forming the P1 helix) could extend to the 3′ exon (P10 helix; see Fig. 1b). The intron sequence, by pairing simultaneously with the 5′ and the 3′ exon, would precisely align the splice sites. For this reason, this sequence has been called the internal guide sequence (IGS). However, and in contrast to what was observed for the 5′ exon, disruption of the intron–3′ exon base pairing (P10) in the *Tetrahymena* rRNA intron was found not to prevent exon ligation. P10 is thus not essential for *in vitro* splicing of this intron (108). None the less, in the presence of potential cryptic sites, the P10 base pairing was shown to increase 3′ splice site specificity in the *Tetrahymena* rRNA intron

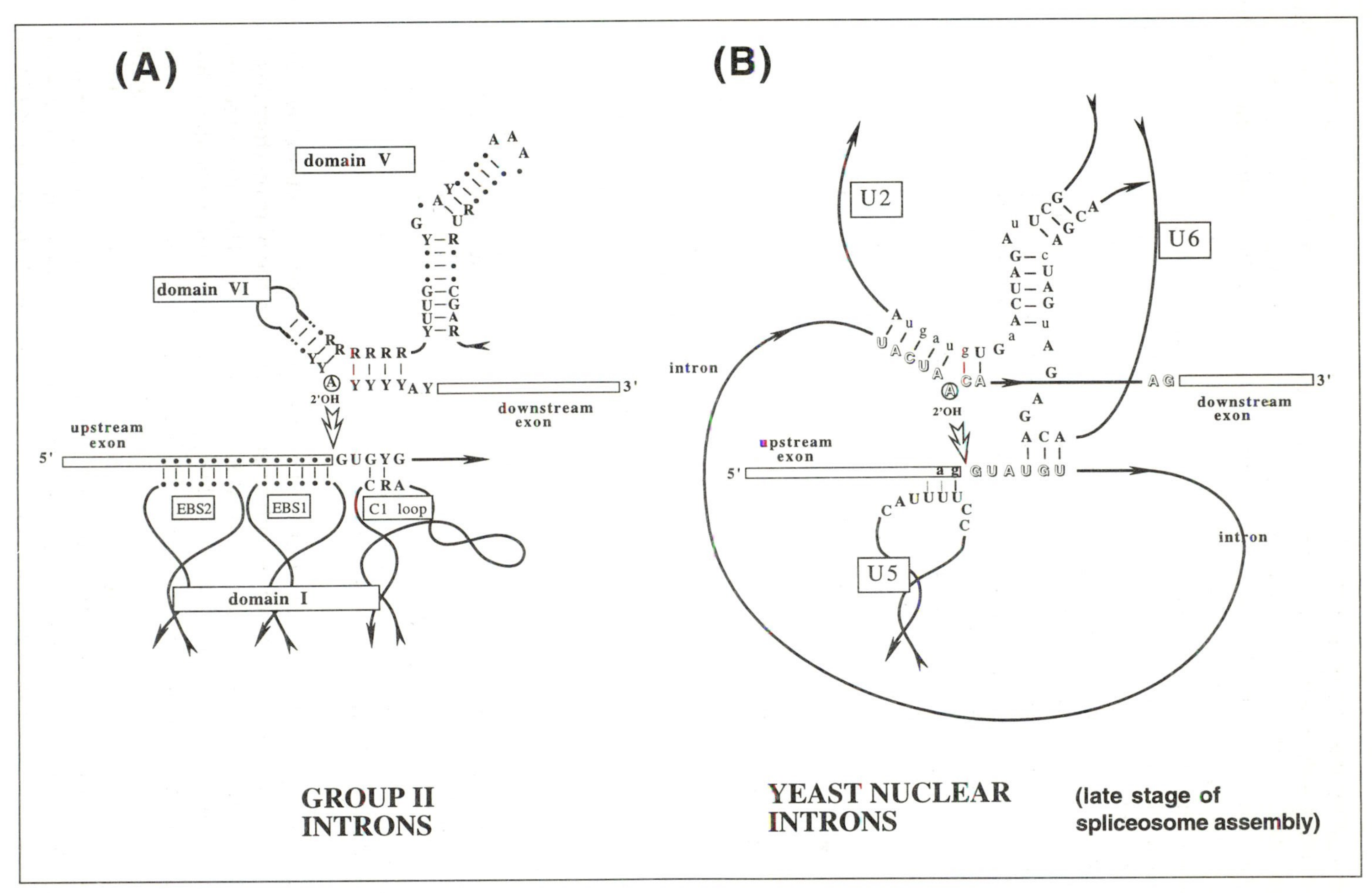

Fig. 4 Currently defined RNA interactions involved in 5′ splice site recognition in group II (a) and nuclear pre-mRNA (b) introns. Exons are shown as thin boxes, and intron or snRNA sequences by thick black lines. The arrowheads indicate the 5′→3′ direction. Upper-case letters represent the most conserved nucleotides. The white arrows indicate the cleavage sites.

(109). Moreover, when other elements required for 3′ splice site recognition are disrupted (see below) (42) or in certain other group I introns (110, 111), the P10 helix was shown to contribute to the efficiency of the second splicing step. Correlated with the relatively minor importance of downstream exon sequences, the terminal intron nucleotides were found to play a major role in 3′ splice site recognition. First, all group I introns end with a G. Alteration of this 3′-terminal guanosine virtually eliminates 3′ splice site activity (112). It was subsequently demonstrated that the terminal G is bound, before the second splicing step, to the same site that binds the exogenous G used as the nucleophile during the first splicing step (see Section 6.3.3). Second, the two nucleotides preceding the terminal G base pair with two nucleotides located between the P7 and P9 helices to form the P9.0 pairing (see Fig. 1b), which is present in most group I introns (113, 114). In conclusion, 3′ splice site recognition seems to be mediated by redundant interactions involving sequences flanking the intron–exon junction.

In a very similar manner, multiple interactions have been implicated in 3′ splice site recognition of group II introns. First, the position of the 3′ splice site relative to domain VI appears to be a major determinant of recognition. The distance between the base of this hairpin and the 3′ junction is highly conserved (two nucleotides in subgroup IIA and three nucleotides in subgroup IIB) (11). Moreover, deletion (40) or alteration of this structure (115) induces the use of cryptic 3′ splice sites. Second, the last nucleotide of the intron, usually a pyrimidine, is involved in an isolated Watson–Crick interaction with an internal core nucleotide (interaction γ/γ; see Figs 1a and 5a), which was shown to be important both for the efficiency (105) and the specificity (40) of the second splicing step. Third, as in group I introns, the base pairing between the intron sequence (exon binding site; EBS1) and the 5′ exon can, in a majority of introns, extend to the 3′ exon. This pairing thus constitutes a potential 'guide' interaction (see Figs 1a and 5a) (11). Mutational analysis of this potential pairing has shown that it barely affects self-splicing efficiency but plays a role in the specificity when cryptic 3′ splice sites are present (40). This situation is remarkably similar to what was observed for group I introns.

6.3.3 The G binding site of group I introns

Guanosine is an exogenous substrate for the first splicing step of group I introns (see Section 3.1). The fact that the initial velocity of this reaction is dependent on the guanosine concentration in an hyperbolic manner, together with the fact that the resulting K_M (about 0.02 mM) is different for different guanosine analogues, show that group I introns contain a saturable binding site specific for guanosine (116).

The first elements of the G binding site were identified by Michel *et al.* (42), who switched the specificity of the enzyme from G to 2-aminopurine by substituting the G^{264}–C^{311} base pair with an A-U base pair within the highly conserved P7 helix. This provided proof for the hydrogen bond between the N_1-H of the exogenous guanosine and the keto oxygen of G^{264} in the P7 helix and led to the proposal that a base triple is formed between the nucleophilic G and the G^{264}–C^{311} base pair. Yarus *et al.* (117) studied the binding of several nucleotides and guanosine analogues to mutant

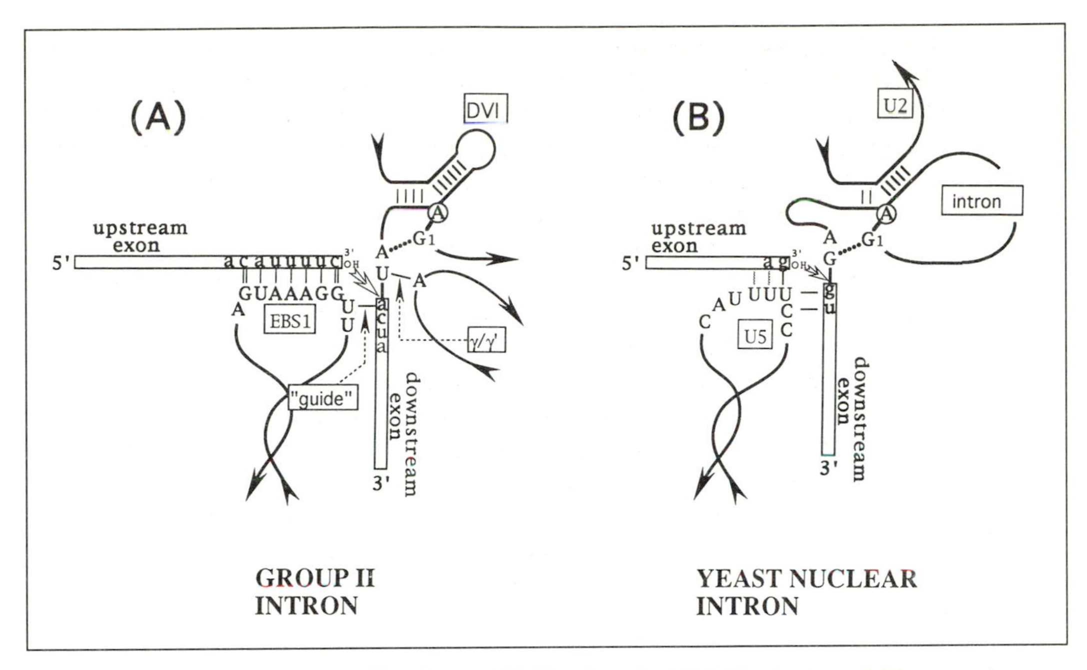

Fig. 5 Currently defined RNA interactions involved in 3′ splice site recognition in group II (a) and nuclear pre-mRNA (b) introns. Exons are shown as thin boxes and intron or snRNA sequences as thick black lines. Upper-case letters are intron or snRNA nucleotides; lower-case letters, intron nucleotides.

ribozymes with substitutions spanning the binding site. Their results were best explained by a model in which the nucleophilic G is non-planar with the G^{264}–C^{311} base pair and is involved in an additional contact with the neighbouring A^{265} (117). The G site, first defined as the binding site for the nucleophilic guanosine involved in the first splicing step, is also responsible for the subsequent binding of the intronic terminal G at the 3′ junction (43). This provides a major argument in favour of a single catalytic site for both splicing steps (see Section 3.2).

6.3.4 Substrate recognition by RNase P

In a cell, a single type of RNase P is responsible for the cleavage of all tRNA precursors, despite their limited primary sequence homology, suggesting that the main recognition features may reside in the tertiary structure of the pre-tRNA molecules. RNase P is also involved in the processing of the precursor to 4.5 S RNA in *Escherichia coli* (118). Moreover, at least in some cases, the K_M of the reaction appears identical with or without the protein subunit, implying that the RNA subunit alone is responsible for most of the substrate recognition process (2) (this is apparently not true for all substrates: see, for examples, 118, 119). Thus, the M1 RNA of RNase P possesses a substrate binding site able to recognize the shape of its substrates, as usual protein enzymes would do. Deletion studies on model pre-tRNA substrates (120, 121) as well as modification interference experiments (122, 123) identified the coaxial T and acceptor stems, including the conserved T loop and the 3′-terminal

NCCA single-stranded sequence, as the main structures involved in pre-tRNA recognition (see Fig. 1e). Even smaller substrates, composed of a short, unpaired 5′ leader sequence, a double helix, and the 3′-terminal NCCA single-stranded sequence on the 3′ side, can be recognized and specifically cleaved by the M1 RNA, although with reduced efficiency (121, 124, 125). The 2′-hydroxyls of at least four nucleotides of the substrate were shown to be important for cleavage (see Fig. 1e) (125). The 2′-hydroxyl of the G at position +1 is required for efficient binding to the M1 RNA, suggesting that it is directly involved in an interaction with the enzyme (G^{+1} is thought to be critical for the specificity of cleavage) (126). In contrast, the 2′-hydroxyls at positions —2, —1, and that of the first C of the CCA 3′-terminal sequence, are not important for binding but are part of an Mg^{2+} binding pocket essential for catalysis. A substrate can bind to the enzyme but is not cleaved if Mg^{2+} has not been bound before the substrate–enzyme interaction. Thus, the substrate–Mg^{2+} complex is the true substrate for the M1 RNA (127).

Some clues about the regions of the M1 RNA that interact with the substrate were provided by cross-linking studies (128, 129).

7. Are group II and nuclear mRNA introns evolutionarily related?

The discovery of self-splicing group I introns suggested that other RNA maturation events, such as nuclear pre-mRNA splicing, might be catalysed by RNAs rather than proteins. The finding that group II introns can also self-splice, and use the same splicing pathway as pre-mRNAs introns, further strengthened this idea. According to this hypothesis, the catalyst of nuclear pre-mRNA splicing would be provided *in trans* by the spliceosomal snRNAs. These snRNAs are the RNA components of the snRNPs (small nuclear ribonucleoprotein particles) which are major constituents of the pre-mRNA splicing machinery; this machinery assembles into a large complex called the spliceosome (see Chapters 4 and 5). A more speculative version of this model is that group II introns and spliceosomal RNAs are evolutionarily related. If so, one could expect that some structural features will be conserved between the group II introns and the structures formed by the RNA–RNA interactions within the spliceosome. As described below, the elucidation of some of the RNA interactions in both systems seems to provide clues consistent with this prediction. However, closer examination of the results leads us to moderate this conclusion.

7.1 Comparison of the RNA–RNA interactions involved in group II and nuclear pre-mRNA splicing

The first structural similarity observed between both types of introns concerns the sequences at the intron boundaries. The 5′ splice site consensus sequence in group II introns (/GUGYG) is reminiscent of the 5′ consensus sequence in nuclear introns (/GURAGU in mammals and /GUAUGU in yeast). The penultimate intron nucleotide is also strongly conserved as an A in both types of introns (the 3′ consensus is AY/ in group II introns and AG in nuclear introns). It is tempting to speculate that this rough primary sequence resemblance is indicative of similar interactions involved in splice site recognition.

7.1.1 Interactions at the 5′ splice site

As seen in Section 6.3.2, recognition of the 5′ splice site in group II introns involves discontinuous base pairing on each side of the splice site with internal loop sequences located in domain I of the intron (sequences EBS1 and EBS2 and loop C1; see Fig. 4a). Since all three base pairings are important for both splicing steps, there is no reason to think that there is a temporal rearrangement of these structures along the splicing pathway.

In nuclear pre-mRNA introns, the recognition of the 5′ splice site seems to be more complex, since recent studies suggest that this process requires a temporal pathway of action by snRNAs, which act in several steps before the cleavage reaction itself. First, both genetic and biochemical studies indicate that U1 base pairs with sequences at the 5′ exon–intron junction (reviewed in 49). Cross-linking performed along a splicing time course showed that the U1 RNA–5′ splice site interaction is an early event, being predominant when splicing does not yet take place and disappearing while the spliced products begin to accumulate (130, 131). It is thus proposed that the 5′ splice site/U1 pairing, required for spliceosome formation, is disrupted when the splicing reaction actually takes place. Consistent with this idea, the U6 RNA was found associated with the 5′ splice site at the time of the splicing reaction (130, 131) and this U6 pairing partially overlaps the U1 pairing (132).

Genetic studies also implicated U5 in 5′ splice site recognition: mutations in a highly conserved loop of U5 can activate cryptic 5′ splice sites by base pairing with 5′ exon sequences in a way that resembles the 5′ exon–intron pairing in group II introns (see Fig. 4b) (133, 134). Cross-linking studies are also consistent with that pairing and suggest that it occurs throughout the splicing reaction (135).

Figure 4 summarizes the known interactions involved in 5′ splice site recognition in both group II and yeast nuclear pre-mRNAs. In conclusion, the U5–5′ exon interaction resembles the EBS1–5′ exon interaction in group II, and the U6–intron interaction may be the equivalent of the C1 loop–5′ intron sequences (ε/ε′) interaction in group II.

7.1.2 Interactions at the 3′ splice site

Figure 5 summarizes the interactions at the 3′ splice site in group II and *Saccharomyces cerevisiae* nuclear pre-mRNA introns. Two types of interactions appear similar.

First, the extension of the EBS1–5′ exon base pairing to the 3′ exon, called 'guide' interaction because it aligns the two exons prior to their ligation, has found its equivalent in yeast nuclear pre-mRNA introns with the conserved loop of U5, as shown in genetic experiments by Newman and Norman (133). However, one U of the U5 loop seems involved in base pairing with both the 5′ and the 3′ exon indicating that part of the two interactions are mutually exclusive and thus may occur in a sequential manner.

Second, genetic experiments indicate that the first nucleotide of the intron (G1) interacts with the 3′ splice site after branch formation in both group II (51) and nuclear pre-mRNA introns (52). However, G1 interacts with the last intron nucleotide in nuclear pre-mRNA introns, whereas it interacts with the penultimate nucleotide in group II introns.

In conclusion, a very similar organization can be recognized between the two systems but the details of the interactions appear different.

7.1.3 Interactions around the branch site

The branch site nucleotide is conserved as an A in both group II and pre-mRNA introns. Strikingly, this A is found bulged out of a helix in both cases. In group II introns, it is bulged out of the domain 6 helix (11) (Figs 1a and 4a), whereas in pre-mRNA introns, it is bulged out of a helix formed between a highly conserved sequence of the U2 snRNA and the sequence surrounding the branch site (see 49 and Fig. 4b). This conserved feature remains the most convincing example of structural similarity between the two systems.

This observation was recently strengthened by the demonstration that U2 sequences just upstream of those base paired with the branch site structure are involved in an extensive interaction with the U6 snRNA (see Fig. 4b) (136). This new interaction brings highly conserved nucleotides of U6 (potentially involved in catalysis) (137) next to the branch site and thus to the site of catalysis in the first splicing step. Several criteria suggest that this structure could be the equivalent of domain V of group II introns. First, it is located directly upstream of the branch site/U2 helix which is the equivalent of the group II domain VI. Second, this helix possesses a two nucleotide bulge which makes it structurally resemble domain V. Third, this helix, like domain V for group II introns, carries some of the most conserved nucleotides of the spliceosomal RNAs. However, these conserved nucleotides differ from those in domain V, which seems surprising if these two structures are true homologues.

In summary, a detailed comparison of the RNA interactions involved in both systems does not necessarily lead to the conclusion that they are directly evolutionarily related. Rather, common strategies seem to have been employed in both cases to bring essential nucleotides to the active site, but with differences in the details of the interactions. For this reason, it has been suggested that the limited structural similarities observed would be the result of chemical determinism rather than phylogenetic relationship (138). The problem, however, is that even interactions that involve invariant nucleotides are not necessarily conserved. For example, in *Schizosaccharomyces pombe*, the pairing between U1 and the 5′ intron consensus sequence in nuclear pre-mRNA extends to the universally conserved 3′ terminal AG/ (139). But this interaction does not appear to be conserved in *S. cerevisiae*, where the same U1 nucleotides interact instead with the 5′ exon terminal sequence (140). So, even in two homologous systems, some crucial, highly conserved nucleotides, which should be at the heart of the spliceosome, seem to be involved in different interactions, at least at some stages of the splicing pathway. This indicates that during evolution, the details of snRNA–intron interactions may have taken different routes. It might be expected, then, that even if the group II and pre-mRNA introns are indeed evolutionarily related, they may be so phylogenetically distant that it will be impossible to recognize conserved detailed structural features. Moreover, if the only conserved structures correspond to the most efficient ones, the same structures could have been independently selected during convergent evolution. For example, the 'guide' structure, which seems conserved between group II and nuclear pre-mRNA introns, is also found in the unrelated group I introns. In conclusion, structural features might not be sufficient to determine whether the obvious similarities and differences found between the group II and nuclear pre-mRNA splicing pathways are the result of convergent or divergent evolution. One way to answer this question would be to find evolutionary intermediates that incorporate features characteristic of both systems.

7.2 Potential evolutionary links between group II and nuclear pre-mRNA introns

If group II introns are the ancestors of nuclear pre-mRNA introns, one would expect that the former are ancient objects and that it might be possible to find intermediates between the two. The first point has recently found a positive answer since group II introns have been found in cyanobacteria and proteobacteria (12). Now, could one find candidates for evolutionary links? The first stage of evolution from group II introns toward present nuclear pre-mRNA introns would have been the fragmentation of a group II intron into distinct transcripts able to reassociate as a functional group II intron. Such transcripts have, in fact, been found. For example, some genes in chloroplasts and plant mitochondria are split in the middle of either domains III or VI of group II introns and now consist of widely separated transcripts (see, for review, 11, 141). In *Chlamydomonas* chloroplasts, splicing of the *psa*A exons 1 and 2 requires a third transcript, encoded by the *tscA* locus. The tripartite association of the *tscA*-RNA and the two intron portions flanking exons 1 and 2 reconstitutes a prototype group II intron (142). Thus, like the spliceosomal snRNAs, the *tscA* transcript represents a *trans*-splicing factor composed of RNA. *Trans*-splicing has also been found in certain nuclear pre-mRNA introns (see Chapter 10). All mRNAs in trypanosomes have an identical short non-coding sequence at their 5′ end. These short sequences were found to be encoded elsewhere in the genome, transcribed separately, and then spliced in *trans* to the 5′ end of precursor RNAs. A similar organization has been found for most coding transcripts of the *Euglena gracilis* nucleus and about 10–15% of the mRNAs in nematodes (reviewed in 141). These 90–140 nucleotide long '*trans*-spliced leader transcripts' (SL RNAs) are composed of a 22–39 nucleotide mini-exon, followed by a 5′ splice site, and sequences eliminated during the splicing process and thus corresponding to 5′ terminal intron sequences. Most interestingly, these latter intronic sequences also share characteristic features of spliceosomal snRNAs. Thus, like self-splicing introns, SL RNAs are composite molecules containing both substrate sequences (the 5′ exon and 5′ splice site) and sequences involved in the splicing mechanism (the snRNA type of sequences).

These introns in pieces, however, can unambiguously be recognized as belonging to either the group II or nuclear pre-mRNA class of introns and do not show more recognizable homologies to each other than regular group II and nuclear pre-mRNA. Therefore, they are not very likely to represent true evolutionary transitions between the two types of introns. They give, however, an idea of the types of intermediates that might have existed during this hypothetical evolutionary process.

The *Euglena gracilis* chloroplast genes contain at least 6 group II introns that possess the highly conserved domains V and VI but are shorter than other group II introns and have abbreviated versions of domains I–IV (11, 143). In addition, *Euglena* chloroplast genes also contain more than 47 peculiar introns, called group III introns, characterized by a very constrained length (around 100 nucleotides). They appear to be streamlined versions of group II introns because their degenerate 5′ consensus sequence retains only the U at position two and the G at position five of the group II (or nuclear pre-mRNA) sequences; they possess a domain VI-like structure; and are excised as lariat molecules (143). An attractive hypothesis is that most of the *cis*-elements normally required for group II intron excision may be supplied in *trans* for these abbreviated versions of group II introns. It will thus be extremely interesting to identify these *trans*-splicing factors and to determine if they contain RNA molecules showing similarities to spliceosomal snRNAs.

In conclusion, true evolutionary intermediates between group II and nuclear pre-mRNA introns remain to be found.

7.3 Is nuclear pre-mRNA splicing catalysed by RNA?

One can ask whether, independently of its potential relationship with group II intron splicing, the excision of nuclear pre-mRNA introns is an RNA-catalysed reaction. As already discussed, enzymatic catalysis involves very high substrate specificity and great acceleration of the rate of the chemical reaction. We have already seen that snRNAs are directly involved in 5′ and 3′ splice site recognition and specificity by directly contacting these substrates. Therefore, one can consider as an established fact that snRNAs act at least at this level of splicing catalysis.

Let us consider now the rate enhancement of the reaction. In enzyme catalysis, a great part of this rate enhancement usually results from the correct positioning and orientation of the different substrates relative to each other. Again, while it has not been formally proven, it is most likely that the binding of the U2 snRNA to the branch site sequence is directly involved in the correct positioning of the attacking 2′-hydroxyl relative to the 5′ splice site during the first reaction step (50). Likewise, one can guess that the binding of the 3′ exon sequences by the U5 snRNA at least contributes to the correct relative positioning and orientation of the substrates during the second reaction step (50). Finally, stabilization of the transition state structure is a determining factor in catalysis. There are no direct indications so far that spliceosomal RNAs are directly involved in that process. However, some highly conserved nucleotides of the U6 snRNA sequence, which are brought into the vicinity of the active site by the U2–U6 pairing, are thought to be directly involved in catalysis (136). It has been proposed that this sequence could be involved in the binding of magnesium, which may be essential for the activation of the chemical step (50).

In conclusion, while the RNA components of the spliceosome have not yet been demonstrated to carry catalytic activity in the absence of proteins, they are certainly directly involved in at least some aspects of splicing catalysis. The real question to be asked, then, should not be whether nuclear pre-mRNA splicing is an RNA-catalysed reaction but, rather, whether RNA is the only component of the catalytic machinery (see Chapters 6 and 7).

8. Conclusions

Are there other potential RNA catalysts? Nuclear pre-mRNA splicing is not the only example of a biological process for which RNA catalysis is not demonstrated but nevertheless suspected. In fact, since the discovery of ribozymes, any RNA found as an enzyme subunit or part of a biological machinery should be considered as a potential RNA catalyst. These include:

(1) a group of small nucleolar RNAs (snoRNAs) involved in pre-rRNA processing (reviewed in 144);

(2) the small nuclear RNA U7 which is involved in the endonucleolytic processing of histone mRNA 3′ ends (reviewed in 144; see Chapter 9);

(3) the RNA component of the RNase MRP (mitochondrial RNA processing) ribonucleoproteins—enzymes involved in the cleavage of the mammalian mitochondrial primer RNA used during mitochondrial replication (some structural similarities between these RNAs, and the RNA component of RNase P have been proposed (145));

(4) the RNA component of the telomerase (reviewed in 146);

(5) the ribosomal RNAs. There are several recent findings suggesting that the peptidyl transferase reaction is catalysed by the 23S RNA. First, antibiotics that are specific inhibitors of the peptidyl transferase reaction have been shown to protect a set of bases in the 23S RNA in a region where mutations conferring resistance toward these antibiotics are located. Second, the 23S RNA interacts with the CCA end of tRNA, a substrate of the peptidyl transferase reaction. Third, the peptidyl transferase activity of *Thermus aquaticus* remains present after extensive protein extraction leaving only about 5% of the initial protein content. All these results indicate that the 23S RNA is intimately involved in the peptidyl transferase reaction, even though the definitive demonstration that it is involved in catalysis remains to be done (reviewed in 147). It should be added

that a derivative of the *Tetrahymena* group I intron was shown to accelerate (modestly) an aminoacyl esterase reaction, demonstrating that ribozymes have indeed the potential to catalyse reactions at carbon rather than at phosphorus centres (see 65 for review).

In addition to these potential natural RNA catalysts, powerful *in vitro* selection techniques (see Section 5.2) have been used to screen for new ribozyme functions. This can be done by screening a large pool of variants of a given RNA. For example, variants of the *Tetrahymena* group I intron have been selected for efficient cleavage of single-stranded DNA (148). The *in vitro* selection technique was also used on very large pools of totally random sequences. Selection for tight and specific binding to a variety of ligands has been highly successful (reviewed in 76, 149). Recently, a new class of ribozymes, able to catalyse a template-directed RNA ligation reaction, has been selected from an unusually large pool of totally random sequences (74). There is no doubt that the use of these very large pools of molecules combined with new selection procedures will expand the repertoire of RNA catalysis in the near future.

The discovery of catalytic RNAs has had an important impact in several areas of biology. First, in enzymology, of course, since biological catalysis was long thought to be restricted to proteins. But, above all, it revealed that RNA has potentially the unique property of being both informational and catalytically active. This gave fuel to theories on the origin of life, which speculate on the 'chicken and egg' problem: who came first—nucleic acids, which require proteins for synthesis, or proteins, which are encoded by nucleic acids? In principle, RNA could do without proteins and would thus be the best candidate for having been the support of some kind of primitive life, simply consisting of self-replicating molecules. Although the laboratory design of a truly self-replicating RNA molecule remains to be done, recent progress in this direction suggests that this goal is not out of reach (see 150). One problem of this theory, however, is that the existence of an 'RNA world' assumes the availability of a sufficient supply of nucleotides, the existence of which cannot yet be simply explained by any of the known abiotic processes (for review, see 151).

Acknowledgements

I thank B. Dujon and all members of his laboratory for many fruitful discussions and support. I am particularly grateful to G. Chanfreau, C. Fairhead, and P. Legrain for their critical reading of the manuscript and their highly constructive comments. The author apologizes for not acknowledging the work of many colleagues who contributed to this field since the final version of this review was submitted.

References

1. Kruger, K., Grabowski, P. J., Zaug, A. J., Sands, J., Gottschling, D. E., and Cech, T. R. (1982) Self-splicing RNA: autoexcision and autocyclization of the ribosomal RNA intervening sequence of *Tetrahymena*. *Cell*, **31**, 147.

2. Guerrier-Takada, C., Gardiner, K., Marsh, N., and Altman, S. (1983) The RNA moiety of ribonuclease P is the catalytic subunit of the enzyme. *Cell*, **35**, 849.
3. Cech, T. R. and Bass, B. L. (1986) Biological catalysis by RNA. *Annu. Rev. Biochem.*, **55**, 599.
4. Peebles, C. L., Perlman, P. S., Mecklenburg, K. L., Petrillo, M. L., Tabor, J. H., Jarrell, K. A., and Cheng, H. L. (1986) A self-splicing RNA excises an intron lariat. *Cell*, **44**, 213.
5. Van der Veen, R., Arnberg, A. C., van der Horst, G., Bonen, L., Tabak, H. F., and Grivell, L. A. (1986) Excised group II introns in yeast mitochondria are lariats and can be formed by self-splicing in vitro. *Cell*, **44**, 225.
6. Lambowitz, A. M. and Perlman, P. S. (1990) Involvement of aminoacyl-tRNA synthetases and other proteins in group I and group II intron splicing. *Trends Biochem. Sci.*, **15**, 440.
7. Michel, F. and Westhof, E. (1990) Modelling of the three-dimensional architecture of group I catalytic introns based on comparative sequence analysis. *J. Mol. Biol.*, **216**, 585.
8. Reinhold-Hurek, B. and Shub, D. A. (1992) Self-splicing introns in tRNA genes of widely divergent bacteria. *Nature*, **357**, 173.
9. Sogin, M. L., Ingold, A., Karlok, M., Nielsen, H., and Engberg, J. (1986) Phylogenetic evidence for the acquisition of ribosomal RNA introns subsequent to the divergence of some of the major *Tetrahymena* groups. *EMBO J.*, **5**, 3625.
10. Belfort, M. (1991) Self-splicing introns in prokaryotes: migrant fossils? *Cell*, **64**, 9.
11. Michel, F., Umesono, K., and Ozeki, H. (1989) Comparative and functional anatomy of group II catalytic introns—a review. *Gene*, **82**, 5.
12. Ferat, J-L. and Michel, F. (1993) Group II self-splicing introns in bacteria. *Nature*, **364**, 358.
13. Lambowitz, A. M. and Belfort, M. (1993) Introns as mobile genetic elements. *Annu. Rev. Biochem.*, **62**, 587.
14. Darr, S. C., Brown, J. W., and Pace, N. R. (1992) The varieties of ribonuclease P. *Trends Biochem. Sci.*, **17**, 178.
15. Altman, S., Kirsebom, L., and Talbot, S. (1993) Recent studies of ribonuclease P. *FASEB J.*, **7**, 7.
16. Prody, G. A., Bakos, J. T., Buzayan, J. M., Schneider, I. R., and Bruening, G. (1986) Autocatalytic processing of dimeric plant virus satellite RNA. *Science*, **231**, 1577.
17. Hutchins, C. J., Rathjen, P. D., Forster, A. C., and Symons, R. H. (1986) Self-cleavage of plus and minus RNA transcripts of avocado sunblotch viroid. *Nucleic Acids Res.*, **14**, 3627.
18. Forster, A. C. and Symons, R. H. (1987) Self-cleavage of plus and minus RNAs of a virusoid and a structural model for the active sites. *Cell*, **49**, 211.
19. Buzayan, J. M., Gerlach, W. L., and Bruening, G. (1986) Satellite tobacco ringspot virus RNA: A subset of the RNA sequence is sufficient for autolytic processing. *Proc. Natl Acad. Sci. USA*, **83**, 8859.
20. Epstein, L. M. and Gall, J. G. (1987) Self-cleaving transcripts of satellite DNA from the newt. *Cell*, **48**, 535.
21. Buzayan, J. M., Gerlach, W. L., and Bruening, G. (1986) Non-enzymatic cleavage and ligation of RNAs complementary to a plant virus satellite RNA. *Nature*, **323**, 349.
22. Hampel, A., Tritz, R., Hicks, M., and Cruz, P. (1990) 'Hairpin' catalytic RNA model: evidence for helices and sequence requirement for substrate RNA. *Nucleic Acids Res.*, **18**, 299.
23. Berzal-Herranz, A., Joseph, S., and Burke, J. M. (1992) *In vitro* selection of active hairpin ribozymes by sequential RNA-catalysed cleavage and ligation reactions. *Genes Dev.*, **6**, 129.

24. Berzal-Herranz, A., Joseph, S., Chowrira, B. M., Butcher, S. E., and Burke, J. M. (1993) Essential nucleotide sequences and secondary structure elements of the hairpin ribozyme. *EMBO J.*, **12**, 2567.
25. Kuo, M. Y-P., Sharmeen, L., Dinter-Gottlieb, G., and Taylor, J. (1988) Characterization of self-cleaving RNA sequences on the genome and antigenome of human hepatitis delta virus. *J. Virol.*, **62**, 4439.
26. Perrotta, A. T. and Been, M. D. (1991) A pseudoknot-like structure required for efficient self-cleavage of hepatitis delta virus RNA. *Nature*, **350**, 434.
27. Saville, B. J. and Collins, R. A. (1990) A site-specific, self-cleavage reaction performed by a novel RNA in *Neurospora* mitochondria. *Cell*, **61**, 685.
28. Saville, B. J. and Collins, R. A. (1991) RNA-mediated ligation of self-cleavage products of a *Neurospora* mitochondrial plasmid transcript. *Proc. Natl Acad. Sci. USA*, **88**, 8826.
29. Pyle, A. M. (1993) Ribozymes: a distinct class of metalloenzymes. *Science*, **261**, 709.
30. Perreault, J. P., Wu, T. F., Cousineau, B., Ogilvie, K. K., and Cedergren, R. (1990) Mixed deoxyribo- and ribo-oligonucleotides with catalytic activity. *Nature*, **344**, 565.
31. Chowrira, B. M. and Burke, J. M. (1991) Binding and cleavage of nucleic acids by the 'hairpin' ribozyme. *Biochemistry*, **30**, 8518.
32. Guerrier-Takada, C., Haydock, K., Allen, L., and Altman, S. (1986) Metal ion requirements and other aspects of the reaction catalyzed by M1 RNA, the RNA subunit of ribonuclease P from *Escherichia coli*. *Biochemistry*, **25**, 1509.
33. Woodson, S. A. and Cech, T. R. (1989) Reverse self-splicing of the *Tetrahymena* group I intron: implication for the directionality of splicing and for intron transposition. *Cell*, **57**, 335.
34. Inoue, T., Sullivan, F. X., and Cech, T. R. (1986) New reactions of the ribosomal RNA precursor of *Tetrahymena* and the mechanism of self-splicing. *J. Mol. Biol.*, **189**, 143.
35. Herschlag, D. and Cech, T. R. (1990) Catalysis of RNA cleavage by the *Tetrahymena thermophila* ribozyme. 1. Kinetic description of the reaction of an RNA substrate complementary to the active site. *Biochemistry*, **29**, 10159.
36. Morl, M. and Schmelzer, C. (1990) Integration of group II intron bI1 into a foreign RNA by reversal of the self-splicing reaction *in vitro*. *Cell*, **60**, 629.
37. Augustin, S., Muller, M. W., and Schweyen, R. J. (1990) Reverse self-splicing of group II intron RNAs *in vitro*. *Nature*, **343**, 383.
38. Jarrell, K. A., Peebles, C. L., Dietrich, R. C., Romiti, S. L., and Perlman, P. S. (1988) Group II intron self-splicing: Alternative reaction conditions yield novel products. *J. Biol. Chem.*, **263**, 3432.
39. Jacquier, A. and Rosbash, M. (1986) Efficient *trans*-splicing of a yeast mitochondrial RNA group II intron implicates a strong 5′ exon–intron interaction. *Science*, **234**, 1099.
40. Jacquier, A. and Jacquesson-Breuleux, N. (1991) Splice site selection and role of the lariat in a group II intron. *J. Mol. Biol.*, **219**, 415.
41. Kay, P. and Inoue, T. (1987) Catalysis of splicing-related reactions between dinucleotides by a ribozyme. *Nature*, **327**, 343.
42. Michel, F., Hanna, M., Green, R., Bartel, D. P., and Szostak, J. W. (1989) The guanosine binding site of the *Tetrahymena* ribozyme. *Nature*, **342**, 391.
43. Been, M. D. and Perrota, A. T. (1991) Group I intron self-splicing with adenosine: evidence for a single nucleoside-binding site. *Science*, **252**, 434.
44. Herschlag, D., Piccirilli, J. A., and Cech, T. R. (1991) Ribozyme-catalysed and nonenzymatic reactions of phosphate diesters: rate effects upon substitution of sulfur for a nonbridging phosphoryl oxygen atom. *Biochemistry*, **30**, 4844.

45. Rajagopal, J., Doudna, J. A., and Szostak, J. W. (1989) Stereochemical course of catalysis by the *Tetrahymena* ribozyme. *Science*, **244**, 692.
46. Suh, E-R. and Waring, R. B. (1992) A phosphorothioate at the 3′ splice site inhibits the second splicing step in a group I intron. *Nucleic Acids Res.*, **20**, 6303.
47. Jaeger, L., Westhof, E., and Michel, F. (1993) Monitoring of the cooperative unfolding of the sunY group I intron of bacteriophage T4. The active form of the sunY ribozyme is stabilized by multiple interactions with 3′ terminal introns components. *J. Mol. Biol.*, **234**, 331.
48. Moore, M. J. and Sharp, P. A. (1993) Evidence for two active sites in the spliceosome provided by stereochemistry of pre-mRNA splicing. *Nature*, **365**, 364.
49. Moore, M. J., Query, C. C., and Sharp, P. A. (1993) Splicing of precursors to mRNA by the spliceosome. In *The RNA world*. Gesteland, R. and Atkins, J. (ed.). Cold Spring Harbor Laboratory Press, Cold Spring Harbor, New York, p. 303.
50. Steitz, T. A. and Steitz, J. A. (1993) A general two-metal-ion mechanism for catalytic RNA. *Proc. Natl Acad. Sci. USA*, **90**, 6498.
51. Chanfreau, G. and Jacquier, A. (1993) Interaction of intronic boundaries is required for the second splicing step efficiency of a group II intron. *EMBO J.*, **12**, 5173.
52. Parker, R. and Siliciano, P. G. (1993) Evidence for an essential non-Watson–Crick interaction between the first and last nucleotides of a nuclear pre-messenger RNA intron. *Nature*, **361**, 660.
53. Eckstein, F. (1985) Nucleoside phosphorothioates. *Annu. Rev. Biochem.*, **54**, 367.
54. McSwiggen, J. A. and Cech, T. R. (1989) Stereochemistry of RNA cleavage by the *Tetrahymena* ribozyme and evidence that the chemical step is not rate-limiting. *Science*, **244**, 679.
55. van Tol, H., Buzayan, J. M., Feldstein, P. A., Eckstein, F., and Bruening, G. (1990) Two autolytic processing reactions of a satellite RNA proceed with inversion of configuration. *Nucleic Acids Res.*, **18**, 1971.
56. Dahm, S. C. and Uhlenbeck, O. C. (1991) Role of divalent metal ions in the hammerhead RNA cleavage reaction. *Biochemistry*, **30**, 9464.
57. Grosshans, C. A. and Cech, T. R. (1989) Metal ion requirements for sequence-specific endoribonuclease activity of the *Tetrahymena* ribozyme. *Biochemistry*, **28**, 6888.
58. Pan, T., Long, D. M., and Uhlenbeck, O. C. (1993) Divalent metal ions in RNA folding and catalysis. In *The RNA world*. Gesteland, R. and Atkins, J. (ed.). Cold Spring Harbor Laboratory Press, Cold Spring Harbor, New York, p. 271.
59. Buzayan, J. M., Feldstein, P. A., Segrelles, C., and Bruening, G. (1988) Autolytic processing of a phosphorothioate diester bond. *Nucleic Acids Res.*, **16**, 4009.
60. Piccirilli, J. A., Vyle, J. S., Caruthers, M. H., and Cech, T. R. (1993) Metal ion catalysis in the *Tetrahymena* ribozyme reaction. *Nature*, **361**, 85.
61. Beese, L. S. and Steitz, T. A. (1991) Structural basis for the 3′–5′ exonuclease activity of *Escherichia coli* DNA polymerase I: A two metal ion mechanism. *EMBO J.*, **10**, 25.
62. Fedor, M. J. and Uhlenbeck, O. C. (1990) Substrate sequence effects on 'hammerhead' RNA catalytic efficiency. *Proc. Natl Acad. Sci. USA*, **87**, 1668.
63. Hampel, A. and Tritz, R. (1989) RNA catalytic properties of the minimum (—)sTRSV sequence. *Biochemistry*, **28**, 4929.
64. Perrotta, A. T. and Been, M. D. (1992) Cleavage of oligoribonucleotides by a ribozyme derived from the hepatitis delta virus RNA sequence. *Biochemistry*, **31**, 16.
65. Cech, T. R. (1993) Structure and mechanism of the large catalytic RNAs: Group I and Group II introns and ribonuclease P. In *The RNA world*, Gesteland, R. and Atkins, J. (ed.). Cold Spring Harbor Laboratory Press, Cold Spring Harbor, New York, p. 39.

66. Bevilacqua, P. C., Kierzek, R., Johnson, K. A., and Turner, D. H. (1992) Dynamics of ribozyme binding of substrate revealed by fluorescence detected stopped-flow. *Science*, **258**, 1355.
67. Tallsjö, A. and Kirsebom, L. A. (1993) Product release is a rate-limiting step during cleavage by the catalytic RNA subunit of *Escherichia coli* RNase P. *Nucleic Acids Res.*, **21**, 51.
68. Reich, C., Olsen, G. J., Pace, B., and Pace, N. R. (1988) Role of the protein moiety of ribonuclease P, a ribonucleoprotein enzyme. *Science*, **239**, 178.
69. Heus, H. A. and Pardi, A. (1991) Nuclear magnetic resonance studies of the hammerhead ribozyme domain. Secondary structure formation and magnesium ion dependence. *J. Mol. Biol.*, **217**, 113.
70. Pace, N. R., Smith, D. K., Olsen, G. J., and James, B. D. (1989) Phylogenetic comparative analysis and the secondary structure of ribonuclease P RNA—a review. *Gene*, **82**, 65.
71. Woese, C. R. and Pace, N. R. (1993) Probing RNA structure, function, and history by comparative analysis. In *The RNA world*, Gesteland, R. and Atkins, J. (ed.). Cold Spring Harbor Laboratory Press, Cold Spring Harbor, New York, p. 91.
72. Haas, E. S., Morse, D. P., Brown, J. W., Schmidt, F. J., and Pace, N. R. (1991) Long-range structure in ribonuclease P RNA. *Science*, **254**, 853.
73. Heus, H. A., Uhlenbeck, O. C., and Pardi, A. (1990) Sequence-dependent structural variations of hammerhead RNA enzymes. *Nucleic Acids Res.*, **18**, 1103.
74. Bartel, D. P. and Szostak, J. W. (1993) Isolation of new ribozymes from a large pool of random sequences. *Science*, **261**, 1411.
75. Green, R., Ellington, A. D., and Szostak, J. W. (1990) *In vitro* genetic analysis of the *Tetrahymena* self-splicing intron. *Nature*, **347**, 406.
76. Szostak, J. W. and Ellington, A. D. (1993) *In Vitro* selection of functional RNA sequences. In *The RNA world*, Gesteland, R. and Atkins, J. (ed.). Cold Spring Harbor Laboratory Press, Cold Spring Harbor, New York, p. 511.
77. Ehresmann, C., Baudin, F., Mougel, M., Romby, P., Ebel, J. P., and Ehresmann, B. (1987) Probing the structure of RNAs in solution. *Nucleic Acids Res.*, **22**, 9109.
78. Latham, J. A. and Cech, T. R. (1989) Defining the inside and outside of a catalytic RNA molecule. *Science*, **245**, 276.
79. Pyle, A. M., Murphy, F. L., and Cech, T. R. (1992) RNA substrate binding site in the catalytic core of the *Tetrahymena* ribozyme. *Nature*, **358**, 123.
80. Moore, M. J. and Sharp, P. A. (1992) Specific modification of pre-mRNA: The 2′ hydroxyl groups at the splice sites. *Science*, **256**, 992.
81. Wyatt, J. R., Sontheimer, E. J., and Steitz, J. A. (1992) Site-specific cross-linking of mammalian U5 snRNP to the 5′ splice site before the first step of pre-mRNA splicing. *Genes Dev.*, **6**, 2542.
82. Wang, J. F., Downs, W. D., and Cech, T. R. (1993) Movement of the guide sequence during RNA catalysis by a group I ribozyme. *Science*, **260**, 504.
83. Michel, F., Jaeger, L., Westhof, E., Kuras, R., and Tihy, F. (1992) Activation of the catalytic core of a group I intron by a remote 3′ splice junction. *Genes Dev.*, **6**, 1373.
84. Beaudry, A. A. and Joyce, G. F. (1990) Minimum secondary structure requirements for catalytic activity of self-splicing group I intron. *Biochemistry*, **29**, 6534.
85. Haseloff, J. and Gerlach, W. L. (1988) Simple RNA enzymes with new and highly specific endoribonuclease activities. *Nature*, **334**, 585.
86. Yang, J. H., Perreault, J. P., Labuda, D., Usman, N., and Cedergren, R. (1990) Mixed DNA/RNA polymers are cleaved by the hammerhead ribozyme. *Biochemistry*, **29**, 11156.

87. Perreault, J. P., Labuda, D., Usman, N., Yang, J. H., and Cedergren, R. (1991) Relationship between 2′-hydroxyls and magnesium binding in the hammerhead RNA domain: a model for ribozyme catalysis. *Biochemistry*, **30**, 4020.
88. Feldstein, P. A., Buzayan, J. M., and Bruening, G. (1989) Two sequences participating in the autolytic processing of satellite tobacco ringspot virus complementary RNA. *Gene*, **82**, 53.
89. Haseloff, J. and Gerlach, W. L. (1989) Sequences required for self-catalysed cleavage of the satellite RNA of tobacco ringspot virus. *Gene*, **82**, 43.
90. Michel, F., Jacquier, A., and Dujon, B. (1982) Comparison of fungal mitochondrial introns reveals extensive homologies in RNA secondary structure. *Biochimie*, **64**, 867.
91. Davies, R. W., Waring, R. B., Ray, J. A., Brown, T. A., and Scazzocchio, C. (1982) Making ends meet: A model for RNA splicing in fungal mitochondria. *Nature*, **300**, 719.
92. Perea, J. and Jacq, C. (1986) Role of the 5′ hairpin structure in the splicing accuracy of the fourth intron of the yeast *cob-box* gene. *EMBO J.*, **4**, 3281.
93. Been, M. D. and Cech, T. R. (1986) One binding site determines sequence specificity of *Tetrahymena* pre-rRNA self-splicing, *trans*-splicing, and RNA enzyme activity. *Cell*, **47**, 207.
94. Waring, R. B., Towner, P., Minter, S. J., and Davies, R. W. (1986) Splice-site selection by a self-splicing RNA of *Tetrahymena*. *Nature*, **321**, 133.
95. Inoue, T., Sullivan, F. X., and Cech, T. R. (1985) Intermolecular exon ligation of the rRNA precursor of *Tetrahymena*: oligonucleotides can function as 5′ exons. *Cell*, **43**, 431.
96. Zaug, A. J., Been, M. D., and Cech, T. R. (1986) The *Tetrahymena* ribozyme acts like an RNA restriction endonuclease. *Nature*, **324**, 429.
97. Sugimoto, N., Tomka, M., Kierzek, R., Bevilacqua, P. C., and Turner, D. H. (1989) Effects of substrate structure on the kinetics of circle opening reactions of the self-splicing intervening sequence from *Tetrahymena thermophila*: evidence for substrate and Mg^{2+} binding interactions. *Nucleic Acids Res.*, **17**, 355.
98. Pyle, A. M., McSwiggen, J. A., and Cech, T. R. (1990) Direct measurement of oligonucleotide substrate binding to wild-type and mutant ribozymes from *Tetrahymena*. *Proc. Natl Acad. Sci. USA*, **87**, 8187.
99. Pyle, A. M. and Cech, T. R. (1991) Ribozyme recognition of RNA by tertiary interactions with specific ribose 2′-OH groups. *Nature*, **350**, 628.
100. Bevilacqua, P. C. and Turner, D. H. (1991) Comparison of binding of mixed ribose-deoxyribose analogues of CUCU to a ribozyme and to GGAGAA by equilibrium dialysis: evidence for ribozyme specific interactions with 2′ OH groups. *Biochemistry*, **30**, 10632.
101. Doudna, J. A. and Szostak, J. W. (1989) RNA-catalysed synthesis of complementary-strand RNA. *Nature*, **339**, 519.
102. Doudna, J. A., Cormack, B. P., and Szostak, J. W. (1989) RNA structure, not sequence, determines the 5′ splice-site specificity of a group I intron. *Proc. Natl Acad. Sci. USA*, **86**, 7402.
103. Herschlag, D. (1992) Evidence for processivity and two-step binding of the RNA substrate from studies of J1/2 mutants of the *Tetrahymena* ribozyme. *Biochemistry*, **31**, 1386.
104. Jacquier, A. and Michel, F. (1987) Multiple exon-binding sites in class II self-splicing introns. *Cell*, **50**, 17.
105. Jacquier, A. and Michel, F. (1990) Base-pairing interactions involving the 5′ and 3′-terminal nucleotides of group II introns. *J. Mol. Biol.*, **213**, 437.

106. Wallasch, C., Morl, M., Niemer, I., and Schmelzer, C. (1991) Structural requirements for selection of 5′ and 3′ splice sites of group II introns. *Nucleic Acids Res.*, **19**, 3307.
107. Morl, M., Niemer, I., and Schmelzer, C. (1992) New reactions catalysed by a group II intron ribozyme with RNA and DNA substrates. *Cell*, **70**, 803.
108. Been, M. D. and Cech, T. R. (1985) Sites of circularization of the *Tetrahymena* rRNA IVS are determined by sequence and influenced by position and secondary structure. *Nucleic Acids Res.*, **13**, 8389.
109. Suh, E. R. and Waring, R. B. (1990) Base pairing between the 3′ exon and an internal guide sequence increases 3′ splice site specificity in the *Tetrahymena* self-splicing rRNA intron. *Mol. Cell. Biol.*, **10**, 2960.
110. Partono, S. and Lewin, A. S. (1990) Splicing of *COB* intron 5 requires pairing between the internal guide sequence and both flanking exons. *Proc. Natl Acad. Sci. USA*, **87**, 8192.
111. Winter, A. J., Koerkamp, M. J. A. G., and Tabak, H. F. (1992) Splice site selection by intron aI3 of the *COX1* gene from *Saccharomyces cerevisiae*. *Nucleic Acids Res.*, **20**, 3897.
112. Price, J. V. and Cech, T. R. (1988) Determinants of the 3′ splice site for self-splicing of the *Tetrahymena* pre-rRNA. *Genes Dev.*, **2**, 1439.
113. Burke, J. M. (1989) Selection of the 3′-splice site in group I introns. *FEBS Letters*, **250**, 129.
114. Michel, F., Netter, P., Xu, M.-Q., and Shub, D. A. (1990) Mechanism of 3′ splice site selection by the catalytic core of the *sunY* intron of bacteriophage T4: the role of a novel base-pairing interaction in group I introns. *Genes Dev.*, **4**, 777.
115. Schmelzer, C. and Muller, M. W. (1987) Self-splicing of group II introns *in vitro*: lariat formation and 3′ splice site selection in mutant RNAs. *Cell*, **51**, 753.
116. Bass, B. L. and Cech, T. R. (1984) Specific interaction between the self-splicing RNA of *Tetrahymena* and its guanosine substrate: implications for biological catalysis by RNA. *Nature*, **308**, 820.
117. Yarus, M., Illangesekare, M., and Christian, E. (1991) An axial binding site in the *Tetrahymena* precursor RNA. *J. Mol. Biol.*, **222**, 995.
118. Peck, M. K. and Altman, S. (1991) Kinetics of the processing of the precursor to 4.5 S RNA, a naturally occurring substrate for RNase P from *Escherichia coli*. *J. Mol. Biol.*, **221**, 1.
119. Kirsebom, L. A. and Svärd, S. G. (1992) The kinetics and specificity of cleavage by RNase P is mainly dependent on the structure of the amino acid acceptor stem. *Nucleic Acids Res.*, **20**, 425.
120. McClain, W. H., Guerrier-Takada, C., and Altman, S. (1987) Model substrates for an RNA enzyme. *Science*, **238**, 527.
121. Schlegl, J., Fürste, J-P., Bald, R., Erdmann, V. A., and Harrmann, R. K. (1992) Cleavage efficiencies of model substrates for ribonuclease P from *Escherichia coli* and *Thermus thermophilus*. *Nucleic Acids Res.*, **20**, 5963.
122. Kahle, D., Wehmeyer, U., and Krupp, G. (1990) Substrate recognition by RNase P and by the catalytic M1 RNA: identification of possible contact points in pre-tRNAs. *EMBO J.*, **9**, 1929.
123. Thurlow, D. L., Shilowski, D., and Marsh, T. L. (1991) Nucleotides in precursor tRNAs that are required intact for catalysis by RNase P RNAs. *Nucleic Acids Res.*, **19**, 885.
124. Forster, A. C. and Altman, S. (1990) External guide sequences for an RNA enzyme. *Science*, **249**, 783.
125. Perreault, J. P. and Altman, S. (1992) Important 2′-hydroxyl groups in model substrates for M1 RNA, the catalytic RNA subunit of RNase-P from *Escherichia coli*. *J. Mol. Biol.*, **226**, 399.

126. Svärd, S. G. and Kirsebom, L. A. (1992) Several regions of a tRNA precursor determine the *Escherichia coli* RNase P cleavage site. *J. Mol. Biol.*, **227**, 1019.
127. Perreault, J. P. and Altman, S. (1993) Pathway of activation by magnesium ions of substrates for the catalytic subunit of RNase P from *Escherichia coli*. *J. Mol. Biol.*, **230**, 750.
128. Guerrier-Takada, C., Lumelsky, N., and Altman, S. (1989) Specific interactions in RNA enzyme-substrate complexes. *Science*, **246**, 1578.
129. Burgin, A. B. and Pace, N. R. (1990) Mapping the active site of ribonuclease P RNA using a substrate containing a photoaffinity agent, *EMBO J.*, **9**, 4111.
130. Sawa, H. and Shimura, Y. (1992) Association of U6 snRNA with the 5′ splice site region of pre-messenger RNA in the spliceosome. *Genes Dev.*, **6**, 244.
131. Wassarman, D. A. and Steitz, J. A. (1992) Interactions of small nuclear RNAs with precursor messenger RNA during *in vitro* splicing. *Science*, **257**, 1918.
132. Kandels-Lewis, S. and Séraphin, B. (1993) Involvement of U6 snRNA in 5′ splice site selection. *Science*, **262**, 2035.
133. Newman, A. J. and Norman, C. (1992) U5 snRNA interacts with exon sequences at 5′ and 3′ splice sites. *Cell*, **68**, 743.
134. Cortes, J. J., Sontheimer, E. J., Seiwert, S. D., and Steitz, J. A. (1993) Mutations in the conserved loop of human U5 snRNA generate use of novel cryptic 5′ splice sites *in vivo*. *EMBO J.*, **12**, 5181.
135. Sontheimer, E. J. and Steitz, J. A. (1993) The U5 and U6 small nuclear RNAs as active site components of the spliceosome. *Science*, **262**, 1989.
136. Madhani, H. D. and Guthrie, C. (1992) A novel base-pairing interaction between U2 and U6 snRNAs suggests a mechanism for catalytic activation of the spliceosome. *Cell*, **71**, 803.
137. Guthrie, C. (1991) Messenger RNA splicing in yeast: clues to why the spliceosome is a ribonucleoprotein. *Science*, **253**, 157.
138. Weiner, A. M. (1993) mRNA splicing and autocatalytic introns: distant cousins or the products of chemical determinism? *Cell*, **72**, 161.
139. Reich, C. I., Hoy, R. W. V., Porter, G. L., and Wise, J. A. (1992) Mutations at the 3′ splice site can be suppressed by compensatory base changes in U1 snRNA in fission yeast. *Cell*, **69**, 1159.
140. Séraphin, B. and Kandels-Lewis, S. (1993) 3′ splice site recognition in *S. cerevisiae* does not require base pairing with U1 snRNA. *Cell*, **73**, 803.
141. Bonen, L. (1993) *Trans*-splicing of pre-mRNA in plants, animals, and protists. *FASEB J.*, **7**, 40.
142. Goldschmidt-Clermont, M., Choquet, Y., Girard-Bascou, J., Michel, F., Schirmer-Rahire, M., and Rochaix, J. D. (1991) A small chloroplast RNA may be required for *trans*-splicing in *Chlamydomonas reinhardtii*. *Cell*, **65**, 135.
143. Copertino, D. W. and Hallick, R. B. (1993) Group II and group III introns of twintrons: potential relationships to nuclear pre-mRNA introns. *Trends Biochem. Sci.*, **18**, 467.
144. Mattaj, I. W., Tollervey, D., and Séraphin, B. (1993) Small nuclear RNAs in messenger RNA and ribosomal RNA processing. *FASEB J.*, **7**, 47.
145. Forster, A. C. and Altman, S. (1990) Similar cage-shaped structures for the RNA components of all ribonuclease P and ribonuclease MRP enzymes. *Cell*, **62**, 407.
146. Blackburn, E. H. (1993) Telomerase. In *The RNA world*, Gesteland, R. and Atkins, J. (ed.). Cold Spring Harbor Laboratory Press, Cold Spring Harbor, New York, p. 557.
147. Noller, H. F. (1993) On the origin of the ribosome: Coevolution of subdomains of tRNA and rRNA. In *The RNA world*, Gesteland, R. and Atkins, J. (ed.). Cold Spring Harbor Laboratory Press, Cold Spring Harbor, New York, p. 137.

148. Beaudry, A. A. and Joyce, G. F. (1992) Directed evolution of an RNA enzyme. *Science*, **257**, 635.
149. Gold, L., Tuerk, C., Allen, P., Brinkley, J., Brown, D., Green, L., *et al.* (1993) RNA: the shape of things to come. In *The RNA world*. Gesteland, R. and Atkins, J. (ed.). Cold Spring Harbor Laboratory Press, Cold Spring Harbor, New York, p. 497.
150. Green, R. and Szostak, J. W. (1992) Selection of a ribozyme that functions as a superior template in a self-copying reaction. *Science*, **258**, 1910.
151. Joyce, G. F. and Orgel, L. E. (1993) Prospects for understanding the origin of the RNA world. In *The RNA world*. Gesteland, R. and Atkins, J. (ed.). Cold Spring Harbor Laboratory Press, Cold Spring Harbor, New York, p. 1.
152. Ruffner, D. E., Stormo, G. D., and Uhlenbeck, O. C. (1990) Sequence requirements of the hammmerhead RNA self-cleavage reaction. *Biochemistry*, **29**, 10695.
153. Uhlenbeck, O. C. (1987) A small catalytic oligoribonucleotide. *Nature*, **328**, 596.
154. Jeffries, A. C. and Symons, R. H. (1989) A catalytic 13-mer ribozyme. *Nucleic Acids Res.*, **17**, 1371.

2 Nuclear organization of pre-mRNA splicing factors and substrates

SUI HUANG and DAVID L. SPECTOR

1. Introduction

Pre-mRNA transcripts must be processed and transported to the cytoplasm where the mature mRNAs are translated into proteins. For most RNA polymerase II transcripts, this processing includes addition of a 7-methyl-guanosine cap structure at the 5′ end of the nascent RNA transcripts; hnRNP assembly; removal of non-coding intron regions and ligation of exons; 3′ end cleavage and polyadenylation; and the exchange of hnRNP proteins for mRNP proteins. Splicing occurs in a multicomponent complex termed a spliceosome. Many of the detailed biochemical steps involved in the pre-mRNA splicing reaction have been extensively studied *in vitro* and are well understood (for reviews see Chapters 3–8). However, the integration of the splicing reaction with other related cellular functions, such as transcription and RNA transport, is best studied in intact cells. The cellular approach to understanding the organization of nuclear function has lagged somewhat behind the *in vitro* approach. However, during the past few years, major inroads have been made toward our understanding of where in the nucleus pre-mRNA splicing occurs and how the organization of both splicing factors and substrates are interrelated to the transcriptional activity of the cell. In this chapter, the nuclear organization of both snRNP and non-snRNP RNA processing factors, as well as pre-mRNA substrates, are examined. In addition, their organization throughout the cell cycle, and the relationship of their nuclear distribution to the transcriptional activity of the cell, are discussed.

2. Localization of splicing factors and other RNA processing components

2.1 Small nuclear ribonucleoprotein particles: proteins and RNAs

The earliest components shown to be essential for pre-mRNA splicing were small nuclear ribonucleoprotein particles (snRNPs) (for a review, see 1; Chapter 5). Each of the snRNP particles (U1, U2) contains a single snRNA species with the exception of the U4/U6·U5 tri-snRNP particle which contains three snRNAs. In addition, each snRNP particle contains a common set of core proteins, as well as unique snRNA-specific proteins (for a review, see 2; Chapter 5). The sub-localization of splicing components in the nucleus of mammalian cells was first suggested in the 1980s by immunofluorescence localization studies using antibodies that recognize protein or RNA components of snRNPs (3–5). Although physicians had reported the presence of antibodies in the sera of some patients with a variety of autoimmune disorders as early as the 1960s (6–8), it was not until some 18 years later that the correlation was made between the speckled staining pattern of these antibodies (4–5), their ability to immunoprecipitate small nuclear RNAs from nuclear extracts (9), and the involvement of snRNPs in pre-mRNA splicing (10–12). Localization studies using anti-Sm antibodies directed against a variety of snRNP-specific proteins (13–17), or antibodies specifically directed against U1 (14, 16), U2 snRNP (14, 17), or the trimethylguanosine (m_3G) cap structure of snRNAs (15) have shown these components to be concentrated in 20–50 nuclear speckles, in addition to being distributed diffusely in the nucleoplasm of mammalian cells (Fig. 1). *In situ* hybridization studies using probes to several of the snRNAs have shown an identical localization (18–20). The diffuse nuclear staining of snRNPs may represent an excess soluble population of snRNPs or it may represent snRNPs in transit to speckles from their assembly sites in the cytoplasm (2).

At the electron microscopic level (3, 5, 21, 22), the speckled distribution of snRNPs has been shown to correspond to nuclear regions enriched in interchromatin granules and perichromatin fibrils (Fig. 2). The interchromatin granule regions contain particles with a mean diameter of 20–25 nm, which are linked together by thin fibrils (23). *In situ* autoradiographic studies following [^{3}H]-uridine incorporation have shown little to no labelling over internal regions of interchromatin granule clusters (24, 25). Based on these data and other studies, interchromatin granule clusters are thought to contain RNA species with a slow turnover rate (for a review, see 26). These studies are consistent with the findings that snRNPs are associated with these nuclear regions (5, 18, 19, 21, 22). Perichromatin fibrils are found at the periphery of regions of condensed chromatin and dispersed throughout the interchromatin space (26). These fibrils generally have a diameter of 3–5 nm but can measure up to 20 nm in diameter. In contrast to interchromatin granules, perichromatin fibrils are rapidly labeled with [^{3}H]-uridine, suggesting that they correspond to nascent pre-mRNA (27, 28). In fact, Monneron and Bernhard (23) first suggested a relationship between these fibrils and extranucleolar RNA synthesis. Both snRNP and hnRNP antigens have been localized to these fibrils (21, 22).

Subsequently, three-dimensional reconstruction techniques were used to examine the distribution of snRNPs (29, 30). These studies showed that regions of the speckled pattern were not isolated islands but, in fact, connections were observed between portions of the speckled pattern resulting in a latticework. Others have previously identified a ribonucleoprotein network or interchromatin net in two-

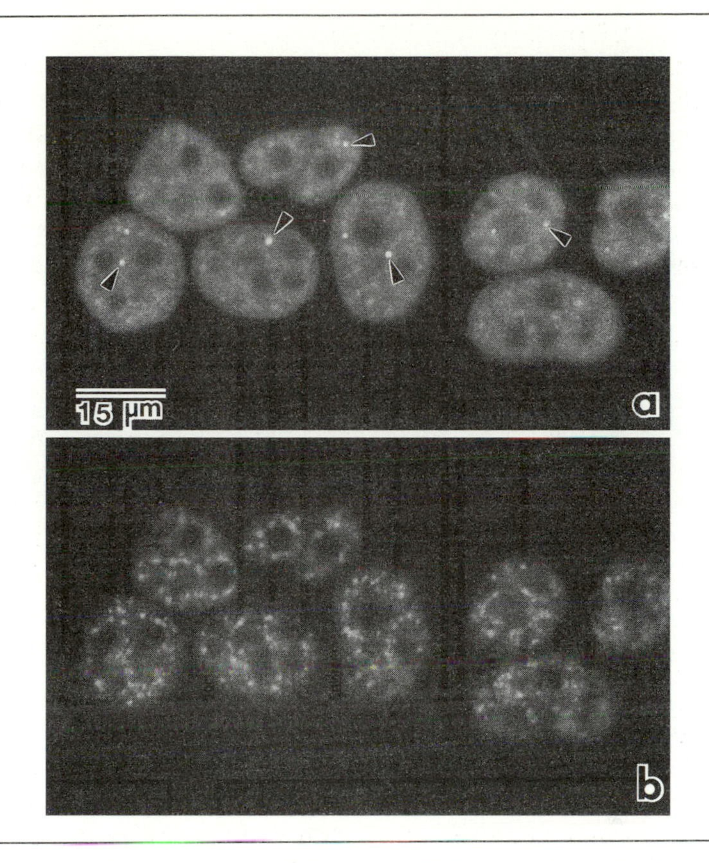

Fig. 1 Double-label immunofluorescence of snRNPs (a) and SC35 (b) in HeLa cells. Both snRNPs and SC35 co-localize in a speckled distribution. In addition, snRNPs are also diffusely distributed throughout the nucleoplasm (a) and are present in coiled bodies (a; arrowheads).

dimensional images by cytochemical staining (31) or in nuclear matrix preparations (32). Using immunofluorescence techniques, the latticework appears to be composed of two components: larger, more intensely stained regions and less intensely stained regions, which in many cases appear to connect two of the larger regions. Immunogold localization studies have shown the larger immunolabelled regions to correspond to interchromatin granule clusters and the less intensely labelled regions to correspond to perichromatin fibrils (30). This organization of splicing factors as a nuclear speckled pattern is, in all likelihood, dynamic and reflects the physiological state of the cell.

In addition to their speckled localization pattern, several studies have also shown snRNPs to be concentrated in coiled bodies (18, 20, 21, 33–37) (Fig. 1). Fakan *et al.* (21) examined the distribution of snRNPs in sections of mouse and rat liver. In addition to labelling interchromatin granules and perichromatin fibrils, which are components of the speckled immunostaining pattern (30), snRNP-specific antibodies were shown to immunolabel coiled bodies in these tissue sections (21). In a

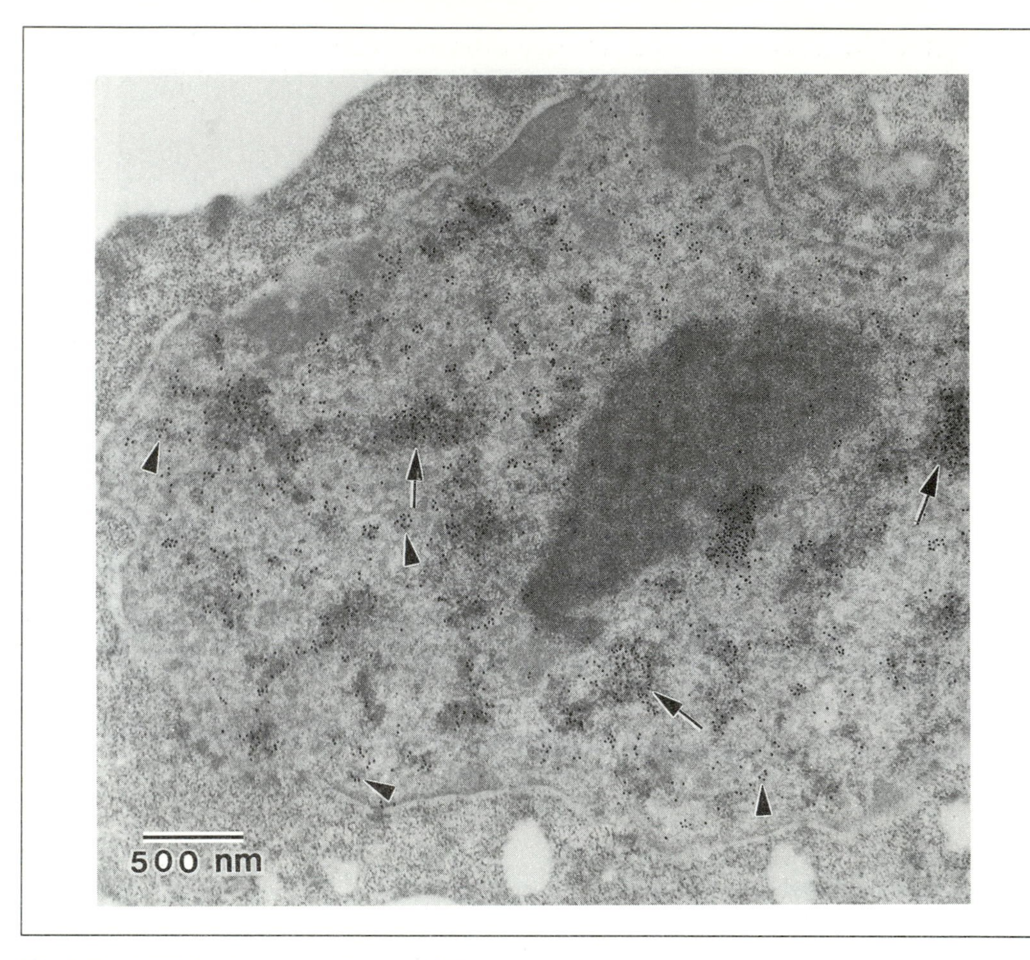

Fig. 2 The speckled staining pattern of SC35 corresponds to interchromatin granule clusters (arrows) and perichromatin fibrils (arrowheads). Connections are observed between many of the immunoreactive regions. Post-stained using the EDTA-regressive method.

second study, Eliceiri and Ryerse (33) immunolabelled HeLa cells with anti-Sm antibodies and described staining of a novel intranuclear structure. Examination of the micrographs in their paper suggests that the structures they observed were coiled bodies. In two studies at the light microscopic level, Leser *et al.* (38) described bright foci in HeLa cells which were immunolabelled with anti-Sm antibodies, and Carmo-Fonseca *et al.* (34, 35) described snRNAs to be present in foci. Based on studies by Raska *et al.* (36), Carmo-Fonseca *et al.* (18), and Spector *et al.* (20), it appears that these bright foci are coiled bodies.

Coiled bodies were first identified at the light microscopic level in 1903 by Ramón y Cajal, who called them 'accessory bodies' in neuronal cells (39). These structures are generally round and measure 0.5–1.0 μm in diameter and consist of coiled fibrillar strands (23, 40). Cytochemical staining of ribonucleoproteins by the

EDTA regressive method (41) demonstrated that coiled bodies contain RNPs (23). In addition, coiled bodies have been shown to contain orthophosphate ions and acid phosphatase activity (40). However, the lack of DNA (23), [^{3}H]-uridine incorporation (42), hnRNP proteins (35, 36), and several non-snRNP splicing factors (see below) in coiled bodies supports the idea that these inclusions are not essential for pre-mRNA splicing.

Raska *et al.* (36) have identified a novel protein, which is localized to the coiled body. The 80 kDa autoantigen, called p80-coilin, has been localized using sera from individuals with various autoimmune disorders (36, 43, 44). Partial cDNA clones coding for the p80-coilin antigen have been characterized (44). The sequence of these clones did not reveal any of the traditional RNA-recognition motifs (45, 46). However, the protein contains two stretches rich in arginine and lysine, which could be involved in RNA binding (47). p80-coilin is the first antigen identified that is restricted to the coiled body. Several other nuclear antigens have been localized to the coiled body in addition to being present in other nuclear regions; these include DNA topoisomerase I and the nucleolar U3 snRNP protein fibrillarin, as well as snRNPs (36). Other nuclear components such as DNA, nucleolin, nucleolar protein B23, 5S rRNP, hnRNP L protein, and SC35 have not been detected in coiled bodies (19, 36). Using the p80-coilin antibody, Raska *et al.* (36) determined that the number of coiled bodies varies among different cell types. Spector *et al.* (20) identified differences in the presence of coiled bodies in transformed cells, immortal cells, and cells of defined passage number. Coiled bodies were found in a low percentage (2–3%) of fibroblasts with defined passage number, in intermediate numbers of immortal cells (4–40%), but were most abundant in transformed cells (81–99%). Therefore, a direct correlation was found between the percentage of cells containing coiled bodies and oncogenic transformation. Based on these findings, it was proposed that the presence of coiled bodies is a reflection of cell physiology.

In frozen sections from rat and mouse brain, a frequent association between coiled bodies and nucleoli was observed (36). A relationship between coiled bodies and nucleoli has been suggested, since these structures are often found in close proximity (23, 48–50). At the cytochemical level, Seite *et al.* (49) and Raska *et al.* (43) have shown that coiled bodies are stained by the nucleolar-organizing region (NOR) silver-staining technique in a way similar to the fibrillar regions of the nucleolus. The association between coiled bodies and nucleoli was investigated further in cycling cells treated with actinomycin D or 5,6-dichloro-1-β-D-ribofuranosylbenzimidazole (DRB) at a dose that results in nucleolar segregation. This study demonstrated that when nucleoli segregate, the p80-coilin antigen co-localizes with fibrillarin-positive nucleolar components (43). Fibrillarin is a 34 kDa antigen associated with the nucleolar U3 snRNP particle (51–53), which is involved in pre-rRNA processing (54). In cells incubated with lower concentrations of actinomycin D (0.02 μg/ml), nucleoli did not segregate. However, p80-coilin was observed to be associated with paranucleolar structures similar to what was observed in untreated primary neurone cultures. Furthermore, Lafarga *et al.* (55) have shown an increase in the number of coiled bodies in supraoptic neurones of the rat, after stimulation of nucleolar transcription. A more recent study by Ochs *et al.* (56) demonstrated coiled bodies to be present within the nucleoli of several cell lines derived from breast cancer tissues. Together, these findings suggest that coiled bodies may be involved in the processing, transport, and storage of nucleolar metabolites (43). However, the presence of other nuclear components not involved in nucleolar function has suggested a more general role for this nuclear organelle (36).

2.2 Non-snRNP splicing factors

A number of non-snRNP splicing activities have been identified by *in vitro* complementation experiments (57, 58 for reviews; Chapter 6), and several have been purified to homogeneity. One of these factors, designated SF2 (59, 60) or alternative splicing factor (ASF) (61), is required for spliceosome assembly, and plays a role in 5′ splice site selection. A second factor, U2-auxiliary factor (U2AF), facilitates U2 snRNP binding to the pre-mRNA branch site (62, 63). A third factor, an 88 kDa protein, was identified by a monoclonal antibody directed against large nuclear ribonucleoprotein particles (64). A fourth factor, spliceosomal protein SC35, was identified using a monoclonal antibody directed against partially purified spliceosomes (65). Extracts depleted of this protein fail to carry out the first step of the splicing reaction and do not form spliceosomes (65). However, these depleted extracts can be complemented with extracts containing SC35 (65) or with purified SC35 (30), but not by extracts that do not contain this antigen (65).

Antibodies to these four factors have been generated and their nuclear organization has been examined. The subnuclear distribution of SC35 was examined in HeLa cells by immunofluorescence labelling. The nuclear distribution of SC35 appears as a speckled pattern that occupies a portion of the nucleoplasm excluding the nucleoli (30, 65) (Fig. 1). However, unlike the snRNP staining, SC35 was not detected diffusely throughout the nucleoplasm (65) nor was it detected in coiled bodies (30, 35) (Fig. 1). This immunolabelling pattern was identical regardless of the fixation used (formaldehyde, glutaraldehyde, methanol) and even in cells which were not fixed prior to immunolabelling. Similar to SC35, SF2/ASF (Spector and Krainer, unpublished) and p88 (G. Ast, personal communication) both exhibit a speckled nuclear distribution but do not stain coiled bodies.

In contrast to the localization of snRNPs and the non-snRNP splicing factors discussed above, Zamore and Green (62) have used peptide antibodies to both subunits of U2AF, which facilitates U2 snRNP binding to the branch site (62, 63, 66), to show this protein to be localized to coiled bodies as well as being diffusely distributed in the nucleoplasm. This reported localization is surprising, since U2AF contains the type of arginine/serine-rich (RS-rich) domain (67, 68) that targets certain proteins to the speckled region (see below), and both U2 snRNA and U2 snRNP localize in nuclear speckles in normal human cells (17, 20). The significance of this unique localization of U2AF, as compared with other splicing factors, remains to be determined.

2.3 An arginine/serine-rich domain targets proteins to nuclear speckles

Insights into the mechanism of the sub-localization of splicing components to speckled nuclear regions have been obtained by the identification of a specific targeting signal (69). An RS-rich domain composed of approximately 120 amino acids from two different *Drosophila* pre-mRNA splicing regulators, *suppressor-of-white-apricot* (*SWAP*) and *transformer* (*tra*), was determined to target proteins into the nucleus and direct them to the speckled regions. Furthermore, the fusion of an RS domain to β-galactosidase directs this protein to the speckle region, suggesting that the RS domain is essential and sufficient to target proteins to this subnuclear region (69). More recently, Hedley *et al.* (70) have re-examined the *tra* RS domain and have identified a nucleoplasmin-like bipartite sequence adjacent to a short stretch of basic amino acids that contains the minimally required targeting sequence. Multiple splicing factors including *tra2* (71) in *Drosophila* and SC35 (72), SF2/ASF (73, 74), U2AF (67, 68), and the U1 snRNP 70 kDa polypeptide (75) in mammalian cells also contain an RS-rich domain. In fact, Roth and co-workers (76–78) had previously identified a family of nuclear phosphoproteins with RS-rich carboxy-terminal domains. The most abundant of these proteins in human (SRp30) and *Drosophila* (SRp55) cells can replace one another or SF2/ASF as an essential splicing factor in a human cell-free system (78). A monoclonal antibody that recognizes the entire family of proteins through a shared phosphoepitope labels mammalian cells in a speckled distribution (76). Although the functional significance of the targeting of splicing factors to the nuclear speckles remains to be determined, it has been suggested to play a role in the regulation of pre-mRNA splicing in the cell nucleus.

2.4 Reorganization of splicing factors during cell division

Several groups have examined the reorganization of splicing factors during mitosis (16, 30, 38, 79, 80, 83). As cells enter prophase, prior to nuclear envelope breakdown, the chromosomes begin to condense (Fig. 3d) and both snRNP antigens (Fig. 3c) and SC35 (Fig. 3b) are more diffusely distributed than in interphase. During metaphase the chromosomes are maximally condensed, and they align along the equatorial plane of the spindle (Fig. 3h). At this phase of mitosis, both SC35 (Fig. 3f) and snRNP antigens (Fig. 3g) are detected as diffuse staining throughout the cytoplasm. While snRNP antigens and SC35 redistribute in a similar manner during prophase and metaphase, a dramatic difference in their localization is observed during the later phases of mitosis. During late anaphase, when chromosomes arrive at opposite poles, SC35 is found in a small number of clusters (Fig. 3j), while snRNP antigens remain diffusely distributed throughout the cell (Fig. 3k). Thus, the assembly of SC35 into concentrated speckles is initiated before the formation of the daughter cell nucleus. A similar distribution has been reported for a nuclear matrix-

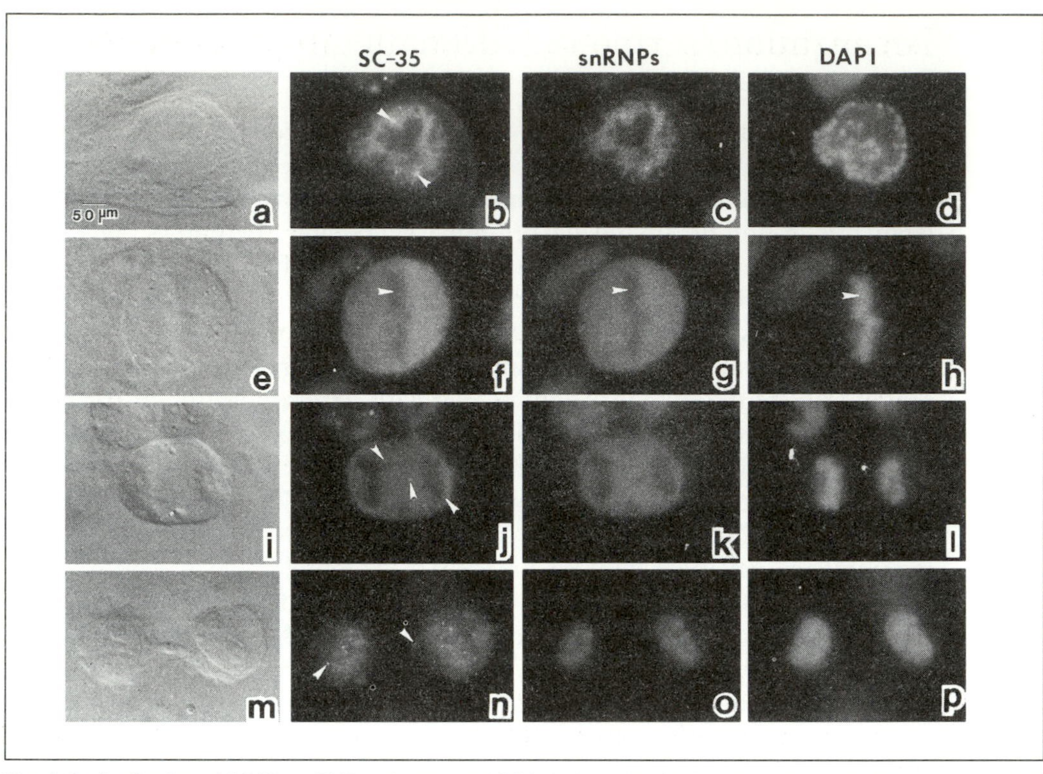

Fig. 3 Redistribution of SC35, snRNP antigens, and DNA during mitosis. As cells enter prophase (a–d) most of the speckles break up, SC35 (b) and snRNP antigens (c) are more uniformly distributed between the condensing chromosomes seen by immunofluorescence in (d). In addition, some diffuse immunoreactivity appears to be present in the cytoplasm of prophase cells (b). During metaphase (e–h) both SC35 (f) and snRNP antigens (g) are uniformly distributed throughout the cytoplasm. However these antigens are not associated with the interior of the chromosomes in the metaphase plate (arrowheads). During anaphase (i–l) SC35 begins to reassociate into speckles (j; arrowheads), while snRNP antigens are still uniformly distributed throughout the cytoplasm (k). During telophase (m–p) the nuclear envelope is reformed and while all of the snRNP immunoreactivity is contained within the nucleus (o) in many cells, several SC35 speckles (n; arrowheads) appear in the cytoplasm in addition to the typical nuclear immunostaining.

associated protein (81) that is thought to be a splicing factor (82), suggesting that the localization of SC35 in clusters during anaphase may be part of nuclear matrix reorganization. In telophase, when the chromosomes decondense and the nuclear envelope is reformed, snRNPs appear to re-establish the speckled distribution in the cell nucleus earlier than SC35 (83). In late telophase, SC35 (Fig. 3n) and snRNPs (Fig. 3o) are once again co-localized within speckled nuclear domains in the newly formed daughter nuclei (Fig. 3p), although some SC35 clusters appear to remain in the cytoplasm after the nuclear envelope is reformed (Fig. 3n). The reformation of nuclear speckles late in telophase seems to correlate closely with the onset of transcription in the newly-formed daughter cell nuclei.

2.5 Localization of splicing factors in amphibian germinal vesicles

Gall and co-workers have examined the localization of a variety of splicing components in amphibian germinal vesicles (84–86). These studies have shown that lampbrush chromosome loops from germinal vesicles of the newt *Notophthalmus viridescens* are uniformly stained with antibodies against snRNPs, SC35, and hnRNP proteins. Thus, the loops appear to be packaged into a ribonucleoprotein complex that includes all the components of the spliceosome for which probes are available. This situation would be comparable to the staining of perichromatin fibrils in mammalian cells by anti-snRNP (21, 22), anti-hnRNP (21), and anti-SC35 (30) antibodies. Therefore, it is likely that pre-mRNA splicing occurs in close proximity to the sites of transcription, in both mammalian cells and amphibian germinal vesicles, as has been previously suggested (87–89). Anti-spliceosome antibodies also stain large extrachromosomal particles in the germinal vesicle, designated snurposomes (86). Snurposomes have been divided into three classes, designated A, B, and C. The A snurposomes vary in size from 1–4 μM in diameter and appear to contain exclusively U1 snRNPs, while the B snurposomes are 4 μM in diameter, and contain U1, U2, U4, U5, and U6, as well as SC35. C snurposomes or spheres can be as large as 10 μM in diameter, and appear to be homologous to the coiled bodies of somatic nuclei. C snurposomes contain U7 snRNA, which is involved in histone 3′ end processing, and are localized extrachromosomally as well as attached to the histone gene loci (90).

2.6 Localization of splicing factors in yeast

The immunolocalization of several nuclear constituents associated with pre-mRNA processing has been reported in the budding yeast *Saccharomyces cerevisiae*. Last and Woolford (91) produced antibodies against fusion proteins that contain portions of the precursor RNA processing (PRP) 2 or PRP3 open reading frames. The PRP2 protein is thought to be associated with spliceosomes (92, 93) and the PRP3 protein with the U4/U6 snRNP (94; Chapter 7). These antibodies were used at the light microscopic level to show that polypeptides expressed from high-copy number plasmids are localized to the cell nucleus in *S. cerevisiae*. However, the precise subnuclear localization was not determined in this study. The PRP11 protein was shown to be specifically associated with the 40S spliceosome and a 30S complex (95). Using immunogold labelling, the PRP11 protein was localized to the non-nucleolar portion of the *S. cerevisiae* nucleus, in a predominantly perinuclear distribution. Potashkin *et al.* (96), using an antibody against the m_3G cap structure of snRNAs, showed that 85% of the m_3G-labelled colloidal gold particles are localized to the nucleolar portion of the nucleoplasm of the fission-yeast *Schizosaccharomyces pombe*. Brennwald *et al.* (97) have shown that this same m_3G antibody immunoprecipitates snRNPs U1 to U5 from *S. pombe* extracts and that U1 and U2 snRNPs are the predominant snRNPs in this organism. A similar localization was observed with an antibody that recognizes a protein component of U1 snRNP. On the basis of these data, it was suggested that the classically designated nucleolar portion of the yeast nucleus is organized differently from the mammalian cell nucleolus and contains functional domains in addition to those associated with rRNA transcription and processing. In support of this, Tani *et al.* (98) recently localized poly(A)$^+$ RNA in the nucleolar as well as the non-nucleolar portions of the *S. pombe* nucleus. The PRP6 protein, which is thought to be associated with the U4/U6 snRNP, has been localized to the nucleoplasm in *S. cerevisiae* (99). Deconvolved immunofluorescent images derived from a digital imaging microscope have shown PRP6 to be localized in what appears to be a speckled nuclear distribution pattern with both more intense and weaker sites of staining within the nuclei. In many cases the less intensely stained regions appear to connect regions that are more intensely stained. These results suggest that at least some components of the yeast splicing machinery may have a similar spatial organization to that of higher eukaryotes (99).

2.7 Localization of hnRNP proteins

Pre-mRNA or heterogeneous nuclear RNA (hnRNA) is present in the cell nucleus in the form of a complex with a discrete set of proteins. These hnRNP particles have been visualized as a linear array of 20–25 nm particles along each pre-mRNA molecule (reviewed in 100, 101; Chapter 3). Individual monoparticles, 40S structures, can be generated by cleavage of the linker RNA. Each monoparticle contains approximately 20 different proteins including the core hnRNP proteins A1 (34 kDa), A2 (36 kDa), B1 (37 kDa), B2 (38 kDa), C1 (41 kDa), and C2 (43 kDa) (102, 103), as well as additional proteins (for a review, see Chapter 3). HnRNP proteins are thought to be involved in the packaging, post-transcriptional processing, and/or transport of all pre-mRNAs (reviewed in 101, 104). Using chromatin-spreading methods, Beyer *et al.* (105) have found RNP particles, averaging 24 nm in diameter, to be associated with the majority of hnRNA transcription units of *Drosophila melanogaster* embryos. Immunodepletion and immunoinhibition studies have suggested a direct or indirect involvement for the hnRNP C-proteins during *in vitro* splicing (106, 107). Subsequently, hnRNP protein A1 was shown to be identical to SF5, a splicing factor involved in the regulation of alternative 5′ splice site selection (108; Chapters 6 and 8). In addition, the hnRNP I protein was shown to be identical to the previously described polypyrimidine tract-binding protein (PTB) (109). Therefore, in some cases the distinction between hnRNP proteins and splicing factors may be artificial.

Antibodies have been generated to several hnRNP proteins and their organization within the nucleus has been examined by immunofluorescence and immunoelectron microscopy. The hnRNP core proteins were shown to be diffusely distributed throughout the nucleoplasm but absent from the nucleoli in a variety of cell types (110–114). In nuclear matrix preparations these proteins were associated with fibrogranular material enmeshed in the core filament network (115). In addition, the abundant hnRNP K (66 kDa) and J (64 kDa) proteins, which are the major poly(C) binding proteins in HeLa cells (116), localized in a diffuse nuclear distribution. This distribution of hnRNP proteins appears to overlap with a portion of the speckled pattern observed with anti-splicing factor antibodies, since Fakan *et al.* (21) have shown, using immunogold electron microscopy, that the C proteins localize to perichromatin fibrils, but not to interchromatin granule clusters. The remainder of the diffuse staining observed with these antibodies may represent the localization of transcripts in transit to the nuclear envelope and/or may represent a soluble pool of hnRNP proteins in transit to new sites of transcription.

Two other hnRNP proteins, the L and I proteins, have been shown to localize in novel nuclear distributions (109, 116). The L protein is a 64–68 kDa protein associated with the majority of non-nucleolar nascent transcripts from the loops of lampbrush chromosomes in the newt, and it is the first hnRNP protein found to be associated with giant loops of lampbrush chromosomes. In mammalian cells this protein is distributed diffusely throughout the nucleoplasm. However, in addition, it stains one to three discrete non-nucleolar regions. The identity of these regions is unknown. The hnRNP I protein is a 58 kDa protein, identical to the previously reported polypyrimidine tract-binding protein (PTB) (117, 118). In addition to its general nucleoplasmic distribution, a high concentration of the I protein appears to form a cap-like structure in HeLa cells, which is associated with the surface of a portion of the nucleolus (109). The identity of this structure is also unknown at the present time.

During mitosis, the C and A1 proteins were shown to be distributed throughout the cell (119). However, hnRNP proteins return to cell nuclei after mitosis by two mechanisms: a transcription-dependent and a transcription-independent mode. In recently divided cells, the C proteins accumulate in the daughter nuclei as soon as the nuclei are formed, while most of the A1 protein remains in the cytoplasm. Therefore, hnRNP complexes appear to dissociate prior to the transport of individual hnRNP proteins back to the nucleus after mitosis. Inhibition of RNA polymerase II transcription does not affect the return of the C proteins but inhibits the transport of A1 to the nucleus. Therefore, transcription by RNA polymerase II appears to be required for the return of the A1 protein, but not the C proteins, to the nucleus after mitosis. Other hnRNP proteins, including A2, B1, B2, E, H, and L, were also reported to be transported to the nucleus in a transcription-dependent mode (119).

Recently, the hnRNP A1 protein was found to shuttle between the nucleus and cytoplasm of interphase cells (120). In interphase cells in which transcription of RNA polymerase II is inhibited by actinomycin D, the A1 protein accumulates in the cytoplasm, whereas the C proteins and the U protein remain restricted to the nucleus. Cross-linking studies have shown that some of the A1 that accumulates in the cytoplasm is bound to poly(A)$^+$ RNA (120). In addition to A1, other hnRNP proteins, such as A2 and E, were found to accumulate in the cytoplasm in the presence of RNA polymerase II inhibitors. Cells treated with translation inhibitors or with actinomycin D, at a concentration which would only inhibit RNA polymerase I transcription, show a normal intracellular distribution of the A1 or C proteins. However, similar results demonstrating a shuttling of the A1 protein were observed in cells treated with the RNA polymerase II inhibitor DRB and in human—*Xenopus laevis* interspecies heterokaryons in the absence of inhibitors of transcription (120). More recently, mutagenesis has defined the amino acid sequence required for the shuttling of the A1 protein between the nucleus and cytoplasm (121, 122).

2.8 Localization of other pre-mRNA processing factors

Most mammalian mRNAs are post-transcriptionally modified by the addition of a poly(A) tail to their 3′ ends. Addition of the poly(A) tail occurs by a two-step reaction in which the transcript is endonucleolytically cleaved at the polyadenylation site and then a stretch of 200–300 nucleotides is added to the 3′ end of the upstream cleavage product (for reviews, see 123, 124; Chapter 9). It has been shown that the AAUAAA consensus sequence, located 10–30 nucleotides upstream of the polyadenylation site, is necessary for both cleavage and polyadenylation and that a set of *trans*-acting factors is required to catalyse accurate polyadenylation. One of these factors, cleavage-stimulation factor (CstF), is composed of three distinct polypeptide subunits of 64, 55, and 50 kDa and has been purified (125). Indirect immunofluorescence microscopy using antibodies specific for the 64 kDa or the 50 kDa subunits has revealed a diffuse nuclear distribution of CstF. The localization of CstF is similar to that observed for the hnRNP core proteins (see above section). Based on the electron microscopic localization of hnRNP proteins to perichromatin fibrils (21), one can deduce that CstF may also be localized to these fibrils, which are thought to represent nascent pre-mRNA transcripts (for a review, see 26). Future studies of the localization of poly(A) polymerase and nuclear poly(A)-binding proteins will add to our overall understanding of the spatial positioning of RNA processing factors.

3. RNA transcription, localization, and pre-mRNA splicing in the mammalian cell nucleus

Transcription and localization of RNA transcripts in the cell nucleus have been examined by a variety of methodologies, including [^{3}H]-uridine incorporation followed by autoradiography; Br-UTP incorporation followed by fluorescent detection; fluorescence *in situ* hybridization (FISH) localization of total poly(A)$^+$ RNA; FISH localization of nascent cellular transcripts; and FISH localization of transcripts from integrated, episomal, or linear non-integrated viral genomes.

3.1 Localization of transcriptionally active nuclear regions by tritiated-uridine, Br-UTP incorporation, or *in situ* nick-translation

High resolution electron microscopic *in situ* autoradiography has been used by several groups to localize sites of active transcription within the cell nucleus (for a review, see 26). After pulses of [^{3}H]-uridine for as short as two minutes, non-nucleolar labelling was first observed over perichromatin fibrils, which are distributed throughout the nucleoplasm adjacent to regions of condensed chromatin (24, 25, 28). These structures were shown to be RNase sensitive, and their appearance was inhibited by actinomycin D (126) or α-amanitin (127) pre-treatment. Based on these data, it was suggested that these fibrils may represent sites of pre-mRNA synthesis (28). When RNAs are extracted from nuclear fractions enriched in perichromatin fibrils, a heterogeneous high molecular weight distribution of RNA is observed, which is distinct from ribosomal RNA (27). Based on chain-length distribution and actinomycin D treatment, these RNAs were reported to represent hnRNA or pre-mRNA (27). The localization of pre-mRNA splicing factors to these fibrils (21, 30) strongly supports their identification as pre-mRNA. Subsequently, Br-UTP incorporation has been used to identify sites of active transcription using fluorescence microscopy. A five-minute pulse of Br-UTP shows scattered labelling throughout the nucleoplasm (128, 129).

In contrast to these studies, two reports using *in situ* nick-translation to visualize DNase I-sensitive nuclear DNA (active genes) have suggested that active genes are preferentially localized around the nuclear periphery (130, 131). However, the possibility that nucleotides or other components of the reaction mix were titrated out or trapped by the peripheral chromatin were not addressed in these studies. These possibilities are supported by the fact that when DNAse I incubation is performed in a separate reaction, prior to the incubation with polymerase and labeled nucleotide, it is possible to demonstrate a decrease in staining of the nuclear periphery and a concomitant increase in staining of an internal subset of chromatin (131). In addition, heterochromatin or inactive chromatin (132), including non-transcribed α-satellite DNA sequences (133), have been shown to be localized to the nuclear periphery.

3.2 Distribution of nuclear polyadenylated RNA

Since almost all messenger RNAs contain a polyadenylated tail, localization of nuclear poly(A)$^{+}$ RNA provides a logical means of identifying the majority of this RNA population in the cell. Carter *et al.* (134) showed that nuclear poly(A)$^{+}$ RNA has a speckled distribution that co-localizes with snRNPs, and which these authors referred to as 'transcript domains'. More recently, examination at the electron microscopic level demonstrated that poly (A)$^{+}$ RNAs are localized to both interchromatin granules clusters and perichromatin fibrils (135, 136). Since interchromatin granule clusters incorporate little or no [^{3}H]-uridine (for a review, see 26), these RNAs are unlikely to represent nascent transcripts. This possibility is supported by the observation that a substantial amount of poly (A)$^{+}$ RNAs remain in the interchromatin granule clusters after a long inhibition of RNA transcription by α-amanitin (135). Thus, the poly(A)$^{+}$ RNA localization may represent both nascent and stable species. Perhaps the portion of the poly(A)$^{+}$ RNA associated with perichromatin fibrils is destined to be spliced and exported, whereas other more stable RNAs localized in the interchromatin granule clusters serve a structural role and/or have other functions in the cell nucleus.

3.3 Localization of specific RNA transcripts

Advances in the ability to localize RNA molecules by FISH and the development of techniques to fluorescently label RNAs and then microinject them into living cells have set the stage for significant progress in our understanding of the nuclear organization of pre-mRNA metabolism. Wang *et al.* (137) microinjected fluorescently tagged β-globin pre-mRNA into interphase nuclei and showed that these RNA molecules localize in a speckled distribution that is coincident with the speckled pattern enriched in pre-mRNA splicing factors (138). In contrast, microinjection of transcripts lacking an intron, or with a deleted polypyrimidine tract and 3′ splice site, resulted in a diffuse distribution of the injected transcripts. These results show that intron-containing transcripts have the ability to associate with nuclear speckles enriched in pre-mRNA splicing factors.

Numerous studies have used FISH to examine the localization of endogenous pre-mRNA and mRNA transcripts in the cell nucleus. When the distribution of neurotensin (139) or β-actin (140, 141) RNAs was examined, the RNA localized as a dot at the site of each allele. Each dot represents the transcription site of the respective gene. Cells stably transfected with a plasmid containing the *neu* oncogene also showed a highly localized concentration of *neu* RNA. However, in many cases the distribution appeared as clusters of dots (142). A more recent examination of the distribution of collagen Iα1 RNA showed the RNA signal to be significantly larger than the dot observed for the respective DNA signal, and the RNA signal did not extend to the nuclear envelope (141). A dot-like distribution has also been shown for the *string* RNA in cells of a cycle-14 *Drosophila* embryo (143). As these RNA localization signals, based upon fluorescence detection methods, for the most part do not reveal RNA contacting the nuclear envelope, it is unclear how the RNAs are transported from their sites of transcription to the cytoplasm.

In several cases, the RNA distribution appeared as a track, rather than a dot, raising the possibility that RNAs are vectorially transported from their sites of synthesis to the nuclear envelope. The first observation of RNA distributed in a track in mammalian cells came from a study by Lawrence *et al.* (142) that examined Namalwa cells, which contain two copies of the Epstein-Barr virus (EBV) genome closely integrated on chromosome 1. EBV *Bam W* RNA transcripts were detected as a track, which averaged 5 μm in length, in the nuclei of Namalwa cells. The tracks appeared to extend between the nuclear interior and the nuclear periphery. However, as very few EBV *Bam W* transcripts are transported to the cytoplasm (144), the observed track may represent RNA accumulation sites, rather than areas of active RNA export. Subsequently, Huang and Spector (145) stimulated c-*fos* transcription in NIH 3T3 cells and observed nascent c-*fos* RNA localized as two dots in interphase nuclei. Hybridization with an intron-specific probe demonstrated that pre-mRNA was present at these sites. When this distribution was examined further by confocal laser scanning microscopy and high voltage electron microscopy, the dot-like distribution of *c-fos* RNA transcripts was found to extend as an elongated path of 0.75–1.5 μm through the depth of the nucleus, from the site of synthesis of the RNA to the nuclear envelope. Electron microscopic *in situ* hybridization confirmed this observation and showed these transcripts to exit the nucleus at a very limited area that might be related to a group of nuclear pores and thus support the gene gating hypothesis (146). More recently, Xing *et al.* (139) localized fibronectin mRNA by FISH and found it to be highly concentrated at one or more sites per nucleus that frequently appear as tracks up to 6 μm long. The gene is positioned at or near one end of the track. Since hybridization to cDNA probes produces longer tracks than with intron probes, it was suggested that splicing occurs within a portion of the tracks. Intron probes also localize as a dispersed signal throughout the nucleoplasm and do not concentrate around the nuclear periphery, (139) as was reported previously for the acetylcholine receptor intron (147).

Perhaps the most intriguing case in which vectorial transport of transcripts has been suggested relates to the *Drosophila* pair-rule genes (148), whose striped expression in the blastoderm embryo is critical in establishing the segmentation pattern. Pair-rule transcripts accumulate exclusively on the apical side of the peripheral blastoderm nuclei. Transcript localization is determined by 3′ sequences in the transcripts. Davis and Ish-Horowicz (149) identified several 3′ pair-rule sequences that confer apical localization on reporter transcripts, whereas the 3′ sequence of human β-globin was found to direct transcripts basally in the blastoderm embryo. These authors have suggested that the cytoplasmic localization of pair-rule transcripts may occur by a nuclear mechanism, whereby transcripts are vectorially exported directly to the apical cytoplasm (148). Selective transcript exit from the nucleus could be due to export through an apical subset of nuclear pore complexes that recognize 3′ localization signals. Alternatively, the RNA signals could be recognized by components of an intranuclear transport machinery and could actively direct transcripts to apical nuclear pore complexes (148).

Recently, Dirks *et al.* (150) examined the localization of three different RNA species and suggested that the observation of nuclear RNA tracks versus dots may be related to splicing activity, rather than to transport pathways. The EBV *Bam W* transcripts in Namalwa cells and the human cytomegalovirus immediate early (HCMV-IE) transcripts in rat 9G cells are extensively spliced, while in the case of the luciferase transcript in X1 cells, only one small intron (66 nucleotide SV40 small-t intron) must be removed. Similar to the *Bam W* transcripts (142, 150), the majority of HCMV-IE nuclear transcripts localize in a main track or elongated dot. However, in the case of HCMV-IE transcripts many small spots of hybridization signal were also observed radiating from the main nuclear RNA signal (150). Analysis of many tracks after double hybridization with intron- and exon-specific probes revealed complete co-localization, indicating that primary transcripts are present all along the nuclear track. This finding is in contrast to the observations by Xing *et al.* (139, 141) with the fibronectin and collagen Iα1 RNAs. In these cases, cDNA probes produce longer tracks than intron probes, suggesting that splicing occurs within a portion of the track. In contrast to RNAs that contain many introns (HCMV-IE; *Bam W*), the nuclear RNA signal from luciferase transcripts (one small intron) mainly consists of a bright fluorescent dot with many small fluorescent spots (150). The density of the small spots decreases away from the major dot and closer to the nuclear periphery. Elongated dots or tracks are never observed. The additional small spots that radiate from the transcription domains were interpreted by Dirks *et al.* (150) as mRNAs in transport to the cytoplasm and spliced-out introns. On the basis of these data, Dirks *et al.* (150) suggested that when the extent of splicing is high, unspliced or partially spliced mRNAs begin to occupy elongated dot or track-like domains in the vicinity of the gene. When the extent of splicing is low, splicing is completed co-transcriptionally, leading to a bright dot-like signal. Therefore, the nuclear RNA tracks do not represent defined RNA transport routes but, more likely, nascent transcription and/or accumulation of precursor RNA (150). These authors further suggested that processed RNA transcripts radiate from this site of transcription and pre-mRNA splicing towards the cytoplasm without following a specific route.

The above interpretation is consistent with several recently proposed models that argue that transcripts are transported to the cytoplasm of cells via an extrachromosomal channel network (151) or interchromosome domain compartment (152, 153), within which transcription and pre-mRNA processing are thought to occur. Zirbel *et al.* (152) proposed that the surfaces of chromosome territories and a space formed between them provide a network-like three-dimensional nuclear compartment for gene expression, mRNA splicing, and transport, termed the interchromosome domain compartment. More than 92% of the nuclear speckles enriched in splicing factors were found at the surface of chromosome territories (152). Zachar *et al.* (151) observed a similar nuclear compartment in *Drosophila* polytene nuclei while examining the localization of an RNA produced from a hybrid gene containing the first three introns of the *SWAP* gene. The RNA was localized in a small area called the primary zone that is thought to represent the gene locus. In addition, pre-mRNA was distributed in a network-like fashion throughout the extranucleolar nucleoplasm. As total poly(A)$^+$ RNA and splicing factors are also observed in the channels, pre-mRNA metabolism is thought to occur within this region of the nucleus. Based upon the channelled distribution of RNA, Zachar *et al.* (151) proposed a channelled diffusion model for mRNA transport, suggesting that mRNAs move from their sites of transcription to the nuclear surface at rates that are consistent with diffusion. However, it remains to be proven whether RNAs are transported to the nuclear envelope by diffusion or active transport.

The clearest studies of RNP transport come from the work done with Balbiani ring (BR) granules in *Chironomus tentans*. These granules are synthesized in BRs 1 and 2, two giant puffs on chromosome IV in the larval salivary glands. The particles are released from the chromatin axis and are found in the nucleoplasm with a higher frequency in the vicinity of the nuclear envelope, in particular, close to the entrance of the nuclear pores (154, 155). The BR granules represent mRNP particles that encode information for secretory polypeptides with molecular masses around

10^3 kDa (for a review, see 156). Since BR granules found free in the nucleoplasm do not immunolabel with anti-snRNP antibodies, it was suggested that splicing takes place prior to the formation of these mature particles (157). The 37 kb transcripts (75S RNA) (158) are packaged into 50 nm RNP particles (159, 160), which have been characterized as a ribbon bent into a ring (161). Electron microscopic tomographic studies have shown that the BR RNP particle positions itself in front of the central channel of the nuclear pore complex (154, 155). The particle then becomes elongated and finally rod-shaped with a diameter of 25 nm and a length of approximately 135 nm (162, 163). The unfolded particle is translocated through the nuclear pore in an oriented fashion, with the 5′ end of the transcript in the lead (154). The RNP particle is not reformed as a globular structure on the cytoplasmic side of the pore complex. Instead, the RNP fibre appears to associate with a row of putative ribosomes. It remains to be determined whether this mode of transport is specialized for the BR granule or ubiquitous to all mRNP particles.

3.4 Transcription and pre-mRNA splicing are temporally and spatially linked

The possibility that transcription and pre-mRNA splicing are closely linked has been suggested by several studies. Ultrastructural analysis of actively transcribing genes from *Drosophila* embryos demonstrated that introns are removed co-transcriptionally (87). In mammalian cells, Fakan *et al.* (88) localized snRNPs to elongating RNP fibrils. More recently, spliceosomes were found to be associated with nascent RNAs transcribed from the Balbiani ring genes of *Chironomus tentans* (164) and splicing factors were localized on the loops of lampbrush chromosomes in amphibian germinal vesicles (85).

The ability to simultaneously detect intracellular RNA and protein components has allowed a direct analysis of the temporal and spatial relationship between nascent RNA and splicing factors in mammalian cell nuclei. Double-labelling of c-*fos* RNA transcripts and the splicing factor SC35 revealed a close association between the RNA transcripts and splicing factors (145). Fibronectin and neurotensin RNA transcripts were also usually found to be associated with splicing factors (139). Most recently, Xing *et al.* (141) examined the nuclear localization of two transcriptionally active and three inactive genes. The inactive genes localized as small fluorescent dots which were largely not associated with nuclear regions enriched in the pre-mRNA splicing factor SC35. However, RNA produced from the active β-actin gene localized as small dots only slightly larger than the DNA signal and 89% of the β-actin RNA signals were associated with SC35 domains. This finding is in sharp contrast to an earlier study (140) that found only 40% of the RNA signals associated with SC35, although both pre-mRNA and mRNA were localized to the same nuclear region. Another transcriptionally active gene, the collagen Iα1 gene, was localized at the periphery of an SC35 domain with the RNA signal co-localizing with the domain (141). All the collagen RNA signals associate with large SC35 domains and they do not extend to the nuclear envelope. These studies clearly demonstrated that transcription and pre-mRNA splicing are spatially and temporally linked.

3.5 Splicing factors are recruited to the sites of intron-containing RNA transcription

Two different approaches have been used to determine whether splicing factors are recruited to the sites of transcription or if active genes and/or transcripts move to pre-existing regions in the nucleus that are enriched in splicing factors. One approach used cells infected with adenovirus (165, 166) and another approach examined the localization of specific transcripts in transiently transfected cells (167).

When HeLa cells are infected with adenovirus, the newly established viral replication and transcription centres form ring-shaped structures, which have been morphologically characterized at both the light and electron microscopic levels (166, 168–173). When the localization of splicing factors such as snRNPs was examined in the infected cells, the distribution of these factors was found to be co-localized with the adenoviral RNA in the ring structures (165, 166). Concomitant with the increased localization of splicing factors at the viral replication and transcription centres, the nucleoplasmic speckled distribution of splicing factors was reduced. In addition to splicing factors, RNA polymerase II, Br-UTP incorporation, and hnRNP C proteins have also been shown to localize to these structures with the viral RNA, confirming that the ring structures represent the sites of viral RNA transcription (165, 166). These findings suggest that splicing factors are recruited to the sites of new RNA synthesis in the infected cells.

Additional support for the recruitment of splicing factors to sites of active transcription comes from recent studies using transiently transfected cells (167). When cells express transiently transfected templates that encode intron-containing RNAs, splicing factors are localized to the sites of RNA transcription and RNAs are spliced at these loci. The localization of the splicing factors appears to be in areas that are similar in size and shape as the localization of the RNA when examined by optical sectioning using confocal laser scanning microscopy. When the RNAs are expressed at very high levels, the nuclear regions occupied by both splicing factors and RNA are much larger than a typical speckle observed in non-transfected cells. Such unusually large clusters of splicing factors at the sites of RNA transcription suggest that splicing factors are recruited from elsewhere in the nucleus to the active sites of new RNA synthesis. Splicing factors do not co-localize with sites of transcription of intron-less RNAs.

Two potential mechanisms, a scanning mechanism and a recruiting mechanism, were proposed to account for the shuttling of both transcription and pre-mRNA processing factors to sites of transcription (165, 167). In the scanning mechanism, factors would continuously diffuse throughout the nucleus either in a soluble form or moving along components of a nuclear matrix. When these factors reach a potential active site they would dock and transcription and processing would occur. Evidence has been provided that both active sites of transcription (174), splicing components (5, 175–177), pre-mRNA (178–180), as well as poly(A)$^+$ RNA (135, 181) are associated with the nuclear matrix. However, the continuous movement of splicing factors in living cells has not yet been demonstrated. In the recruiting mechanism model, factors would be associated with specific storage and/or assembly sites in the nucleus. Prior to, or at the initiation of, transcription, these factors would be recruited to the active sites of transcription by another factor or chaperon molecule. Evidence to support this model comes from previous studies, which showed that a subpopulation of splicing factors are localized to interchromatin granule clusters in mammalian cell nuclei (21, 80). These clusters contain little labelled RNA after short pulses with [^{3}H]uridine (25), suggesting that they may not represent active sites of transcription. Therefore, splicing factors are localized to both sites of active transcription (perichromatin fibrils) and storage and/or reassembly sites (interchromatin granule clusters). Furthermore, upon introduction of new transcription sites into the cell nucleus, there is a concomitant decrease in the signal intensity of splicing factors at host cell speckles with an increase at new active sites of viral transcription, while the overall level of snRNP proteins remains constant throughout the infection process (165). These findings strongly suggest that there are signals generated in the nucleus that regulate the compartmentalization of factors to nuclear regions where they will be functioning. Identification of these signalling mechanisms will be key to understanding the integration of a variety of functional events which occur within the boundaries of the nuclear envelope.

3.6 Effect of transcription by RNA polymerase II on the organization of the speckled pattern

Several studies have used RNA polymerase II inhibitors to examine whether the speckled nuclear organization of splicing factors depends on active transcription. When cells are treated with α-amanitin at a concentration that specifically inhibits RNA polymerase II, the nuclear speckles round up into larger and fewer clusters and no interconnections are observed between them (182). The specificity of the inhibitory activity of the α-amanitin has been clearly demonstrated by using control cells with a mutant RNA polymerase II that is no longer sensitive to the drug (135). A similar change was observed with the adenosine analogue DRB (5, 183). However, unlike α-amanitin, the effect of DRB on the organization of splicing factors was reversible within 30 minutes of removal of the drug (5). In addition, Carmo-Fonseca *et al.* (18) showed that snRNPs are no longer concentrated in coiled bodies after α-amanitin or actinomycin D treatment. These results demonstrate that the organization of splicing factors in the cell nucleus is dependent on RNA polymerase II activity. When drug-treated cells were examined at the electron microscopic level, a significant decrease in the number of perichromatin fibrils was observed, as compared to control cells (128, 184–186). These data support the idea that the connections observed between the larger speckles represent snRNPs associated with perichromatin fibrils or sites of nascent transcripts. They are also in agreement with previous studies that showed [^{3}H]uridine incorporation, as well as hnRNP and snRNP antigens, to be associated with perichromatin fibrils (21, 22, 27, 28). Therefore, the organization of splicing factors in the nucleus is a reflection of the transcriptional activity of the cell.

3.7 Analysis of RNA splicing *in vivo* with antisense probes

To assess the effect of inhibition of pre-mRNA splicing on the organization of splicing factors, antisense DNA probes were microinjected into living cells (187). A 20-base oligonucleotide probe (5′-CTCCCCTGCCAGGTAAGTAT-3′), complementary to the region of U1 snRNA that base pairs with the 5′ splice site of pre-mRNA, and encompassing sequences previously shown to inhibit pre-mRNA splicing *in vitro* (188–190), was used. When this oligonucleotide probe was microinjected into the cytoplasm of HeLa cells, the oligonucleotide entered the nucleus and, within 1 hour, changes were observed in the organization of the speckled pattern. Similar to what was observed using drugs that inhibit RNA polymerase II, antisense oligonucleotides targeted to U1 snRNA caused the speckles to round up and the connections between speckles were no longer visible (187). When these cells were examined at the ultrastructural level, interchromatin granule clusters appeared to be round and larger than usual, and a significant reduction was observed in the amount of perichromatin fibrils in the nuclei. However, when a control oligonucleotide with no complementarity to splice sites or regions of interaction between snRNAs (5′-TCCGGTACCACGACG-3′) was microinjected into cells, no change in the organization of splicing factors was observed. Therefore, interfering with the interaction of U1 snRNP with the 5′ splice site of pre-mRNAs resulted in a reorganization of splicing factors. This reorganization was reversible over time. Microinjected DNA oligonucleotides that hybridize to RNA sequences have been shown to stimulate RNase H activity, which results in degradation of the RNA strand of the RNA-DNA hybrid (191). Upon RNase H cleavage of the RNA portion of the U1–oligo hybrid, splicing is inhibited. Splicing factors may return to storage and/or reassembly sites (interchromatin granule clusters) awaiting the initiation of new pre-mRNA synthesis, or synthesis of new snRNAs and snRNPs. Based upon these data, it was suggested that perichromatin fibrils represent sites of active splicing of nascent pre-mRNAs and that factors shuttle between storage and/or reassembly sites and sites of active transcription.

4. A model for the functional organization of the pre-mRNA splicing apparatus

On the basis of the studies described above, we present a unifying model attempting to explain the functional organization of pre-mRNA splicing in the cell nucleus. In a typical mammalian cell nucleus, the splicing factors that are concentrated in the majority of the interchromatin granule clusters are likely to be stored or reassembled at these sites, since interchromatin granule clusters do not actively incorporate

[^{3}H]uridine. In contrast, the factors at the perichromatin fibrils are likely to be actively engaged in splicing, as these structures represent the sites of RNA synthesis. Splicing factors are constantly shuttling between the sites of active splicing and the sites of storage and/or reassembly, in response to transcriptional activity. When the level of transcription is low or moderate, the splicing factors and nascent RNAs are probably visualized as perichromatin fibrils throughout the nucleoplasm and on the surface of some of the interchromatin granule clusters. When the level of transcription is very high, the large amount of splicing factors together with nascent RNAs at the transcription site can result in a morphological structure that closely resembles, but is functionally different from, a typical interchromatin granule cluster. Inhibition of either transcription or splicing abolishes the need for the shuttling of splicing factors, resulting in an accumulation of these factors in the interchromatin granule clusters, which become larger and more uniform in shape. Therefore, we conclude that the distribution of splicing factors and the organization of interchromatin granule clusters is highly regulated and dynamic in response to the transcriptional activity of the cell.

Acknowledgements

We wish to thank Scott Henderson and Raymond O'Keefe for helpful suggestions on the manuscript. D.L.S. is supported by grants from the National Institutes of Health (GM42694 and 5P30 CA45508) and The Council for Tobacco Research, Inc. (3295).

References

1. Green, M. R. (1991) Biochemical mechanisms of constitutive and regulated pre-mRNA splicing. *Annu. Rev. Cell Biol.*, **7**, 559.
2. Zieve, G. and Sauterer, R. A. (1990) Cell biology of the snRNP particles. *Crit. Rev. Biochem. Mol. Biol.*, **25**, 1.
3. Perraud, M., Gioud, M., and Monier, J. C. (1979) Structures intranucleaires reconnues par les autoanticorps anti-ribonucleoprotéines: étude sur cellules de rein de singe en culture par les techniques d'immunofluorescence et d'immunomicroscopie electronique. *Ann. Immunol.*, **130**, 635.
4. Lerner, E. A., Lerner, M. R., Janeway, L. A., and Steitz, J. A. (1981) Monoclonal antibodies to nucleic acid containing cellular constituents: probes for molecular biology and autoimmune disease. *Proc. Natl Acad. Sci. USA*, **78**, 2737.
5. Spector, D. L., Schrier, W. H., and Busch, H. (1983) Immunoelectron microscopic localization of snRNPs. *Biol. Cell*, **49**, 1.
6. Beck, J. S. (1961) Variations in the morphological patterns of autoimmune nuclear fluorescence. *Lancet*, **6**, 1203.
7. Tan, E. M. and Kunkel, H. G. (1966) Characteristics of a soluble nuclear antigen precipitating with sera of patients with systemic lupus erythematosus. *J. Immunol.*, **96**, 464.
8. Tan, E. M. (1989) Antinuclear antibodies: diagnostic markers for autoimmune diseases and probes for cell biology. *Adv. Immunol.*, **44**, 93.

9. Lerner, M. R. and Steitz, J. A. (1979) Antibodies to small nuclear RNAs complexed with proteins are produced by patients with systemic lupus erythematosus. *Proc. Natl Acad. Sci. USA*, **76**, 5495.
10. Yang V. W., Lerner, M. R., Steitz, J. A., and Flint, S. J. (1981) A small ribonucleoprotein is required for splicing of adenoviral early RNA sequences. *Proc. Natl Acad. Sci. USA*, **78**, 1371.
11. Padgett, R. A., Mount, S. M., Steitz, J. A., and Sharp, P. A. (1983) Splicing of messenger RNA precursors is inhibited by antisera to small nuclear ribonucleoprotein. *Cell*, **35**, 101.
12. Hernandez, N. and Keller, W. (1983) Splicing of in vitro synthesized messenger RNA precursors in HeLa cell extracts. *Cell*, **35**, 89.
13. Nyman, U., Hallman, H., Hadlaczky, G., Pettersson, I., Sharp, G., and Ringertz, N. R. (1986) Intranuclear localization of snRNP antigens. *J. Cell Biol.*, **102**, 137.
14. Spector, D. L. (1984) Colocalization of U1 and U2 small nuclear RNPs by immunocytochemistry. *Biol. Cell.*, **51**, 109.
15. Reuter, R., Appel, B., Bringmann, P., Rinke, J., and Lührmann, R. (1984) 5′-terminal caps of snRNAs are reactive with antibodies specific for 2,2,7-trimethylguanosine in whole cells and nuclear matrices. *Exp. Cell Res.*, **154**, 548.
16. Verheijen, R., Kuijpers, H., Vooijs, P., van Venrooij, W., and Ramaekers, F. (1986) Distribution of the 70K U1 RNA-associated protein during interphase and mitosis. Correlation with other U RNP particles and proteins of the nuclear matrix. *J. Cell Sci.*, **86**, 173.
17. Habets, W. J., Hoet, M. H., De Jong, B. A. W., Van der Kemp, A., and Van Venrooij, W. J. (1989) Mapping of B cell epitopes on small nuclear ribonucleoproteins that react with human autoantibodies as well as with experimentally-induced mouse monoclonal antibodies. *J. Immunol.*, **143**, 2560.
18. Carmo-Fonseca, M., Pepperkok, R., Carvalho, M. T., and Lamond, A. I. (1992) Transcription-dependent colocalization of the U1, U2, U4/U6, and U5 snRNPs in coiled bodies. *J. Cell Biol.*, **117**, 1.
19. Huang, S. and Spector, D. L. (1992) U1 and U2 small nuclear RNAs are present in nuclear speckles. *Proc. Natl Acad. Sci. USA*, **89**, 305.
20. Spector, D. L., Lark, G., and Huang, S. (1992) Differences in snRNP localization between transformed and nontransformed cells. *Mol. Biol. Cell*, **3**, 555.
21. Fakan, S., Leser, G., and Martin, T. E. (1984) Ultrastructural distribution of nuclear ribonucleoproteins as visualized by immunocytochemistry on thin sections. *J. Cell Biol.*, **98**, 358.
22. Puvion, E., Viron, A., Assens, C., Leduc, E. H., and Jeanteur, P. (1984) Immunocytochemical identification of nuclear structures containing snRNPs in isolated rat liver cells. *J. Ultrastruct. Res.*, **87**, 180.
23. Monneron, A. and Bernhard, W. (1969) Fine structural organization of the interphase nucleus in some mammalian cells. *J. Ultrastruct. Res.*, **27**, 266.
24. Fakan, S. and Bernhard, W. (1971) Localisation of rapidly and slowly labelled nuclear RNA as visualized by high resolution autoradiography. *Exp. Cell Res.*, **67**, 129.
25. Fakan, S. and Nobis, P. (1978) Ultrastructural localization of transcription sites and of RNA distribution during the cell cycle of synchronized CHO cells. *Exp. Cell Res.*, **113**, 327.
26. Fakan, S. and Puvion, E. (1980) The ultrastructural visualization of nucleolar and extranucleolar RNA synthesis and distribution. *Int. Rev. Cytol.*, **65**, 255.

27. Bachellerie, J-P., Puvion, E., and Zalta, J-P. (1975) Ultrastructural organization and biochemical characterization of chromatin RNA protein complexes isolated from mammalian cell nuclei. *Eur. J. Biochem.*, **58**, 327.
28. Fakan, S., Puvion, E., and Spohr, G. (1976) Localization and characterization of newly synthesized nuclear RNA in isolated rat hepatocytes. *Exp. Cell Res.*, **99**, 155.
29. Spector, D. L. (1990) Higher order nuclear organization: three-dimensional distribution of small nuclear ribonucleoprotein particles. *Proc. Natl Acad. Sci. USA*, **87**, 147.
30. Spector, D. L., Fu, X-D., and Maniatis, T. (1991) Associations between distinct pre-mRNA splicing components and the cell nucleus. *EMBO J.*, **10**, 3467.
31. Puvion, E. and Bernhard, W. (1975) Ribonucleoprotein components in liver cell nuclei as visualized by cryoultramicrotomy. *J. Cell Biol.*, **67**, 200.
32. Smetana, K., Steele, W. J., and Busch H. (1963) A nuclear ribonucleoprotein network. *Exp. Cell Res.*, **31**, 198.
33. Eliceiri, G. L. and Ryerse, J. S. (1984) Detection of intranuclear clusters of Sm antigens with monoclonal anti-Sm antibodies by immunoelectron microscopy. *J. Cell. Physiol.*, **121**, 449.
34. Carmo-Fonseca, M., Tollervey, D., Barabino, S. M. L., Merdes, A., Brunner, C., Zamore, P. D., *et al.* (1991) Mammalian nuclei contain foci which are highly enriched in components of the pre-mRNA splicing machinery. *EMBO J.*, **10**, 195.
35. Carmo-Fonseca, M., Pepperkok, R., Sproat, B. S., Ansorge, W., Swanson, M., and Lamond, A. I. (1991) *In vivo* detection of snRNP-rich organelles in the nuclei of mammalian cells. *EMBO J.*, **10**, 1863.
36. Raska, I., Andrade, L. E. C., Ochs, R. L., Chan, E. K. L., Chang, C.-M., Roos, G., *et al.* (1991) Immunological and ultrastructural studies of the nuclear coiled body with autoimmune antibodies. *Exp. Cell Res.*, **195**, 27.
37. Huang, S. and Spector, D. L. (1992) Will the real splicing sites please light up? *Curr. Biol.*, **2**, 188.
38. Leser, G. P., Fakan, S., and Martin, T. E. (1989) Ultrastructural distribution of ribonucleoprotein complexes during mitosis. snRNP antigens are contained in mitotic granule clusters. *Eur. J. Cell Biol.*, **50**, 376.
39. Ramón y Cajal, S. (1903) Un sencillo método de coloración selectiva del retículo protoplásmico y sus efectos en los diversos órganos nerviosos. *Trab. Lab. Invest. Biol.*, **2**, 129.
40. Moreno Diaz de la Espina, S., Sanchez Pina, A., and Risueno, M. C. (1982) Localization of acid phosphatase activity, phosphate ions and inorganic cations in plant nuclear coiled bodies. *Cell Biol. Int. Rep.*, **6**, 601.
41. Bernhard, W. (1969) A new staining procedure for electron microscopical cytology. *J. Ultrastruct. Res.*, **27**, 250.
42. Moreno Diaz de la Espina, S., Sanchez Pina, A., Risueno, M. C., Medina, F. J., and Fernandez-Gomez, M. E. (1980) The role of plant coiled bodies in the nuclear RNA metabolism. *Electron Micros.*, **2**, 240.
43. Raska, I., Ochs, R. L., Andrade, L. E. C., Chan, E. K. L., Burlingame, R., Peebles, C., *et al.* (1990) Association between the nucleolus and the coiled body. *J. Struct. Biol.*, **104**, 120.
44. Andrade, L. E. C., Chan, E. K. L., Raska, I., Peebles, C. L., Roos, G., and Tan, E. M. (1991) Human autoantibody to a novel protein of the nuclear coiled body: immunological characterization and cDNA cloning of p80-coilin. *J. Exp. Med.*, **173**, 1407.
45. Chan, E. K. L., Sullivan, K. F., and Tan, E. M. (1989) Ribonucleoprotein SS-B/La belongs to a protein family with consensus sequences for RNA-binding. *Nucleic Acids Res.*, **17**, 2233.

46. Dreyfuss, G., Swanson, M., and Piñol-Roma, S. (1988) Heterogeneous nuclear ribonucleoprotein particles and the pathway of mRNA formation. *Trends Biochem. Sci.*, **13**, 86.
47. Zamore, P. D., Zapp, M. L., and Green, M. R. (1990) RNA binding: βs and basics. *Nature*, **348**, 485.
48. Hardin, J. H., Spicer, S. S., and Greene, W. B. (1969) The paranucleolar structure, accessory body of Cajal, sex chromatin, and related structures in nuclei of rat trigeminal neurons: a cytochemical and ultrastructural study. *Anat. Rec.*, **164**, 403.
49. Seite, R., Pebusque, M-J., and Vio-Cigna, M. (1982) Argyrophilic proteins on coiled bodies in sympathetic neurons identified by Ag-NOR procedure. *Biol. Cell*, **46**, 97.
50. Schultz, M. C. (1989) Ultrastructural study of the coiled body and a new inclusion, the 'mykaryon', in the nucleus of the adult rat sertoli cell. *Anat. Rec.*, **225**, 21.
51. Lischwe, M. A., Ochs, R. L., Reddy, R., Cook, R. G., Yeoman, L. C., Tan, E. M., *et al.* (1985) Purification and partial characterization of a nucleolar scleroderma antigen (Mr=34,000; pI 8.5) rich in N^G,N^G-dimethylarginine. *J. Biol. Chem.*, **260**, 14304.
52. Ochs, R. L., Lischwe, M. A., Spohn, W. H., and Busch, H. (1985) Fibrillarin: a new protein of the nucleolus identified by autoimmune sera. *Biol. Cell*, **54**, 123.
53. Tyc, K. and Steitz, J. A. (1989) U3, U8 and U13 comprise a new class of mammalian snRNPs localized in the cell nucleolus. *EMBO J.*, **8**, 3113.
54. Kass, S., Tyc, K., Steitz, J. A., and Sollner-Webb, B. (1990) The U3 small nucleolar ribonucleoprotein functions in the first step of preribosomal RNA processing. *Cell*, **60**, 897.
55. Lafarga, M., Hervas, J. P., Santa-Cruz, M. C., Villegas, J., and Crespo, D. (1983) The 'accessory body' of Cajal in the neuronal nucleus. A light and electron microscopic approach. *Anat. Embryol.*, **166**, 19.
56. Ochs, R. L., Stein, T. W. J., and Tan, E. M. (1994) Coiled bodies in the nucleolus of breast cancer cells. *J. Cell Sci.*, **107**, 385.
57. Krainer, A. and Maniatis, T. (1988) RNA splicing. In *Frontiers in molecular biology: transcription and splicing*. Hames, B. D. and Glover, D. M. (ed.). IRL Press, Oxford, p. 131.
58. Bindereif, A. and Green, M. (1990) Identification and functional analysis of mammalian splicing factors. In *Genetic engineering: principles and methods*. Vol. 12 Setlow, J. (ed.). Plenum Press, New York, p. 201.
59. Krainer, A. R. and Maniatis, T. (1985) Multiple factors, including the small nuclear ribonucleoproteins U1 and U2, are necessary for pre-mRNA splicing *in vitro*. *Cell*, **42**, 725.
60. Krainer, A., Conway, G. C., and Kozak, D. (1990) The essential pre-mRNA splicing factor SF2 influences 5′ splice site selection by activating proximal sites. *Cell*, **62**, 35.
61. Ge, H. and Manley, J. L. (1990) A protein factor, ASF, controls alternative splicing of SV40 early pre-mRNA *in vitro*. *Cell*, **62**, 25.
62. Zamore, P. D. and Green, M. R. (1991) Biochemical characterization of U2 snRNP auxiliary factor: an essential pre-mRNA splicing factor with a novel intranuclear distribution. *EMBO J.*, **10**, 207.
63 Ruskin, B., Zamore, P. D., and Green, M. R. (1988) A factor, U2AF, is required for U2 snRNP binding and splicing complex assembly. *Cell*, **52**, 207.
64. Ast, G., Goldblatt, D., Offen, D., Sperling, J., and Sperling, R. (1991) A novel splicing factor is an integral component of 200S large nuclear ribonucleoprotein (lnRNP) particles. *EMBO J.*, **10**, 425.
65. Fu, X-D. and Maniatis, T. (1990) Factor required for mammalian spliceosome assembly is localized to discrete regions in the nucleus. *Nature*, **343**, 437.

66. Zamore, P. D. and Green, M. R. (1989) Identification, purification and biochemical characterization of U2 small nuclear ribonucleoprotein auxiliary factor. *Proc. Natl Acad. Sci. USA*, **86**, 9243.
67. Zamore, P. D., Patton, J. G., and Green, M. R. (1992) Cloning and domain structure of the mammalian splicing factor U2AF. *Nature*, **355**, 609.
68. Zhang, M., Zamore, P. D., Carmo-Fonseca, M., Lamond, A. I., and Green, M. R. (1992) Cloning and intracellular localization of the U2 small nuclear ribonucleoprotein auxiliary factor small subunit. *Proc. Natl Acad. Sci. USA*, **89**, 8769.
69. Li, H. and Bingham, P. M. (1991) Arginine/serine-rich domains of the *su(w^a)* and *tra* RNA processing regulators target proteins to a subnuclear compartment implicated in splicing. *Cell*, **67**, 335.
70. Hedley, M. L., Amrein, H., and Maniatis, T. (1995) An amino acid sequence motif sufficient for subnuclear localization of an arginine/serine-rich splicing factor. *Proc. Natl Acad. Sci. USA*, **92**, 11524.
71. Amrein, H., Gorman, M., and Nothiger, R. (1988) The sex-determining gene *tra*-2 of *Drosophila* encodes a putative RNA binding protein. *Cell*, **55**, 1025.
72. Fu, X-D. and Maniatis, T. (1992) Isolation of a complementary DNA that encodes the mammalian splicing factor SC35. *Nature*, **256**, 535.
73. Krainer, A. R., Mayeda, A., Kozak, D., and Binns, G. (1991) Functional expression of cloned human splicing factor SF2: homology to RNA-binding proteins, U1 70K, and *Drosophila* splicing regulators. *Cell*, **66**, 383.
74. Ge, H., Zuo, P., and Manley, J. L. (1991) Primary structure of the human splicing factor ASF reveals similarities with *Drosophila* regulators. *Cell*, **66**, 373.
75. Theissen, H., Etzerodt, M., Reuter, R., Schneider, C., Lottspeich, F., Argos, P., *et al.* (1986) Cloning of the human cDNA for the U1 RNA-associated 70K protein. *EMBO J.*, **5**, 3209.
76. Roth, M. B., Murphy, C., and Gall, J. G. (1990) A monoclonal antibody that recognizes a phosphorylated epitope stains lampbrush chromosome loops and small granules in the amphibian germinal vesicle. *J. Cell Biol.*, **111**, 2217.
77. Zahler, A. M., Lane, W. S., Stolk, J. A., and Roth, M. B. (1992) SR proteins: a conserved family of pre-mRNA splicing factors. *Genes & Dev.*, **6**, 837.
78. Mayeda, A., Zahler, A. M., Krainer, A. R., and Roth, M. B. (1992) Two members of a conserved family of nuclear phosphoproteins are involved in pre-mRNA splicing. *Proc. Natl Acad. Sci. USA*, **89**, 1301.
79. Reuter, R., Appel, B., Rinke, J., and Lührmann, R. (1985) Localization and structure of snRNPs during mitosis: immunofluorescent and biochemical studies. *Exp. Cell Res.*, **159**, 63.
80. Spector, D. L. and Smith, H. C. (1986) Redistribution of U-snRNPs during mitosis. *Exp. Cell Res.*, **163**, 87.
81. Smith, H. C., Spector, D. L., Woodcock, C. L. F., Ochs, R. L., and Bhorjee, J. (1985) Alterations in chromatin conformation are accompanied by reorganization of nonchromatin domains that contain U-snRNP protein p28 and nuclear protein p107. *J. Cell Biol.*, **101**, 560.
82. Smith, H. C., Harris, S. G., Zillmann, M., and Berget, S. M. (1989) Evidence that a nuclear matrix protein participates in premessenger RNA splicing. *Exp. Cell Res.*, **182**, 521.
83. Ferreira, J. A., Carmo-Fonseca, M., and Lamond, A. I. (1994) Differential interaction of splicing snRNPs with coiled bodies and interchromatin granules during mitosis and assembly of daughter cell nuclei. *J. Cell Biol.*, **126**, 11.

84. Gall, J. G. and Callan, H. G. (1989) The sphere organelle contains small nuclear ribonucleoproteins. *Proc. Natl Acad. Sci. USA*, **86**, 6635.
85. Wu, Z., Murphy, C., Callan, H. G., and Gall, J. G. (1991) Small nuclear ribonucleoproteins and heterogeneous nuclear ribonucleoproteins in the amphibian germinal vesicle: loops, spheres, and snurposomes. *J. Cell Biol.*, **113**, 465.
86. Gall, J. G. (1991) Spliceosomes and snurposomes. *Science*, **252**, 1499.
87. Beyer, A. L. and Osheim, Y. M. (1988) Splice site selection, rate of splicing, and alternative splicing on nascent transcripts. *Genes Dev.*, **2**, 754.
88. Fakan, S., Leser, G., and Martin, T. E. (1986) Immunoelectron microscope visualization of nuclear ribonucleoprotein antigens within spread transcription complexes. *J. Cell Biol.*, **103**, 1153.
89. Sass, H. and Pederson, T. (1984) Transcription-dependent localization of U1 and U2 small nuclear ribonucleoproteins at major sites of gene activity in polytene chromosomes. *J. Mol. Biol.*, **180**, 911.
90. Wu, C-H. H. and Gall, J. G. (1993) U7 small nuclear RNA in C snurposomes of the *Xenopus* germinal vesicle. *Proc. Natl Acad. Sci. USA*, **90**, 6259.
91. Last, R. L. and Woolford, J. L., Jr (1986) Identification and nuclear localization of yeast pre-messenger RNA processing components: RNA2 and RNA3 proteins. *J. Cell Biol.*, **103**, 2103.
92. Lin, R-J., Lustig, A. J., and Abelson, J. (1987) Splicing of yeast nuclear pre-mRNA *in vitro* requires a functional 40S spliceosome and several extrinsic factors. *Genes Dev.*, **1**, 7.
93. King, D. S. and Beggs, J. D. (1990) Interactions of PRP2 protein with pre-mRNA splicing complexes in *Saccharomyces cerevisiae*. *Nucleic Acids Res.*, **18**, 6559.
94 Ruby, S. W. and Abelson, J. (1991) Pre-mRNA splicing in yeast. *Trends Genet.*, **7**, 79.
95. Chang, T-H., Clark, M. W., Lustig, A. J., Cusick, M. E., and Abelson, J. (1988) RNA11 protein is associated with the yeast spliceosome and is localized in the periphery of the cell nucleus. *Mol. Cell. Biol.*, **8**, 2379.
96. Potashkin, J. A., Derby, R. J., and Spector, D. L. (1990) Differential distribution of factors involved in pre-mRNA processing in the yeast cell nucleus. *Mol. Cell. Biol.*, **10**, 3524.
97. Brennwald, P., Porter, G., and Wise, J. A. (1988) U2 small nuclear RNA is remarkably conserved between *Schizosaccharomyces pombe* and mammals. *Mol. Cell. Biol.*, **8**, 5575.
98. Tani, T., Derby, R. J., Hiraoka, Y., and Spector, D. L. (1995) Nucleolar accumulation of poly (A)+ RNA in heat-shocked yeast cells: implication of nucleolar involvement in mRNA transport. *Mol. Biol. Cell*, **6**, 1515.
99. Elliott, D. J., Bowman, D. S., Abovich, N., Fay, F. S., and Rosbash, M. (1992) A yeast splicing factor is localized in discrete subnuclear domains. *EMBO J.*, **11**, 3731.
100. LeStourgeon, W. M., Barnett, S. F., and Northington, S. J. (1990) Tetramers of the core proteins of 40S nuclear ribonucleoprotein particles assemble to package nascent transcripts into a repeating array of regular particles. In *The eukaryotic nucleus*. Vol. 2, Strauss, P. R. and Wilson, S. H. (ed.). Telford Press, Caldwell, NJ, p. 477.
101. Chung, S. Y. and Wooley, J. (1986) Set of novel, conserved proteins fold premessenger RNA into ribonucleosomes. *Proteins*, **1**, 195.
102. Beyer, A. L., Christensen, M. E., Walker, B. W., and LeStourgeon, W. M. (1977) Identification and characterization of the packaging proteins of core 40S hnRNP particles. *Cell*, **11**, 127.
103. Piñol-Roma, S., Choi, Y. D., Matunis, M. J., and Dreyfuss, G. (1988) Immunopurification of heterogeneous nuclear ribonucleoprotein particles reveals an assortment of RNA-binding proteins. *Genes Dev.*, **2**, 215.

104. Dreyfuss, G. (1986) Structure and function of nuclear and cytoplasmic ribonucleoprotein particles. *Annu. Rev. Cell Biol.*, **2**, 459.
105. Beyer, A. L., Miller, O. L., Jr., and McKnight, S. L. (1980) Ribonucleoprotein structure in nascent hnRNA is non-random and sequence-dependent. *Cell*, **20**, 75.
106. Choi, Y. D., Grabowski, P. J., Sharp, P. A., and Dreyfuss, G. (1986) Heterogeneous nuclear ribonucleoproteins: role in RNA splicing. *Science*, **231**, 1534.
107. Sierakowska, H., Szer, W., Furdon, P. J., and Kole, R. (1986) Antibodies to hnRNP core proteins inhibit *in vitro* splicing of human β-globin pre-mRNA. *Nucleic Acids Res.*, **14**, 5241.
108. Mayeda, A. and Krainer, A. R. (1992) Regulation of alternative pre-mRNA splicing by hnRNP A1 and splicing factor SF2. *Cell*, **68**, 365.
109. Ghetti, A., Piñol-Roma, S., Michael, W. M., Morandi, C., and Dreyfuss, G. (1992) hnRNP I, the polypyrimidine tract-binding protein: distinct nuclear localization and association with hnRNAs. *Nucleic Acids Res.*, **20**, 3671.
110. Jones, R. E., Okamura, C. S., and Martin, T. E. (1980) Immunofluorescent localization of the proteins of nuclear ribonucleoprotein complexes. *J. Cell Biol.*, **86**, 235.
111. Choi, Y. D. and Dreyfuss, G. (1984) Monoclonal antibody characterization of the C proteins of heterogeneous nuclear ribonucleoprotein complexes in vertebrate cells. *J. Cell Biol.*, **99**, 1997.
112. Dreyfuss, G., Choi, Y. D., and Adam, S. A. (1984) Characterization of hnRNA–protein complexes *in vivo* with monoclonal antibodies. *Mol. Cell. Biol.*, **4**, 1104.
113. Leser, G. P., Escara-Wilke, J., and Martin, T. E. (1984) Monoclonal antibodies to heterogeneous nuclear RNA-protein complexes. The core proteins comprise a conserved group of related polypeptides. *J. Biol. Chem.*, **259**, 1827.
114. Piñol-Roma, S., Swanson, M. S., Gall, J. G., and Dreyfuss, G. (1989) A novel heterogeneous nuclear RNP protein with a unique distribution on nascent transcripts. *J. Cell Biol.*, **109**, 2575.
115. He, D., Martin, T., and Penman, S. (1991) Localization of heterogeneous nuclear ribonucleoprotein in the interphase nuclear matrix core filaments and on perichromosomal filaments at mitosis. *Proc. Natl Acad. Sci. USA*, **88**, 7469.
116. Matunis, M. J., Michael, W. M., and Dreyfuss, G. (1992) Characterization and primary structure of the polyC-binding heterogeneous nuclear ribonucleoprotein complex K protein. *Mol. Cell. Biol.*, **12**, 164.
117. Gil, A., Sharp, P. A., Jamison, S. F., and Garcia-Blanco, M. (1991) Characterization of cDNAs encoding the polypyrimidine tract-binding protein. *Genes Dev.*, **5**, 1224.
118. Paton, J. G., Mayer, S. A., Tempst, P., and Nadal-Ginard, B. (1991) Characterization and molecular cloning of polypyrimidine tract-binding protein: a component of a complex necessary for pre-mRNA splicing. *Genes Dev.*, **5**, 1237.
119. Piñol-Roma, S. and Dreyfuss, G. (1991) Transcription-dependent and transcription-independent nuclear transport of hnRNP proteins. *Science*, **253**, 312.
120. Piñol-Roma, S. and Dreyfuss, G. (1992) Shuttling of pre-mRNA binding proteins between nucleus and cytoplasm. *Nature*, **355**, 730.
121. Siomi, H. and Dreyfuss, G. (1995) A nuclear localization domain in the hnRNP A1 protein. *J. Cell Biol.*, **129**, 551.
122. Michael, W. M., Choi, M., and Dreyfuss, G. (1995) A nuclear export signal in hnRNP A1: a signal-mediated, temperature-dependent nuclear protein export pathway. *Cell*, **83**, 415.
123. Humphrey, T. and Proudfoot, N. J. (1988) A beginning to the biochemistry of polyadenylation. *Trends Genet.*, **4**, 243.

124. Manley, J. L. (1988) Polyadenylation of mRNA precursors. *Biochim. Biophys. Acta*, **950**, 1.
125. Takagaki, Y., Manley, J. L., MacDonald, C. C., Wilusz, J., and Shenk, T. (1990) A multisubunit factor, CstF, is required for polyadenylation of mammalian pre-mRNAs. *Genes Dev.*, **4**, 2112.
126. Miyawaki, H. (1974) Extranucleolar pyroninophilic substances in the liver cell nuclei of starve–refed mice as revealed by nonaqueous negative staining. *J. Ultrastruct. Res.*, **47**, 255.
127. Petrov, P. and Sekeris, C. E. (1971) Early action of α-amanitin on extranucleolar ribonucleoproteins, as revealed by electron microscopic observation. *Exp. Cell Res.*, **69**, 393.
128. Jackson, D. A., Hassan, A. B., Errington, R. J., and Cook, P. R. (1993) Visualization of focal sites of transcription within human nuclei. *EMBO. J.*, **12**, 1059.
129. Wansink, D. G., Schul, W., van der Kraan, I., van Steensel, B., van Driel, R., and de Jong, L. (1993) Fluorescent labeling of nascent RNA reveals transcription by RNA polymerase II in domains scattered throughout the nucleus. *J. Cell Biol.*, **122**, 283.
130. Hutchison, N. and Weintraub, H. (1985) Localization of DNAase I-sensitive sequences to specific regions of interphase nuclei. *Cell*, **43**, 471.
131. Krystosek, A. and Puck, T. T. (1990) The spatial distribution of exposed nuclear DNA in normal, cancer, and reverse-transformed cells. *Proc. Natl Acad. Sci. USA*, **87**, 6560.
132. Comings, D. E. (1980) Arrangement of chromatin in the nucleus. *Hum. Genet.*, **53**, 131.
133. Manuelidis, L. (1984) Different central nervous system cell types display distinct and nonrandom arrangements of satellite DNA sequences. *Proc. Natl Acad. Sci. USA*, **81**, 3123.
134. Carter, K. C., Taneja, K. L., and Lawrence, J. B. (1991) Discrete nuclear domains of poly(A) RNA and their relationship to the functional organization of the nucleus. *J. Cell Biol.*, **115**, 1191.
135. Huang, S., Deerinck, T. J., Ellisman, M. H., and Spector, D. L. (1994) *In vivo* analysis of the stability and transport of nuclear poly(A)$^+$ RNA. *J. Cell Biol.*, **126**, 877.
136. Visa, N., Puvion-Dutilleul, F., Harper, F., Bachellerie, J. P., and Puvion, E. (1993) Intranuclear distribution of poly(A) RNA determined by electron microscope *in situ* hybridization. *Exp. Cell Res.*, **208**, 19.
137. Wang, J., Cao, L-G., Wang, Y-L., and Pederson, T. (1991) Localization of pre-mRNA at discrete nuclear sites. *Proc. Natl Acad. Sci. USA*, **88**, 7391.
138. Spector, D. L. (1993) Macromolecular domains within the cell nucleus. *Annu. Rev. Cell Biol.*, **9**, 265.
139. Xing, Y., Johnson, C. V., Dobner, P. R., and Lawrence, J. B. (1993). Higher level organization of individual gene transcription and RNA splicing: integration of nuclear structure and function. *Science*, **259**, 1326.
140. Zhang, G., Taneja, K. L., Singer, R. H., and Green, M. R. (1994) Localization of pre-mRNA splicing in mammalian nuclei. *Nature*, **372**, 809.
141. Xing, Y., Johnson, C. V., Moen, P. T., Jr, McNeil, J. A., and Lawrence, J. B. (1995) Nonrandom gene organization: Structural arrangements of specific pre-mRNA transcription and splicing with SC-35 domains. *J. Cell Biol.*, **131**, 1635.
142. Lawrence, J. B., Singer, R. H., and Marselle, L. M. (1989) Highly localized tracks of specific transcripts within interphase nuclei visualized by *in situ* hybridization. *Cell*, **57**, 493.
143. O'Farrell, P. H., Edgar, B. A., Lakich, D., and Lehner, C. F. (1989) Directing cell division during development. *Science*, **246**, 635.

144. Dambaugh, T., Hennessy, K., Fennewald, S., and Kieff, E. (1986) The virus genome and its expression in latent infection. In *The Epstein–Barr virus: recent advances*, Epstein M. A. and Achong, B. G. (ed.). John Wiley and Sons, New York, p. 13.
145. Huang, S. and Spector, D. L. (1991) Nascent pre-mRNA transcripts are associated with nuclear regions enriched in splicing factors. *Genes Dev.*, **5**, 2288.
146. Blobel, G. (1985) Gene gating: a hypothesis. *Proc. Natl Acad. Sci. USA*, **82**, 8527.
147. Berman, S. A., Bursztajn, S., Bowen, B., and Gilbert, W. (1990) Localization of an acetylcholine receptor intron to the nuclear membrane. *Science*, **247**, 212.
148. Davis, I., Francis-Lang, H., and Ish-Horowicz, D. (1993) Mechanisms of intracellular transcript localization and export in early *Drosophila* embryos. *Cold Spring Harbor Symp. Quant. Biol.*, **58**, 793.
149. Davis, I. and Ish-Horowicz, D. (1991) Apical localization of pair-rule transcripts requires 3′ sequences and limits protein diffusion in the *Drosophila* blastoderm embryo. *Cell*, **67**, 927.
150. Dirks, R. W., Daniël, K. C., and Raap, A. K. (1995) RNAs radiate from gene to cytoplasm as revealed by fluorescence *in situ* hybridization. *J. Cell Sci.*, **108**, 2565.
151. Zachar, Z., Kramer, J., Mims, I. P., and Bingham, P. M. (1993) Evidence for channeled diffusion of pre-mRNAs during nuclear RNA transport in metazoans. *J. Cell Biol.*, **121**, 729.
152. Zirbel, R. M., Mathieu, U. R., Kurz, A., Cremer, T., and Lichter, P. (1993) Evidence for a nuclear compartment of transcription and splicing located at chromosome domain boundaries. *Chromosome Res.*, **1**, 93.
153. Cremer, T., Kurz, A., Zirbel, R., Dietzel, S., Rinke, B., Schrock, E., *et al.* (1993) Role of chromosome territories in the functional compartmentalization of the cell nucleus. *Cold Spring Harbor Symp. Quant. Biol.*, **58**, 777.
154. Mehlin, H., Daneholt, B., and Skoglund, U. (1992) Translocation of a specific premessenger ribonucleoprotein particle through the nuclear pore studied with electron microscope tomography. *Cell*, **69**, 605.
155. Olins, A. L., Olins, D. E., and Bazett-Jones, D. P. (1992) Balbiani ring hnRNP substructure visualized by selective staining and electron spectroscopic imaging. *J. Cell Biol.*, **117**, 483.
156. Daneholt, B. (1982) Structural and functional analysis of Balbiani ring genes in the salivary glands of *Chironomus tentans*. In *Insect ultrastructure*, Vol. 1, King, R. and Akai, H. (ed.). Plenum Press, New York, p. 382.
157. Vazquez-Nin, G. H., Echeverria, O. M., Fakan, S., Leser, G., and Martin, T. E. (1990) Immunoelectron microscope localization of snRNPs in the polytene nucleus of salivary glands of *Chironomus thummi*. *Chromosoma*, **99**, 44.
158. Wurtz, T., Lonnroth, A., Ovchinnikov, L., Skoglund, U., and Daneholt, B. (1990) Isolation and initial characterization of a specific premessenger ribonucleoprotein particle. *Proc. Natl Acad. Sci. USA*, **80**, 6436.
159. Lamb, M. M. and Daneholt, B. (1979) Characterization of active transcription units in Balbiani rings of *Chironomus tentans*. *Cell*, **17**, 835.
160. Olins, A. L., Olins, D. E., and Franke, W. W. (1980) Stereo-electron microscopy of nucleoli, Balbiani rings and endoplasmic reticulum in *Chironomus* salivary gland cells. *Eur. J. Cell Biol.*, **22**, 714.
161. Skoglund, U., Andersson, K., Strandberg, B., and Daneholt, B. (1986) Three-dimensional structure of a specific pre-messenger RNP particle established by electron microscope tomography. *Nature*, **319**, 560.

162. Stevens, B. J. and Swift, H. (1966) RNA transport from nucleus to cytoplasm in *Chironomus* salivary glands. *J. Cell Biol.*, **31**, 55.
163. Mehlin, H., Skoglund, U., and Daneholt, B. (1991) Transport of Balbiani ring granules through nuclear pores in *Chironomus tentans*. *Exp. Cell Res.*, **193**, 72.
164. Kiseleva, E., Wurtz, T., Visa, N., and Daneholt, B. (1994) Assembly and disassembly of spliceosomes along a specific pre-messenger RNP fiber. *EMBO J.*, **13**, 6052.
165. Jiménez-García, L. F. and Spector, D. L. (1993) Pre-mRNA processing components reorganize and colocalize with adenovirus RNA transcripts: evidence that transcription and splicing are coordinated by a cellular recruiting mechanism. *Cell*, **73**, 47.
166. Pombo, A., Ferreira, J., Bridge, E., and Carmo-Fonseca, M. (1994) Adenovirus replication and transcription sites are spatially separated in the nucleus of infected cells. *EMBO J.*, **13**, 5075.
167. Huang, S. and Spector, D. L. (1996) Intron-dependent recruitment of pre-mRNA splicing factors to sites of transcription. *J. Cell Biol.*, **133**, 719.
168. Martinez-Palomo, A., LeBuis, J., and Bernhard, W. (1967) Electron microscopy of adenovirus 12 replication. 1. Fine structural changes in the nucleus of infected KB cells. *J. Virol.*, **1**, 817.
169. Puvion-Dutilleul, F., Rousseu, R. and Puvion, E. (1992) Distribution of viral RNA molecules during the adenovirus 5 infectious cycle in HeLa cells. *J. Struct. Biol.*, **108**, 209.
170. Puvion-Dutilleul, F. and Puvion, E. (1991) Sites of transcription of adenovirus type 5 genomes in relation to early viral DNA replication in infected HeLa cells. A high resolution *in situ* hybridization and autoradiographical study. *Biol. Cell.*, **71**, 135.
171. Reich, N. C., Sarnow, P., Duprey, E., and Levine, A. J. (1983) Monoclonal antibodies which recognize native and denatured forms of the adenovirus DNA-binding protein. *Virology*, **128**, 480.
172. Sugawara, K., Gilead, Z., Wold, M., and Green, M. (1977) Immunofluorescence study of the adenovirus type 2 single-stranded DNA-binding protein in infected and transformed cells. *J. Virol.*, **22**, 527.
173. Voelkerding, K. and Klessig, D. F. (1986) Identification of two nuclear subclasses of the adenovirus type 5-encoded DNA-binding protein. *J. Virol.*, **60**, 353.
174. Berezney, R. (1984) Organization and functions of the nuclear matrix. In *Chromosomal nonhistone proteins*, vol. 4, Hnilica, L. S. (ed.). CRC Press, Boca Raton, Florida, p. 119.
175. Vogelstein, B. and Hunt, B. F. (1982) A subset of small nuclear ribonucleoprotein particle antigens is a component of the nuclear matrix. *Biochem. Biophys. Res. Commun.*, **105**, 1224.
176. Smith, H. C., Ochs, R. L., Fernandez, E. A., and Spector, D. L. (1986) Macromolecular domains containing nuclear matrix protein p107 and U-snRNP protein p28: further evidence for an *in situ* nuclear matrix. *Mol. Cell. Biochem.*, **70**, 151.
177. Zeitlin, S., Parent, A., Silverstein, S., and Efstratiadis, A. (1987) Pre-mRNA splicing and the nuclear matrix. *Mol. Cell. Biol.*, **7**, 111.
178. Ciejek, E. M., Nordstrom, J. L., Tsai, M-J., and O'Malley, B. W. (1982) Ribonucleic acid precursors are associated with the chick oviduct nuclear matrix. *Biochemistry*, **21**, 4945.
179. Mariman, E. C., van Eekelen, C. A. G., Reinders, R. J., Berns, A. J. M., and van Venrooij, W. J. (1982) Adenoviral heterogeneous nuclear RNA is associated with the host nuclear matrix during splicing. *J. Mol. Biol.*, **154**, 103.
180. Xing, Y. and Lawrence, J. B. (1991) Preservation of specific RNA distribution within the chromatin-depleted nuclear substructure demonstrated by *in situ* hybridization coupled with biochemical fractionation. *J. Cell Biol.*, **112**, 1055.

181. Carter, K. C., Bowman, D., Carrington, W., Fogarty, K., McNeil, J. A., Fay, F. S., *et al.* (1993) A three-dimensional view of precursor messenger RNA metabolism within the mammalian nucleus. *Science*, **259**, 1330.
182. Spector, D. L., O'Keefe, R. T., and Jiménez-García, L. F. (1993) Dynamics of transcription and pre-mRNA splicing within the mammalian cell nucleus. *Cold Spring Harbor Symp. Quant. Biol.*, **58**, 799.
183. Sehgal, P. B., Darnell J. E., Jr, and Tamm, I. (1976) The inhibition by DRB (5,6-dichloro-1-β-d-ribofuranosylbenzimidazole) of hnRNA and mRNA production in HeLa cells. *Cell*, **9**, 473.
184. Marinozzi, V. and Fiume, L. (1971) Effects of α-amanitin on mouse and rat liver cell nuclei. *Exp. Cell Res.*, **67**, 311.
185. Sinclair, G. D. and Brasch, K. (1978) The reversible action of α-amanitin on nuclear structure and molecular composition. *Exp. Cell Res.*, **111**, 1.
186. Kedinger, C. and Simard, R. (1974) The action of α-amanitin on RNA synthesis in Chinese hamster ovary cells. *J. Cell Biol.*, **63**, 831.
187. O'Keefe, R. T., Mayeda, A., Sadowski, C. L., Krainer, A. R., and Spector, D. L. (1994) Disruption of pre-mRNA splicing in vivo results in reorganization of splicing factors. *J. Cell Biol.*, **124**, 249.
188. Krämer, A., Keller, W., Appel, B., and Lührmann, R. (1984) The 5′ terminus of the RNA moiety of U1 small nuclear ribonucleoprotein particles is required for the splicing of messenger RNA precursors. *Cell*, **38**, 299.
189. Krainer, A. and Maniatis, T. (1985) Multiple factors including the small nuclear ribonucleoproteins U1 and U2 are necessary for pre-mRNA splicing in vitro. *Cell*, **42**, 725.
190. Black, D. L., Chabot, B., and Steitz, J. A. (1985) U2 as well as U1 small nuclear ribonucleoproteins are involved in pre-mRNA splicing. *Cell*, **42**, 737.
191. Akhtar, S. and Juliano, R. L. (1992) Cellular uptake and intracellular fate of antisense oligonucleotides. *Trends Cell Biol.*, **2**, 139.

3 | The packaging of pre-mRNA

JAMES G. MCAFEE, MAE HUANG, SYRUS SOLTANINASSAB, JANE E. RECH, SUNITA IYENGAR, and WALLACE M. LESTOURGEON

1. Introduction

In eukaryotes, pre-mRNA is bound by protein either simultaneously with the initiation of transcription or very soon thereafter (1–3), and it is not possible to isolate pre-mRNA molecules free from a unique set of abundant nuclear proteins unless protein denaturants are used in the purification scheme. The packaging of DNA into nucleosomes, and their association in the 30 nm chromatin fibre, greatly foreshortens the DNA substrate and provides some protection of the genome from nuclease activity (4). However, the major pre-mRNA binding proteins possess well-characterized, helix-unwinding activities (5–8) and, in comparison with nucleosome packaging, provide less protection from nuclease activity (9, 10). It seems then, that the pre-mRNA packaging mechanism functions primarily to maintain the transcript in a single-stranded state. This generally fits with the short-lived nature of pre-mRNA, and with the fact that nascent transcripts must be accessed by factors involved in intron removal, polyadenylation, and transport. Unlike rRNA, tRNA, and snRNA, pre-mRNAs are usually transcribed from single-copy genes and the pool of total nuclear pre-mRNA reflects great length heterogeneity (hence the term, heterogeneous nuclear RNA or hnRNA). More importantly to the RNA packaging mechanism, pre-mRNA looks like randomly generated sequence (11, 12). It is therefore not surprising that the major pre-mRNA binding proteins (the A-, B- and C-group proteins; see Fig. 1) bind *in vitro* to any and all RNAs and even to single-stranded DNA (13–16). The exclusion of pre-mRNA packaging proteins from unique abundant transcripts like rRNA, tRNA, and snRNAs may be due largely to the fact that these RNAs are bound by transcript-specific proteins (17, 18) that can recognize unique RNA secondary structures.

Students new to the field of pre-mRNA metabolism sometimes assume that a sequence-independent packaging mechanism may preclude the events of RNA processing. On this point it helps to remember that *trans*-acting factors access highly packaged sequences in the heterochromatic nuclei of differentiated cells and quickly activate host-specific transcription following heterokaryon formation (19). Perhaps through analogous mechanisms, pre-mRNAs that are packaged in a sequence-independent manner *in vitro* by the major pre-mRNA binding proteins are efficiently spliced when added to nuclear splicing extracts (20). It should also be remembered that numerous transcripts possess hundreds and even thousands of nucleotides 5′ to the first 3′ splice site and this RNA is clearly packaged by the A-,

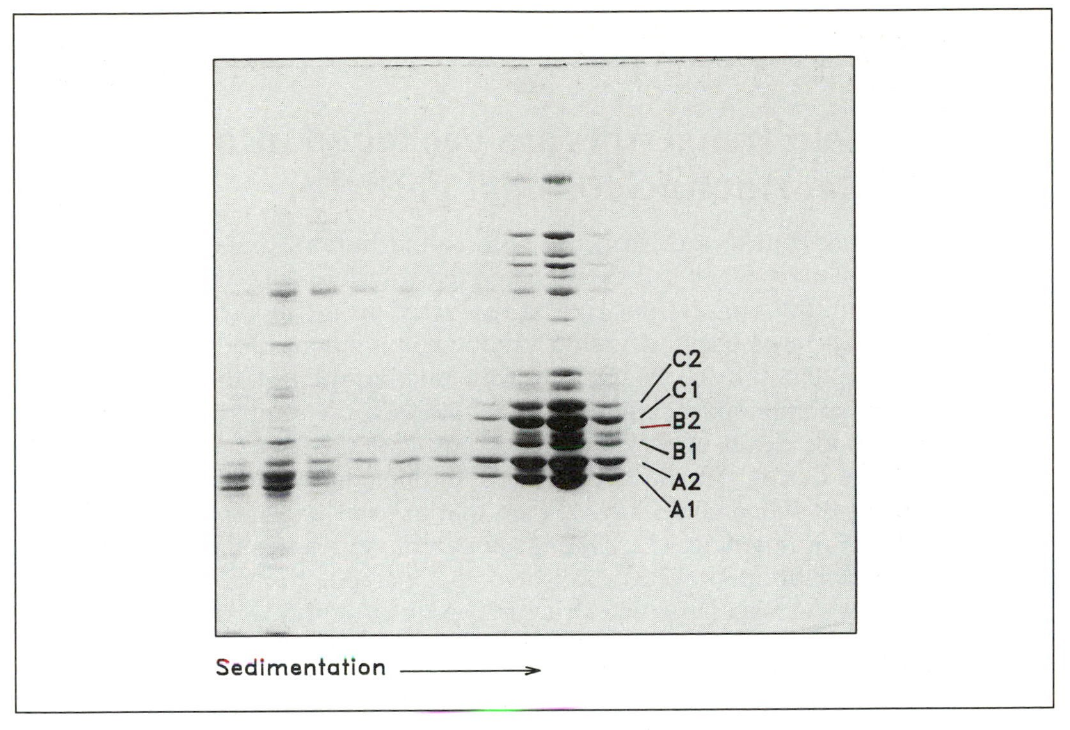

Fig. 1 Coomassie-stained SDS-PAGE showing the proteins present in successive fractions of a 15–30% glycerol gradient following the sedimentation of HeLa 40S hnRNP particles reconstituted on a 726 nucleotide transcript of the phage T4 genome (14). The lane showing the highest protein load corresponds to the 40S region of the gradient. Note the symmetrical distribution of the major core particle proteins above and below the peak fraction. Note also that proteins A1, A2, and C1 are present in approximately equal molar amounts. In this *in vitro* assembly experiment, varying amounts of protein A2 and B1 can be seen to sediment in a heterodisperse manner between the pool of soluble protein near the top to regions near the bottom of the gradient. This is due to the artifactual assembly of RNA-free fibres composed entirely of $(A2)_3B1$ tetramers (10). These fibres spontaneously assemble when intact monoparticles are dissociated by digesting the RNA substrate with nuclease. Fibre formation can be retarded but not completely inhibited by conducting the assembly reactions at 0°C.

B- and C-group proteins. It is likely, however, that splicing factors may compete *in vivo* with packaging proteins for functionally important sites during transcription and that packaging proteins may be phased by the sequence-specific association of snRNP complexes.

The information presented in this chapter is focused on the evidence indicating that, in mammalian cells, nascent pre-mRNAs are mostly packaged into repeating arrays of regular ribonucleoprotein particles, and on the function of the A-, B- and C-group proteins in 40S hnRNP core particle assembly. These proteins were initially termed the 'core' hnRNP proteins (21, 22) because they comprise 80–90% of the total protein present in gradient-isolated hnRNP particles; they are present in stoichiometric ratios (3, 20, 23, 24); and they possess the intrinsic ability to package RNA *in vitro* into a repeating array of 20–25 nm particles (Figs. 2–4) (13, 14, 20).

Observations indicating that pre-mRNAs may not be similarly packaged in invertebrates will also be discussed.

2. Nascent transcripts are packaged into an array of regular ribonucleosomal particles

If isolated nuclei are extracted with saline buffers under conditions that inhibit nuclease activity, or if they are subjected to brief sonic disruption, then 70–90% of the pre-mRNA molecules can be recovered from the 30S–300S region of glycerol gradients in association with a unique complement of abundant nuclear proteins (see Fig. 2a) (21, 22, 25–27). Electron micrographs of the 30–40S material reveal 20–25 nm monoparticles while the faster-sedimenting material exists either as a contiguous array of 20–25 nm polyparticles (28, 29) or as apparent clusters of particles (30, 31). In addition to these findings, numerous ultrastructural studies on spreads of active genes have shown that nascent transcripts exist either as an array of 20–25 nm particles (32, 33) or as apparent aggregates thereof (30, 31). These were the initial findings which indicated that nascent transcripts are packaged into a repeating array of regular structural entities or into a ribonucleosomal structure (34). More recently, antibodies against the major hnRNP core particle proteins were used to identify their *in situ* distribution in mildly dispersed rat neurone nuclei. Electron micrographs reveal nascent transcripts as polyparticle complexes corresponding in morphology and protein composition to isolated polyparticle complexes (35). Individual 30–40S hnRNP particles (monoparticles), as well as polyparticle complexes, are termed hnRNP complexes because the packaged RNA substrate is hnRNA or pre-mRNA. Summarized below are some of the findings

Fig. 2 (a) The sedimentation in glycerol gradients of hnRNP complexes stabilized by the aggressive inhibition of nuclease activity during hnRNP extraction from isolated HeLa nuclei. The top of the gradients is at the left. Electron micrographs of material taken from increasingly dense regions of the gradients (multiple insets) reveal monoparticles, dimers, and polyparticle complexes. The insets are positioned over gradient regions from which the samples were obtained. The solid tracing shows the absorbance of 260 nm light (representing RNA) throughout 15—30% glycerol gradients. The hnRNP complexes were freed via sonication from isolated HeLa nuclei at 0°C in the presence of RNasin and tRNA. The procedure used to preserve polyparticle complexes during extraction and sedimentation is similar to that of Sperling (31). The electron micrographs of samples taken from fast-sedimenting regions of the gradients reveal that about two-thirds of the polyparticle complexes appear as clumps of particles as described by Sperling (31). The more spread polyparticle complexes (insets in panel A) were chosen to better reveal the uniformity of particle morphology and to suggest the presence of a continuous RNA substrate. Electrophoresis of the RNA recovered throughout the gradient clearly reveals that the RNA increases in length proportional to the size of the polyparticle complex (not shown). (b) A sample centrifuged under the same conditions as in panel a, but without RNasin or tRNA. It was also incubated at 37° for 15 minutes to facilitate nuclease activity prior to loading on the gradient. In this sample almost all of the RNA is recovered as a single peak. The peak corresponds to the 40S region of the gradient. Electron micrographs of material taken from this gradient region reveal monoparticles. Below the A_{260} tracing in each panel is shown the distribution of the major packaging proteins (see Figure 1). Note that the composition and stoichiometry of the major core particle proteins is essentially the same for polyparticle complexes and for monoparticles. These experiments, conducted in the authors' laboratory, confirm the earliest findings on pre-mRNA packaging obtained almost 30 years ago (see 28, 30, 32, 33).

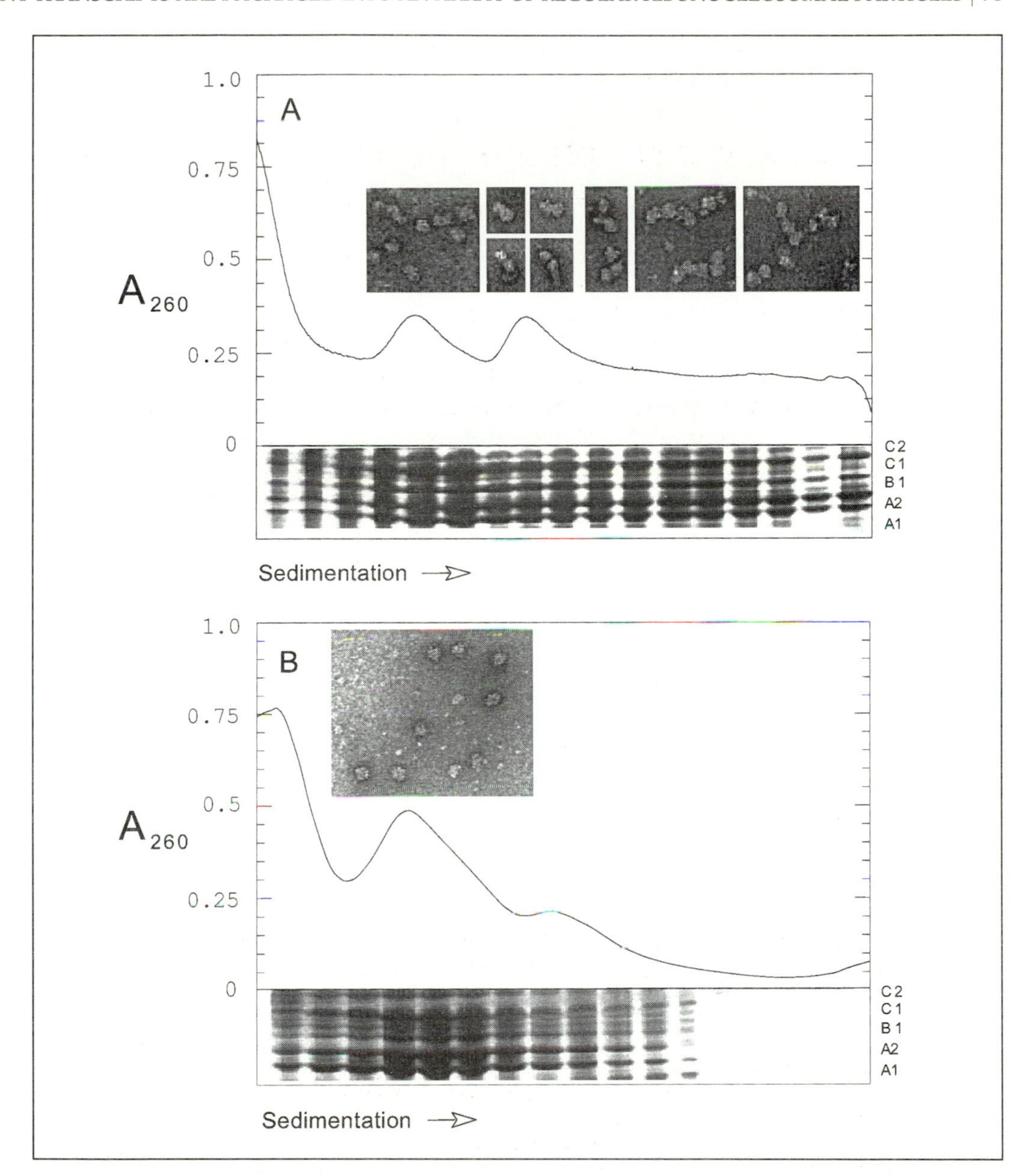

demonstrating that pre-mRNA is packaged into repeating arrays of regular particles (polyparticles). The first five observations are direct evidence for a ribonucleosome structure while the last three observations are confirmatory in nature, but obligatory, if nascent transcripts are packaged in an array of regular particles.

1. The smallest pre-mRNA-containing entity that can be recovered from crude nuclear sonicates or extracts is a ribonucleoprotein particle which sediments from 30–40S in glycerol or sucrose gradients (3, 29). In HeLa cells 80–90% of total

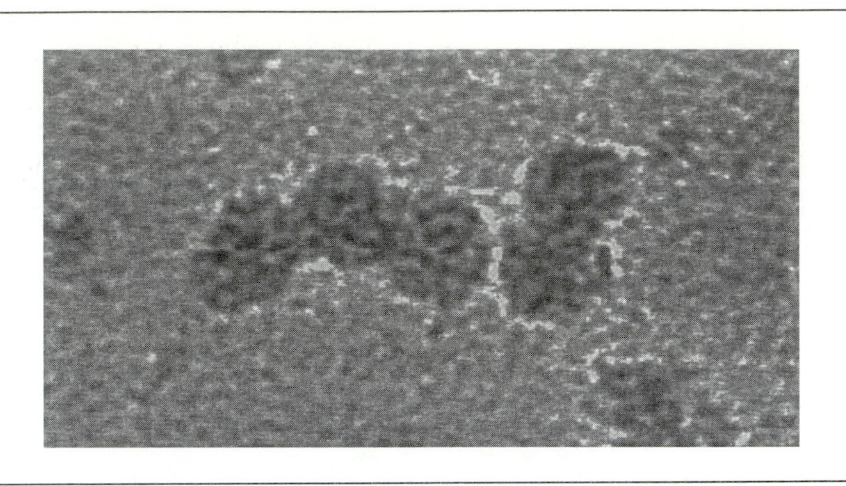

Fig. 3 Electron micrograph of a negatively stained polyparticle complex obtained as described in the legend to Fig. 2. The sample was taken from the lower third of the 15–30% (w/v) glycerol gradient after centrifugation in a Beckman Sw 28.1 rotor for 15 hours at 20 krpm. This print reveals the general uniformity of particles in the complex. This complex should contain a packaged RNA substrate 3500 nucleotides in length (about 700 nucleotides per monoparticle). The procedure used in the authors' laboratory for sample preparation usually does not reveal gaps between particles as shown here between the grouping of three- and two-particle clusters. It is not likely that the two-particle cluster is a free dimer of hnRNP particles because this 'fivemer' complex was recovered from gradient regions considerably below the dimer (small peak) and trimer region of the gradient.

pre-mRNA sequences are recovered with the 30–40S material, which is composed of the proteins shown in Figs 1 and 2.

2. Electron micrographs reveal that the 30–40S pre-mRNA-containing material recovered from glycerol gradients exists as 20–25 nm particles and that the fast-sedimenting material, containing long RNA molecules, exists either as an array of particles or as apparent particle clusters (Fig. 2a and Fig. 3) (3, 29, 31).

3. Brief nuclease cleavage converts polyparticle complexes into 30–40S monoparticles possessing the same protein composition (Fig. 2b) (28–33). The pre-mRNA fragments recovered from crude preparations of nuclease-generated monoparticles range in length from 600 to about 1000 nucleotides (21).

4. The most definitive evidence for a ribonucleosomal structure has come from the finding that the A-, B-, and C-group proteins possess the intrinsic ability to package monoparticle-lengths of RNA (700 ± 20 nucleotides) *in vitro* into 40S particles and to package multiples of this RNA length into dimers, trimers, and polyparticle complexes (13, 14, 20) (Fig. 4). Reconstituted polyparticle complexes possess the same physical–chemical properties as native hnRNP complexes. The reader is reminded here that 12–15 years ago, the conclusive evidence that histones package DNA into a repeating array of regular particles did not come from electron micrographs of spread chromatin, but through protein cross-linking studies (36) and especially

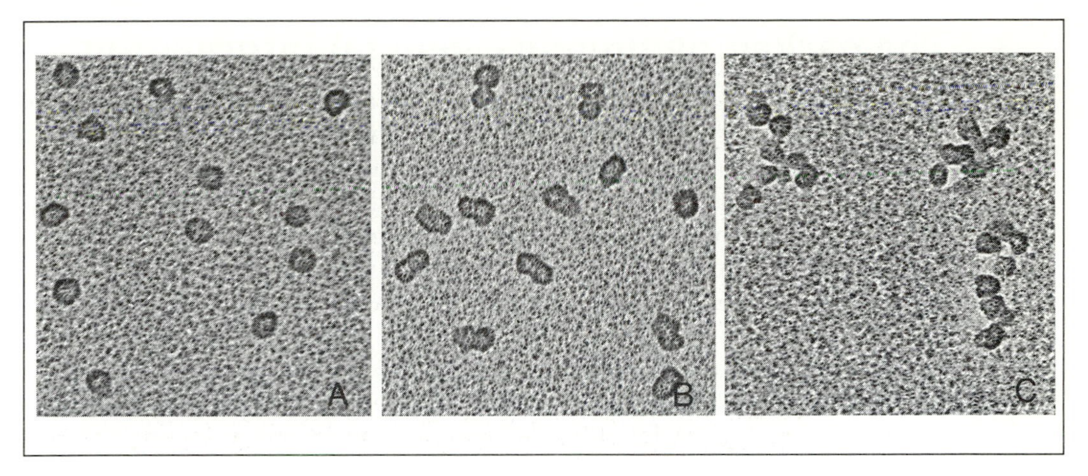

Fig. 4 Negatively stained and rotary shadowed electron micrographs showing the monoparticles (a) which reconstitute *in vitro* on a 709 nucleotide transcript of the human β-globin gene, dimer particles which assemble on a 1509 nucleotide transcript of the adenovirus major late promoter (b), and polyparticle clusters which assemble on the linearized 5386 nucleotide ϕX 174 ssDNA genome (c). The average diameter of the individual particles seen in these micrographs is 22 nm. The particles shown in these figures are composed of the proteins seen in Figs 1 and 2.

through the demonstration that purified histone tetramers spontaneously package DNA *in vitro* into nucleosomes with the same physical–chemical properties as native chromatin (for example, see 37, 38, and 39). Spontaneous self-assembly (reconstitution) is the assay for structural protein function (nucleosomes, ribosomes, snRNP particles, thick and thin filaments, microtubules, viral capsids, etc.) just as catalytic activity is the assay for enzyme function.

5. The C protein tetramer has an RNA binding site size near 230 nucleotides and binds RNA through a self-cooperative binding mode (consistent with a regular packaging function) (16). Each successive group of three tetramers folds monoparticle-lengths of RNA (about 700 nucleotides) into a 19S triangular complex that is present in native 40S ribonucleosomes and that is required to nucleate the stepwise assembly of the hnRNP core particle *in vitro* (20).

6. The major A, B, and C-group proteins (Figs. 1 and 2), which comprise 70–90% of the total protein in gradient isolated 30–40S particles, exist *in vivo* as three different tetramers—$(A1)_3B2$, $(C1)_3C2$, and $(A2)_3B1$ (10, 23, 24, 40, 41)—that are recovered from polyparticle complexes and from monoparticles in the same stoichiometric ratio.

7. In rapidly dividing cells in tissue culture, or in transcriptionally active cells such as those of the brain, the major A-, B-, and C-group proteins are present at very high intranuclear concentrations; about one-third the amount of the histones in HeLa cell nuclei (9), or about 2 mM (42).

8. The major A-, B-, and C-group proteins are generally located along the length of nascent transcripts (20, 35, 43–45).

3. The packaging proteins: general information

The A-, B-, and C-group proteins (A1 and A2; B1 and B2; C1 and C2) probably exist in the 40S ribonucleosomes isolated from actively dividing HeLa cells, mostly as three different heterotetramers $(A1)_3B2$, $(A2)_3B1$, and $(C1)_3C2$) (10, 20, 23, 24, 40, 41) (see Fig. 5). If true, this would explain why proteins A1, A2, and C1 are present at a 1:1:1 molar ratio and why these proteins are present at three times the molar ratio of B1, B2, and C2. Recently, however, we have shown that bacterially expressed C1 and C2 spontaneously oligomerize to form homotetramers in bacterial cells (16). This suggests that the core proteins in HeLa cells might exist as unique homotetramers. The protein cross-linking studies performed to date indicate that homotrimers of A1, A2, and C1 are the major species recovered from cross-linked intact 40S monoparticles but definitive heterotypic contacts between the homotrimers and their alternatively spliced forms B1, B2, and C2 could not be confirmed (10, 24, 40, 41). The existence of homooligomers of B1, B2, or C2 also could not be confirmed. The chemical cross-linking studies performed on purified native C protein demonstrate the presence of the heterotetramer but they do not exclude the existence of low amounts of C2 homotetramer (24). While the evidence to date argues for the existence of heterotetramers, the oligomerization mechanism that functions to ensure this assembly scheme is not readily apparent nor is the significance of one monomer in each tetramer being derived via an alternative splicing event (discussed below). Should they exist, the biological significance of unique homotetramers, for example $C1_4$ or $C2_4$, in a 3:1 molar ratio (generated by alternative splicing) is also not immediately apparent.

Each 40S monoparticle contains three copies of each tetramer such that the core-particle protein composition is $3[(A1)_3B2, (A2)_3B1), (C1)_3C2]$, yielding an expected core-particle protein mass of 1.25 million. Adding the mass of a 700 nucleotide fragment of pre-mRNA (235 kDa) yields a ribonucleosome mass near 1.48 million. This is consistent with its sedimentation coefficient and size (about 20–25 nm in negatively stained preparations). For example, the 30S and 50S ribosomal subunits of *E. coli* have respective masses of 1.0 and 1.8 million (46). In addition, preliminary mass determinations via scanning transmission electron microscopy place the mass of the ribonucleosome at 1.3–1.5 million (J. Wooley, personal communication). As discussed below, the best evidence that each 40S monoparticle contains three copies of each tetramer lies in the finding that only three C protein tetramers will bind to monoparticle lengths of RNA (about 700 nucleotides) and that this binding event precedes and directs the binding of three $(A2)_3B1$ and three $(A1)_3B2$ tetramers to complete monoparticle assembly (20). Support for this argument can be seen in chemical cross-linking studies which reveal that the C proteins in native hnRNP preparations exist as oligomeric complexes corresponding to three tetramers (40, 41).

The $(A2)_3B1$ and $(C1)_3C2$ tetramers have been isolated and partially characterized (20, 23, 24). The $(A1)_3B2$ tetramer has not been isolated but the cross-linking studies reveal that A1 exists at least as homotrimers in hnRNP complexes (10, 40, 41) and in a 3:1 ratio with B2. Of considerable interest from both an evolutionary and a

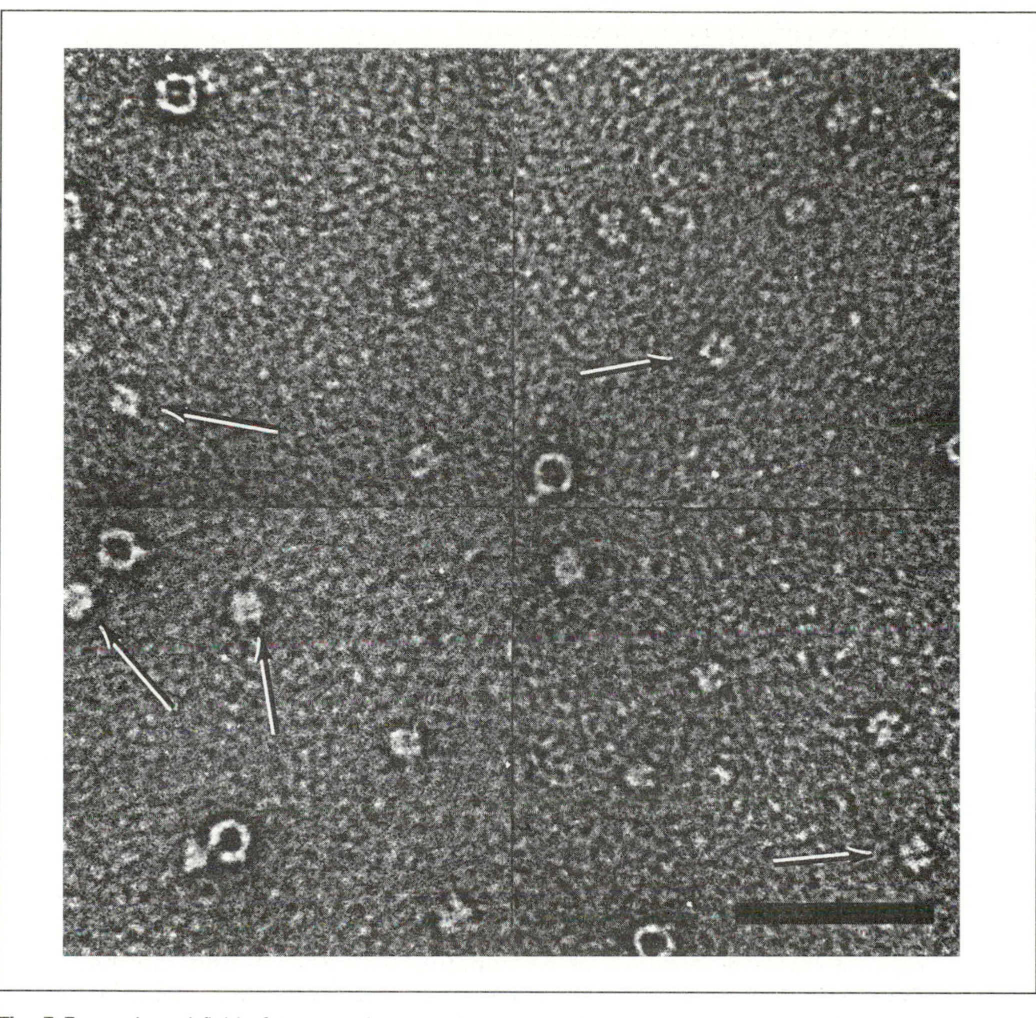

Fig. 5 Four enlarged fields from two electron micrographs taken from different areas of a negatively stained formvar-coated and glow-discharged grid. The holes in the doughnut-like horse spleen ferritin molecules added as an internal size standard are about 8 nm (138). The purified C protein tetramers appear mostly as four-lobed structures (denoted by arrows). In about 30% of the structures one of the four lobes is poorly resolved (see text).

functional perspective has been the finding that proteins C1 and C2, which form the $(C1)_3C2$ tetramer, proteins A2 and B1 which form the $(A2)_3B1$ tetramer, and proteins A1 and B2 which form the $(A1)_3B2$ tetramer are encoded by single genes. More specifically, protein C2 differs from C1 by the presence of a 13-residue insert at Gly106 (47, 48), B1 differs from A2 by the presence of a 12-residue insert at the amino terminus (48), and B2 differs from A1 by the presence of a 52-amino-acid insert near the carboxy-terminal third of A1 (49). In the original characterization of B2 (21) it was misidentified in a two-dimensional gel as an acidic polypeptide. B2 is basic like A1 with a pI near 9.5 (22, 50). The alternative splicing event that yields the two mRNAs for each type of tetramer is a constitutive homeostatic event, in that it happens in all somatic cells. Like the histones (reviewed in 4), the core particle proteins are transcribed from multigene families (51–53), and they reveal charge and nonallelic variants (49, 53, 54) that can be resolved in various two-dimensional electrophoretic systems (22, 55–59).

The C proteins are acidic (pI 5.9) (largely due to their 130-residue carboxy terminus, which contains 41 acidic and 21 basic residues) (60) whereas the A and B group proteins are basic (pI 8.4–9.5) (22, 48, 50). The C protein tetramers, present in native and reconstituted ribonucleosomes, dissociate completely from RNA at salt concentrations of 400–500 mM (20, 21). C1 and C2 apparently are phosphorylated at multiple serine and/or threonine residues and hypophosphorylation has been reported to correlate with enhanced RNA binding affinity (61). The A- and B-group proteins dissociate from RNA at 100–200 mM salt (21). On this point it is interesting to note that the salt concentration that dissociates the $(A2)_3B1$ tetramer from RNA is also the concentration that dissociates the tetramer into A2–A2 and A2–B1 dimers (24). Not only does the C protein tetramer bind RNA in a salt-resistant manner but the tetramer itself does not dissociate into soluble polypeptides even in 1 M salt, in 0.1% deoxycholate, in EDTA-containing buffers, or in the presence of reducing agent (23). The C protein tetramer does dissociate in 6 M urea in the absence of reducing agent.

4. A1–RNA binding and the question of binding specificity

hnRNP A1 was initially characterized as a glycine-rich basic polypeptide containing several residues of the unusual amino acid N^G-N^G-dimethyl arginine (21, 56, 62). The A- and B-group proteins are all similar in these regards (48, 63). The 135-residue carboxy terminus of A1 contains 45% glycine and the repeating sequence GN(F/Y)GG(S/G)RG, within which is found a regular spacing of both positively charged and aromatic residues (64, 65). Also within the carboxy terminus are six Arg-Gly-Gly (RGG) repeats termed the RGG box (66). The RGG box is thought to be a general RNA-binding motif and is often found in proteins that possess other RNA-binding motifs (63). Many RGG box-containing proteins also contain dimethyl arginine (67). The sites of arginine methylation in A1 have recently been mapped to residues 205, 217, and 224 (numbering based on the aminoterminal Ser as residue 1) and are located within the RGG motif. The methylase consensus sequence is Phe/Gly-Gly-Gly-Arg-Gly-Gly-Gly/Phe (K. Williams, personal communication; 68). The RNA-binding elements of the carboxy terminus appear to drive A1's co-operative binding mode (69) and its strand annealing activity (70, 71). The amino terminal 1–195 residue UP1, or A-domain (6, 7), contains two RNA-recognition motifs (RRMs) (63, 72, 73). The RRM is the most commonly observed structural motif of RNA-binding proteins and is characterized by the β1-α1-β2-β3-α2-β4 secondary structure (74, 75). The RRMs of A1 display a clear preference for single-stranded nucleic acids and bind RNA through a non-cooperative mode (69, 76). The two RRMs of A1 contribute about 50% of A1's free energy for RNA binding (76) while the remaining binding energy is derived from cooperative A1–A1 and A1–nucleic acid interactions located in the carboxy-terminal domain (15, 70, 77). For more information concerning the type, distribution, and possible functions of these and other RNA-binding elements, see references 63, 69, 73 and 78.

The recovery of A1 as a major stoichiometric component of a regular hnRNP core particle; the contiguous distribution of hnRNP complexes (and A1) along the length of nascent transcripts (see Figs 2 and 3); and the remarkably high intranuclear concentration of A1 in actively growing cells (about 2 mM) (42) all seemed initially to indicate that hnRNP A1 is strictly a structural component of 40S monoparticles possessing no physiologically relevant sequence-specific RNA-binding activity. However, when hnRNP-specific antibodies were used to isolate hnRNP-RNA complexes resistant to RNase T1 digestion, it was observed that A1-recovered complexes were enriched in the 19-nucleotide polypyrimidine tract containing the 3′ terminal intron sequence of the β-globin pre-mRNA (79). In addition, when the intron-terminal AG dinucleotide was changed to GG, a 10-fold reduction in A1 recovery was reported. It was also observed in these studies that when a 494-nucleotide bacterial transcript was used as the RNA substrate, no RNase T1 fragments could be immunopurified, a finding interpreted as further evidence that A1 binds specifically to intron-terminal elements of the 3′ splice sequence. In further support of this possibility, it was shown through gel retardation assays that when the synthetic oligo ribonucleotide $r(UUAGGG)_4$, a 24-nucleotide repeat suggested to match the 3′ splice site consensus sequence (Y_nNYAG/G), is used as a binding substrate, proteins A1, A2–B1, D, and E are major components of the oligonucleotide-bound complex (80). Finally, using the SELEX method of Tuerk and Gold (77), it was shown that recombinant A1 binds with high affinity to 20-nucleotide 'winner' sequences that contain two copies of the sequence UAGGGU/A separated by a CU dinucleotide (81). In these studies, it was reported that A1's affinity for the winner sequence is about 300-fold greater than for a 20-nucleotide sequence of β-globin intron lacking a single occurrence of the UAGGGU/A element. In addition to these studies, it was previously shown that recombinant A1 preferentially cross-links (induced via UV-irradiation) through the aminoterminal RRMs to short thymidine-rich ssDNA sequences corresponding to the 3′ terminal pyrimidine-rich elements of human β-globin intron 1 and to the same intron 1 region of the adenovirus type 2 major late transcript (82).

The non-equilibrium RNA and DNA binding studies described above have been interpreted (and widely cited) as evidence that hnRNP A1 possesses physiologically relevant RNA-binding specificity for the 3′ consensus sequence. Recently, however, other findings raise serious questions about these interpretations. For example, it is possible to monitor the interaction of RNA with the amino-terminal RRMs of A1 under equilibrium conditions by quantifying the binding-induced quench of tryptophan 36 fluorescence. Using this spectrophotometric method, the laboratory of K. R. Williams has failed to observe under physiological conditions that A1 possesses significant binding specificity for the 3′ splice site of IVSI of β-globin pre-mRNA.

Moreover, no significant difference in A1 affinity was observed upon mutating or randomizing the sequence (83). In fact, it was observed that A1 binds two to three fold more tightly to an unrelated intron sequence. Furthermore, the amino-terminal domain of A1 showed little RNA binding activity for the winner sequence (described above) but bound with reasonable affinity to a randomized version of the winner sequence. However, in 'forward' titrations, where the winner sequence was tested for its ability to compete as a binding substrate with the fluorescent probe poly[r(εA)], it was observed that a 50-fold molar excess of winner to probe essentially eliminated A1 binding to the probe. When this titration protocol was used under equilibrium conditions, A1 bound about 100-fold more tightly to the winner sequence than to a β-globin intron sequence or to a randomized construct of the winner sequence. These surprising results suggested that A1 binds the winner sequence differently than other RNAs and prompted these investigators to further characterize the winner sequence. It was discovered that the winner sequence exists as a unique dimeric/tetrameric complex formed through G-quartet pairings among the three adjacent guanosines. More importantly, it was shown that the winner sequence is actually bound by the glycine-rich carboxy-terminal domain and not by the RRMs of the amino terminus (the binding domains thought to confer binding specificity) (78, 82). In this context the $r(UUAGGGG)_4$ oligonucleotide (described above) seems as likely to form the G-tetrad structure as the SELEX-identified winner sequence. These findings serve as reminders that the SELEX procedure identifies RNAs for which a given protein binds with high affinity. It does not yield information about the structure of the RNA selected, the nature of the interaction, nor whether the high-affinity RNAs have biological relevance (84–86). The significance of the finding that recombinant A1 cross-links upon UV-irradiation preferentially to thymidine-rich ssDNA sequences corresponding to the polypyrimidine tract of introns is difficult to interpret in light of the fact that thymidine and uridine are more than 100-fold more photoreactive than the purines (87–91). In other words, during the period of irradiation, brief interactions between A1 and the thymidine-rich oligonucleotides are likely to be covalently trapped and preferentially accumulate in difference to oligonucleotides not enriched in UV-labile bases. This same caveat holds for short RNA sequences enriched in uridine. For example, it has been reported that A1 specifically binds (as detected via UV-cross-linking) to non-consensus-like but uridine-rich 3′ untranslated regions of several cytoplasmic RNAs (92).

5. C-protein–RNA binding and the question of binding specificity

hnRNP C1 and C2 were initially characterized as acidic polypeptides that bind pre-mRNA through a salt-resistant interaction (21). As in the case of hnRNP A1, it was assumed that the C proteins did not bind pre-mRNA in a sequence-specific manner because they are present at very high intranuclear concentrations (about 2 mM) (42), they are localized along the length of pre-mRNA molecules, and they are stoichiometric components of 40S monoparticles. Also, purified native C protein binds bacterial, plasmid, and pre-mRNA to form the same unique triangular complex that is recovered from native hnRNP particles following low salt extraction (discussed below). Finally, C protein nucleates spontaneous *in vitro* reconstitution of 40S monoparticles whether or not the RNA substrates contain splicing elements (20). However, findings suggesting that the C proteins may preferentially bind pre-mRNA at specific sites first came with the report that the C proteins bind sepharose-linked poly(U) with a remarkably high affinity (resistant to 2.0 M salt) (50). On the basis of this result, the C proteins were classified as 'poly(U)-binding proteins' and it was suggested that the C proteins may possess binding specificity for the poly-pyrimidine tract at the 3′ terminus of introns (50). In a follow-up study, it was in fact reported that if pre-mRNAs are added to nuclear extracts containing the hnRNP proteins, then monoclonal antibodies specific for the C proteins could be used to recover C protein–polypyrimidine tract complexes following aggressive digestion with RNase T1 (79). These experiments have been referred to as 'nuclease-protection experiments' but this description is not applicable in this case because the polypyrimidine tracts of the synthetic RNAs possessed no T1-cut sites. More recently, the SELEX procedure was used to identify high affinity 'winner sequences' for the C proteins (93). In support of the finding that the C proteins have a higher intrinsic affinity for uridine-rich regions than for other sequences, the selected sequences were found to be enriched in uridine and to possess at least one region of five contiguous uridylates.

As in the case of protein A1, the findings described above were interpreted (and have been widely cited) as evidence that the C proteins function in the early events of 3′ splice site recognition. In support of this interpretation, the finding that monoclonal antibodies against C protein block *in vitro* splicing (94) is often cited. More recently, however, equilibrium binding studies on purified native and recombinant C protein demonstrate that C protein binds RNA through a mode that seems to preclude an ability to specifically recognize polypyrimidine tracts in mRNA-sized substrates (16). Regarding the question of RNA-binding specificity, it is important first to know a ligand's binding site size (n), its state of oligomerization, its intrinsic affinity, and its predominant binding mode (i.e. cooperative or non-cooperative). If substrates significantly smaller than n are used in binding assays, the binding equilibrium under investigation is not likely to reflect the biologically relevant equilibrium. This is especially relevant if a ligand has multiple independent-acting RNA-binding domains (i.e. if the ligand is an oligomer). If a ligand binds a lattice cooperatively then a unique mechanism (a remarkably high intrinsic affinity or low off rate) must attenuate or block this activity until specific interactions have occurred. Regarding n, ultrastructural and hydrodynamic studies have indicated that the C-protein tetramer binds about 230 nucleotides of RNA (20). The recent equilibrium binding isotherms generated under near stoichiometric conditions using the fluorescent probe poly[r(εA)] independently confirm this approximate site size (16). Regarding oligomerization, the C proteins bind RNA as preformed tetramers, not as monomeric polypeptides (discussed below). In addition to the sedimentation, cross-linking, and ultrastructural studies that establish this fact, recombinant C1 and C2 spontaneously oligomerize in the absence of pre-mRNA to form tetramers. Regarding the intrinsic affinity of C protein for various homoribopolymers, recent equilibrium binding studies reveal that C protein binds poly(G) with an approximate intrinsic affinity (K_i) of 10^9 M^{-1} which is 100-fold higher than its equilibrium-determined affinity for poly(U). In opposition to reports that C protein does not bind poly(A) and poly(C), the equilibrium binding studies show that C protein binds these substrates with moderate K_i, but with high cooperativity (ω). The overall affinity ($K\omega$) for binding these substrates is greater than 10^8 but less than 10^9 M^{-1}. These intrinsic affinities do not readily reveal a mechanism through which the C protein tetramers might bind specifically to polypyrimidine tracts when coupled with its remarkably high intranuclear concentration and its cooperative binding activity (16).

In the SELEX experiments performed with C protein, the randomized sequence was about 20 nucleotides and, together with the flanking sequences, the binding substrates were about 60 nucleotides in length. This corresponds to a length that may not contact more than one RRM of the tetramer (discussed below). The equilibrium binding studies indicate that G-rich sequences should have been amplified in the SELEX experiments. Their absence from the amplified and selected pool might occur if, due to their remarkable affinity, they were not quantitatively removed from C protein prior to the amplification step or perhaps if G-tetrad secondary structures attenuate reverse transcriptase or *Taq* polymerase function. This was not the case, however, in the SELEX studies performed with A1 (see above). Clearly, C protein's intrinsic affinity for short poly(U) substrates is higher than its intrinsic affinity for short poly(A) or poly(C) substrates. In this regard, the selection of short U-rich RNAs is consistent with the equilibrium binding studies.

The highly cooperative binding of C protein to single-stranded RNAs provides a mechanistic basis for its distribution along the length of nucleic acid substrates. Given this binding mode, it is difficult to understand how C-protein tetramers, at an intranuclear concentration near 2 mM, with an occluded site size of 230 nucleotides and with at least four RNA binding sites, might select short polypyrimidine tracts in a physiologically relevant manner. To explore this possibility further, the consensus winner sequence for C protein was tested under equilibrium binding conditions for its ability to attenuate native C-protein binding to the fluorescent probe poly[r(εA)] (about 700 nucleotides). At an equal molar ratio of total nucleotides (winner to non-intron-containing RNAs), no attenuation in C-protein binding to the probe was observed (in preparation). This and related findings indicate that binding studies that neglect a ligand's site size, its intrinsic potential for strong cooperative interactions, and its state of oligomerization may not correctly reflect the physiologically relevant binding mode of a particular protein. The caveat (described above for the A1 studies) associated with the use of UV-irradiation and the high photoreactivity of uridine also holds for studies on C-protein–RNA binding.

Regarding the general question of protein specificity for pre-mRNA molecules, it is conceptually and physically impossible for every transcript to be bound by transcript-specific proteins. If true, this would open a 'hall of mirrors' in that the mRNAs for each transcript-specific protein would themselves be bound by unique proteins. This process would continue *ad infinitum*. The suggestion that this may occur came initially from the finding that when several short RNAs (< 550 nucleotides) were added to nuclear extracts competent for directing *in vitro* splicing, unique complements of protein were recovered with each and every RNA (95). Also, unique RNP complexes were observed to assemble on ecdysone-responsive sites in *Drosophila* (96). The results observed after addition of RNAs to nuclear extracts are not unexpected given the large number of biologically relevant RNA-binding and non-physiologically relevant RNA-binding proteins (as shown by the SELEX procedure) (84–86). As an example, when RNAs shorter than about 700 nucleotides (the length required for monoparticle assembly) are added to nuclease-dissociated hnRNP preparations, a very large number of non-stoichiometric RNP complexes assemble that sediment in a heterodisperse manner throughout density gradients (14). As pointed out above, without knowledge of a particular protein's site size or whether binding reflects a linear-sequence or unique-structure driven event, it is difficult to interpret the physiological relevance of each protein in an *in vitro*-assembled RNP complex. It is also true that most procedures for preparing splicing extracts do not yield preparations containing the normal composition nor stoichiometry of the major RNA packaging proteins (our unpublished observation).

6. Ribonucleosomes are regular structures

If nascent transcripts are packaged into a ribonucleosomal structure then, to be consistent with this terminology, the three different tetramers must interact to form a single particle type rather than three different particles or a mix of particles with differing protein compositions and packaged RNA lengths. In addition to the ultrastructural evidence for a single ribonucleosome size and morphology, there is biochemical evidence for a single ribonucleosomal architecture:

1. The core proteins co-sediment in sucrose and glycerol gradients with a single peak of pre-mRNA fragments (reviewed in 3), and it has not been possible to resolve fixed RNP complexes with differing protein compositions and densities in isopycnic gradients (56, 97–99).

2. As specific proteins dissociate from RNA upon increasing ionic strength, the RNA sediments more slowly in gradients but always in association with the remaining bound protein (21). In these experiments, the C-protein tetramers are the last to dissociate from the RNA (between 0.4–0.5 M salt). If the A-, B-, and C-group proteins exist in different particles that possess the same sedimentation coefficient then, upon dissociation of the A and B particle types, the C particles should continue to

sediment at 40S. As discussed below, when the A- and B-group proteins are dissociated from ribonucleosomes at 400 mM salt, the C proteins sediment as a 19S triangular complex in association with the initially packaged RNA.

3. The polyparticle complexes that spontaneously assemble *in vitro* are morphologically homogeneous (13, 14) and look like native polyparticle complexes (Fig. 4).

4. The *in vitro* assembly of stoichiometric hnRNP particles is an RNA length-dependent phenomenon (13, 20). More specifically, lengths less than 700 nucleotides support the assembly of artifactual complexes composed mostly of proteins A2 and B1. Stoichiometric 40S monoparticles assemble on 700 nucleotide lengths of RNA and multiples of this length are packaged into dimers, trimers, and polyparticle complexes (13, 14, 20) (Fig. 4).

5. Antibodies against the individual core proteins (i.e. against the A-, B-, or C-group proteins) immunoprecipitate 30–40S complexes possessing the same core protein composition as gradient-purified 40S particles (27, 100).

6. Finally, as discussed below, the most definitive evidence for a regular ribonucleosome structure is seen in the fact that the *in vitro* assembly of the 40S core particle is dependent on a stepwise interaction of all three tetramers (20).

These observations argue that hnRNP core particles are regular structures and that there is a default mechanism involving the A-, B-, and C-group proteins for pre-mRNA packaging. In crude preparations of isolated hnRNP complexes, numerous additional proteins can be detected in gels (Fig. 1) and many are RNA-binding proteins (22, 101). These non-packaging proteins are likely to be distributed amongst the pool of monoparticles because they are present at sub-stoichiometric levels in Coomassie-stained gels of gradient-isolated monoparticle preparations. When specific antibodies against the core proteins are used in immunopurification schemes, particles containing typical core-protein composition are obtained (27). It is likely that RNA-binding proteins with unique specificity, as well as snRNP complexes, can displace the core proteins at functionally important sites along the transcript. While these events could lead to a non-stoichiometric ratio of the core proteins for a particular particle, they do not argue against the existence of a regular packaging mechanism. Evidence that snRNP particles and other *trans*-acting factors can displace the packaging proteins from RNA is seen in the fact that various transcripts can be efficiently spliced *in vitro* following their *in vitro* packaging into ribonucleosomes (20).

7. The stepwise assembly of the ribonucleosome

7.1 Binding of the C-protein tetramer: the establishment of ribonucleosome architecture

Purified C-protein tetramers $(C1)_3C2$ appear in negatively stained electron micrographs as spherical or slightly ellipsoidal structures in the size range 8.5–10 nm. In some preparations, as many as 30% of the structures appear to be more dispersed on the grids and these reveal a four-lobed morphology (Fig. 5). One of the four subunits is often poorly resolved. This might be due to the C2 polypeptide with its additional 13-amino-acid insert (47, 48). The structures that reveal the four-lobed morphology appear about as large as ferritin in negatively stained electron micrographs. While these more extended structures only represent about one-third of the tetramers resolved in electron micrographs, they appear to reflect more accurately the morphology of the tetramer in solution. This statement is based on the observation that the C protein tetramer elutes from gel filtration columns as a symmetrical peak (390–400 kDa) (20, 23, 102) with a Stokes radius of 6.8 nm. The large Stokes radius is probably due to asymmetry and to the presence of 41 acidic and 21 basic residues in the 130 residue carboxy terminus and to the salt and hydration shell associated with these four highly charged domains. The C-protein tetramer may be structurally similar to the p53 tetramer, which is asymmetric and elutes from size exclusion columns with an anomalously large apparent mass (103, 104).

Through a series of protein–RNA binding studies using highly purified RNA-free C protein tetramers and various pre-mRNA molecules, it has been demonstrated via mass and density analyses and by electron microscopy that a single C protein tetramer binds 230–240 nucleotides of RNA (20) and recent equilibrium binding studies independently confirm this occluded site size range (Fig. 6) (16). Two tetramers bind twice this RNA length (Fig. 7) and, when a third tetramer binds to monoparticle lengths of RNA, an RNA folding event occurs (Fig. 8). This folding event leads to the formation of a unique triangular 19S C protein–RNA complex. The folding event which occurs when the third tetramer binds RNA is also apparent in the sedimentation properties of the one-, two-, and three-tetramer C protein–RNA complexes. The one- and two-tetramer complexes have very similar sedimentation rates in 15–30% glycerol gradients but the 19S complex sediments much faster than would be expected from the addition of the third tetramer alone. The folding event leads to a significant reduction in the relative frictional coefficient of the three-tetramer C-protein–RNA complex (20, 105, 106).

The intrinsic activity of the three C-protein tetramers to fold 700-nucleotide lengths of RNA into a unique triad complex, suggests that on long lengths of RNA each 700-nucleotide increment should also be so folded. If true, then this event would provide mechanistic insight into the finding that when long RNAs are added to preparations of nuclease or salt-dissociated hnRNP complexes a contiguous array of 40S ribonucleosomes spontaneously assemble *in vitro* (see Fig. 5) (14). In Fig. 9 it can be seen that purified C-protein tetramers spontaneously package two-triangle lengths of RNA (1452 nucleotides) into two-triangle complexes. Three-triangle complexes form when the C proteins bind 2087-nucleotide lengths of RNA (20). In the numerous C-protein–RNA binding studies conducted in our laboratory, purified native C protein has been observed to bind RNA regardless of RNA base composition or the presence or absence of eukaryotic RNA processing signals. In addition, C protein binds homoribopolymers of uridine (average length 268 nucleotides) to form structures like those shown in Fig. 5. Unlike the binding of C protein to

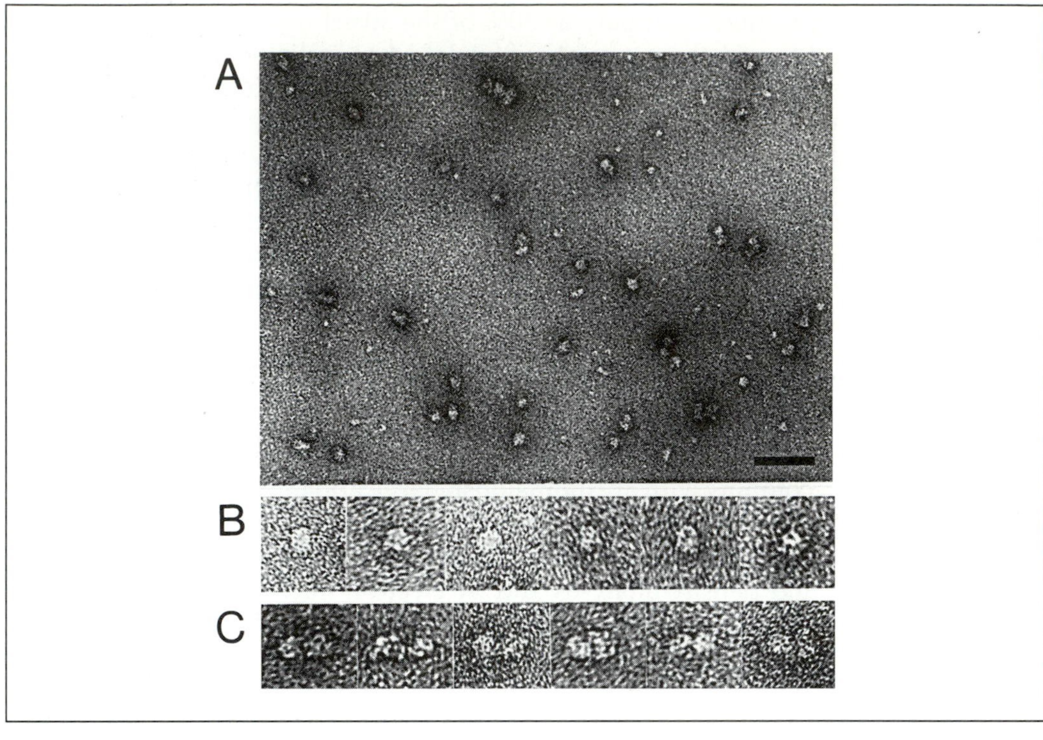

Fig. 6 Electron micrographs of negatively stained C-protein–RNA complexes which form when a single tetramer binds 230 nucleotides of RNA. A typical field of single-tetramer–RNA complexes (bar = 60 nm). Two-fold enlargements of the single-tetramer–RNA complexes, that represent about 70% of the observed structures. Structures representing about 30% of the observed structures, and appearing in the two-fold enlargements to be aggregates of two single-tetramer RNP complexes.

sepharose-linked poly(U) (50), C-protein–poly(U) complexes formed under equilibrium conditions readily dissociate at 500 mM salt into free poly(U) and protein, as do native 40S particles (Fig. 10).

7.2 The triangular 19S complex is present in isolated hnRNP particles

The findings discussed above demonstrate that purified C-protein tetramers possess the intrinsic ability to fold 700-nucleotide increments of RNA into 19S triangular complexes *in vivo*. Previous studies have shown that 700-nucleotide increments of RNA are packaged in each 40S monoparticle (14). If the C-protein-driven folding event exists as a mechanism for measuring monoparticle lengths of RNA and is an *in vivo* step in the assembly pathway leading to core particle assembly, then these structures should be present in native hnRNP complexes. In this context, if native or reconstituted ribonucleosomes are exposed to buffers of increasing ionic strength, the A- and B-group proteins dissociate at salt concentrations of 150–250 mM (21). At

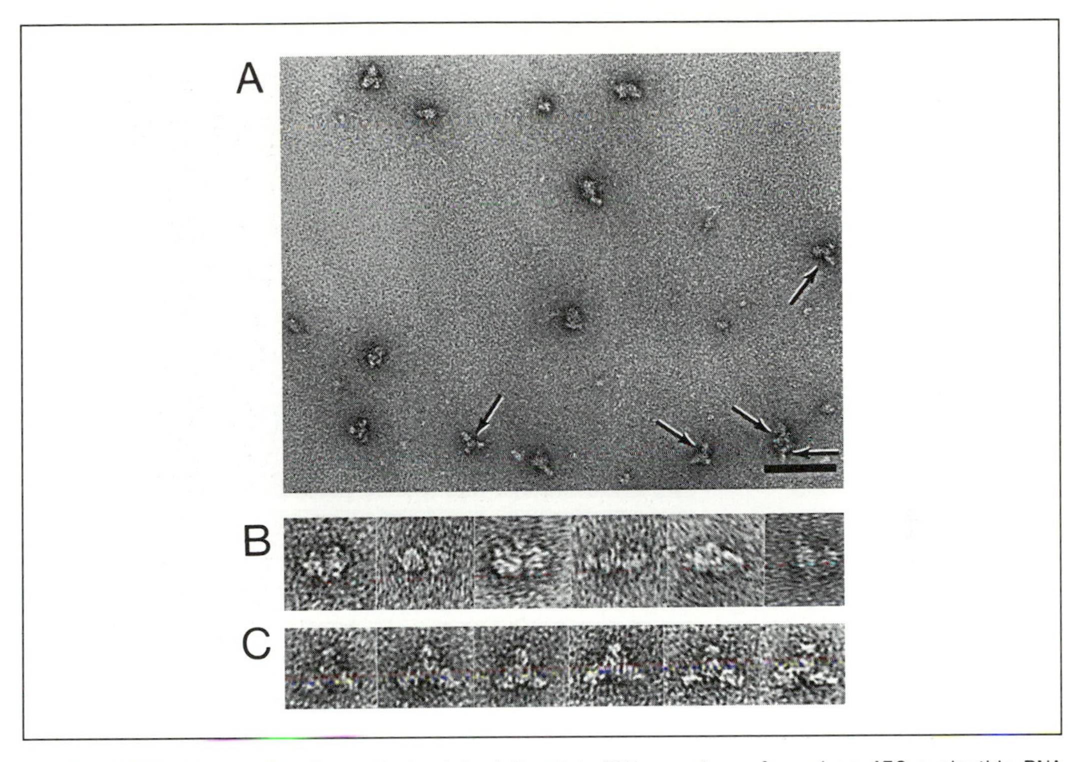

Fig. 7 Electron micrographs of negatively stained C-protein–RNA complexes formed on 456 nucleotide RNA molecules. (a) A typical field of the two-tetramer RNP structures (bar = 60 nm). (b) A two-fold enlargement of the RNP structures appearing to possess two C-protein–RNA complexes. (c) These structures are termed 'broken propeller' structures since they appear as three-blade propellers with one broken blade. The enlarged complexes, arranged with the broken blade up. About 20% of the observed complexes appear as poorly formed triangles (seen in upper and lower left of panel a).

350 mM salt the C proteins are the only core proteins bound to RNA. This salt-resistant complex sediments as 19S and electron micrographs of material taken from these gradient regions reveal the characteristic triangular 19S complex (20). Furthermore, when isolated or reconstituted hnRNP complexes are irradiated with UV light (to covalently link the C proteins to the RNA substrate), triangular C-protein–RNA complexes are again recovered as a remnant structure following hnRNP dissociation (105, 106). This argues that the triangular complex is not an artifact due to C-protein rearrangements along the RNA substrate, following A- and B-group protein dissociation. Triangular structures (thought to be assembly intermediates) with the dimensions of the 19S C-protein–RNA complex were previously observed by T. E. Martin in electron micrographs of crude hnRNP complexes (107). The recovery of the 19S triangular C-protein complex from native hnRNP complexes is another demonstration that it is the tetramer, and not monomeric C1 and C2, that initially binds RNA. This must be true because recombinant C1 and C2 tetramers spontaneously form tetramers in the absence of RNA and because purified native tetramers bind RNA *in vitro* to form the triangular complex, and also because the

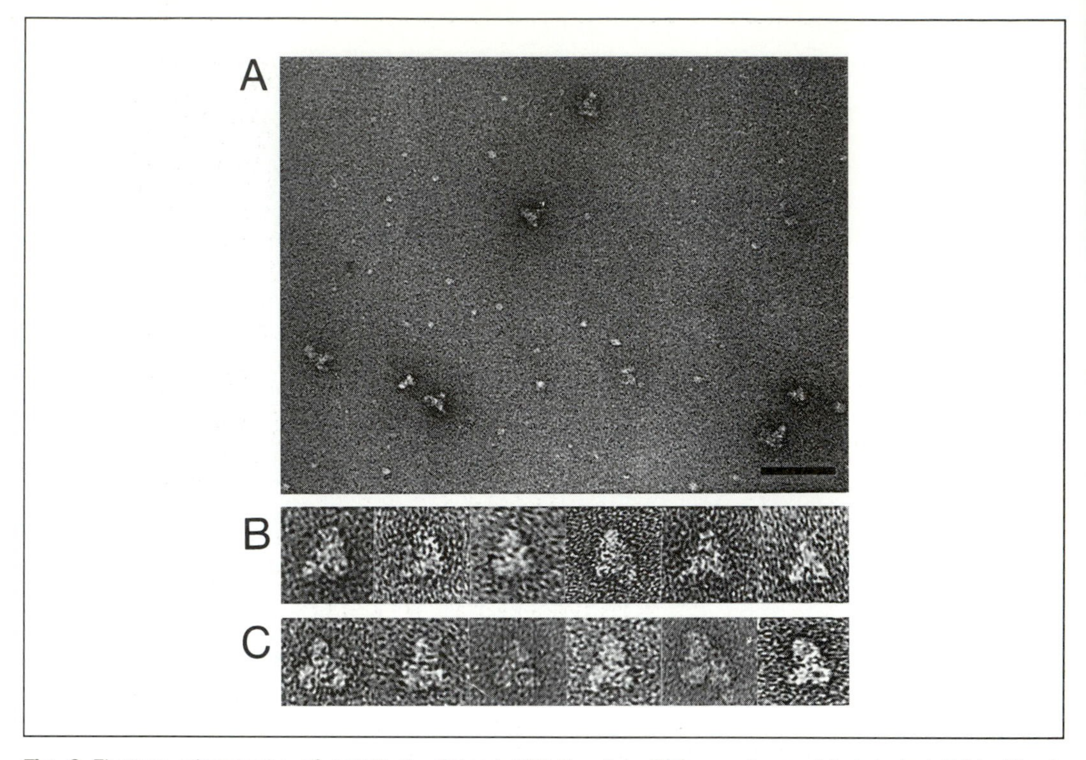

Fig. 8 Electron micrographs of negatively stained 19S C-protein–RNA complexes. (a) A typical field with six triangular 19S complexes (bar = 60 nm) (b and c) Two-fold enlargements of 19S complexes. The structures shown in panel b are typical of about 80% of the complexes and those in panel c are typical of about 20% of the complexes. The structures in panel c appear less compacted such that the three individual tetramers with their associated RNA are resolved. The 19S complexes shown in panel b and c are arranged with the shorter side as the base. Note that in many of the complexes the stain appears to penetrate the base more than the two equal sides.

tetramer does not dissociate into monomeric C1 and C2 in the absence of 6 M urea, guanidinium hydrochloride, or SDS-containing solutions.

7.3 Binding of the packaging proteins to RNA is physiologically relevant

Stated briefly below are the observations which demonstrate that purified C protein tetramers bind RNA *in vitro* as they do *in vivo*:

1. On the basis of their mass each ribonucleosome should contain three C-protein tetramers, and it is exactly three C-protein tetramers that fold monoparticle lengths of RNA (700 nucleotides) into triangular 19S RNP complexes.

2. Triangular 19S C-protein–RNA complexes are recovered from native and reconstituted 40S hnRNP complexes as remnant structures following salt-induced particle dissociation. Thus, the spontaneous assembly of this structure *in vitro* mimics an *in vivo* event.

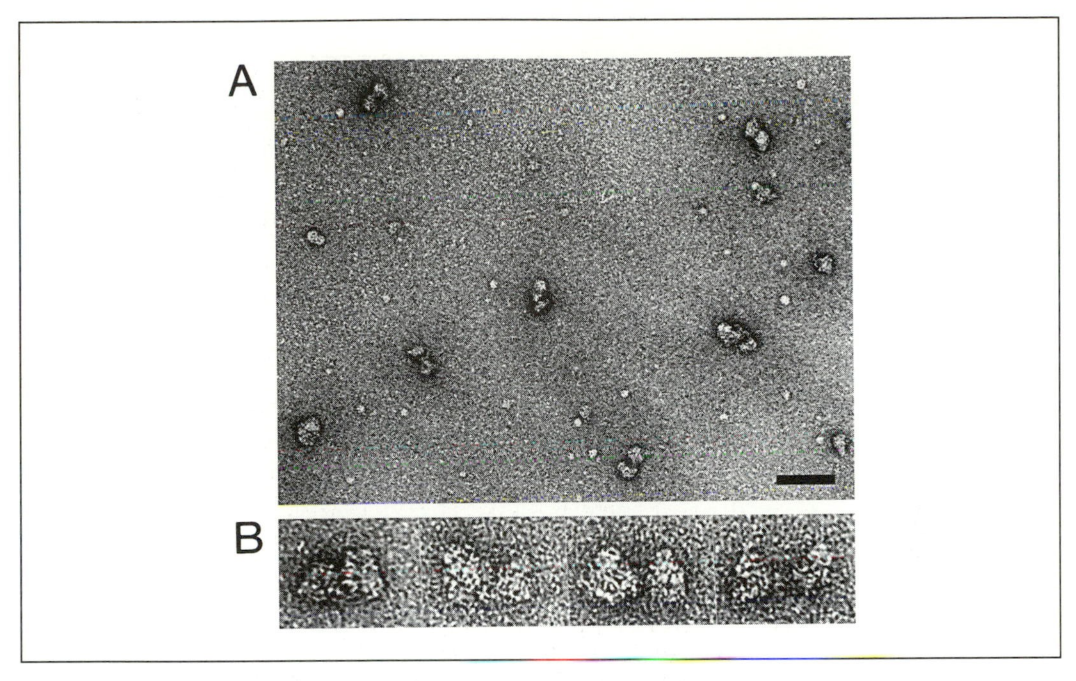

Fig. 9 Electron micrographs of negatively stained C-protein–RNA complexes which form on transcripts long enough (1452 nucleotides) to support the binding of six C-protein tetramers. (a) A typical field of RNP complexes (bar = 60 nm). (b) Two-fold enlargements reveal two contiguous triangular RNP complexes possessing the same size and general morphology as the 19S triangular C-protein–RNA complexes which form on 700-nucleotide transcripts.

3. Like the C proteins of native hnRNP complexes, the three C-protein tetramers that form the 19S C-protein–RNA complex dissociate from the RNA substrate at 400–500 mM salt.

4. Chemical cross-linking studies have shown that the C proteins in native hnRNP preparations exist as an oligomer of three tetramers (40, 41).

5. The C-protein tetramers bind long lengths of RNA through a self-cooperative binding mode to form contiguous arrays of triangular 19S complexes (16, 105).

6. Purified 19S C-protein–RNA complexes nucleate the spontaneous assembly of ribonucleosomes when added to nuclease-dissociated hnRNP preparations (20). Through the use of radiolabelled C protein in these experiments, it was found that the C protein in the 19S complex does not exchange with the free C protein during ribonucleosome assembly *in vitro*. This is an additional finding showing that the C proteins bind RNA as tetramers and it demonstrates the stability of the triangular complex, once formed.

7. The triangular 19S C-protein–RNA complex possesses the intrinsic ability to direct the binding of three $(A2)_3B1$ tetramers to form a 35S assembly intermediate (discussed below).

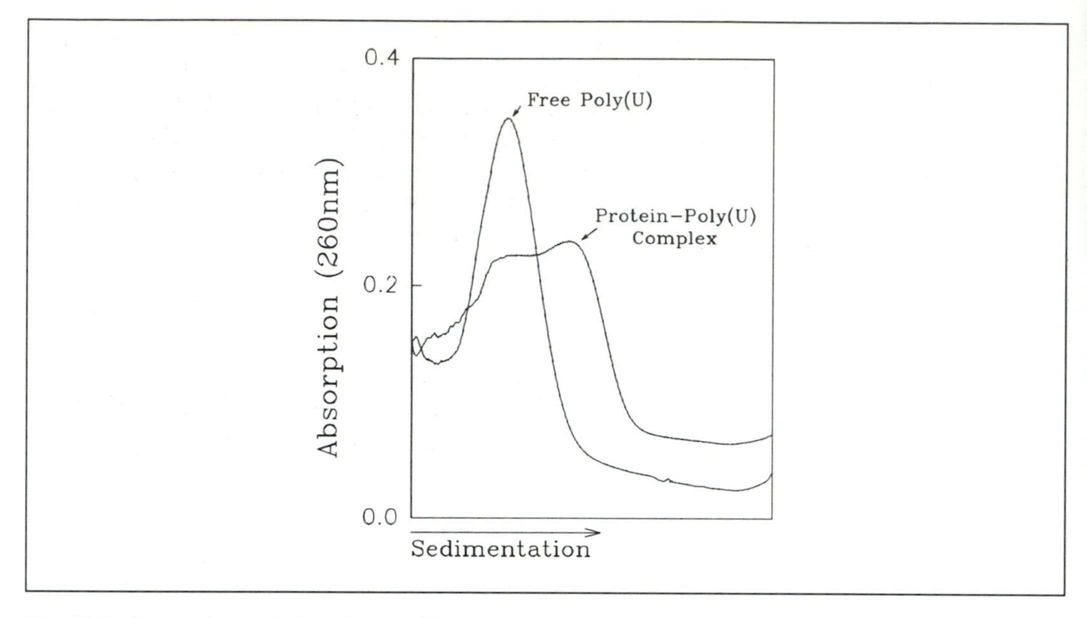

Fig. 10 Sedimentation and dissociation of C-protein–poly(U) complexes. Purified C protein was allowed to bind a two-fold molar excess of poly(U) with a length average of 268 nucleotides and half of the assembly mix was sedimented on 15–30% glycerol gradients containing 90 mM STM (salt-Tris-magnesium) buffer (21). The sedimentation of the protein-bound poly(U) and the excess free RNA (5.7S) was monitored by scanning the gradients at 260 nm. This tracing is denoted as the protein–poly(U) complex. The second half of the assembly mix was loaded on 15–30% glycerol gradients containing 500 mM STM buffer (500 mM NaCl). This tracing is denoted in the figure as free poly(U). Note that at 500 mM salt the C-protein–poly(U) complex has dissociated and that all the poly(U) sediments as protein-free RNA at 5.7S. Because a two-fold molar excess of poly(U) was used in the binding–dissociation assay, the unbound RNA can be seen to sediment in the 90 mM gradient as protein-free RNA.

7.4 The C proteins bind RNA first in the stepwise assembly of the core particle

Two early observations suggested that the C proteins may bind RNA first during core particle assembly. First, the C proteins bind RNA at salt concentrations which completely dissociate the basic A- and B-group proteins (21) and it is possible that the C proteins could play a 'first on, last off' role in RNA packaging. Secondly, in the absence of the C proteins, purified $(A2)_3B1$ tetramers were known to bind RNA in an artifactual manner to form a 43S complex which is not present in native monoparticle preparations (10). Summarized below are three recent observations indicating that the C proteins bind RNA first during ribonucleosome assembly.

1. If purified proteins A2 and B1 are allowed to bind monoparticle-length transcripts, the resulting 43S $(A2)_3$ B1–RNA complex does not function as a substrate for the *in vitro* assembly of core particles or even the binding of the other core particle proteins (20).

2. When purified 19S C-protein–RNA complexes are added to purified $(A2)_3B1$

tetramers *in vitro*, such that equal molar amounts of the two tetramer types are present, a 35S RNP complex spontaneously forms which, like native 40S hnRNP particles, contains a 1:1 molar ratio of the $(A2)_3B1$ and $(C1)_3C2$ tetramers (20). The protein composition and stoichiometry of the 35S RNP complex is like that of core particles that are missing the major core protein A1 and B2. As in the case of native and reconstituted particles, the 19S triangular complex can be recovered from the 35S complex upon salt-induced dissociation of the $(A2)_3B1$ tetramers. These findings demonstrate that the C proteins direct the correct *in vitro* association of proteins A2 and B1. If a twofold excess of $(A2)_3B1$ tetramers exists prior to the addition of 19S C-protein–RNA complexes then an $(A2)_3B1$-rich complex forms, which sediments at 40S. This suggests that the additional $(A2)_3B1$ tetramers may bind at sites normally occupied by $(Al)_3B2$ tetramers. In physical–chemical terms, as well as in terms of the amino acid sequence of the constituent polypeptides, these two tetramers are very similar.

3. The 19S triangular C-protein–RNA complex is required for the correct *in vitro* assembly of the 40S core particle. This was shown through the observation that only the triad structure would nucleate correct hnRNP assembly. In these experiments, one, two, three, and four tetramer-containing C-protein–RNA complexes (formed on transcripts of 230, 456, 709, and 962 nucleotides were added to aliquots of dissociated hnRNP complexes to determine if spontaneous assembly is dependent on the presence of the three-tetramer 19S complex. It was found that the 230-nucleotide-containing single-tetramer complex nucleates the assembly of fast-sedimenting artifactual structures which are A2-rich. This was not a surprising result because proteins A2 and B1 alone bind short lengths of RNA to form the artifactual 43S complexes, which sediment faster than 40S particles (10, 14). The two-tetramer 456 nucleotide C-protein–RNA complex was found to nucleate the assembly of a 'two-thirds' core particle (i.e two copies of each tetramer type and two-thirds the monoparticle length of RNA). Only the three-tetramer 19S C-protein–RNA complex, which assembles on monoparticle-length transcripts, could nucleate the correct assembly of stoichiometric 40S core particles.

8. A model for ribonucleosome structure and assembly

The most data-consistent mechanism through which about 700-nucleotide increments of pre-mRNA are packaged into a repeating array of regular particles is shown in Fig. 11. The packaging mechanism is largely explained by the intrinsic ability of three C-protein tetramers to bind RNA first, and fold successive monoparticle lengths of RNA into stable 19S triad structures. Support for this function is seen in the obligate requirement for the triangular 19S complex in the *in vitro* assembly of the 40S core particle and in the recovery of this unique structure from native hnRNP complexes, and from reconstituted core particles following dissoci-

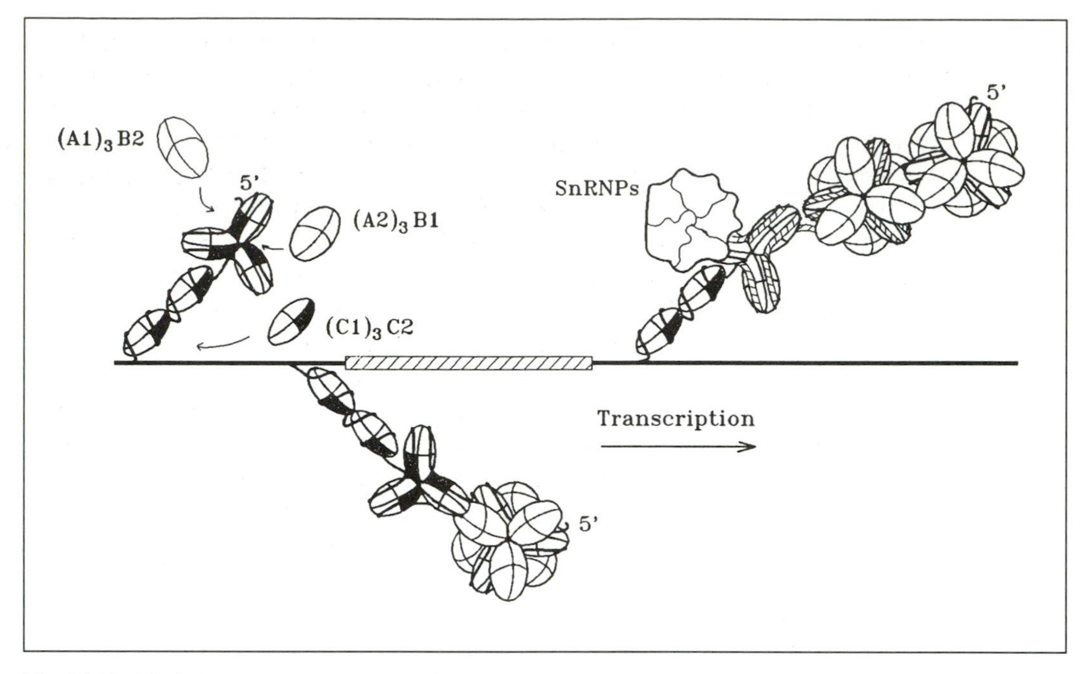

Fig. 11 Model of the events involved in pre-mRNA packaging. Note that the C-protein tetramers bind RNA first, that each successive group of three tetramers folds to form the 19S triangular C-protein–RNA complex, which establishes the architecture of the hnRNP core particle. Next, the $(A2)_3B1$ and $(A1)_2B2$ tetramers bind to complete core particle assembly and perhaps charge-neutralize the acidic carboxy terminus of C protein. This model for hnRNP core particle assembly on nascent transcripts implies a sequence-independent packaging mechanism, but one which allows *trans*-acting factors, such as snRNPs, access to the packaged substrate. The model assumes that splicing factors displace or loosen the packaging proteins at functionally important sites. The model argues that large ribonucleosomal complexes should exist that possess most or all of the factors required for intron removal and that spliceosomes assembled *in vitro* are likely to possess varying amounts of packaging proteins depending on intron and flanking-sequence length. See text for details.

ation. The stability of the tetramer in solution, its salt-resistant binding affinity for RNA, and the ability of the tetramers to form the 19S complex in the presence of the basic A- and B-group proteins indicate that the C-protein tetramer is well suited for this packaging function.

Further evidence that the C proteins function to direct correct ribonucleosome assembly is seen in the ability of the triangular 19S C-protein–RNA complex to direct the binding of three $(A2)_3B1$ tetramers to form a 35S assembly intermediate, and in the artifactual complexes which form when the A- and B-group proteins bind RNA either in the absence of, or before, C protein binding. The spontaneous assembly of the 35S complex which occurs when RNA is added to a mixture of $(A2)_3B1$ and $(C1)_3C2$ tetramers, or when purified $(A2)_3B1$ tetramers are added to the 19S C-protein–RNA complex, argues for the existence of a stepwise pathway toward ribonucleosome assembly. Namely, three C protein tetramers bind first to form the 19S complex, the triad structure forms and, after or simultaneously with $(A2)_3B1$ binding, proteins A1 and B2 complete monoparticle assembly (Fig. 11). The ability of $(A2)_3B1$ tetramers to bind at sites normally bound by $(A1)_3B2$ tetramers may explain in part why erythroleukemia cells that produce only trace amounts of A1 are viable (108, 109). It further suggests that the intranuclear concentration of these and other proteins may ultimately determine hnRNP composition, stability, and function. In this context, it has been observed that the mix of A1 isoforms differs with the state of cellular differentiation (58) and it has been shown that protein A1 can influence alternative splice site selection *in vitro* and *in vivo* in a concentration-dependent manner (see Chapters 6 and 7) (109–111).

Consistent with its RNA packaging role, numerous findings have demonstrated that C-protein–RNA binding is a highly self-cooperative process. For example, in the presence of a several-fold molar excess of long RNA molecules, complexes possessing only one or two tetramers were never observed either in gradients or in electron micrographs (20). Rather, under conditions of limited protein, all the protein was observed to saturate a limited number of RNA molecules, leaving the excess RNA free of protein. Also, when RNA molecules that are too short to support cooperative binding (less than 700 nucleotides) are used in an attempt to nucleate core particle assembly, artifactual A–B-rich structures are formed. This indicates that the intrinsic RNA-binding affinity of a single C-protein tetramer (though high) is not sufficient to preclude the artifactual binding of the basic A- and B-group proteins. An *in vitro* finding in support of this interpretation is the assembly of a C-protein-rich ribonucleosome when transcripts sufficient for the binding of a fourth tetramer (962 nucleotides) are used as the assembly substrate. In other words, cooperative interactions between the third and fourth tetramer along the RNA substrate appear sufficient to prevent A- and B-group protein binding to the extra 250-nucleotide length of RNA. The folding event that leads to the formation of the triangular 19S complex could not occur if tetramer–tetramer interactions were absent. A finding of particular interest in our reconstitution studies has been the observation that, if a significant excess of any core protein is present during assembly, then hnRNP complexes are formed that are enriched in the high concentration species. This indicates that RNA-binding proteins can add on to, or bind, peripheral RNA in hnRNP complexes. It is likely that such an event could occur *in vivo* upon the over-expression of a particular core particle protein and that this could influence unpackaging rates or splicing factor access to important sites.

The observations discussed above demonstrate that the C proteins pre-exist in tetrameric form before binding RNA, that the tetramers bind through a self-cooperative mode to single stranded RNAs, and that a single tetramer binds about 230 nucleotides of RNA. The tetramer's large RNA-binding site size is consistent with its large Stokes radius (6.8 nm) and suggests that the four amino-terminal RNA-binding domains probably bind RNA independently to regularly spaced sites along the 230-nucleotide substrate. NMR (112) and crystallographic studies (74) on the RRMs of the C proteins and related proteins argue that each polypeptide binds RNA indpendently. This binding mode complicates suggestions that the C proteins bind preferentially at specific sites in pre-mRNA molecules (discussed above). If the C-protein tetramer binds to the polypyrimidine sequence of introns with a biologically relevant enhanced affinity, then the intrinsic binding constant must be remarkably high in order to:

(1) negate the binding interactions of the three other binding domains in the tetramer;

(2) negate the self-cooperative binding mode established prior to the synthesis of a 3′ splice site in pre-mRNAs with a long first exon;

(3) negate the protein–protein interactions which exist in the triangular three-tetramer complex within each ribonucleosome. As described above, equilibrium binding studies do not support the suggestion that C protein first binds to, or localizes specifically, on polypyrimidine tracts (16).

Electron microscopy, together with the sedimentation properties of C-protein–RNA complexes of increasing tetramer number, suggest two possibilities for tetramer folding to form the 19S triangular structure. The tetramers may associate along the length of the RNA in a head-to-tail fashion (the 13 amino acid insert in C2 confers compositional polarity to the tetramer) and 'wrap up' such that the head of the third tetramer associates with the tail of the first to form a regular triangular structure. Through a mechanism more consistent with the observations to date, the tetramers could bind through 'limited' self-cooperativity along the length of the RNA and coalesce in groups of three through head-to-head-to-head interactions, as shown in Fig. 11. The association of three C-protein tetramers with monoparticle lengths of RNA (700 nucleotides) must occur in such a manner that RNA-binding sites are available for the binding of three $(A2)_3B1$ and three $(A1)_3B2$ tetramers to complete ribonucleosome assembly. If the 19S complex exists as an open triad complex, as shown, then the other tetramers may interdigitate in an ordered manner and charge-neutralize the acidic carboxy terminus of the C proteins. Most reports place the diameter of 30–40S particles at between 20–25 nm (3). In negatively stained preparations, the isosceles sides of the 19S C-protein-RNA complex average 23 nm and the height 20 nm. Thus, if the $(A2)_3B1$ and $(A1)_3B2$ tetramers interdigitate within an open triad structure (as shown in Fig. 11), the similarity in dimensions between 40S hnRNP particles and the 19S complex indicates that the triangular complex could function to establish the basic topology of protein and RNA in the core particle. A packaging mechanism based on an open arrangement of tetramers in the 19S complex (and thus in monoparticles) is favoured by the lability of the substrate to nuclease and by the fact that RNA is available for the binding of splicing factors. The inhibition of RNA splicing by antibodies against the A-, B-, and C-group proteins (94, 113) could result from the inability of *trans*-acting factors to access the RNA in antibody-bound hnRNP complexes and is not necessarily evidence that these proteins play a direct role in the biochemical events of intron removal.

If the C proteins associate with RNA *in vivo* as *in vitro* (a contiguous array of tetramers) then, unless excluded by other factors, pre-mRNA molecules should be bound along their entire length by these highly abundant nuclear proteins. The finding that the C proteins are cross-linked by UV-irradiation to all regions of adenoviral transcripts (corresponding to 16.2–91.5 map units) (44), the recovery of oligomeric arrays of C-protein tetramers after cross-linking splicing-competent native ribonucleosome complexes (40, 41), and the self-cooperative RNA-binding mode of C protein (16) strongly support this binding mechanism. The obligate C-protein-directed stepwise assembly of the core particle demonstrates that the core proteins interact together to package RNA into an array of regular structures. An irregular packaging system, compositionally and structurally unique to each transcript, would seem to place almost impossible evolutionary demands on the *trans*-acting factors involved in the basic events of RNA processing due to substrate variability.

9. Pre-mRNA packaging in non-mammalian cells

Proteins possessing motifs common to mammalian core particle proteins have been found in distantly related organisms (i.e. cyanobacteria (114), yeast (115–19), plants (120, 121), crustaceans (122, 123), insects (124–28), amphibians (129), and birds and reptiles (130)). However, this does not argue that the pre-mRNAs of all eukaryotes are packaged into regular hnRNP particles because the major motifs (the RRMs and the RGG box) exist in a wide variety of RNA-binding proteins (63). If C-protein tetramers function as protein rulers and direct monoparticle assembly every 700 nucleotides, then the existence of C protein at high intranuclear concentrations in a given species might argue for the presence of a regular packaging system. In this regard, the monoclonal antibody 4F4 (specific for human C1 and C2) has revealed the presence of doublet bands with similar mobilities to human C1 and C2 in gels of nuclear proteins from the chicken and lizard (130). In addition, a C-protein homologue (30916 Da, 78.5% identity) has been demonstrated in the frog (129) and some of the most definitive electron micrographs of polyparticle complexes have been obtained on lampbrush transcripts from developing amphibian oocytes (32, 33, 131, 132). However, neither C protein nor definitive polyparticle complexes have yet been demonstrated in more distantly related species.

A deduction that C protein may be a prerequisite for a mammalian-like ribonucleosomal packaging mechanism in a particular species may be premature, as follows: isolated poly(A)$^+$ RNP complexes from the brine shrimp *Artemia* (an arthropod) possess a single major basic protein (hd40) which appears to be a true homologue to the human A/B core particle proteins (123). At low protein:RNA ratios, highly purified hd40 removes secondary structure from long RNAs *in vitro* and at higher concentrations the RNAs are packaged into a repeating array of regular hnRNP-like particles (122). It is not, however, known whether or not pre-mRNAs in *Artemia* are packaged *in vivo* into polyparticle complexes. This deserves mention because a large number of A/B-like proteins may possess an RNP particle-generating activity *in vitro*. For example, purified human $(A2)_3B1$ tetramers package RNAs into a repeating array of particles, but the individual particles are larger than native hnRNP particles; they do not accommodate the binding of the other core particle proteins, and they are not present in crude hnRNP preparations (24). In addition, apparent A/B homologues in wheat mimic the activity of hd40 from *Artemia* but the *in vitro*-assembled polyparticle complexes are larger and irregular (120).

The ultrastructural and biochemical evidence for polyparticles in *Drosophila* (a dipteran and, like *Artemia*, an arthropod) mostly argues against a ribonucleosomal structure. As pointed out by Beyer in a previous review (2), the often cited studies that 'demonstrate' arrays of regular 20–25 nm particles on *Drosophila* pre-mRNA are likely to be in error. More recent studies, that also use the Miller spreading technique, mostly reveal a 5 nm RNP fibril, not arrays of regular particles. Also, in *Chironomus*, a second dipteran species, the large 75S Balbiani ring transcript (35–40 kb in length) is a 7 nm fibril that folds into a large 50 nm BR granule (133). The granule unfolds at the nuclear pore complex to facilitate RNA transport to the cytoplasm (134). Although electron spectroscopic imaging reveals the presence of 10–12 'particles' (or localized concentrations of RNA) within the 50 nm BR granule (135), for several reasons, these are clearly not mammalian-like hnRNP particles. First, the RNA present in the 50 nm BR granule cannot be an aggregate of hnRNP-like particles because 37 kb of RNA would exist in 50–53 hnRNP particles and, even if highly aggregated, only about 12 such particles could be present in a 50 nm granule. Second, the RNA foreshortening ratio in the 40S hnRNP particle is about 14:1 and in the BR granule it would be about 2500:1. Third, unlike large polyparticle complexes from mammalian cells, a population of regular 30–40S monoparticles are not liberated upon brief exposure of the BR granule to nuclease (136). Finally, it is hard to imagine how any protein can be present in the granule together with 35–40 kb of RNA. For example, a single C-protein tetramer has a diameter of about 9 nm which is larger than the 5 nm *Drosophila* or the 7 nm *Chironomus* transcripts (20, 105, 106). If 35–40 kb of RNA is folded into the 50 nm BR granule it is likely to exist as a paracrystalline complex. As pointed out by Beyer, it is possible that the detergent-based Miller spreading technique may dissociate hnRNP proteins from *Drosophila* pre-mRNAs and generate 5 nm thick fibres possessing little bound RNP. This does not seem likely because the same spreading technique reveals very distinct polyparticle complexes on pre-mRNAs but not on pre-rRNAs of amphibians (for example, see 132). In strong support of a 5 nm fibril in *Drosophila* is the mode of transcript packaging in the BR granule of *Chironomus*, which is different from that observed in vertebrates.

Although the dipterans may lack C protein and functional homologues thereof, they do possess several proteins that are homologous to the basic A and B group proteins of the mammalian core particle. For example, monoclonal antibodies to the *Drosophila* hnRNP homologue (hrp40) cross-react with human A/B and G proteins (124). In addition the Hrb98DE gene of *Drosophila* encodes four basic 38–41 kDa proteins that, together with another homologous gene product (Hrb87F), reveal functional as well as sequence homology to the mammalian A/B proteins (125, 126). More specifically, these proteins co-isolate with *Drosophila* poly(A)$^+$ RNA and can be recovered from 40–80S regions of density gradients (128); also, like their mammalian homologues, they are basic; they are distributed along the length of nascent transcripts (43); their specific antibodies strongly cross react with human A/B proteins; and, like human A1, they induce internal exon skipping when overproduced *in vivo* (137). While these proteins may not package pre-mRNAs into a regular core particle structure, it seems likely that they would exist in an ordered arrangement, driven by protein–protein interactions, along the length of *Drosophila* pre-mRNAs.

References

1. Fakan, S., Leser, G., and Martin, T. E. (1986) Immunoelectron microscope visualization of nuclear ribonucleoprotein antigens within spread transcription complexes. *J. Cell Biol.*, **103**, 1153.
2. Beyer, A. L. and Osheim, Y. N. (1990) Ultrastructural analysis of the ribonucleoprotein substrate for pre-mRNA processing. In *The eukaryotic nucleus: molecular biochemistry and macromolecular assemblies*. Vol. 2, Strauss, P. R. and Wilson, S. H. (ed.). Telford Press, Caldwell, N.J., p. 431.
3. LeStourgeon, W. M., Barnett, S. F., and Northington, S. J. (1990) Tetramers of the core proteins of 40S nuclear ribonucleoprotein particles assemble to package nascent transcripts into a repeating array of regular particles. In *The eukaryotic nucleus: molecular biochemistry and macromolecular assemblies*, Vol. 2, Strauss, P. R. and Wilson, S. H. (ed.). Telford Press, Caldwell, N.J., p. 477.
4. van Holde, K. E. (1989) The proteins of chromatin. I. Histones. In *Chromatin*, Rich, A. (ed.). Springer-Verlag, New York, p. 69.
5. Cobianchi, F., SenGupta, D. N., Zmudzta, B. Z., and Wilson, S. H. (1986) Structure of rodent helix-destabilizing protein revealed by cDNA cloning. *J. Biol. Chem.*, **261**, 3536.
6. Merrill, B. M., LoPresti, B., Stone, K. L., and Williams, K. R. (1986) High pressure liquid chromatography purification of UP1 and UP2: two related single-stranded nucleic acid binding proteins from calf thymus. *J. Biol. Chem.*, **261,** 878.
7. Merrill, B. M., Lopresti, M. B., Stone, K. L., and Williams, K. R. (1987) Amino acid sequence of UP1, an hnRNP-derived single-stranded nucleic acid binding protein from calf thymus. *Int. J. Peptide Protein Res.*, **29**, 21.
8. Merrill, B. M. and Williams, K. R. (1990) Structure/function relationships in hnRNP proteins. In *The eukaryotic nucleus: molecular biochemistry and macromolecular assemblies*. Vol. 2, Strauss, P. R. and Wilson, S. H. (ed.). Telford Press, Caldwell, N. J., p. 579.
9. LeStourgeon, W. M., Lothstein, L., Walker, B. W., and Beyer, A. L. (1981) The composition and general topology of RNA and protein in monomer 40S ribonucleoprotein particles. In *The cell nucleus*. Vol. 9, Busch, H. (ed.). Academic Press, New York, p. 49.
10. Lothstein, L., Arenstorf, H. P., Chung, S., Walker, B. W., Wooley, J. C., and LeStourgeon, W. M. (1985) General organization of protein in HeLa 40S nuclear ribonucleoprotein particles. *J. Cell Biol.*, **100**, 1570.
11. Senapathy, P. (1986) Origin of eukaryotic introns: A hypothesis based on codon distribution statistics in genes and its implications. *Proc. Natl Acad. Sci. USA*, **83**, 2133.
12. Senapathy, P. (1995) Introns and the origin of protein-coding genes. *Science*, **268**, 1366.
13. Wilk, H. E., Angeli, G., and Schafer, K. P. (1983) *In vitro* reconstitution of 35S ribonucleoprotein complexes. *Biochemistry*, **22**, 4592.
14. Conway, G., Wooley, J., Bibring, T., and LeStourgeon, W. M. (1988) Ribonucleoproteins package 700 nucleotides of pre-mRNA into a repeating array of regular particles. *Mol. Cell. Biol.*, **8**, 2884.

15. Nadler, S. G., Merrill, B. M., Roberts, W. J., Keating, K. M., Lisbin, M. J., Barnett, S. F., *et al.* (1991) Interactions of the A1 heterogeneous nuclear ribonuclearprotein and its proteolytic derivative, UP1, with RNA and DNA: evidence for multiple RNA binding domains and salt-dependent binding mode transitions. *Biochemistry*, **30**, 2968.
16. McAfee, J. G., Soltaninassab, S. R., Lindsay, M. E., and LeStourgeon, W. M. (1996) Proteins C1 and C2 of heterogeneous nuclear ribonucleoprotein complexes bind RNA in a highly cooperative fashion: support for their contiguous deposition on pre-mRNA during transcription. *Biochemistry*, **35**, 1212.
17. Herrera, A. H. and Olson, M. O. J. (1986) Association of protein C23 with rapidly labeled nucleolar RNA. *Biochemistry*, **25**, 6258.
18. Lührmann, R. (1988) snRNP Proteins, in *Structure and function of major and minor small nuclear ribonucleoprotein particles*, Birnstiel, M. L. (ed.). Springer, New York, p. 71.
19. Bergman, M. and Ringertz, N. (1990) Gene expression pattern of chicken erythrocyte nuclei in heterokaryons. *J. Cell Sci.*, **97**, 167.
20. Huang, M., Rech, J. E., Northington, S. J., Flicker, P. F., Mayeda, A., Krainer, A. R., *et al.* (1994) The C protein tetramer binds 230 to 240 nucleotides of pre-mRNA and nucleates the assembly of 40S heterogeneous nuclear ribonucleoprotein particles. *Mol. Cell. Biol.*, **14**, 518.
21. Beyer, A. L., Christensen, M. E., Walker, B. W., and LeStourgeon, W. M. (1977) Identification and characterization of the packaging proteins of core 40S hnRNP particles. *Cell*, **11**, 127.
22. Wilk, H. E., Werr, H., Friedrich, D., Kiltz, H. H., and Schaefer, K. P. (1985) The core proteins of 35S hnRNP complexes: characterization of nine different species. *Eur. J. Biochem.*, **146**, 71.
23. Barnett, S. F., Friedman, D. L., and LeStourgeon, W. M. (1989) The C proteins of HeLa 40S nuclear ribonucleoprotein particles exist as anisotropic tetramers of (C1)3C2. *Mol. Cell. Biol.*, **9**, 492.
24. Barnett, S. F., Theiry, T. A., and LeStourgeon, W. M. (1991) The core proteins A2 and B1 exist as (A2)3B1 tetramers in 40S nuclear ribonucleoprotein particles. *Mol. Cell. Biol.*, **11**, 864.
25. Samarina, O. P., Krichevskaya, A. A., and Georgiev, G. P. (1966) Nuclear ribonucleoprotein particles containing messenger ribonucleic acid. *Nature*, **210**, 1319.
26. Martin, T. E., Billings, P. B., Levy, A., Ozarslan, S., Quinlan, T., Swift, H., *et al.* (1974) Some properties of RNA: protein complexes from the nucleus of eucaryotic cells. *Cold Spring Harbor Symp. Quant. Biol.*, **38**, 921.
27. Choi, Y. D. and Dreyfuss, G. (1984) Isolation of the heterogeneous nuclear RNA-ribonucleoprotein complex (hnRNP): A unique supramolecular assembly. *Proc. Natl Acad. Sci. USA*, **81**, 7471.
28. Samarina, O. P., Lukanidin, E. M., Molnar, J., and Georgiev, G. P. (1968) Structural organization of nuclear complexes containing DNA-like RNA. *J. Mol. Biol.*, **33**, 251.
29. Samarina, O. and Krichevskaya, A. (1981) Nuclear 30S RNP particles. In *The cell nucleus*. Vol. 9, H. Busch (ed.). Academic Press, Inc., New York, p. 1.
30. Tsanev, R. G. and Djondurov, L. P. (1982) Ultrastructure of free ribonucleoprotein complexes in spread mammalian nuclei. *J. Cell Biol.*, **94**, 662.
31. Sperling, J. and Sperling, R. (1990) Large nuclear ribonucleoprotein particles of specific RNA polymerase II transcripts. In *The eukaryotic nucleus: molecular biochemistry and macromolecular assemblies*. Vol. 2, Strauss, P. R. and Wilson, S. H. (ed.). Telford Press, Caldwell, N.J., p. 453.

32. Malcolm, D. B. and Sommerville, J. (1977) The structure of nuclear ribonucleoprotein of amphibian oocytes. *J. Cell Sci.*, **24**, 143.
33. Malcolm, D. B. and Sommerville, J. (1977) Structural organization of ribonucleoproteins in amphibian oocytes. *Biochem. Soc. Trans.*, **5**, 627.
34. Chung, S. Y. and Wooley, J. (1986) Set of novel, conserved proteins fold pre-messenger RNA into ribonucleosomes. *Proteins: Struct. Funct. Genet.*, **1**, 195.
35. Vazquez-Nin, G. H., Echeverria, O. M., Martin, T. E., Lührmann, R., and Fakan, S. (1994) Immunocytochemical characterization of nuclear ribonucleoprotein fibrils in cells of the central nervous system of the rat. *European J. Cell Biol.*, **65**, 291.
36. Kornberg, R. D. (1974) Chromatin structure: a repeating unit of histones and DNA. *Science*, **184**, 868.
37. Oudet, P., Gross-Bellard, M., and Chambon, P. (1975) Electron microscopic and biochemical evidence that chromatin structure is a repeating unit. *Cell*, **4**, 281.
38. Jorcano, J. L. and Ruiz-Carrillo, A. (1979) H3.H4 tetramer directs DNA and core histone octamer assembly in the nucleosome core particle. *Biochemistry*, **18**, 768.
39. Tatchell, K. and van Holde, K. E. (1979) Nucleosome reconstitution: effect of DNA length on nucleosome structure. *Biochemistry*, **18**, 2871.
40. Harris, S. G., Martin, T. E., and Smith, H. C. (1988) Reversible chemical cross-linking and ribonuclease digestion analysis of the organization of proteins in ribonucleoprotein particles. *Mol. Cell. Biochem.*, **84**, 17.
41. Harris, S. G., Hoch, S. O., and Smith, H. C. (1988) Chemical cross-linking of Sm and RNP antigenic proteins. *Biochemistry*, **27**, 4595.
42. Kiledjian, M., Burd, C. G., Gorlach, M., Portman, D. S., and Dreyfuss, G. (1994) Structure and function of hnRNP proteins. In *RNA–protein interactions*, Mattaj, I. W. and Nagai, K. (ed.). Oxford University Press. p. 127.
43. Amero, S. A., Raychaudhuri, G., Cass, C. L., vanVenrooij, W. J., Habets, W. J., Krainer, A. R., *et al.* (1992) Independent deposition of heterogeneous nuclear ribonucleoproteins and small nuclear ribonucleoprotein particles at sites of transcription. *Proc. Natl Acad. Sci. USA*, **89**, 8409.
44. van Eekelen, C., Ohlsson, R., Philipson, L., Mariman, E., van Beek, R., and van Venrooij, W. (1982) Sequence dependent interaction of hnRNP proteins with late adenoviral transcripts. *Nucleic Acids Res.*, **10**, 7115.
45. Steitz, J. A. and Kamen, R. (1981) Arrangement of 30S heterogeneous nuclear ribonucleoprotein on polyoma virus late nuclear transcripts. *Mol. Cell. Biol.*, **1**, 21.
46. van Holde, K. E. and Hill, W. E. (1974) General properties of ribosomes. In *Ribosomes* Nomura, M., Tissieres, A., and Lengyel P. (ed.). Cold Spring Harbor Laboratory, Cold Spring Harbor, New York, p. 53.
47. Merrill, B. M., Barnett, S. F., LeStourgeon, W. M., and Williams, K. R. (1989) Primary structure differences between proteins C1 and C2 of HeLa 40S nuclear ribonucleoprotein particles. *Nucleic Acids Res.*, **17**, 8441.
48. Burd, C. G., Swanson, M. S., Gorlach, M., and Dreyfuss, G. (1989) Primary structures of the heterogeneous nuclear ribonucleoprotein A2, B1, and C2 proteins: a diversity of RNA binding proteins is generated by small peptide inserts. *Proc. Natl Acad. Sci. USA*, **86**, 9788.
49. Buvoli, M., Cobianchi, F., Bestagno, M. G., Mangiarotti, A., Bassi, M. T., Biamonti, G., *et al.* (1990) Alternative splicing in the human gene for the core protein A1 generates another hnRNP protein. *EMBO J.*, **9**, 1229.
50. Swanson, M. S. and Dreyfuss, G. (1988) Classification and purification of proteins of

heterogeneous nuclear ribonucleoprotein particles by RNA-binding specificities. *Mol. Cell. Biol.*, **8**, 2237.

51. Buvoli, M., Biamonti, G., Tsoulfas, P., Bassi, M. T., Ghetti, A., Riva, S., *et al.* (1988) cDNA cloning of human hnRNP protein A1 reveals the existence of multiple mRNA isoforms. *Nucleic Acids Res.*, **16**, 3751.
52. Nakagawa, T. Y., Swanson, M. S., Wold, B. J., and Dreyfuss, G. (1986) Molecular cloning of cDNA for the nuclear ribonucleoprotein particle C proteins: a conserved gene family. *Proc. Natl Acad. Sci. USA*, **83**, 2007.
53. Cobianchi, F., Biamonti, G., Bassi, M. T., Buvoli, M., and Riva, S. (1990) The protein A1 of mammalian hnRNP complex: A new insight into protein complexity and gene family structure. In *The eukaryotic nucleus: molecular biochemistry and macromolecular assemblies.* Vol. 2, Strauss, P. R. and Wilson, S. H. (ed.). Telford Press, Caldwell, N.J., p. 561.
54. Buvoli, M., Cobianchi, F., Bestagno, M., Bassi, M. T., Biamonti, G., and Riva, S. (1990) A second A1-type protein is encoded by the human hnRNP A1 gene. *Mol. Biol. Rep.*, **14**, 83.
55. Celis, J. E., Bravo, R., Arenstorf, H. P., and LeStourgeon, W. M. (1986) Identification of proliferation-sensitive human proteins amongst components of the 40S hnRNP particles. Identity of hnRNP core proteins in the HeLa protein catalogue. *FEBS Lett.*, **194**, 101.
56. Karn, J., Vidali, G., Boffa, L., and Allfrey, V. (1977) Characterization of the non-histone nuclear proteins associated with rapidly labeled heterogeneous nuclear RNA. *J. Biol. Chem.*, **252**, 969.
57. Leser, G. P. and Martin, T. E. (1987) Changes in heterogeneous nuclear RNP core polypeptide complements during the cell cycle. *J. Cell Biol.*, **105**, 2083.
58. Minoo, P., Martin, T. E., and Riehl, R. M. (1991) Nucleic acid binding characteristics of group A/B hnRNP proteins. *Biochem. Biophys. Res. Commun.*, **176**, 747.
59. Brunel, C. and Lelay, M. (1979) Two-dimensional analysis of proteins associated with heterogeneous nuclear RNA in various animal cell lines. *Eur. J. Biochem.*, **99**, 273.
60. Swanson, M. S., Nakagawa, T. Y., LeVan, K., and Dreyfuss, G. (1987) Primary structure of human nuclear ribonucleoprotein particle C proteins: conservation of sequence and domain structures in heterogeneous nuclear RNA, mRNA, and pre-RNA binding proteins. *Mol. Cell. Biol.*, **7**, 1731.
61. Mayrand, S. H., Dwen, P., and Pederson, T. (1993) Serine/threonine phosphorylation regulates binding of C hnRNP proteins to pre-mRNA. *Proc. Natl Acad. Sci. USA*, **90**, 7764.
62. Christensen, M. E., Beyer, A. L., Walker, B., and LeStourgeon, W. M. (1977) Identification of N^G, N^G-dimethylarginine in a nuclear protein from the lower eukaryote physarum polycephalum homologous to the major proteins of mammalian 40S ribonucleoprotein particles. *Biochem. Biophys. Res. Commun.*, **74**, 621.
63. Burd, C. G. and Dreyfuss, G. (1994) Conserved structures and diversity of functions of RNA-binding proteins. *Science*, **265**, 615.
64. Cobianchi, F., Karpel, R. L., Williams, K. R., Notario, V., and Wilson, S. H. (1988) Mammalian heterogeneous nuclear ribonucleoprotein complex protein A1. Large-scale overproduction in *Escherichia coli* and cooperative binding to single-stranded nucleic acids. *J. Biol. Chem.*, **263**, 1063.
65. Wilson, S. H., Cobianchi, F., and Guy, H. R. (1987) In *DNA, protein interaction and gene regulation,* Thompson, E. B. and Papaconstatinou, J. (ed.). University of Texas Press, Austin, p. 129.

66. Kiledjian, M. and Dreyfuss, G. (1992) Primary structure and binding activity of the hnRNP U protein: binding RNA through RGG box. *EMBO J.*, **11**, 2655.
67. Christensen, M. E. and Fuxa, K. P. (1988) The nucleolar protein, B-36, contains a glycine and dimethylarginine-rich sequence conserved in several other nuclear RNA-binding proteins. *Biochem. Biophys. Res. Commun.*, **155**, 1278.
68. Kim, S., Merrill, B. M., Rajpurohit, R., Kumar, A., Stone, K. L., Szer, W., *et al.* (1997) Identification of N^G-methylarginine residues in human heterogeneous RNP protein A1: Phe/Gly-Gly-Gly-Arg-Gly-Gly-Gly/Phe is a major motif. *Biochem.*, in press.
69. Casas-Finet, J. R., Smith, J. D., Jr, Kumar, A., Kim, J. G., Wilson, S. H., and Karpel, R. L. (1993) Mammalian heterogeneous ribonucleoprotein A1 and its constituent domains. Nucleic acid interaction, structural stability and self-association. *J. Mol. Biol.*, **229**, 873.
70. Kumar, A., Casas-Finet, J. R., Luneau, C. J., Karpel, R. L., Merrill, B. M., Williams, K. R., *et al.* (1990) Mammalian heterogeneous nuclear ribonucleoprotein A1: nucleic acid binding properties of the COOH-terminal domain. *J. Biol. Chem.*, **265**, 17094.
71. Munroe, S. H. and Dong, X. (1992) Heterogeneous nuclear ribonucleoprotein A1 catalyzes RNA–RNA annealing. *Proc. Natl Acad. Sci. USA*, **89**, 895.
72. Query, C. C., Bently, R. C., and Keene, J. D. (1989) A common RNA recognition motif identified within a defined U1 RNA binding domain of the 70K U1 snRNP protein. *Cell*, **57**, 89.
73. Keene, J. D. and Query, C. C. (1991) Nuclear RNA-binding proteins. *Prog. Nucleic Acid Res.*, **41**, 179.
74. Nagai, K., Oubridge, C., Jessen, T. H., Li, J., and Evans, P. R. (1990) Crystal structure of the RNA-binding domain of the U1 small nuclear ribonucleoprotein A. *Nature*, **348**, 515.
75. Wittekind, M., Gorlach, M., Friedrichs, M., Dreyfuss, G., and Mueller, L. (1992) ^{1}H, ^{13}C, and ^{15}N NMR assignments and global folding patterns of the RNA-binding domain of the human hnRNP C protein tetramer. *Biochemistry*, **31**, 6254.
76. Shamoo, Y., Abdul-Manan, N., Patten, A. M., Crawford, J. K., and Pellegrini, M. C. (1994) Both RNA-binding domains in heterogeneous nuclear ribonucleoprotein A1 contribute toward single-stranded-RNA binding. *Biochemistry*, **33**, 8272.
77. Tuerk, C. and Gold, L. (1990) Systematic evolution of ligands by exponential enrichment: RNA ligand to bacteriophage T4 DNA polymerase. *Science*, **249**, 505.
78. Biamonti, G. and Riva, S. (1994) New insights into the auxiliary domains of eukaryotic RNA binding proteins. *FEBS Lett.*, **340**, 1.
79. Swanson, M. S. and Dreyfuss, G. (1988) RNA binding specificity of hnRNP proteins: a subset bind to the 3′ end of introns. *EMBO J.*, **7**, 3519.
80. Ishikawa, F., Matunis, M. J., Dreyfuss, G., and Cech, T. R. (1993) Nuclear proteins that bind the pre-mRNA 3′ splice site sequence r(UUAG/G) and the human telomeric DNA sequence d(TTAGGG)n. *Mol. Cell. Biol.*, **13**, 4301.
81. Burd, C. G. and Dreyfuss, G. (1994) RNA binding specificity of hnRNP A1: significance of hnRNP A1 high-affinity binding sites in pre-mRNA splicing. *EMBO J.*, **13**, 1197.
82. Buvoli, M., Cobianchi, F., Biamonti, G., and Riva, S. (1990) Recombinant hnRNP protein A1 and its N-terminal domain show preferential affinity for oligodeoxynucleotides homologous to intron/exon acceptor sites. *Nucleic Acids Res.*, **18**, 6595.
83. Abdul-Manan, N., O'Malley, S. M., and Williams, K. R. (1996) Origins of binding specificity of the A1 heterogeneous nuclear ribonucleoprotein. *Biochemistry*, **35**, 3545.
84. Klug, S. J. and Famulok, M. (1994) All you wanted to know about SELEX. *Mol. Biol. Rep.*, **20**, 97.

85. Jellinek, D., Lynott, C. K., Rifkin, D. B., and, Janjic, N. (1993) High-affinity RNA ligands to basic fibroblast growth factor inhibit receptor binding. *Proc. Natl Acad. Sci. USA*, **90**, 11227.
86. Kubik, M. F., Stephens, A. W., Schneider, D., Marlar, R. A., and Tasset, D. (1994) High-affinity RNA ligands to human α-thrombin. *Nucleic Acids Res.*, **22**, 2619.
87. Shetlar, M. D. (1980) Cross-linking of proteins to nucleic acids by ultraviolet light. *Photochem. Photobiol. Reviews*, **5**, 105.
88. Shetlar, M. D., Carbone, J., Steady, E., and Hom, K. (1984) Photochemical addition of amino acids and peptides to polyuridylic acid. *Photochem. Photobiol.*, **39**, 141.
89. Nikogosyan, D. N. (1990) Two-quantum UV photochemistry of nucleic acids: comparison with conventional low-intensity UV photochemistry and radiation chemistry. *Int. J. Radiat. Biol.*, **57**, 233.
90. Hockensmith, J. W., Kubasek, W. L., Vorachek, W. R., and von-Hippel, P. H. (1986) Laser cross-linking of nucleic acids to proteins: methodology and first applications to the phage T4 DNA replication system. *J. Biol. Chem.*, **261**, 3512.
91. Hockensmith, J. W., Kubasek, W. L., Vorachek, W. R., Evertsz, E. M., and von-Hippel, P. H. (1991) Laser cross-linking of protein-nucleic acid complexes. *Methods Enzymol.*, **208**, 211.
92. Hamilton, B. J., Nagy, E., Malter, J. S., Arrick, B. A., and Rigby, W. F. (1993) Association of heterogeneous nuclear ribonucleoprotein A1 and C proteins with reiterated AUUUA sequences. *J. Biol. Chem.*, **1268**, 8881.
93. Gorlach, M., Burd, C. G., and Dreyfuss, G. (1994) The determinants of RNA-binding specificity of the heterogeneous nuclear ribonucleoprotein C proteins. *J. Biol. Chem.*, **269**, 23074.
94. Choi, Y. D., Grabowski, P. J., Sharp, P. A., and Dreyfuss, G. (1986) Heterogeneous nuclear ribonucleoproteins: role in RNA splicing. *Science*, **231**, 1534.
95. Bennett, M., Piñol-Roma, S., Staknis, D., Dreyfuss, G., and Reed, R. (1992) Differential binding of heterogeneous nuclear ribonucleoproteins to mRNA precursors prior to spliceosome assembly *in vitro*. *Mol. Cell. Biol.*, **12**, 3165.
96. Amero, S. A., Matunis, M. J., Matunis, E. L., Hockensmith, J. W., Raychaudhuri, G., and Beyer, A. L. (1993) A unique ribonucleoprotein complex assembles preferentially on ecdysone-responsive sites in *Drosophila melanogaster*. *Mol. Cell. Biol.*, **13**, 5323.
97. Kish, V. M. and Pederson, T. (1978) Isolation and characterization of ribonucleoprotein particles containing heterogeneous nuclear RNA. *Methods Cell. Biol.*, **17**, 377.
98. Pederson, T. and Munroe, S. H. (1981) Ribonucleoprotein organization of eukaryotic RNA. XV. Different nucleoprotein structures of globin messenger RNA sequences in nuclear and polyribosomal ribonucleoprotein particles. *J. Mol. Biol.*, **150**, 509.
99. Gattoni, R., Stévenin, J., and Jacob, M. (1980) Comparison of the nuclear ribonucleoproteins containing the transcripts of adenovirus-2 and HeLa cell DNA. *Eur. J. Biochem.*, **108**, 203.
100. Dreyfuss, G., Swanson, M. S., and Piñol-Roma, S. (1990) The composition, structure, and organization of proteins in heterogeneous nuclear ribonucleoprotein complexes. In *The eukaryotic nucleus: molecular biochemistry and macromolecular assemblies*. Vol. 2, Strauss, P. R. and Wilson, S. H. (ed.). Telford Press, Caldwell, N.J., p. 503.
101. Piñol-Roma, S., Choi, Y. D., Matunis, M. J., and Dreyfuss, G. (1988) Immunopurification of heterogeneous nuclear ribonucleoprotein particles reveals an assortment of RNA-binding proteins. *Genes Dev.*, **2**, 215.
102. Barnett, S. F., LeStourgeon, W. M., and Friedman, D. L. (1988) Rapid purification of native C protein from nuclear ribonucleoprotein particles. *J. Biochem. Biophys. Methods*, **16**, 87.

103. Friedman, P. N., Chen, X., Bargonetti, J., and Prives, C. (1993) The p53 protein is an unusually shaped tetramer that binds directly to DNA. *Proc. Natl Acad. Sci. USA*, **90**, 3319.
104. Wang, P., Reed, M., Wang, Y., Gregory, M., Stenger, J. E., Anderson, M. E., *et al.* (1994) p53 domains: structure, oligomerization, and transformation. *Mol. Cell. Biol.*, **14**, 5182.
105. Rech, J. E., Huang, M. H., LeStourgeon, W. M., and Flicker, P. F. (1995) An ultrastructural characterization of *in vitro*-assembled hnRNP C protein–RNA complexes. *J. Struct. Biol.*, **114**, 84.
106. Rech, J. E., LeStourgeon, W. M., and Flicker, P. F. (1995) Ultrastructural morphology of the hnRNP C protein tetramer. *J. Struct. Biol.*, **114**, 77.
107. Martin, T. E., Billings, P. B., Pullman, J. M., Stevens, B. J., and Kinniburgh, A. J. (1978) Substructures of nuclear ribonucleoprotein complexes in chromosome structure and function. *Cold Spring Harbor Symp. Quant. Biol.*, **42**, 899.
108. Ben-David, Y., Bani, M. R., Chabot, B., Koven, A. D., and Bernstein, A. (1992) Retroviral insertions downstream of the heterogeneous nuclear ribonucleoprotein A1 gene in erythroleukemia cells: evidence that A1 is not essential for cell growth. *Mol. Cell. Biol.*, **12**, 4449.
109. Yang, X., Bani, M. R., Lu, S. J., Rowna, S., Ben-David, Y., and Chabot, B (1994) The A1 and A1B proteins of heterogeneous nuclear ribonucleoprotein modulate 5′ splice site selection *in vivo*. *Proc. Natl Acad. Sci. USA*, **91**, 6924.
110. Mayeda, A. and Krainer, A. R. (1992) Regulation of alternative pre-mRNA splicing by hnRNP A1 and splicing factor SF2. *Cell*, **68**, 365.
111. Mayeda, A., Munroe, S. H., Cáceres, J. F., and Krainer, A. R. (1994) Function of conserved domains of hnRNP A1 and other hnRNP A/B proteins. *EMBO J.*, **13**, 5483.
112. Gorlach, M., Wittekind, M., Beckman, R. A., Mueller, L., and Dreyfuss, G. (1992) Interaction of the RNA-binding domain of the hnRNP C proteins with RNA. *EMBO J.*, **11**, 3289.
113. Sierakowska, H., Szer, W., Furdon, P. J., and Kole, R. (1986) Antibodies to hnRNP core proteins inhibit *in vitro* splicing of human β-globin pre-mRNA. *Nucleic Acids Res.*, **14**, 4475.
114. Mulligan, M. E., Jackman, D. M., and Murphy, S. T. (1994) Heterocyst-forming filamentous cyanobacteria encode proteins that resemble eukaryotic RNA-binding proteins of the RNP family. *J. Mol. Biol.*, **235**, 1162.
115. Anderson, J. T., Wilson, S. M., Datar, K. V., and Swanson, M. S. (1993) Nab2: a yeast nuclear polyadenylated RNA-binding protein essential for cell viability. *Mol. Cell. Biol.*, **13**, 2730.
116. Matunis, M. J., Matunis, E. L., and Dreyfuss, G. (1993) PUB1: a major yeast poly(A) RNA-binding protein. *Mol. Cell. Biol.*, **13**, 6114.
117. Cusick, M. E. (1994) RNP1: a new ribonucleoprotein gene of the yeast *Saccharomyces cerevisiae*. *Nucleic Acids Res.*, **22**, 869.
118. Wilson, S. M., Kshama, V. D., Paddy, M. R., Swedlow, J. R., and Swanson, M. S. (1994) Characterization of nuclear polyadenylated RNA-binding proteins in *Saccharomyces cerevisiae*. *J. Cell Biol.*, **127**, 1173.
119. Gamberi, C., Contreas, G., Romanelli, M. G., and Morandi, C. (1994) Analysis of the yeast NSR1 gene and protein domain comparison between Nsr1 and human hnRNP type A1. *Gene*, **148**, 59.
120. Raziuddin, Thomas, J. O., and Szer, W. (1982) Nucleic acid binding properties of major proteins from the heterogeneous nuclear ribonucleoproteins of wheat. *Nucleic Acids Res.*, **10**, 7777.

121. Soulard, M., Della Valle, V., Siomi, M. C., Piñol-Roma, S., Codogno, P., Bauvy, C., *et al.* (1993) hnRNP G: sequence and characterization of a glycosylated RNA-binding protein. *Nucleic Acids Res.*, **21** 4210.
122. Thomas, J. O., Glowacka, S. K., and Szer, W. (1983) Structure of complexes between a major protein of heterogeneous nuclear ribonucleoprotein particles and polyribonucleotides. *J. Mol. Biol.*, **171**, 439.
123. Cruz-Alvarez, M., Szer, W., and Pellicer, A. (1985) Cloning of cDNA sequences for an Artemia salina hnRNP protein: evidence for conservation through evolution. *Nucleic Acids Res.*, **13**, 3917.
124. Matunis, E. L., Matunis, M. J., and Dreyfuss, G. (1992) Characterization of the major hnRNP proteins from *Drosophila melanogaster*. *J. Cell Biol.*, **116**, 257.
125. Haynes, S. R., Raychaudhuri, G., and Beyer, A. L. (1990) The *Drosophila* Hrb98DE locus encodes four protein isoforms homologous to the A1 protein of mammalian heterogeneous nuclear ribonucleoprotein complexes. *Mol. Cell. Biol.*, **10**, 316.
126. Haynes, S. R., Johnson, D., Raychaudhuri, G., and Beyer, A. L. (1991) The *Drosophila* Hrb87F gene encodes a new member of the A and B hnRNP protein group. *Nucleic Acids Res.*, **19**, 25.
127. Haynes, S. R., Raychaudhuri, G., Johnson, D., Amero, S., and Beyer, A. L. (1990) The *Drosophila* Hrb loci: a family of hnRNA binding proteins. *Mol. Biol. Reports*, **14**, 93.
128. Raychaudhuri, G., Haynes, S. R., and Beyer, A. L. (1992) Heterogeneous nuclear ribonucleoprotein proteins and complexes in *Drosophila melanogaster*. *Mol. Cell. Biol.*, **12**, 847.
129. Preugschat, F. and Wold, B. (1988) Isolation and characterization of a *Xenopus laevis* C protein cDNA: structure and expression of a heterogeneous nuclear ribonucleoprotein core protein. *Proc. Natl Acad. Sci. USA*, **85**, 9669.
130. Choi, Y. D. and Dreyfuss, G. (1984) Monoclonal antibody characterization of the C proteins of heterogeneous nuclear ribonucleoprotein complexes in vertebrate cells. *J. Cell Biol.*, **99**, 1997.
131. Malcolm, D. B. and Sommerville, J. (1974) The structure of chromosome derived ribonucleoprotein of amphibian oocytes. *Chromosoma*, **48**, 137.
132. Angelier, N. and Lacroix, J. C. (1975) Complexes de transcription d'origines nucleolaire et chromosomique d'ovocytes de Pleurodeles waltii et P. poireti (amphibiens, urodeles). *Chromosoma*, **51**, 323.
133. Lonnroth, A., Alexciev, K., Mehlin, H., Wurtz, T., Skoglund, U., and Daneholt, B. (1992) Demonstration of a 7-nm RNP fiber as the basic structural element in a premessenger RNP particle. *Exp. Cell Res.*, **199**, 292.
134. Mehlin, H., Daneholt, B., and Skoglund, U. (1995) Structural interaction between the nuclear pore complex and a specific translocating RNP particle. *J. Cell Biol.*, **129**, 1205.
135. Olins, A. L., Olins, D. E., and Bazett-Jones, D. P. (1992) Balbiani ring hnRNP substructure visualized by selective staining and electron spectroscopic imaging. *J. Cell Biol.*, **117**, 483.
136. Wurtz, T., Lonnroth, A., and Daneholt, B. (1990) Biochemical characterization of Balbiani ring premessenger RNP particles. *Mol. Biol. Rep.*, **14**, 95.
137. Shen, J., Zu, K., Cass, C. L., Beyer, A. L., and Hirsh, J. (1995) Exon skipping by overexpression of a *Drosophila* heterogeneous nuclear ribonucleoprotein *in vivo*. *Proc. Natl Acad. Sci. USA*, **92**, 1822.
138. de Haen, C. (1987) Molecular weight standards for calibration of gel filtration and sodium dodecyl sulfate-polyacrylamide gel electrophoresis: ferritin and apoferritin. *Anal. Biochem.*, **166**, 235.

4 Spliceosome assembly

ROBIN REED and LEON PALANDJIAN

1. Introduction

Pre-mRNA splicing takes place within a large, highly dynamic complex designated the spliceosome. Spliceosomes are 50–60S, have a calculated molecular weight on the order of 3–5 × 10^6 kDa, and are estimated to be 40–60 nm in diameter (see 1, 2 for reviews). Among the best characterized of the spliceosomal components are the small nuclear RNAs (snRNAs) U1, U2, U4, U5, and U6, which are thought to play central roles not only in spliceosome assembly, but also in the two catalytic steps of the splicing reaction (reviewed in 1, 3, 4; see Chapter 5). Less is known about the protein components of the splicing machinery. Genetic studies in yeast have identified over 50 protein-coding genes essential for splicing (reviewed in 2; see Chapter 7), and an equal number of proteins have been detected in highly purified mammalian spliceosomes (5, 6). In addition to obtaining direct evidence for RNA-mediated catalysis, major challenges of the splicing field include cloning, characterizing, and determining the functions of the protein components of the spliceosome and achieving a detailed understanding of spliceosome assembly.

1.1 The spliceosomal complexes

Both the snRNAs and spliceosomal proteins assemble on pre-mRNA in a stepwise pathway. Four distinct spliceosomal complexes have been detected during incubation in HeLa cell nuclear extracts, and these assemble in the order E→A→B→C. The E, A, and B complexes contain unspliced pre-mRNA, and the C complex contains the products of catalytic step I of the splicing reaction (exon 1 and the lariat-exon intermediate). Additional complexes, i and D, which contain the excised intron and spliced exons respectively, are generated from catalytic step II of the splicing reaction. Although there are multiple intermediate forms of each of the complexes, resulting from dynamic interactions of spliceosomal components, we will only distinguish between the complexes (E, A, B, C, D and i) that are readily detected under normal splicing conditions. In addition to assembling into spliceosomal complexes, a portion of the pre-mRNA incubated in splicing extracts assembles into a discrete complex, designated the H complex. Unlike the spliceosomal complexes, the H complex has not been shown to be a functional intermediate in the spliceosome assembly pathway.

Below, the H complex and the spliceosomal complexes are considered individually, with descriptions of what has been learned about the pre-mRNA sequence requirements for complex assembly, the protein and snRNA composition of the complex, and the functional requirement for individual factors in complex assembly. The review is focused primarily on recent developments in understanding the protein components of mammalian spliceosomal complexes (summarized in Table 1), and we refer the reader to excellent reviews that emphasize the multiple roles of RNA–RNA interactions in spliceosome assembly, including both snRNA–snRNA and snRNA–pre-mRNA interactions (1, 7; see Chapter 5). Yeast spliceosomal complexes are mentioned in cases where there are important differences between yeast and mammals or in cases where more is known about the yeast complex (reviewed in 2, 8, 9; see Chapter 7).

2. The H complex

The H complex assembles immediately when pre-mRNA is added to HeLa cell nuclear extracts, does not require functional 5′ or 3′ splice sites for assembly, forms in the presence or absence of ATP, and assembles on ice or at 30°C. The H complex has been detected by native gel electrophoresis, gel filtration, and density gradients and has been designated nonspecific or unspecific complex in some studies (10–14).

Table 1 Protein components of the splicing machinery

Protein	snRNP	Cross-link	cDNA/ab	Structure	Yeast
Proteins that bind in the E complex					
U2AF^{65}		py tract	+/p	3 RRMs, RS domain	
U2AF^{35}		–	+/p	RS domain	
SR		+	+/m	1 or 2 RRMs, RS domain	
SAP 115		5′ portion			
SAP 92		–			
SAP 88		–			
SAP 72		–			
SAP 42		–			
U1 70K	U1	–	+/p,m	1 RRM, RS domain	SNP1
U1 A	U1	–	+/p	2 RRMs, proline-rich domain	MUD1
U1 C	U1	–	+/	Zn finger, proline-rich domain	
B	core	–	+/p,m	proline-rich domain	
B′	core	–	+/p,m	proline-rich domain	
D	core	–	+/p,m		SMD1
Proteins that first bind in the A complex					
SAP 155	U2	3′ portion			
SAP 145	U2	3′ portion	+/		
SAP 130	U2	–			
SAP 114 (SF3a^{110})	U2	3′ portion			PRP21 (SPP91)
SAP 62 (SF3a^{66})	U2	3′ portion	+/m	Zn finger, 22 proline-rich repeats	PRP11
SAP 61 (SF3a^{60})	U2	3′ portion	+/p	Zn finger	PRP9

Table 1 (continued)

Protein	snRNP	Cross-link	cDNA/ab	Structure	Yeast
SAP 49	U2	3′ portion	+/	2 RRMs, proline-glycine rich domain	
SAP 33		–			
B″	U2	–	+/p,m	2RRMs	
A′	U2	–	+/p		
Proteins that first bind in the B complex					
220	U5	5′ splice site			PRP8
116	U5	–			
110/112	U5	–			
SAP 90	U4/U6	–			
SAP 82		–			
SAP 60	U4/U6	–			
SAP 55		–			
Proteins that first bind in the C complex					
SAP 165		–			
SAP 102 (PSF)*+		+	+/p	2 RRMs, proline-rich domain	
SAP 95		–			
SAP 75		–			
SAP 70		–			
SAP 68 (PSF)*+		+	+/p		
SAP 65		–			
SAP 58		–			
SAP 57*		–			
SAP 52		–			
SAP 48		–			
SAP 45		–			
40*	U5	–			
SAP 36		+			
SAP 35		–			
SAP 30		–			

Each protein is listed only in the complex in which it first binds and in the complex in which the protein is the most enriched. Except for the SR proteins, all proteins are present in splicing complexes affinity-purified in 250 mM salt. All of the components remain on the pre-mRNA as spliceosome assembly proceeds, except for U2AF^{65}, U2AF^{35}, U1 snRNP, and U4 snRNP (see Section 5.4).

snRNP: indicates the snRNP-association of the protein; core refers to proteins common to all snRNPs.

Cross-link: indicates proteins that can be UV cross-linked to pre-mRNA (py tract, pyrimidine tract at the 3′ splice site; 5′ portion, portion of pre-mRNA containing exon 1 and the 5′ splice site; 3′ portion, portion of pre-mRNA containing the BPS, 3′ splice site and exon 2; +, cross-links to unknown site; –, cross-linking not detected.

cDNA/ab: indicates mammalian proteins that have been cloned and for which antibodies are available (p, polyclonal; m, monoclonal; blank, not available).

structure: indicates major structural motifs in mammalian protein.

Yeast: indicates likely yeast homologue of mammalian protein.

* four proteins that were previously identified in the B complex, but were subsequently found to be significantly more abundant in the C complex.

+ SAP 68 is thought to be a breakdown product of SAP 102.

Analysis of the composition of affinity-purified H complex by two-dimensional gel electrophoresis revealed that it consists almost exclusively of hnRNP proteins (15), the more than 20 abundant nuclear proteins that package newly transcribed pre-mRNA *in vivo* (reviewed in 16; see Chapter 3). Significantly, however, there are differences in the composition of the H complex assembled on different pre-mRNAs, indicating that the hnRNP proteins have distinct sequence preferences (15; see 16 and Chapter 3 for review).

There is no evidence that the H complex is a functional intermediate in spliceosome assembly, nor that any of the hnRNP proteins are essential components of the basic splicing machinery. However, the observation that pre-mRNAs are packaged in a transcript-dependent manner raises the possibility that hnRNP proteins play a role in initial splice-site recognition. For example, the specific manner in which pre-mRNA is packaged by hnRNP proteins could result in the masking of cryptic splice sites or in the exposure of authentic splice sites to the splicing machinery. In support of a role for hnRNP proteins in initial splice-site recognition, hnRNP A1 can antagonize the activity of SR proteins, which are a family of serine/arginine-rich proteins involved in early splice-site recognition (see Chapters 6 and 8; 17). Specifically, in pre-mRNAs containing duplicated 5′ splice sites, the proximal 5′ splice site is preferentially used in extracts supplemented with SR proteins alone (18, 19), whereas use of the distal site increases when hnRNP A1 is also present (17).

The observation that hnRNP proteins are the first factors that bind to pre-mRNA *in vitro* parallels the observation that these proteins bind immediately to nascent pre-mRNA *in vivo* (for review, see 16, 20; Chapter 3). hnRNP proteins have long been known to cross-link to RNA both *in vivo* and *in vitro* (reviewed in 16). Not unexpectedly, the hnRNP proteins also cross-link to pre-mRNA in the H complex (21). Significantly, however, the levels of the hnRNP proteins, detected either by cross-linking or silver staining, are dramatically lower in the spliceosomal complexes than in the H complex (21). This observation suggests that any role of hnRNP–pre-mRNA interactions in affecting splice-site recognition or spliceosome assembly is restricted to the earliest steps in spliceosome assembly.

3. The early complex

The earliest detectable functional intermediate in spliceosome assembly is an ATP-independent complex that requires incubation at 30°C for assembly (22–25; see 26 for review). The early (E) complex in mammals and commitment complex (CC) in yeast are defined as 'committing' pre-mRNA to spliceosome assembly based on the observation that the pre-mRNA in these complexes can be chased into spliced products in the presence of a vast excess of competitor pre-mRNA (22–25). The E complex was originally detected as a discrete peak by gel filtration, and is not resolved from the H complex on native gels or density gradients (24, 25, 27). In yeast, native gel conditions were established to resolve the commitment complex (23), but no other fractionation method has so far been used to characterize this complex.

The E complex assembles on all pre-mRNAs tested, and the efficiency of its assembly correlates with the efficiency of spliceosome assembly on a particular pre-mRNA (24, 25, 27). The E complex accumulates only in splicing extracts lacking ATP and is rapidly converted to the first ATP-dependent spliceosomal complex (the A complex) under normal splicing conditions. In yeast, low levels of the commitment complex can be detected in splicing reactions containing ATP (9). However, in yeast as well as in mammals, it is not clear whether the E complex assembled in the absence of ATP is identical to the complex that precedes A complex assembly in the presence of ATP.

3.1 Pre-mRNA sequence requirements for E complex assembly

Several sequence elements are required for the splicing reaction, and each of these elements is recognized multiple times throughout spliceosome assembly. The critical sequence elements in the intron are located at the 5′ splice site, the branch site and the 3′ splice site. In addition, exon sequences play an important role in splicing. Recently, specific sequence elements were identified within exons and shown to be required for excision of the upstream intron (28–31). These elements, designated exonic enhancers, have been characterized as purine-rich sequences, though examples of non-purine-rich enhancers have also been identified (31, 32) (for additional discussion, see Section 3.4). The consensus sequences of elements recognized during splicing are shown in Fig. 1.

Mutation of either the 5′ or the 3′ splice site in pre-mRNA does not abolish E complex assembly completely. Instead, specific complexes assemble on pre-mRNAs containing only a 5′ or 3′ splice site, and these are designated the E5′ and E3′ complexes, respectively (27). The situation is similar in yeast where CC1 is a complex that assembles on pre-mRNA containing only a 5′ splice site (9, 23). Although a yeast equivalent of the E3′ complex has not been detected, this particular complex

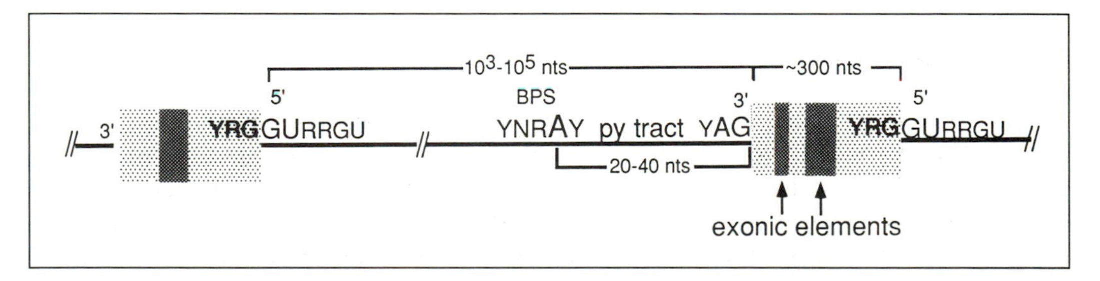

Fig. 1 Sequence elements in metazoan pre-mRNAs. Boxes indicate exon sequences and lines indicate intron sequences. Typical intron and exon lengths for naturally occurring metazoan pre-mRNAs are shown (not drawn to scale). The consensus sequences at the 5′ splice site, branchpoint sequence (BPS) and 3′ splice site are indicated (R, purine; Y, pyrimidine; N, any nucleotide). The highly conserved residues at each site are shown in capitals. Py tract is a pyrimidine-rich tract located between the BPS and AG dinucleotide. The 5′ splice site on the downstream end of the 2nd exon and/or exonic elements are required for splicing upstream introns in some pre-mRNAs (see Section 3.4).

may not be stable in the native gel system used. The 5′ splice site is required for E5′ assembly while the pyrimidine tract at the 3′ splice site is essential for E3′ assembly (27). The context of the 5′ and 3′ splice sites also appears to play a role at the time of E complex assembly, as the E5′ complex does not assemble on a random-sequence RNA containing a consensus 5′ splice site. Similarly, the E3′ complex assembles much less efficiently when exon sequences are altered (27).

Because specific complexes assemble independently on the 5′ or 3′ splice sites, a sensitive substrate competition assay was used to identify the sequences involved in E complex assembly on intact pre-mRNA. In the yeast system, wild-type pre-mRNA competes for commitment complex assembly more efficiently than pre-mRNAs with mutations in the 5′ splice site or branchpoint sequence (BPS) (9, 23). Similarly, in mammals, pre-mRNA containing wild type 5′ and 3′ splice sites assembles the E complex more efficiently than pre-mRNAs containing either a mutation in the 5′ splice site or in the pyrimidine tract (27). For the early steps in spliceosome assembly, the BPS in yeast appears to be functionally equivalent to the pyrimidine tract in mammals (2). Thus, in both mammals and yeast, the 5′ and 3′ ends of the intron are functionally associated with each other at the time of E complex assembly (9, 23, 27). Despite this notable similarity between yeast and mammals, it is likely that mammalian E complex assembly across the intron only occurs on the short model pre-mRNAs used for *in vitro* studies. Unlike yeast pre-mRNAs, which contain short introns, mammalian pre-mRNAs usually contain very large introns, typically ranging from 10^3 nucleotides up to greater than 10^5 nucleotides. As discussed below, in pre-mRNAs containing large introns, E complex assembly may occur across the exon (see Section 3.4).

3.2 snRNA and protein composition of the E complex

3.2.1 The snRNAs

U1 is the only snRNA present at about stoichiometric levels in affinity-purified E and E5′ complexes (24, 27). U1 snRNA is also present in the yeast commitment complex (9, 23). These observations are consistent with earlier snRNA binding studies which showed that U1 snRNA interacts with the 5′ splice site in the absence of ATP and before the other snRNAs (33–36). Moreover, the 5′ end of U1 snRNA can be psoralen-crosslinked to the 5′ splice site before the binding of the other snRNAs or in the absence of ATP (37). Low levels of U2 snRNA are present in affinity-purified E complex (27), and antibodies to the U2 snRNPs specifically immunoprecipitate the E complex (Bennett and Reed, unpublished data). Whether or not the association of U2 snRNP components with the E complex is functionally significant remains to be determined.

3.2.2 The proteins

Affinity-purified E complex (5, 27) contains the essential non-snRNP splicing factor $U2AF^{65}$, which binds specifically to the pyrimidine tract at the 3′ splice site (see Table 1) (37–39) (Chapter 6). $U2AF^{35}$, which tightly associates with $U2AF^{65}$ (40), is also present in the E complex (5, 21). In addition, the E complex contains U1 snRNP proteins A, C, and 70K, the snRNP core proteins B, B′, and D, and five spliceosome-associated proteins (SAPs) of 42, 72, 88, 92, and 115 KDa (see Table 1) (5). It is not yet known whether the low molecular weight snRNP core proteins (E, F, G) are present in the E or other spliceosomal complexes. SR proteins (18, 19, 41, 42) are present in the E complex, but are not as tightly bound as the other E complex proteins. The SR proteins can only be detected in gel filtration-purified spliceosomal complexes, in which case UV cross-linking was used to facilitate their detection (32). SR proteins are not detected in spliceosomal complexes purified by the more stringent biotin/avidin affinity chromatography (5).

3.3 Functional activities required for E complex assembly

U1 snRNP is required for commitment (or E) complex assembly (9, 23, 25). Although there is no direct evidence that $U2AF^{65}$ is required for E complex assembly, observations that $U2AF^{65}$ is bound to the pyrimidine tract in the E complex, and that the pyrimidine tract is essential for E complex assembly suggest this requirement. Moreover, $U2AF^{65}$ is required for U2 snRNP binding and for A complex formation (38, 40, 43).

SR proteins are also thought to play a role in E complex assembly because the addition of purified SR proteins to normal nuclear extracts can stimulate E complex assembly, including the binding of both U1 snRNP to the 5′ splice site, and $U2AF^{65}$ to the 3′ splice site (32). Moreover, using purified factors, it has been shown that SR proteins can promote the binding of U1 snRNP to the 5′ splice site (44). This effect is thought to be mediated by a direct interaction between the SR protein and the 70K component of U1 snRNP (44, 45; see next section for further discussion).

3.4 General implications: E complex formation across introns or exons

For mammalian pre-mRNAs containing short introns and consensus splice sites, substrate competition assays have shown that the 5′ and 3′ splice sites first functionally interact in the E complex. U1 snRNP is bound to the 5′ splice site, and $U2AF^{65}$ is bound to the 3′ splice site in this complex (see Fig. 2). Thus, the 5′ and 3′ splice sites are likely to be brought together via an interaction between U2AF and U1 snRNP. Likely candidates for mediating this interaction are members of the SR protein family (32, 45). SR proteins have been shown to interact specifically with $U2AF^{35}$ and with the snRNP protein, U1-70K, in farwestern assays, by co-immunoprecipitation, and in the yeast two-hybrid system (44, 45). In addition, SR proteins can interact directly with one another (45). SR proteins also directly contact pre-mRNA in the E complex (32), and it is possible that they bind to exon sequences (see below for further discussion). Thus, the 5′ and 3′ splice sites may be brought together in the E complex via a network of RNA–protein and protein–protein interactions involving U2AF, U1 snRNP, and SR proteins (see Fig. 2).

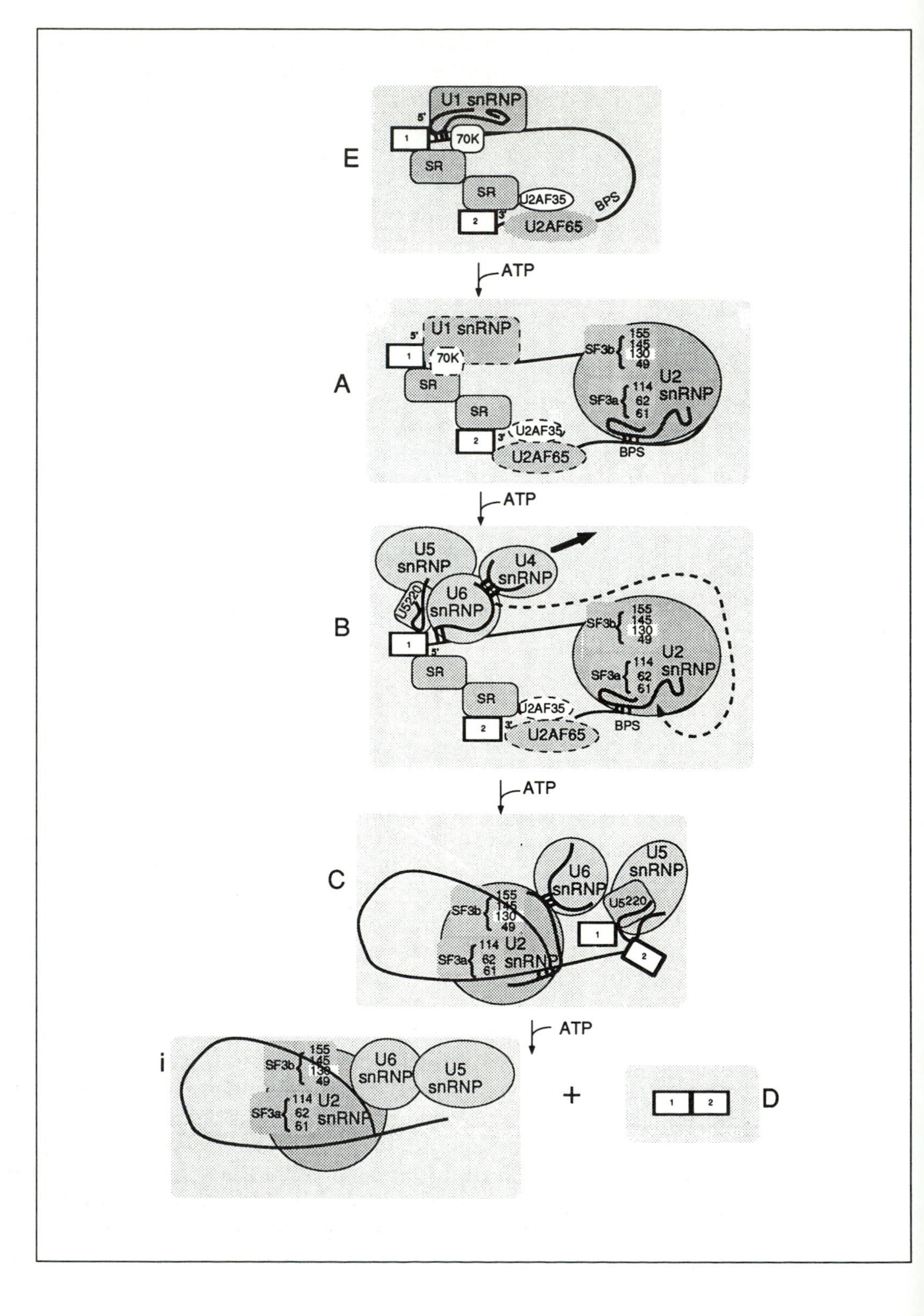
E
U1 snRNP
5'
1
70K
SR
SR
U2AF35
BPS
2
3'
U2AF65
ATP
A
U1 snRNP
5'
1
70K
SR
SR
U2AF35
2
3'
U2AF65
SF3b
155
145
130
49
SF3a
114
62
61
U2 snRNP
BPS
ATP
B
U5 snRNP
U4 snRNP
U6 snRNP
U5220
1
5'
SR
SR
U2AF35
2
3'
U2AF65
SF3b
155
145
130
49
SF3a
114
62
61
U2 snRNP
BPS
ATP
C
U6 snRNP
U5 snRNP
U5220
SF3b
155
145
130
49
SF3a
114
62
61
U2 snRNP
1
2
ATP
i
SF3b
155
145
130
49
SF3a
114
62
61
U2 snRNP
U6 snRNP
U5 snRNP
+
1
2
D

Fig. 2 Model for spliceosome assembly on pre-mRNAs containing short introns. Symbols are the same as in Fig. 1. Spliceosomal complexes E, A, B, C, D, and i are shown. Only the best characterized splicing factors are depicted in each complex. Components that interact directly with pre-mRNA are shaded. Dashes around U2AF and U1 snRNP indicate that they become less tightly bound during the E to A transition in B complex, the solid arrow indicates U4 snRNP dissociation, and the dashed arrow indicates the association of U2 and U6 snRNAs. SR proteins, which are thought to be present in all of the spliceosomal complexes, are not shown after the B complex for simplicity. See Section 3.4 for discussion.

In contrast to the pre-mRNAs used for *in vitro* splicing studies, most mammalian pre-mRNAs contain very large introns and weakly conserved splice sites. 'Cryptic' splice sites are abundant in such long introns, and the authentic splice sites contain minimal sequence information. Thus, specific mechanisms must exist for recognition of the authentic splice sites in complex metazoan pre-mRNAs. The first important insight into this problem came from the observation that a 5′ splice site on the downstream end of an exon promotes splicing of the intron immediately upstream (46). This result, considered together with the observation that mammalian exons are generally small (about 300 nucleotides) relative to introns (10^3–10^5 nucleotides), led to the 'exon definition' model for splice-site recognition (46). According to this model, an interaction is first established between the 5′ and 3′ splice sites across the short exon, followed by an interaction across the long intron. A great deal of recent data support the exon definition model, and a molecular explanation of it has begun to emerge. In the simplest model, the interactions that occur in the E complex across short introns are instead formed across the exon (see Fig. 3). In this case, U2AF is bound at the 3′ splice site, SR proteins are bound to the exon, and U1 snRNP is bound to the downstream 5′ splice site. Among the many observations that support this model, a downstream 5′ splice site promotes use of a weak upstream 3′ splice site, and U1 snRNP binding to the downstream 5′ splice site is required for this effect (47). Significantly, the presence of the downstream 5′ splice site was also found to promote U2AF binding to the weak 3′ splice site upstream (48). In addition, specific exonic sequences, designated splicing enhancers, have been shown to promote splicing of the intron upstream. Importantly, members of the SR family of splicing factors bind to exonic splicing enhancers and are essential for enhancer activity (30, 31, 49; Chapters 6 and 8). Finally, E3′ complex, which assembles on pre-mRNA containing only a 3′ splice site and exon 2, contains SR proteins, U1 snRNP, and U2AF (32). Thus, this complex contains all the components expected for E complex assembly across the exon (Fig. 3). Functional studies show that the E3′ complex can be chased into the A3′ complex (described below), which contains U2 snRNP bound at the branch site. Thus, assembly of the E3′ and A3′ complexes may reflect events on normal pre-mRNAs containing long introns. Ultimately, the upstream 5′ splice site across the long intron must be paired with the downstream 3′ splice site. How and when this pairing occurs is not known. However, U5 and U6 snRNPs are thought to replace U1 snRNP on the 5′ splice site during B complex assembly. Thus, it is possible that these snRNPs are involved in 5′ and 3′ splice site pairing across the intron.

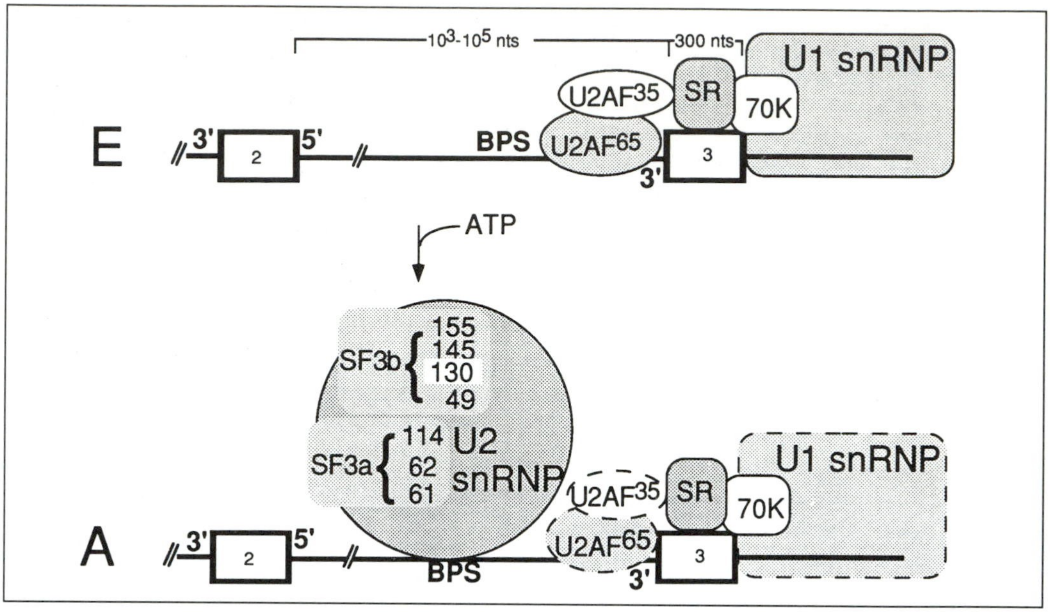

Fig. 3 Early steps in spliceosome assembly on pre-mRNAs containing long introns. Symbols are the same as in Fig. 1 and 2. Typical intron and exon lengths for naturally occurring metazoan pre-mRNAs are indicated above E complex; figure is not drawn to scale. For simplicity, complex assembly around exon 3 only is shown. See Section 3.4 for discussion.

4. The A complex

The A complex, which was originally detected on native gels, is the first ATP-dependent complex in spliceosome assembly (12, 50, 51). The A complex can also be detected by density gradient sedimentation and by gel filtration (24, 25, 52). Accumulation of the A complex begins after about one minute of incubation at 30°C, and within five to ten minutes the B complex is detected (note, however, that the kinetics of complex assembly vary depending on the particular pre-mRNA substrate and the preparation of nuclear extract).

4.1 Pre-mRNA sequence requirements for A complex assembly

As observed with the E complex (see above), mutation of the 5′ or 3′ splice site does not abolish complex assembly in the presence of ATP, but instead results in formation of the A5′ and A3′ complexes (12, 27). The A3′ complex can be detected on native gels, as well as by gel filtration, whereas the A5′ complex can only be detected by gel filtration. *In vitro* complementation assays show that the E3′ complex is a precursor to the A3′ complex (27). In yeast, the 5′ and 3′ splice sites are essential for ATP-dependent complex assembly; no complexes analogous to the mammalian A5′ and A3′ complex have been detected (50, 53, 54).

The pyrimidine tract is essential for A complex and A3′ complex formation (reviewed in 55). As the E complex is a precursor to the A complex, the requirement for the pyrimidine tract may result from its prior requirement for E complex assembly. The AG dinucleotide at the 3′ splice site is required for efficient A and A3′ complex assembly for some introns (56), but is dispensable for others (57, 58). The AG-independent introns require a long pyrimidine tract for efficient complex assembly while AG-dependent introns require a significantly shorter pyrimidine tract (57). Mutations in the BPS reduce the efficiency of A complex assembly (59). However, significant levels of both the A and B complexes do assemble on pre-mRNAs with mutations at the branch-site adenosine or with BPS mutations that disrupt the potential for base-pairing with U2 snRNA (60; for further discussion, see Section 4.2).

4.2 snRNA and protein composition of the A complex

4.2.1 The snRNAs

U2 snRNA forms an essential base-pairing interaction with the BPS in the A complex (37, 61–63). This duplex is thought to bulge the branch-site adenosine and position it for nucleophilic attack on the 3′ splice site (see 1 for review; see also Chapter 1). U1 snRNA can also be detected in the A complex but appears to be substoichiometric (based on ethidium bromide staining) relative to both U2 snRNA and to the pre-mRNA (27). Thus, it appears that U1 snRNA is stably associated with the pre-mRNA in the E complex, but then becomes less tightly bound, or dissociates completely, as spliceosome assembly proceeds. It is difficult to establish precisely when U1 snRNA becomes less tightly bound to the pre-mRNA because spliceosome assembly is not synchronous on the pre-mRNA added to the splicing reaction.

4.2.2 The proteins

The proteins that first bind to pre-mRNA at the time of A complex assembly have been determined for the A3′ complex (the A complex was not used because it is so short-lived) (5). At least nine of the ten proteins that first bind in the A3′ complex are components of 17S U2 snRNP (B″, A′ and SAPs 155, 145, 130, 114, 62, 61, and 49) (see Table 1, where, for simplicity, our nomenclature is used for these proteins); (5, 21). The 17S form of U2 snRNP contains all seven of these A complex-specific SAPs, as well as 92 and 35 kDa proteins (64). Whether these two proteins are also present in the A complex remains to be established. $U2AF^{65}$ is detected at lower levels in affinity-purified A3′ complex, than in E3′ or E complexes. Thus, similar to U1 snRNP, U2AF appears to become less tightly bound to pre-mRNA as spliceosome assembly proceeds. SR proteins are present in the A complex, as this complex can be immunoprecipitated by antibodies to these proteins; SR proteins are thought to bridge the 5′ and 3′ splice sites in this complex (65). SR proteins have also been shown to cross-link to pre-mRNA in the E and B complexes, and thus they probably also cross-link in the A complex (32). A model showing the interactions of the best characterized of the A complex components is presented in Fig. 2 (see also Fig. 3).

4.3 Functional activities required for A complex assembly

Assembly of the A complex requires U1 and U2 snRNPs, U2AF^{65}, the SR proteins, SF1, SF3a, and SF3b (38, 40, 43, 66–71). Of these factors, the requirements for U1 snRNP, U2AF^{65}, and the SR proteins may be due to their prior requirement for E complex assembly, while SF3a and SF3b are required for the E to A complex transition. SF3a and SF3b are multimeric complexes that consist of the U2 snRNP proteins, SAPs 61, 62, and 114, and SAPs 155, 145, 130, and 49, respectively (see Table 1) (67, 68). SF3a, SF3b, and the 12S U2 snRNP associate to yield a particle similar, if not identical, to 17S U2 snRNP (67). SF1 is a 75 kDa protein required for A complex assembly, but it is not yet known if there is a prior requirement for SF1 in E complex assembly (72).

A number of observations in both mammals and yeast indicate that PRPs 9, 11, and 21 are the yeast homologues of SAPs 61, 62, and 114, and hence of SF3a. The first evidence for this homology was the observation that antibodies to PRP9 recognize the 60 kDa 17S U2 snRNP protein (same as SAP 61) (73). Subsequently, cDNAs encoding SAPs 61 and 62 were isolated and found to encode proteins bearing significant amino acid identity (~25%) to PRPs 9 and 11, respectively (74, 75). In addition, using the yeast two-hybrid system, PRPs 9, 11, and 21 were shown to form a heterotrimer in which PRP21 interacts with both PRPs 9 and 11 (76, 77). Likewise in mammals, SAPs 61 and 62 interact specifically with SAP 114 in farwestern assays (74, 75). Together, these studies provide strong evidence that the SF3a complex has been conserved from yeast to humans. All of the SF3a subunits, and all but one (SAP 130) of the SF3b subunits, can be specifically UV cross-linked to pre-mRNA in A3′ and B complexes (21). These proteins are also thought to associate with the 5′ portion of U2 snRNA, which contains the sequences that base pair to the BPS (64). Thus, it is possible that SF3a/b functions to mediate interactions between U2 snRNA and the BPS.

5. The B complex

Conversion of the A to the B complex requires ATP and incubation at 30°C (24, 25). The B complex can be detected on native gels, density gradients, and by gel filtration (11, 12, 51, 78).

5.1 Sequence requirements for B complex assembly

Both the 5′ and 3′ splice sites are essential for B complex assembly (12, 79). Because of the prior requirement for the 3′ splice site for E and A complex assembly, it has not been possible to determine whether this sequence element plays an additional role in the B complex. In contrast, a distinct role for the 5′ splice site in B complex assembly is suggested by the observation that mutation of the 5′ splice site abolishes B complex assembly, whereas sensitive substrate competition assays were required to uncover the role of the 5′ splice site in E and A complex assembly. Based on studies of factor interactions at the 5′ splice site, it is possible that the involvement of the 5′ splice site in E and A complex assembly is due to U1 snRNP interactions, whereas the essential role of the 5′ splice site in B complex assembly is due to U5 and/or U6 snRNP interactions (see below for further discussion). That factors other than U1 snRNA interact with the 5′ splice site is supported further by the observation that some mutations in the 5′ splice site are not suppressed by the expected compensatory changes in U1 snRNA (54, 80).

5.2 snRNA and protein composition of the B complex

5.2.1 The snRNAs

U2, U4, U5, and U6 snRNAs are detected in the B complex (13, 51, 56). U2, U5, and U6 snRNAs are present at about a 1:1 stoichiometry whereas the relative levels of U4 snRNA are lower (27). Thus, similar to U1 snRNP, U4 snRNP appears to associate only transiently with the assembling spliceosome (see Section 5.4 for further discussion of U4 snRNP dynamics) (50, 51, 56, 106).

Direct interactions between pre-mRNA and U2, U5, and U6 snRNAs have been detected in UV cross-linking studies (37, 82–84). U2 snRNA can be cross-linked to the branchpoint sequence in the B complex (37, 83). This cross-linking interaction is probably indicative of the essential U2 snRNA–BPS base-pairing interaction (37, 61–63). In addition, site-specific cross-linking studies showed that U5 snRNA cross-links to a photoactivatable thiouridine residue located either one or two nucleotides upstream from the 5′ splice junction in the B complex (82, 85). Finally, cross-linking studies showed that U6 snRNA interacts with the 5′ splice site prior to the first catalytic step of the splicing reaction (37, 83, 84).

5.2.2 The proteins

More than 30 distinct protein components have been identified in affinity-purified B complex (5). Among these are all of the U2 snRNP-specific SAPs identified in the A complex. In addition, 10 new proteins bind to the pre-mRNA in the B complex (see Table 1) (5). Four of these proteins were subsequently found to be present at much higher levels in the C complex, and are therefore now designated as C complex-specific components (indicated by asterisks in Table 1) (6). Five of the proteins that first bind in the B complex (220 (doublet), 112/110, and 116 kDa) are components of a 20S form of U5 snRNP which was purified under gentle conditions and found to contain eight proteins (220 (doublet), 116, 112/110, 52, 40, and 15 kDa) in addition to the common snRNP core proteins previously identified in 10S U5 snRNP (86, 87; see Chapter 5). The 52 and 15 kDa U5 snRNP proteins have not been detected in the purified B or C complexes. Two of the proteins that first bind in the B complex—SAPs 60 and 90—were identified as U4/U6 snRNP components (see Table 1) (6).

Finally, SR proteins, though not present in affinity-purified complexes, are detected in gel-filtration isolated B complex (32, 41).

The U4, U5, and U6 snRNPs associate with one another to form a tri-snRNP particle (13, 51). This particle was immunopurified from nuclear extracts and found to contain, in addition to all of the 20S U5 snRNP proteins, at least five U4/U6·U5 tri-snRNP-specific proteins (90, 60, 27, 20, and 15.5 kDa) (88). The 60 and 90 kDa proteins are likely to correspond to the U4/U6 snRNP-specific SAPs 60 and 90 (mentioned above). It is not yet known whether the other U4/U6·U5 proteins are also spliceosome components. A kinase activity associated with the purified U4/U6·U5 particle phosphorylates the 52 kDa U5 snRNP protein, but the role of this phosphorylation is not known (88).

SAPs 155, 145, 114, 62, 61, and 49, which UV cross-link to the pre-mRNA in the affinity-purified A3′ complex, also crosslink in affinity-purified B complex (21). The only new protein that cross-links in the B complex is the 220 kDa U5 snRNP protein which requires the 5′ splice site for cross-linking (21, 89). This protein also cross-links to a thiouridine residue located two nucleotides upstream of the 5′ splice site (82). A model of spliceosome assembly is presented in Fig. 2, which shows the B complex along with the factors that have been best characterized.

5.3 Functional activities required for B complex assembly

The B complex does not assemble in extracts specifically depleted of U5 snRNP (90, 91), of the 200 kDa U5 snRNP protein (PRP8) (92), or of U4/U6 snRNP (71). A large body of evidence indicates that U4, U5, and U6 snRNPs enter the spliceosome as part of the U4/U6·U5 tri-snRNP particle, including the observations that all three snRNPs bind to pre-mRNA at the same time, and that the U4/U6 snRNP requires U5 snRNP for binding to pre-mRNA and vice versa (13, 51, 71, 90–92) (see Chapter 5). Furthermore, isolated U4/U6·U5 snRNP can restore splicing activity to extracts specifically depleted of U4/U6 and U5 snRNPs (93). Assembly of the triple snRNP does not occur with isolated 12S U4/U6 snRNP and 20S U5 snRNP, but requires one or more of the proteins that are specific to the 25S U4/U6·U5 snRNP particle (the 90, 60, 27, 20, and 15.5 kDa proteins) (88). Evidence that these U4/U6·U5-specific components are required for B complex assembly comes from the observation that one or more of these proteins corresponds to HSLF (heat shock-labile splicing factor), a factor that restores spliceosome assembly and splicing to heat-shocked inactivated HeLa cell extracts (94).

Other than the snRNP components, little is known about mammalian factors required for B complex assembly. In yeast, a major class of splicing proteins contains an ATP-dependent RNA helicase motif; the hallmark of this motif is the amino acid sequence DEAD or DEAH. The DEAD/H box family members essential for splicing include PRPs2, 5, 16, 22, and 28. Of these, PRP5 is required for A complex assembly, PRP2 is required for catalytic step I, PRP16 is essential for catalytic step II, and PRP22 is necessary for intron turnover. Finally, genetic evidence indicates that PRP28 functions after PRP5, possibly at about the time that

U4/U6·U5 snRNP binds to the B complex (reviewed in 2) (see Chapter 7). None of the mammalian splicing proteins that have been purified or cloned are DEAD/H box proteins. However, given that these proteins play such a critical role in splicing in yeast, it is likely that mammalian counterparts have thus far eluded detection.

The 220 kDa U5 snRNP protein is the only protein having an apparent yeast homologue (PRP8) that first binds to pre-mRNA in the B complex. Although similarity between these proteins has not yet been demonstrated at the amino acid level, both PRP8 and the mammalian 220 kDa protein are U5 snRNP components, and are present in the U4/U6·U5 particle and in the spliceosome (5, 86, 95, 96). Moreover, antibodies to PRP8 detect the 220 kDa mammalian protein, both PRP8 and the mammalian protein cross-link to pre-mRNA, and the proteins are similar to each other in size (21, 82, 89, 96, 97).

5.4 General implications: dynamic interactions during spliceosome assembly

Dynamic interactions between spliceosomal components and the pre-mRNA, and among spliceosomal components themselves, are emerging as central features of spliceosome assembly. One of the first dynamic changes takes place when U1 snRNP and U2AF, which bind tightly to pre-mRNA in E complex, are either significantly destabilized or completely dissociate from pre-mRNA during subsequent steps of spliceosome assembly (5, 14, 21, 27). Whether U2AF remains loosely bound to pre-mRNA as depicted in the model (Fig. 2), or is replaced by another factor is not yet known. In contrast, it appears that U1 snRNP is replaced by other factors as spliceosome assembly proceeds. Cross-linking of U1 snRNA to a thiouridine residue located one or two nucleotides upstream of the 5′ splice junction peaks after about five minutes of spliceosome assembly and is then replaced by cross-linking of U5 snRNA (82, 85). The kinetics of these interactions are consistent with an exchange of the U1 snRNA–5′ splice site interaction for the U5 snRNA–5′ splice site interaction in the B complex. The 220 kDa U5 snRNP protein that first cross-links in the B complex (21, 89) also cross-links to the same thiouridine residue as U5 snRNA and may be involved in mediating the U5 snRNA–5′ splice site interaction (82). The observation that U5 snRNP is positioned at the 5′ splice site is undoubtedly related to yeast genetic suppression studies, which indicate that exon sequences at the 5′ and 3′ splice sites base-pair with sequences in a highly conserved single-stranded loop in U5 snRNA (98, 99). On the basis of these observations, U5 snRNA is thought to play a critical role in the transesterification reactions (85).

Another dynamic interaction that occurs around the time of the B complex formation involves U4 snRNP. In nuclear extracts, U4 snRNA resides in three distinct particles, the U4 snRNP, the U4/U6 snRNP and the U4/U6·U5 snRNP (reviewed in 100) (see Chapter 5). U4 snRNA is extensively base-paired to U6 snRNA in the U4/U6 snRNP; this base-pairing is thought to be present in the U4/U6·U5 particle as well (101–104). Analysis of spliceosomal complexes has suggested that U4 snRNP is initially bound tightly, but then dissociates at or near the time of catalytic step I (50, 51, 56, 105, 106). U4 snRNP is not required for splicing after B complex assembly. This conclusion is based on studies in yeast showing that isolated B complex which lacks U4 snRNP is able to carry out both the first and second catalytic steps of the splicing reaction (106). This observation lends support to the hypothesis that one of the primary functions of U4 snRNA is to be an antisense regulator of U6 snRNA, which has been proposed to participate directly in catalysis (3). Due in part to the fact that spliceosome assembly is not synchronous on the pre-mRNA added to nuclear extracts, it is has not been possible to determine precisely when the destabilization of U4 snRNP occurs. In any case, the unwinding of U4/U6 snRNAs is thought to involve disruption of two strongly base-paired stems, and therefore may require energy and a helicase-type activity. Such a role has been proposed for the above described DEAD box factors, which have ATP-dependent helicase motifs (see 107).

Subsequent to the disruption of the U4–U6 base-pairing interaction, it is possible that U2–U6 base-pairing interactions are established (107). Genetic suppression studies in yeast demonstrated an essential U2–U6 base-pairing interaction that is mutually exclusive with the U4–U6 snRNA base-pairing. The site of the U2–U6 interaction is immediately adjacent to the ACAGAGA sequence, which is thought to be a catalytically active site on U6 snRNA, and also adjacent to the region of U2 snRNA that base-pairs with the BPS (107). Whether the U2–U6 interaction occurs in the spliceosome and the precise timing of this interaction remain to be established.

6. The C complex

The products of catalytic step I of the splicing reaction—exon 1 and lariat-exon 2—are present in the C complex. With most model pre-mRNAs, the C complex is a very short-lived intermediate due to the rapid conversion of the splicing intermediates into spliced products (ligated exons and lariat intron). The spliced mRNA is in turn rapidly released from the C complex, such that no discrete spliceosomal complex has been identified containing the two products of the second step of the reaction. The spliced mRNA alone is detected in the D complex, which fractionates faster than the H complex on a native gel (79). Thus, the D complex most likely consists of RNA-binding proteins, such as the SR proteins, which are thought to interact with exon sequences (see Section 3.4). In contrast, the snRNPs remain associated with the lariat in the i complex (79).

The C complex is detected on native gels (56) and was isolated by gel filtration under conditions in which the second catalytic step of the reaction was blocked by a mild heat treatment (78, 108). Conversion of the B to the C complex, as well as conversion of the splicing intermediates into the spliced products, requires ATP and incubation at 30°C (78, 109, 110). Heat treatment of nuclear extracts has not proven to be a reliable method for accumulating the C complex, due to variability among extracts. However, the C complex does accumulate when assembled on AG-independent introns lacking the AG dinucleotide (6, 14).

6.1 Pre-mRNA sequence requirements for catalytic steps I and II

The 5′ splice site, pyrimidine tract, and AG dinucleotide (in AG-dependent introns) are required for catalytic step I (57, 79, 111–113), and no mutations in any of these elements uncouple B complex assembly from step I (i.e. from C complex formation). Thus, it is not possible to determine whether a distinct role in step I exists for any of these elements. With respect to mutations in the BPS, essentially three different phenotypes have been observed, depending on the particular mutation and on the context of this mutation. First, in several pre-mRNAs, when the branch-site adenosine is substituted with other nucleotides, or when the entire BPS is absent, the lariat forms at a 'cryptic' BPS located within the normal distance (20–40 nucleotides) upstream from the 3′ splice junction (111, 114, 115). As expected, most of the cryptic BPSs have the potential to form at least a few base-pairs with U2 snRNA; in general, the lower the base-pairing potential, the lower the efficiency of catalytic step I (60, 61). Second, when the BPS is substituted with sequences incapable of base-pairing with U2 snRNA (and/or lacking an adenosine), but in the absence of a cryptic branch site, step I is dramatically decreased (60). The third phenotype of BPS mutations was observed with a rabbit β-globin pre-mRNA, in which lariat formation was found to occur fairly efficiently with C, G, or U residues substituted for the branch-site adenosine (116). A naturally occurring example of this phenotype was observed with the human growth hormone pre-mRNA, in which lariat formation occurs via use of a C residue, and sometimes via use of two U residues nearby (117). It is likely that the BPS mutations have variable phenotypes in the different pre-mRNAs due to differences in the strengths of other sequence elements (for example, the pyrimidine tract, exon sequences, 5′ splice site).

In contrast to the B to C complex transition, there are several examples of mutations in the pre-mRNA that uncouple catalytic steps I and II. Mutations in the conserved GU dinucleotide at the 5′ splice site more severely affect step II than I (79, 113, 118–120). Similarly, mutation of the branch-site A to G or U residues results in lariats that are unable to undergo step II (116). As the branch site and the G and U at the 5′ terminus of the intron are not likely to participate directly in catalysis of step II, it is possible that 'proofreading' of the branched nucleotides is a prerequisite for this reaction. The AG dinucleotide is essential for step II (111, 113), and in most pre-mRNAs, the first AG downstream from the BPS is used (56, 111, 121). However, both the distance and the context of the AG play a role in the efficiency of this reaction (57, 121, 122).

6.2 snRNA and protein composition of the C complex

6.2.1 The snRNAs

The C complex contains U2, U5, and U6 snRNAs (14, 56). Significantly, both genetic and cross-linking methods have converged to provide strong evidence that U5 and U6 snRNAs interact with pre-mRNA very near or at the active site of the spliceosome and may be the key snRNAs involved in catalysis (37, 82–85, 123, 124). U5 snRNA interacts with exon sequences immediately upstream of the 5′ splice site before catalytic step I of the splicing reaction and then also interacts with exon sequences downstream of the 3′ splice site after step I (37, 82, 85, 98, 99). In addition, U6 snRNA base-pairs with sequences at the 5′ splice site before catalytic step I and then is detected near the branch site after step I has occurred (37, 83–85, 123, 124). These snRNA-pre-mRNA interactions are discussed below in the context of the model for spliceosome assembly (see Section 7).

6.2.2 The proteins

The C complex contains over 45 specific protein components. These include all the proteins that are present in the B complex plus an additional 15 C complex-specific SAPs (6). Four of the C-complex-specific proteins were originally shown to bind in the B complex, but were subsequently found to be much more abundant in the C complex (5, 6). These are SAPs 57, 68, and 102 and the 40 kDa U5 snRNP protein (SAPs 102 and 68 are thought to correspond to the essential splicing factor PSF and an apparent breakdown product of PSF, respectively). All of the proteins that cross-link in the B complex also cross-link in the C complex (the 220 kDa U5 snRNP protein and SAPs 155, 145, 114, 62, 61, and 49). In addition, the C-complex-specific SAP 36 and two others, possibly SAPs 102 and 68, also cross-link in C complex (6).

6.3 Functional activities required for catalytic steps I and II

In mammals, SF4 is the only splicing factor identified and at least partially purified that is required for catalytic step I, but not for spliceosome assembly (81). The splicing factor PSF and U5 snRNP are the only known mammalian factors required for catalytic step II (6, 125). In yeast, the only well characterized factor required for step I, but not for spliceosome assembly is PRP2 (126) (see Chapter 7). PRP2 is a member of the DEAH box family of splicing factors and has RNA-dependent ATPase activity (127, 128). PRP2 interacts transiently with both the B and C complexes (129), and the addition of ATP alone is sufficient to generate the splicing intermediates from an isolated spliceosome containing bound PRP2 (130). Kim and Lin (130) proposed that PRP2 may cause a conformational change or be involved in a proofreading step required for catalysis.

Several factors involved in the second catalytic step of the reaction have been identified in yeast. The most intensively studied of these, PRP16, is a DEAH family member that bears resemblance to PRP2 (128, 131–133) (Chapter 7). As with PRP2, PRP16 has an RNA-dependent ATPase activity essential for its activity in splicing (133). One class of mutant PRP16 alleles has reduced ATPase activity and is also a dominant suppressor of mutations at multiple positions in the BPS (131, 132). Based on these and other observations, Burgess and Guthrie (134) have proposed that PRP16 functions in an ATP-dependent proofreading step required for step II. The observation, in mammals, that certain BPS and 5′ splice site mutations abolish catalytic step II is consistent with the notion that the structure of the branched lariat is proofread as a prerequisite for step II.

PRP17, PRP18, and SLU7 are also essential for catalytic step II in yeast. These proteins, as well as PRP16, interact genetically with one another and/or with U5 snRNA, and so far at least PRP18 is known to be a U5 snRNP (and/or a U4/U6·U5 snRNP) component (135, 136). Thus, it possible that these four proteins constitute a U5 snRNP-associated complex required for step II (135, 136). This complex may be analogous to the U2 snRNP-associated PRP9/11/21 heterotrimer that is required for pre-spliceosome assembly. In both cases, the complexes interact genetically with a DEAD/H box family member: the U5 snRNP complex interacts with PRP16 and the U2 snRNP complex with PRP15. In mammals, the likely homologues of PRPs 9, 11, and 21 have been identified (as the U2 snRNP-associated SAPs 114, 62, and 61 or SF3a, see above), and it is likely that some of the U5 snRNP proteins enriched in the mammalian C complex are counterparts of PRP17, PRP18, and SLU7.

7. General implications: model for spliceosome assembly

Two of the central functions of spliceosome assembly are to position the critical factors of the splicing machinery at the active sites of the spliceosome and to ensure a high level of accuracy in identifying the bonds in the pre-mRNA to be cleaved and ligated. It is generally believed that the catalytic steps of the splicing reaction are carried out, at least in part, by the snRNAs (reviewed in 1, 3, 4). This notion originated from the discovery that pre-mRNA splicing and group II autocatalytic splicing take place via the same two-step mechanism and from other parallels between pre-mRNA splicing and group I and II autocatalytic splicing (37) (Chapter 1). U6 snRNA was identified as a candidate for a catalytic RNA for several reasons, including the observations that certain regions of U6 snRNA are unexpectedly highly conserved and that these conserved regions of U6 snRNA are exposed after the release of U4 snRNA, before catalytic step I (see 3, 104, 138–140). The central role of U6 snRNA in catalysis is also supported by the observation that mutations in *Ascaris* U6 snRNA upstream of a putative catalytic domain (ACAGAG sequence) result in attachment of the branch site to U6 snRNA rather than to the 5′ splice site (141) (Chapter 10).

A current model for spliceosome assembly on short pre-mRNAs typically used for *in vitro* studies is presented in Fig. 2 and summarized briefly below. In the E complex, U1 snRNP binds to the 5′ splice site, and a duplex is formed between U1 snRNA and the 5′ splice site. In addition, U2AF binds to the pyrimidine tract at the 3′ splice site. U1 snRNP and U2AF are bridged by SR proteins, which can also interact with one another and with the pre-mRNA. In the A complex, U2 snRNP binds to the BPS, an essential duplex is formed between U2 snRNA and the BPS, and all but one (SAP 130) of the SF3a/b subunits can be cross-linked to pre-mRNA. The base-pairing interaction between U2 snRNA and the BPS functions to bulge the branch-site adenosine, specifying it as the nucleophile for attack on the 5′ splice site.

U1 snRNP and U2AF become less tightly bound to pre-mRNA during the ATP-dependent E to A complex transition. The B complex is formed subsequently by the binding of U4/U6·U5 snRNP, and U5 and U6 snRNA base-pairing interactions replace U1 snRNA at the 5′ splice site. The U5 snRNP protein $U5^{220}$ cross-links to exon 1 sequences immediately adjacent to the 5′ splice site and may stabilize the U5 snRNA–pre-mRNA interaction. At some point, a major conformational change occurs, as the U4–U6 snRNA helix is disrupted, U4 snRNP dissociates, and an interaction between U2 and U6 snRNAs is established (indicated by arrows in B complex, Fig. 2). These rearrangements result in the close juxtapositioning of the 5′ splice site–U6 duplex and the BPS–U2 duplex, and these RNA–RNA interactions may establish the active site of the spliceosome (for catalytic step 1). The C complex contains the products of catalytic step I of the splicing reaction. U5 snRNA interacts with both exons in the C complex and may function to hold them together for ligation. After catalytic step II, the spliced exons are released from the spliceosome and are detected in a discrete complex of unknown composition (possibly containing SR proteins). The snRNPs are detected bound to the excised lariat intron in complex i.

Although the snRNAs are likely to play a direct role in both catalytic steps of the splicing reaction, the protein components of the splicing machinery are almost certainly vital for establishing and/or stabilizing interactions between the snRNAs and the pre-mRNA. This function for the protein components of the spliceosome is suggested by the observation that the sequences in the pre-mRNA that interact with the snRNAs are short and often not well conserved. Thus, proteins may be necessary to direct snRNAs to the critical regions of the pre-mRNA and to promote a stable and specific association of the snRNAs with the sequence elements in the pre-mRNA. Candidates for this function for U2 snRNA–BPS interactions are the U2 snRNP-associated SAPs; these proteins can be UV cross-linked to the pre-mRNA (21) and also interact with the 5′ portion of U2 snRNA which contains the sequence that base pairs to the BPS (64). The 220 kDa U5 snRNP component, which cross-links to the terminus of the first exon, is a good candidate for mediating the U5 snRNA–5′ splice site interaction (82). That these U2 and U5 snRNP proteins play crucial roles near the active site of the spliceosome is suggested by the striking conservation of these proteins between yeast and humans (67, 73–76, 89, 96, 97, 142). In addition to promoting snRNA–pre-mRNA interactions, non-snRNP protein factors, such as the DEAD/H box family members, are likely to play key roles in disrupting RNA–RNA interactions during the catalytic steps of the splicing reaction.

References

1. Moore, M. J., Query, C. C., and Sharp, P. A. (1993) Splicing of precursors to messenger RNAs by the spliceosome. In *The RNA World*, Gesteland, R. and Atkins, J. (ed.). Cold Spring Harbor Laboratory Press, New York, p. 303.
2. Rymond, B. C. and Rosbash, M. (1992) Yeast pre-mRNA splicing. In *The Molecular and Cellular Biology of the Yeast Saccharomyces*, Vol. 2, Broach, J. R., Pringle, J., and Jones, E. W. (eds.). Cold Spring Harbor Laboratory Press, New York, p. 143.

3. Guthrie, C. (1991) Messenger RNA splicing in yeast: clues to why the spliceosome is a ribonucleoprotein. *Science*, **253**, 157.
4. Steitz, J. A. (1992) Splicing takes a holliday. *Science*, **257**, 888.
5. Bennett, M., Michaud, S., Kingston, J., and Reed, R. (1992) Protein components specifically associated with prespliceosome and spliceosome complexes. *Genes Dev.*, **6**, 1986.
6. Gozani, O., Patton, J. G., and Reed, R. (1994) A novel set of spliceosome-associated proteins and the essential splicing factor PSF bind stably to pre-mRNA prior to catalytic step II of the splicing reaction. *EMBO J.*, **13**, 3356.
7. Madhani, H. D. and Guthrie, C. (1994) Dynamic RNA–RNA interactions in the spliceosome. *Annu. Rev. Genet.*, **28**, 1.
8. Ruby, S. W. and Abelson, J. (1991) Pre-mRNA splicing in yeast. *Trends Genet.*, **7**, 79.
9. Séraphin, B. and Rosbash, M. (1991) The yeast branchpoint sequence is not required for the formation of a stable U1 snRNA-pre-mRNA complex and is recognized in the absence of U2 snRNA. *EMBO J.*, **10**, 1209.
10. Frendewey, D. and Keller, W. (1985) Stepwise assembly of a pre-mRNA splicing complex requires U-snRNPs and specific intron sequences. *Cell*, **42**, 355.
11. Grabowski, P. J., Seiler, S. R., and Sharp, P. A. (1985) A multicomponent complex is involved in the splicing of messenger RNA precursors. *Cell*, **42**, 345.
12. Konarska, M. M. and Sharp, P. A. (1986) Electrophoretic separation of complexes involved in the splicing of precursors to mRNAs. *Cell*, **46**, 845.
13. Konarska, M. M. and Sharp, P. A. (1987) Interactions between small nuclear ribonucleoprotein particles in formation of spliceosomes. *Cell*, **49**, 763.
14. Reed, R. (1990) Protein composition of mammalian spliceosomes assembled in vitro. *Proc. Natl Acad. Sci. USA*, **87**, 8031.
15. Bennett, M., Piñol-Roma, S., Staknis, D., Dreyfuss, G., and Reed, R. (1992) Differential binding of heterogeneous nuclear ribonucleoproteins to mRNA precursors prior to spliceosome assembly in vitro. *Mol. Cell. Biol.*, **12**, 3165.
16. Dreyfuss, G., Matunis, M. J., Piñol-Roma, S., and Burd, C. G. (1993) hnRNP proteins and the biogenesis of mRNA. *Annu. Rev. Biochem.*, **62**, 289.
17. Mayeda, A. and Krainer, A. R. (1992) Regulation of alternative pre-mRNA splicing by hnRNP A1 and splicing factor SF2. *Cell*, **68**, 365.
18. Krainer, A. R., Conway, G. C., and Kozak, D. (1990) The essential pre-mRNA splicing factor SF2 influences 5′ splice site selection by activating proximal sites. *Cell*, **62**, 35.
19. Ge, H. and Manley, J. L. (1990) A protein factor, ASF, controls cell-specific alternative splicing of SV40 early pre-mRNA *in vitro*. *Cell*, **62**, 25.
20. Chung, S. Y. and Wooley, J. (1986) Set of novel, conserved proteins fold pre-messenger RNA into ribonucleosomes. *Proteins*, **1**, 195.
21. Staknis, D. and Reed, R. (1994) Direct interactions between pre-mRNA and six U2 small nuclear ribonucleoproteins during spliceosome assembly. *Mol. Cell. Biol.*, **14**, 2994.
22. Legrain, P., Séraphin, B., and Rosbash, M. (1988) Early commitment of yeast pre-mRNA to the spliceosome pathway. *Mol. Cell. Biol.*, **8**, 3755.
23. Séraphin, B. and Rosbash, M. (1989) Identification of functional U1 snRNA–pre-mRNA complexes committed to spliceosome assembly and splicing. *Cell*, **59**, 349.
24. Michaud, S. and Reed, R. (1991) An ATP-independent complex commits pre-mRNA to the mammalian spliceosome assembly pathway. *Genes Dev.*, **5**, 2534.
25. Jamison, S. F., Crow, A., and Garcia-Blanco, M. A. (1992) The spliceosome assembly pathway in mammalian extracts. *Mol. Cell. Biol.*, **12**, 4279.

26. Rosbash, M. and Séraphin, B. (1991) Who's on first? The U1 snRNP-5′ splice site interaction and splicing. *Trends Biochem. Sci.*, **16**, 187.
27. Michaud, S. and Reed, R. (1993) A functional association between the 5′ and 3′ splice site is established in the earliest prespliceosome complex (E) in mammals. *Genes Dev.*, **7**, 1008.
28. Watakabe, A., Tanaka, K., and Shimura, Y. (1993) The role of exon sequences in splice site selection. *Genes Dev.*, **7**, 407.
29. Sun, Q., Mayeda, A., Hampson, R. K., Krainer, A. R., and Rottman, F. M. (1993) General splicing factor SF2/ASF promotes alternative splicing by binding to an exonic splicing enhancer. *Genes Dev.*, **7**, 2598.
30. Lavigueur, A., La Branche, H., Kornblihtt, A. R., and Chabot, B. (1993) A splicing enhancer in the human fibronectin alternate ED1 exon interacts with SR proteins and stimulates U2 snRNP binding. *Genes Dev.*, **7**, 2405.
31. Tian, M. and Maniatis, T. (1993) A splicing enhancer complex controls alternative splicing of doublesex pre-mRNA. *Cell*, **74**, 105.
32. Staknis, D. and Reed, R. (1994) SR proteins promote the first specific recognition of pre-mRNA and are present together with the U1 small nuclear ribonucleoprotein particle in a general splicing enhancer complex. *Mol. Cell. Biol.*, **14**, 7670.
33. Black, D. L., Chabot, B., and Steitz, J. A. (1985) U2 as well as U1 small nuclear ribonucleoproteins are involved in premessenger RNA splicing. *Cell*, **42**, 737.
34. Chabot, B. and Steitz, J. A. (1987) Multiple interactions between the splicing substrate and small nuclear ribonucleoproteins in spliceosomes. *Mol. Cell. Biol.*, **7**, 281.
35. Bindereif, A. and Green, M. R. (1987) An ordered pathway of snRNP binding during mammalian pre-mRNA splicing complex assembly. *EMBO J.*, **6**, 2415.
36. Ruby, S. W. and Abelson, J. (1988). An early hierarchic role of U1 small nuclear ribonucleoprotein in spliceosome assembly. *Science*, **242**, 1028.
37. Wassarman, D. A. and Steitz, J. A. (1992) Interactions of small nuclear RNA's with precursor messenger RNA during *in vitro* splicing. *Science*, **257**, 1918.
38. Ruskin, B., Zamore, P. D., and Green, M. R. (1988) A factor, U2AF, is required for U2 snRNP binding and splicing complex assembly. *Cell*, **52**, 207.
39. Zamore, P. D. and Green, M. R. (1991) Biochemical characterization of U2 snRNP auxiliary factor: an essential pre-mRNA splicing factor with a novel intranuclear distribution. *EMBO J.*, **10**, 207.
40. Zamore, P. D. and Green, M. R. (1989) Identification, purification, and biochemical characterization of U2 small nuclear ribonucleoprotein auxiliary factor. *Proc. Natl Acad. Sci. USA*, **86**, 9243.
41. Fu, X. D. and Maniatis, T. (1990) Factor required for mammalian spliceosome assembly is localized to discrete regions in the nucleus. *Nature*, **343**, 437.
42. Zahler, A. M., Lane, W. S., Stolk, J. A., and Roth, M. B. (1992) SR proteins: a conserved family of pre-mRNA splicing factors. *Genes Dev.*, **6**, 837.
43. Krämer, A. (1988) Presplicing complex formation requires two proteins and U2 snRNP. *Genes Dev.*, **2**, 1155.
44. Kohtz, J. D., Jamison, S. F., Will, C. L., Zuo, P., Lührmann, R., Garcia-Blanco, M. A., and Manley, J. L. (1994) Protein–protein interactions and 5′-splice-site recognition in mammalian mRNA precursors. *Nature*, **368**, 119.
45. Wu, J. Y. and Maniatis, T. (1993) Specific interactions between proteins implicated in splice site selection and regulated alternative splicing. *Cell*, **75**, 1061.
46. Robberson, B. L., Cote, G. J., and Berget, S. M. (1990) Exon definition may facilitate splice site selection in RNAs with multiple exons. *Mol. Cell. Biol.*, **10**, 84.

47. Kuo, H. C., Nasim, F. H., and Grabowski, P. J. (1991) Control of alternative splicing by the differential binding of U1 small nuclear ribonucleoprotein particle. *Science*, **251**, 1045.
48. Hoffman, B. E. and Grabowski, P. J. (1992) U1 snRNP targets an essential splicing factor, U2AF65, to the 3′ splice site by a network of interactions spanning the exon. *Genes Dev.*, **6**, 2554.
49. Sun, Q., Mayeda, A., Hampson, R. K., Krainer, A. R., and Rottman, F. M. (1993) General splicing factor SF2/ASF promotes alternative splicing by binding to an exonic splicing enhancer. *Genes Dev.*, **7**, 2598.
50. Pikielny, C. W., Rymond, B. C., and Rosbash, M. (1986) Electrophoresis of ribonucleoproteins reveals an ordered assembly pathway of yeast splicing complexes. *Nature*, **324**, 341.
51. Cheng, S. C. and Abelson, J. (1987) Spliceosome assembly in yeast. *Genes Dev.*, **1**, 1014.
52. Grabowski, P. J. and Sharp, P. A. (1986) Affinity chromatography of splicing complexes: U2, U5, and U4+U6 small nuclear ribonucleoprotein particles in the spliceosome. *Science*, **233**, 1294.
53. Pikielny, C. W. and Rosbash, M. (1986) Specific small nuclear RNAs are associated with yeast spliceosomes. *Cell*, **45**, 869.
54. Séraphin, B., Kretzner, L., and Rosbash, M. (1988) A U1 snRNA:pre-mRNA base pairing interaction is required early in yeast spliceosome assembly but does not uniquely define the 5′ cleavage site. *EMBO J.*, **7**, 2533.
55. Green, M. R. (1986) Pre-mRNA splicing. *Annu. Rev. Genet.*, **20**, 671.
56. Lamond, A. I., Konarska, M. M., Grabowski, P. J., and Sharp, P. A. (1988) Spliceosome assembly involves the binding and release of U4 small nuclear ribonucleoprotein. *Proc. Natl Acad. Sci. USA*, **85**, 411.
57. Reed, R. (1989) The organization of 3′ splice-site sequences in mammalian introns. *Genes Dev.*, **3**, 2113.
58. Smith, C. W., Porro, E. B., Patton, J. G., and Nadal-Ginard, B. (1989) Scanning from an independently specified branch point defines the 3′ splice site of mammalian introns. *Nature*, **342**, 243.
59. Nelson, K. K. and Green, M. R. (1989) Mammalian U2 snRNP has a sequence-specific RNA-binding activity. *Genes Dev.*, **3**, 1562.
60. Reed, R. and Maniatis, T. (1988) The role of the mammalian branchpoint sequence in pre-mRNA splicing. *Genes Dev.*, **2**, 1268.
61. Zhuang, Y. and Weiner, A. M. (1989) A compensatory base change in human U2 snRNA can suppress a branch site mutation. *Genes Dev.*, **3**, 1545.
62. Parker, R., Siliciano, P. G., and Guthrie, C. (1987) Recognition of the TACTAAC box during mRNA splicing in yeast involves base pairing to the U2-like snRNA. *Cell*, **49**, 229.
63. Wu, J. and Manley, J. L. (1989) Mammalian pre-mRNA branch site selection by U2 snRNP involves base pairing. *Genes Dev.*, **3**, 1553.
64. Behrens, S. E., Tyc, K., Kastner, B., Reichelt, J., and Lührmann, R. (1993) Small nuclear ribonucleoprotein (RNP) U2 contains numerous additional proteins and has a bipartite RNP structure under splicing conditions. *Mol. Cell. Biol.*, **13**, 307.
65. Fu, X. D. and Maniatis, T. (1992) The 35-kDa mammalian splicing factor SC35 mediates specific interactions between U1 and U2 small nuclear ribonucleoprotein particles at the 3′ splice site. *Proc. Natl Acad. Sci. USA*, **89**, 1725.
66. Krämer, A. and Utans, U. (1991) Three protein factors (SF1, SF3 and U2AF) function in pre-splicing complex formation in addition to snRNPs. *EMBO J.*, **10**, 1503.

67. Brosi, R., Groning, K., Behrens, S. E., Lührmann, R., and Krämer, A. (1993) Interaction of mammalian splicing factor SF3a with U2 snRNP and relation of its 60-kD subunit to yeast PRP9. *Science*, **262**, 102.
68. Brosi, R., Hauri, H. P., and Krämer, A. (1993) Separation of splicing factor SF3 into two components and purification of SF3a activity. *J. Biol. Chem.*, **268**, 17640.
69. Zillmann, M., Rose, S. D., and Berget, S. M. (1987) U1 small nuclear ribonucleoproteins are required early during spliceosome assembly. *Mol. Cell. Biol.*, **7**, 2877.
70. Zillmann, M., Zapp, M. L., and Berget, S. M. (1988) Gel electrophoretic isolation of splicing complexes containing U1 small nuclear ribonucleoprotein particles. *Mol. Cell. Biol.*, **8**, 814.
71. Barabino, S. M., Blencowe, B. J., Ryder, U., Sproat, B. S., and Lamond, A. I. (1990) Targeted snRNP depletion reveals an additional role for mammalian U1 snRNP in spliceosome assembly. *Cell*, **63**, 293.
72. Krämer, A. (1992) Purification of splicing factor SF1, a heat-stable protein that functions in the assembly of a presplicing complex. *Mol. Cell. Biol.*, **12**, 4545.
73. Behrens, S. E., Galisson, F., Legrain, P., and Lührmann, R. (1993) Evidence that the 60-kDa protein of 17S U2 small nuclear ribonucleoprotein is immunologically and functionally related to the yeast PRP9 splicing factor and is required for the efficient formation of prespliceosomes. *Proc. Natl Acad. Sci. USA*, **90**, 8229.
74. Bennett, M. and Reed, R. (1993) Correspondence between a mammalian spliceosome component and an essential yeast splicing factor. *Science*, **262**, 105.
75. Chiara, M. D., Champion-Arnaud, P., Buvoli, M., Nadal-Ginard, B., and Reed, R. (1994) Specific protein–protein interactions between the essential mammalian spliceosome-associated proteins SAP 61 and SAP 114. *Proc. Natl Acad. Sci. USA*, **91**, 6403.
76. Legrain, P. and Chapon, C. (1993) Interaction between PRP11 and SPP91 yeast splicing factors and characterization of a PRP9-PRP11-SPP91 complex. *Science*, **262**, 108.
77. Legrain, P., Chapon, C., and Galisson, F. (1993) Interactions between PRP9 and SPP91 splicing factors identify a protein complex required in prespliceosome assembly. *Genes Dev.*, **7**, 1390.
78. Reed, R., Griffith, J., and Maniatis, T. (1988). Purification and visualization of native spliceosomes. *Cell*, **53**, 949.
79. Lamond, A. I., Konarska, M. M., and Sharp, P. A. (1987) A mutational analysis of spliceosome assembly: evidence for splice site collaboration during spliceosome formation. *Genes Dev.*, **1**, 532.
80. Siliciano, P. G. and Guthrie, C. (1988) 5′ splice site selection in yeast: genetic alterations in base-pairing with U1 reveal additional requirements. *Genes Dev.*, **2**, 1258.
81. Utans, U. and Krämer, A. (1990) Splicing factor SF4 is dispensable for the assembly of a functional splicing complex and participates in the subsequent steps of the splicing reaction. *EMBO J.*, **9**, 4119.
82. Wyatt, J. R., Sontheimer, E. J., and Steitz, J. A. (1992) Site-specific cross-linking of mammalian U5 snRNP to the 5′ splice site before the first step of pre-mRNA splicing. *Genes Dev.*, **6**, 2542.
83. Sawa, H. and Shimura, Y. (1992) Association of U6 snRNA with the 5′-splice site region of pre-mRNA in the spliceosome. *Genes Dev.*, **6**, 244.
84. Sawa, H. and Abelson, J. (1992) Evidence for a base-pairing interaction between U6 small nuclear RNA and 5′ splice site during the splicing reaction in yeast. *Proc. Natl Acad. Sci. USA*, **89**, 11269.
85. Sontheimer, E. J. and Steitz, J. A. (1993) The U5 and U6 small nuclear RNAs as active site components of the spliceosome. *Science*, **262**, 1989.

86. Bach, M., Winkelmann, G., and Lührmann, R. (1989). 20S small nuclear ribonucleoprotein U5 shows a surprisingly complex protein composition. *Proc. Natl Acad. Sci. USA*, **86**, 6038.
87. Bringmann, P. and Lührmann, R. (1986) Purification of the individual snRNPs U1, U2, U5 and U4/U6 from HeLa cells and characterization of their protein constituents. *EMBO J.*, **5**, 3509.
88. Behrens, S. E. and Lührmann, R. (1991) Immunoaffinity purification of a [U4/U6·U5] tri-snRNP from human cells. *Genes Dev.*, **5**, 1439.
89. Garcia-Blanco, M. A., Anderson, G. J., Beggs, J., and Sharp, P. A. (1990) A mammalian protein of 220 kDa binds pre-mRNAs in the spliceosome: a potential homologue of the yeast PRP8 protein. *Proc. Natl. Acad. Sci. USA*, **87**, 3082.
90. Lamm, G. M., Blencowe, B. J., Sproat, B. S., Iribarren, A. M., Ryder, U., and Lamond, A. I. (1991) Antisense probes containing 2-aminoadenosine allow efficient depletion of U5 snRNP from HeLa splicing extracts. *Nucleic Acids Res.*, **19**, 3193.
91. Séraphin, B., Abovich, N., and Rosbash, M. (1991) Genetic depletion indicates a late role for U5 snRNP during in vitro spliceosome assembly. *Nucleic Acids Res.*, **19**, 3857.
92. Brown, J. D. and Beggs, J. D. (1992) Roles of PRP8 protein in the assembly of splicing complexes. *EMBO J.*, **11**, 3721.
93. Blencowe, B. J., Carmo-Fonseca, M., Behrens, S. E., Lührmann, R., and Lamond, A. I. (1993) Interaction of the human autoantigen p150 with splicing snRNPs. *J. Cell. Sci.*, **105**, 685.
94. Utans, U., Behrens, S. E., Luhrmann, R., Kole, R., and Krämer, A. (1992) A splicing factor that is inactivated during *in vivo* heat shock is functionally equivalent to the [U4/U6·U5] triple snRNP-specific proteins. *Genes Dev.*, **6**, 631.
95. Lossky, M., Anderson, G. J., Jackson, S. P., and Beggs, J. (1987) Identification of a yeast snRNP protein and detection of snRNP–snRNP interactions. *Cell*, **51**, 1019.
96. Pinto, A. L. and Steitz, J. A. (1989) The mammalian analogue of the yeast PRP8 splicing protein is present in the U4/5/6 small nuclear ribonucleoprotein particle and the spliceosome. *Proc. Natl Acad. Sci. USA*, **86**, 8742.
97. Anderson, G. J., Bach, M., Lührmann, R., and Beggs, J. D. (1989) Conservation between yeast and man of a protein associated with U5 small nuclear ribonucleoprotein. *Nature*, **342**, 819.
98. Newman, A. and Norman, C. (1991) Mutations in yeast U5 snRNA alter the specificity of 5′ splice-site cleavage. *Cell*, **65**, 115.
99. Newman, A. J. and Norman, C. (1992) U5 snRNA interacts with exon sequences at 5′ and 3′ splice sites. *Cell*, **68**, 743.
100. Lührmann, R., Kastner, B., and Bach, M. (1990) Structure of spliceosomal snRNPs and their role in pre-mRNA splicing. *Biochim. Biophys. Acta*, **1087**, 265.
101. Bringmann, P., Appel, B., Rinke, J., Reuter, R., Theissen, H., and Lührmann, R. (1984) Evidence for the existence of snRNAs U4 and U6 in a single ribonucleoprotein complex and for their association by intermolecular base pairing. *EMBO J.*, **3**, 1357.
102. Hashimoto, C. and Steitz, J. A. (1984) U4 and U6 RNAs coexist in a single small nuclear ribonucleoprotein particle. *Nucleic Acids Res.*, **8**, 3283.
103. Rinke, J., Appel, B., Digweed, M., and Lührmann, R. (1985) Localization of a base-paired interaction between small nuclear RNAs U4 and U6 in intact U4/U6 ribonucleoprotein particles by psoralen cross-linking. *J. Mol. Biol.*, **185**, 721.
104. Brow, D. A. and Guthrie, C. (1988) Spliceosomal RNA U6 is remarkably conserved from yeast to mammals. *Nature*, **334**, 213.

105. Blencowe, B. J., Sproat, B. S., Ryder, U., Barabino, S., and Lamond, A. I. (1989) Antisense probing of the human U4/U6 snRNP with biotinylated 2′-OMe RNA oligonucleotides. *Cell*, **59**, 531.
106. Yean, S. L. and Lin, R. J. (1991) U4 small nuclear RNA dissociates from a yeast spliceosome and does not participate in the subsequent splicing reaction. *Mol. Cell. Biol.*, **11**, 5571.
107. Madhani, H. D. and Guthrie, C. (1992) A novel base-pairing interaction between U2 and U6 snRNAs suggests a mechanism for the catalytic activation of the spliceosome. *Cell*, **71**, 803.
108. Krainer, A. R. and Maniatis, T. (1985) Multiple factors including the small nuclear ribonucleoproteins U1 and U2 are necessary for pre-mRNA splicing *in vitro*. *Cell*, **42**, 725.
109. Abmayr, S. M., Reed, R., and Maniatis, T. (1988) Identification of a functional mammalian spliceosome containing unspliced pre-mRNA. *Proc. Natl. Acad. Sci. USA*, **85**, 7216.
110. Sawa, H., Ohno, M., Sakamoto, H., and Shimura, Y. (1988) Requirement of ATP in the second step of the pre-mRNA splicing reaction. *Nucleic Acids Res.*, **16**, 3157.
111. Reed, R. and Maniatis, T. (1985) Intron sequences involved in lariat formation during pre-mRNA splicing. *Cell*, **41**, 95.
112. Ruskin, B. and Green, M. R. (1985) Role of the 3′ splice site consensus sequence in mammalian pre-mRNA splicing. *Nature*, **317**, 732.
113. Aebi, M., Hornig, H., Padgett, R. A., Reiser, J., and Weissmann, C. (1986) Sequence requirements for splicing of higher eukaryotic nuclear pre-mRNA. *Cell*, **47**, 555.
114. Ruskin, B., Greene, J. M., and Green, M. R. (1985) Cryptic branch point activation allows accurate *in vitro* splicing of human beta-globin intron mutants. *Cell*, **41**, 833.
115. Padgett, R. A., Konarska, M. M., Aebi, M., Hornig, H., Weissmann, C., and Sharp, P. A. (1985) Nonconsensus branch-site sequences in the *in vitro* splicing of transcripts of mutant rabbit beta-globin genes. *Proc. Natl Acad. Sci. USA*, **82**, 8349.
116. Hornig, H., Aebi, M., and Weissmann, C. (1986) Effect of mutations at the lariat branch acceptor site on beta-globin pre-mRNA splicing *in vitro*. *Nature*, **324**, 589.
117. Hartmuth, K. and Barta, A. (1988) Unusual branch point selection in processing of human growth hormone pre-mRNA. *Mol. Cell. Biol.*, **8**, 2011.
118. Newman, A. J., Lin, R. J., Cheng, S. C., and Abelson, J. (1985) Molecular consequences of specific intron mutations on yeast mRNA splicing *in vivo* and *in vitro*. *Cell*, **42**, 335.
119. Aebi, M., Hornig, H., and Weissmann, C. (1987) 5′ cleavage site in eukaryotic pre-mRNA splicing is determined by the overall 5′ splice region, not by the conserved 5′ GU. *Cell*, **50**, 237.
120. Parker, R. and Guthrie, C. (1985) A point mutation in the conserved hexanucleotide at a yeast 5′ splice junction uncouples recognition, cleavage, and ligation. *Cell*, **41**, 107.
121. Smith, C. W., Chu, T. T., and Nadal-Ginard, B. (1993) Scanning and competition between AGs are involved in 3′ splice site selection in mammalian introns. *Mol. Cell. Biol.*, **13**, 4939.
122. Patterson, B. and Guthrie, C. (1991) A U-rich tract enhances usage of an alternative 3′ splice site in yeast. *Cell*, **64**, 181.
123. Lesser, C. F. and Guthrie, C. (1993) Mutations in U6 snRNA that alter splice site specificity: implications for the active site. *Science*, **262**, 1982.
124. Kandels-Lewis, S. and Séraphin, B. (1993) Role of U6 snRNA in 5′ splice site selection. *Science*, **262**, 2035.

125. Winkelmann, G., Bach, M., and Lührmann, R. (1989) Evidence from complementation assays *in vitro* that U5 snRNP is required for both steps of mRNA splicing. *EMBO J.*, **8**, 3105.
126. Lin, R. J., Lustig, A. J., and Abelson, J. (1987) Splicing of yeast nuclear pre-mRNA in vitro requires a functional 40S spliceosome and several extrinsic factors. *Genes Dev.*, **1**, 7.
127. Kim, S. H., Smith, J., Claude, A., and Lin, R. J. (1992) The purified yeast pre-mRNA splicing factor PRP2 is an RNA-dependent NTPase. *EMBO J.*, **11**, 2319.
128. Chen, J. H. and Lin, R. J. (1990) The yeast PRP2 protein, a putative RNA-dependent ATPase, shares extensive sequence homology with two other pre-mRNA splicing factors. *Nucleic Acids Res.*, **18**, 6447.
129. King, D. S. and Beggs, J. D. (1990) Interactions of PRP2 protein with pre-mRNA splicing complexes in *Saccharomyces cerevisiae*. *Nucleic Acids Res.*, **18**, 6559.
130. Kim, S. H. and Lin, R. J. (1993) Pre-mRNA splicing within an assembled yeast spliceosome requires an RNA-dependent ATPase and ATP hydrolysis. *Proc. Natl Acad. Sci. USA*, **90**, 888.
131. Burgess, S., Couto, J. R., and Guthrie, C. (1990) A putative ATP binding protein influences the fidelity of branchpoint recognition in yeast splicing. *Cell*, **60**, 705.
132. Schwer, B. and Guthrie, C. (1991) PRP16 is an RNA-dependent ATPase that interacts transiently with the spliceosome. *Nature*, **349**, 494.
133. Schwer, B. and Guthrie, C. (1992) A conformational rearrangement in the spliceosome is dependent on PRP16 and ATP hydrolysis. *EMBO J.*, **11**, 5033.
134. Burgess, S. M. and Guthrie, C. (1993) A mechanism to enhance mRNA splicing fidelity: the RNA-dependent ATPase Prp16 governs usage of a discard pathway for aberrant lariat intermediates. *Cell*, **73**, 1377.
135. Horowitz, D. S. and Abelson, J. (1993) A U5 small nuclear ribonucleoprotein particle protein involved only in the second step of pre-mRNA splicing in *Saccharomyces cerevisiae*. *Mol. Cell. Biol.*, **13**, 2959.
136. Frank, D., Patterson, B., and Guthrie, C. (1992) Synthetic lethal mutations suggest interactions between U5 small nuclear RNA and four proteins required for the second step of splicing. *Mol. Cell. Biol.*, **12**, 5197.
137. Sharp, P. A. (1985) Splicing of messenger RNA precursors. *Harvey Lect.*, **81**, 1.
138. Guthrie, C. and Patterson, B. (1988) Spliceosomal snRNAs. *Annu. Rev. Genet.*, **22**, 387.
139. Brow, D. A. and Guthrie, C. (1989) Splicing a spliceosomal RNA. *Nature*, **337**, 14.
140. Fabrizio, P. and Abelson, J. (1990) Two domains of yeast U6 small nuclear RNA required for both steps of nuclear precursor messenger RNA splicing. *Science*, **250**, 404.
141. Yu, Y. T., Maroney, P. A., and Nilsen, T. W. (1993) Functional reconstitution of U6 snRNA in nematode *cis-* and *trans*-splicing: U6 can serve as both a branch acceptor and a 5′ exon. *Cell*, **75**, 1049.
142. Ruby, S. W., Chang, T. H., and Abelson, J. (1993) Four yeast spliceosomal proteins (PRP5, PRP9, PRP11, and PRP21) interact to promote U2 snRNP binding to pre-mRNA. *Genes Dev.*, **7**, 1909.

5 | snRNP structure and function

CINDY L. WILL and REINHARD LÜHRMANN

1. Introduction

In contrast to self-splicing RNA molecules which contain extensive, highly conserved structural elements sufficient for intron excision and exon ligation, nuclear pre-mRNA molecules possess short, conserved sequences which are limited to the 5′ and 3′ splice site and branchpoint regions (see Chapters 1, 4, and 8). While these sequences are necessary for the conversion of the pre-mRNA to mRNA, they are not sufficient. Nuclear pre-mRNA splicing is thus dependent upon the activity of a number of *trans*-acting splicing factors, which in addition to protein factors, include the four small nuclear ribonucleoprotein particles (snRNPs) U1, U2, U5, and U4/U6. These evolutionarily conserved RNA–protein complexes interact stepwise with nuclear pre-mRNA molecules to form the large multicomponent RNP complex (the spliceosome) in which splicing occurs (see Chapter 4). U1, U2, U5, and U4/U6 snRNPs play a central role in the recognition and alignment of splice sites. In addition, snRNPs may directly contribute to the catalytic activity of spliceosomes: snRNA constituents have been proposed to form the catalytic centre responsible for excision–ligation events.

While a clearer picture of the role of snRNPs in the splicing process is slowly emerging, many aspects of snRNP function, for example the contributions of the various snRNAs and snRNP proteins to individual steps of the splicing process at the molecular level, require additional investigation. Much progress has been made in elucidating base-pairing interactions of individual snRNAs with the pre-mRNA molecule and/or with other snRNA species within the spliceosome. This network of RNA–RNA interactions may provide the structural framework required for the catalysis of pre-mRNA splicing. As a prerequisite to ultimately elucidating functional aspects of the snRNPs, much effort has also been focused on the biochemical characterization of snRNP particles and on the investigation of their higher order structure. In recent years, significant advances have been made in identifying and characterizing the functional forms of several spliceosomal snRNPs, including the 17S U2 snRNP and 25S U4/U6·U5 tri-snRNP. The protein composition of these newly characterized snRNP forms has proven unexpectedly complex. In addition, while the function of the majority of the snRNP proteins remains elusive, recent studies have demonstrated essential roles for several snRNP proteins at early stages of the splicing process.

In this chapter we first describe what is currently known about the biochemical composition of the spliceosomal snRNPs. Since U1, U2, U5, and U4/U6 snRNPs are relatively more abundant in higher eukaryotic cells (2×10^5–10^6 particles per cell, compared with 50–100 in yeast), significantly more biochemical and structural information is available and, therefore, results obtained from higher eukaryotes will be emphasized. The interactions between snRNAs and snRNP proteins are subsequently discussed in terms of snRNP morphogenesis and transport, followed by current information regarding snRNP higher order structure. Lastly, we discuss the various roles of the individual snRNP particles in the splicing process, as investigated both in mammalian and yeast splicing systems.

2. Structure of the spliceosomal snRNAs

The spliceosomal snRNPs are named according to their RNA component—U1, U2, U5, or U4 and U6 snRNA; the latter two species are most often found together in the same ribonucleoprotein complex. The snRNA class of RNA molecules includes several additional non-spliceosomal species which are generally much less abundant and appear to be involved in other RNA maturation processes (see 1 for review). The spliceosomal snRNAs are characterized by their small size, metabolic stability, and high degree of sequence conservation. They are also highly modified, containing pseudouridine and methylated bases such as, N^6-methyladenosine, as well as a unique 2,2,7-trimethylguanosine 5′ cap structure (2). U6 snRNA, which possesses a monomethyl phosphate cap, represents an exception to the latter rule, which reflects the difference in the RNA polymerase responsible for its synthesis (3). That is, in contrast to U1, U2, U4, and U5 snRNA genes, which are transcribed by RNA polymerase II, the U6 gene is transcribed by RNA polymerase III (4–6).

The primary sequence and most probable secondary structure of the human spliceosomal snRNAs are shown in Figs 1 and 2. Aside from these predominant forms, several spliceosomal snRNA variants have been described, which include both constitutively expressed and tissue-specific or developmentally regulated forms. For example, at least 10 U5 snRNA variants have been detected in HeLa cells (7–9) and embryonic and adult forms of U1 have been observed in mice (10). The function of these snRNA variants remains unclear. However, it is tempting to speculate that they may play a role in alternative splicing events (for a more detailed discussion see 11).

2.1 Primary structure of snRNAs

The sequences of U1, U2, U4, U5, and U6 snRNAs are highly conserved, with differences of less than 10% observed when comparing mammals and birds. Although greater sequence variations, in particular differences in size, are observed between more evolutionarily distant organisms, sequence conservation remains high in a number of snRNA regions known to be functionally important. Generally, these highly conserved regions are mostly single-stranded in secondary structure models

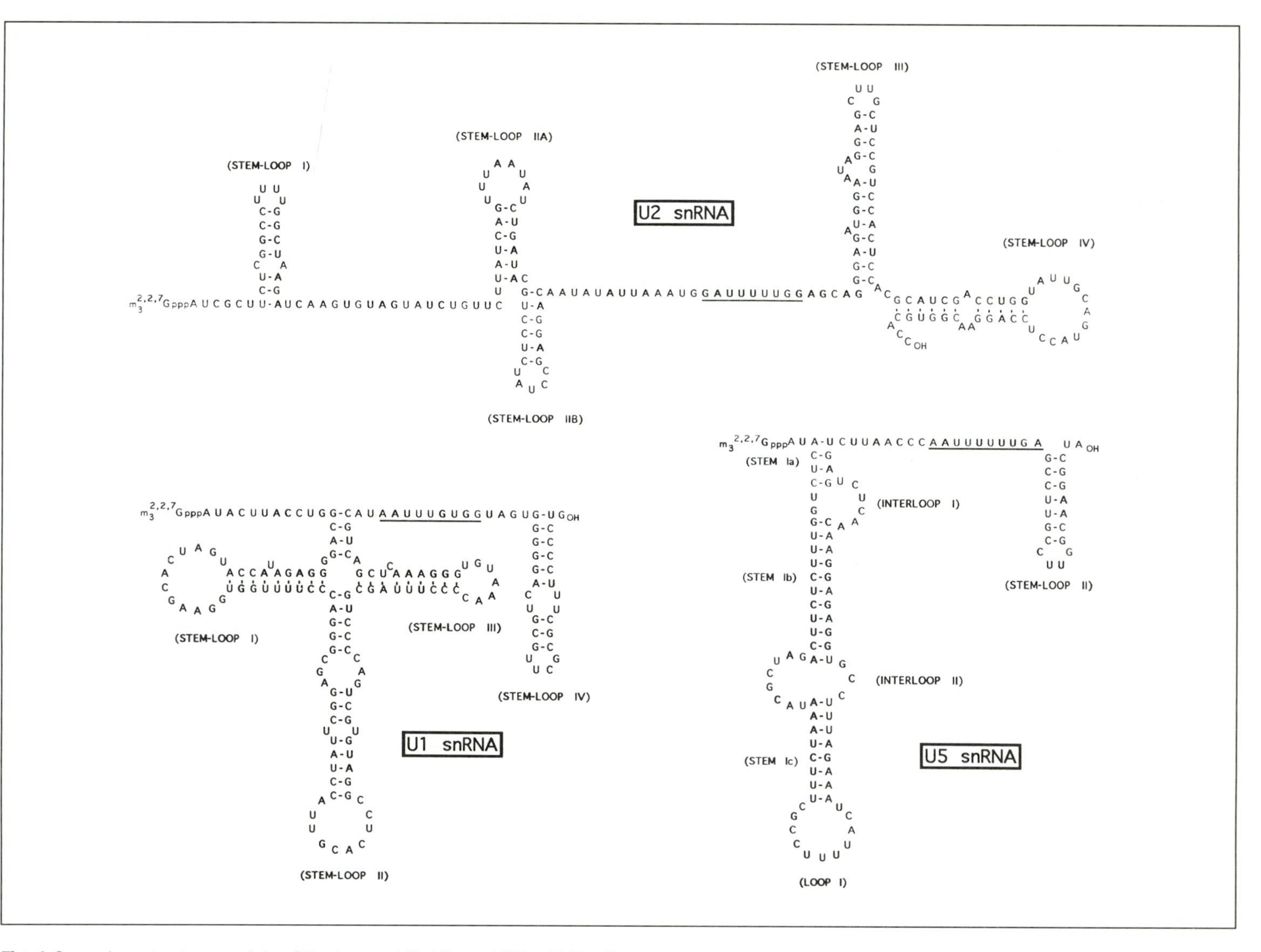

Fig. 1 Secondary structure models of the human U1, U2, and U5 snRNAs. The consensus secondary structures of U1 and U5 are shown according to the models of Guthrie and Patterson (17), and that of U2, as proposed by Ares and Igel (190). The conserved sequences of the Sm site are underlined.

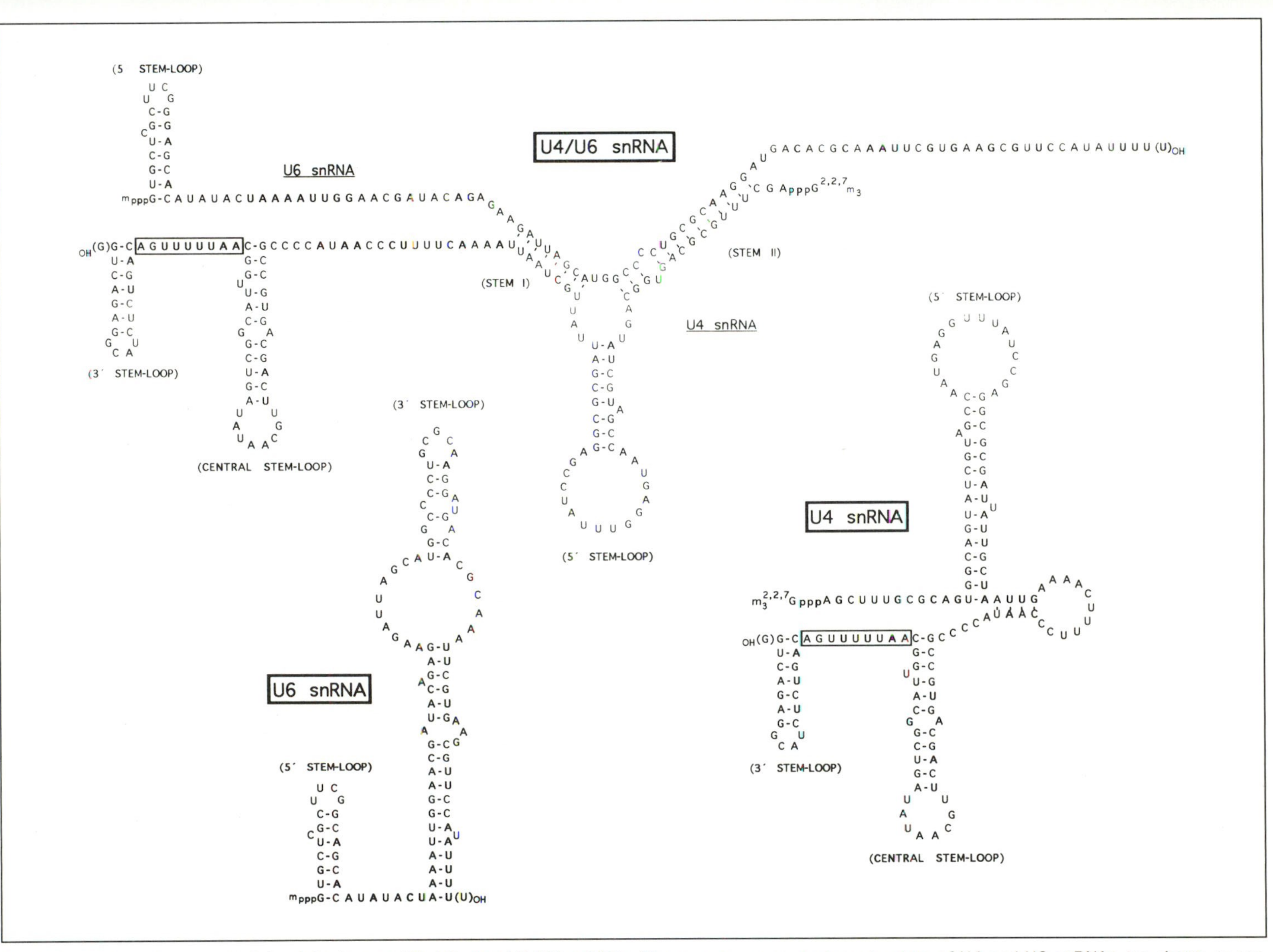

Fig. 2 Secondary structure models of the human U4, U6, and U4/U6 snRNAs. The consensus secondary structure of U4 and U6 snRNAs are shown as proposed by Myslinski and Branlant (191) and Rinke *et al.* (192), respectively, whereas that of U4/U6 is according to the model of Guthrie and Patterson (17). The conserved sequences of the U4 Sm site are boxed.

Table 1 Spliceosomal snRNA sizes

snRNA	Size (nucleotides) Mammals	*S. cerevisiae*
U1	164	568
U2	187	1175
U4	145	160
U5(L)	116	214
(S)	—	179
U6	106	112

and serve either as protein binding sites or to base pair with either pre-mRNA sequences or other snRNAs in the spliceosome (see Section 8). One example of a conserved protein binding region is the Sm site, a single-stranded U-rich sequence, often flanked by two hairpins, which is found in U1, U2, U4, and U5 snRNAs and serves as a binding site for proteins common to each of the spliceosomal snRNPs (see Section 7.1). The conserved, single-stranded Sm site consensus is also present in yeast snRNAs. An example of a conserved single-stranded region that interacts with pre-mRNA is the 5′ end of U1 snRNA; a base-pairing interaction between nucleotides 3–10 of U1 and the 5′ splice site junction contributes to 5′ splice site recognition (discussed in Section 8.1.1). Additional functionally important snRNA sequences are discussed in more detail in the context of snRNA/snRNP function (see Section 8).

Although significant size differences are observed when comparing metazoan U1, U2, and U5 snRNAs with those of the budding yeast, *Saccharomyces cerevisiae* (Table 1), these differences are often accounted for by variations in a limited number of RNA domains. Moreover, most regions exhibiting extensive size variation are functionally dispensable; for example, deletion of the majority of those regions specific to the yeast U1 or U2 snRNA has no effect on cell viability (12–14). In addition, the lethal phenotype of a yeast U2 snRNA gene deletion can be complemented by a minimally altered human U2 gene (15). In contrast to U1, U2, and U5, the length of U4 and U6 snRNA is conserved through evolution. Comparisons of both the sequence and size of U6 snRNA have established that it is one of the most extensively conserved RNA species thus far identified (16); this observation forms the basis for the proposed central role of U6 snRNA in nuclear pre-mRNA splicing (discussed in Section 8.3.3).

2.2 Secondary structure of snRNAs

Despite some primary sequence divergence, the secondary structure of the metazoan spliceosomal snRNAs is highly conserved. Based on extensive sequence comparisons, a consensus secondary structure model has been generated for U1, U2, and U5, as well as for U4 and U6; the general structure of metazoan snRNAs can be inferred from those of the human snRNAs (Figs 1 and 2). While free U4 and U6 snRNAs exist, the bulk of these RNAs interact by extensive base pairing within the U4/U6 snRNP. The base-paired region forms a characteristic Y-shaped structure, consisting of two intermolecular helices (stem I and II) separated by an intramolecular helix of U4 snRNA (17).

Although distinct secondary structures have been presented (Figs 1 and 2), the overall structure of some of the snRNAs should not be viewed as fixed, but rather as dynamic, with potentially multiple configurations. The preferred structure of a given snRNA may be dependent upon its association with snRNP proteins and/or its association with other snRNAs or with the pre-mRNA molecule itself. This principle is best illustrated by U6 snRNA, which undergoes several conformational transitions, first during the assembly of the U4/U6 snRNP particle and later at several distinct stages of the splicing process (see Section 8; Fig. 8). The conformational state of the U4 snRNA, and the 5′ half of the U2 snRNA molecule, are also dynamic in nature (see Fig. 8). (For a more detailed discussion of structural and evolutionary aspects of the snRNAs, see 17.)

3. Organization of snRNAs as RNPs

The spliceosomal snRNAs are not biologically active as such, but rather associate with a number of proteins to form discrete ribonucleoprotein complexes. The RNP nature of the snRNAs was first indicated by their sedimentation coefficients in glycerol gradients, which are significantly greater than naked RNA (18, 19). Their RNP nature was later confirmed by immunoprecipitation studies with autoantibodies obtained from patients suffering from systemic lupus erythematosus (20). The antigenic determinants recognized by such antibodies are located on a subset of the proteins associated with the snRNAs, resulting in co-precipitation of both RNP components. Through the use of immunoaffinity and biochemical techniques, individual spliceosomal snRNPs have since been isolated and their protein composition determined. (For a summary of the approaches used to purify snRNPs in the past, see 21, 22.)

Refinements in snRNP isolation techniques have demonstrated that the protein composition of the spliceosomal snRNPs is highly dependent on ionic strength. This is reflected by salt-dependent alterations in the sedimentation behaviour of several snRNP species during sucrose or glycerol gradient centrifugation. For example, at moderate salt concentrations (300 mM) U1, U2, and U4/U6 sediment as 12S particles, and U5 as a 20S particle. Under low-salt conditions (approaching *in vitro* splicing conditions), a 17S U2 and 25S U4/U6·U5 tri-snRNP complex can be isolated (23–25). The 17S and 25S snRNP complexes are likely to represent the active forms of these spliceosomal snRNAs. That is, U2 snRNA, and U4/U6 and U5 snRNAs, appear to enter the spliceosome in the form of a 17S and 25S RNP particle, respectively. In this sense (by analogy to ribosomes), they may be considered functional spliceosomal subunits.

4. snRNP proteins

4.1 Human snRNP proteins

The protein composition of human (HeLa cell) snRNP particles has been most extensively studied (Table 2). Individual 12S U1, U2, and U4/U6 snRNPs, as well as 20S U5 snRNPs, can be isolated from HeLa nuclear extracts first by immunoaffinity chromatography, utilizing antibodies directed against the unique snRNA cap structure (m_3G), followed by Mono Q ion exchange chromatography (26). However, as the Mono Q resin tends to dissociate weakly bound snRNP proteins, alternative purification schemes, such as multiple immunoaffinity chromatography steps, have been employed to isolate the 17S U2 snRNP and 25S U4/U6·U5 (23, 25).

4.1.1 Common snRNP proteins

snRNP proteins can generally be divided into two classes: those which are common to all spliceosomal species and those which are associated with a given snRNP particle or complex. The common snRNP proteins comprise a group of polypeptides whose interaction with U1, U2, U4, or U5 snRNA is maintained even under relatively harsh conditions. This tightly associated group of proteins includes eight polypeptides denoted B, B′, D1, D2, D3, E, F, and G, whose molecular weights range from 9 to 29 kDa (Table 2). A subset of these proteins—B, B′, D1, D2, and D3—possess the main antigenic determinants recognized by anti-Sm antibodies (27, 28). For this reason, the common snRNP proteins are often also referred to as the Sm proteins.

An additional common protein with an apparent molecular weight of 69 kDa has recently been identified (29). This newly characterized protein is less tightly associated with the snRNP core domain than the other common proteins are, and appears to be transiently associated with a number of snRNP species. The transient nature of the 69 kDa protein interaction was inferred from the following observations. The 69 kDa protein associates with the core structure of all spliceosomal snRNAs, either under *in vitro* reconstitution conditions or upon injection of the snRNAs into the cytoplasm of *Xenopus* oocytes. However, when snRNPs are isolated from HeLa nuclear extracts, significant amounts of the 69 kDa protein are found in association with U1 snRNP, but little or none with 17S U2 or 25S U4/U6·U5 snRNPs. The apparent transient association of this protein has led to speculation that it may play a role in the metabolism of snRNP particles.

4.1.2 Particle-specific proteins

Particle-specific proteins exhibit a wide range of binding affinities and the variations in their association behaviour account for the appearance of the different forms of the spliceosomal snRNPs. The HeLa U1 snRNA is consistently isolated in association with eight common snRNP proteins and three particle-specific proteins denoted 70K, A, and C (Table 2) which are the best characterized of the mammalian snRNP proteins. The Hela U2 snRNP had previously been isolated solely as a 12S

Table 2 Protein composition of HeLa snRNPs

Protein	Apparent M_r (kDa)	Presence in snRNP particles					
		12S (U1)	12S (U2)	17S (U2)	20S (U5)	25S (U4/U6·U5)	12S (U4/U6)
G	9	•	•	•	•	•	•
F	11	•	•	•	•	•	•
E	12	•	•	•	•	•	•
D1	16	•	•	•	•	•	•
D2	16,5	•	•	•	•	•	•
D3	18	•	•	•	•	•	•
B	28	•	•	•	•	•	•
B′	29	•	•	•	•	•	•
C	22	•					
A	34	•					
70K	70	•					
B″	28,5		•	•			
A′	31		•	•			
	33			•			
	35			•			
	53			•			
	60			•			
	66			•			
	92			•			
	110			•			
	120			•			
	150			•			
	160			•			
	15				•	•	
	40				•	•	
	52				•	•	
	100				•	•	
	102				•	•	
	110				•	•	
	116				•	•	
	200				•	•	
	220				•	•	
	60				•		•
	90				•		•
	15,5				•		
	20				•		
	27				•		
	61				•		
	63				•		

particle which contained, in addition to the common proteins, two specific proteins denoted A′ and B″. However, under splicing conditions (100 mM salt), a considerably more complex U2 snRNP particle is observed containing 10 additional proteins with molecular masses ranging from 33 to 160 kDa (23) (Table 2). U4/U6 snRNPs isolated from HeLa nuclear extracts by immunoaffinity chromatography and ion exchange chromatography contain only the common snRNP proteins. The absence of specific proteins in these particles appears to reflect the harshness of the isolation procedure. Studies investigating the fate of proteins upon disruption of the HeLa U4/U6·U5 tri-snRNP complex indicate that several tri-snRNP proteins, including the 60 kDa and 90 kDa proteins, associate preferentially with the U4/U6 snRNP (J. Lauber and R. Lührmann, in preparation). Similarly, recent studies using a monoclonal antibody which specifically precipitates complexes containing U4/U6 snRNA have demonstrated that a 60 kDa and a 90 kDa spliceosomal-associated protein (SAPs 60 and 90; see Chapter 4) are U4/U6-specific proteins (30).

In contrast to immunoaffinity-purified U4/U6 snRNPs, purified 20S U5 RNP particles possess a relatively complex protein composition (31); a total of eight U5-specific proteins with molecular masses ranging from 15 to 220 kDa have been identified (Table 2). These proteins remain associated with U5 upon its interaction with U4/U6 to form the 25S tri-snRNP complex. This complex contains five novel proteins with molecular masses ranging from 15.5 to 63 kDa (25) (Table 2) which are not present in 12S U4/U6 or 20S U5 particles. One or more of the tri-snRNP-specific proteins probably mediates the association of U4/U6 and U5, since isolated 12S U4/U6 and 20S U5 particles do not form a stable 25S complex *in vitro* (25). Interestingly, purified human U4/U6·U5 complexes possess a kinase activity that phosphorylates the 52 kDa U5-specific protein; the identity of the kinase and, more importantly, its role in the morphogenesis or function of the tri-snRNP complex remain to be established (25).

At present, it is not clear whether the majority of the U4/U6 and U5 snRNPs represent nascent particles that subsequently associate to form the tri-snRNP complex. Alternatively, they may represent tri-snRNP dissociation products that occur naturally, or as the result of the snRNP isolation procedure. Although formal proof is lacking, a dissociation–reassociation cycle has been hypothesized for both the tri-snRNP complex and the U4/U6 snRNP particle for each round of splicing (32). Since U4 and U6 snRNAs initially exist as monomeric RNP complexes (33), but subsequently interact with each other and ultimately with U5 to form the tri-snRNP complex, the proteins associated with them may vary depending on their RNP status. The majority of proteins identified in isolated HeLa snRNPs are also present in the spliceosome (discussed in Chapter 4). Although proteins associated with a given isolated particle are operationally defined as particle-specific, it is conceivable that some of these proteins may also associate with other snRNAs within the spliceosome. Thus, their interaction with a given snRNA during splicing may be dynamic in nature.

Several additional snRNP-associated proteins have been reported which are not detected in isolated HeLa snRNPs by Coomassie blue staining. These generally represent species that are present in sub-stoichiometric amounts and are detected solely by immunoprecipitation or by their enzymatic activity. Sub-stoichiometric amounts of a 150 kDa protein are found in association with 17S U2, 20S U5, and 25S U4/U6·U5 particles (34); as yet, it is not clear whether this protein is only transiently associated with this group of snRNPs during their assembly in the nucleus. An unidentified protein bound specifically to U6 snRNA has also been detected in both HeLa and trypanosome cell extracts. This protein appears to interact transiently with U6 (it is not present in U4/U6 complexes) and has been proposed to be involved in U6 snRNP maturation and transport (35). A 120 kDa U4/U6 snRNP-associated protein has also been identified in HeLa cell extracts by immunoprecipitation studies with autoimmune serum (36). Lastly, a minor unidentified protein, which phosphorylates the arginine/serine-rich (RS-rich) region of the U1-70K protein, has been found in association with immunoaffinity purified U1, U2, U5, and U4/U6 snRNPs (37). This kinase appears to be predominantly associated with U1 snRNPs and has been proposed to phosphorylate RS domains common to a number of splicing factors.

4.2 snRNP proteins in other species

The protein composition of the spliceosomal snRNPs appears to be remarkably conserved among eukaryotes. Comparison of those particles containing the more tightly associated, particle-specific proteins (the 20S U5, 12S U1 and U2 snRNPs) from a wide variety of organisms indicates that both the number and molecular weights of the spliceosomal snRNP proteins are evolutionarily conserved. Characterization of purified *Droshophila melanogaster* snRNPs, for example, identified proteins of similar molecular weight corresponding to most of the characterized HeLa 12S U1 and U2, and 20S U5 snRNP proteins (38). Even the protein composition of plant and trypanosomal spliceosomal snRNPs is strikingly similar to that of mammals (see 39, 40 for review). In mice, the majority of identified snRNP proteins even exhibit similar pI values and appear to be post-translationally modified in a manner identical to that of HeLa snRNP proteins (41). One notable exception, however, is the absence of the B′ protein in murine spliceosomal snRNPs (41, 42).

Characterization of snRNP proteins in other organisms has also revealed snRNP-associated proteins thus far not identified in HeLa cells. Significant amounts of a U6 snRNP have been observed in germ cells and early embryos of *Xenopus laevis*, but only a low level of this particle is present in somatic cells. The U6 snRNA present in these RNP particles appears to associate solely with a 50 kDa protein (33).

4.2.1 Yeast snRNP proteins

Compared with mammalian systems, little is known about the protein components of yeast snRNPs. The paucity of biochemical information is mainly due to the relatively low abundance of the spliceosomal snRNPs in yeast. Recently, however, preparative amounts of U1 snRNP and U4/U6·U5 tri-snRNP have been isolated from *S. cerevisiae* cell extracts by a combination of immunoaffinity chromatography and glycerol gradient centrifugation. The initial biochemical characterization of these yeast spliceosomal snRNPs has demonstrated the existence of both common and particle-specific proteins. Between eight and 10 proteins common to both U1 and U4/U6·U5 snRNPs were observed, in addition to several apparent particle-specific proteins (43). Interestingly, the yeast U1 snRNP possesses at least seven particle-specific proteins (in contrast to only three in HeLa cells) and sediments in glycerol gradients as an 18S particle (43). The presence of additional U1-specific proteins may explain apparent differences observed between yeast and mammalian splicing with respect to early steps of spliceosome assembly (see Section 8.1.2).

In yeast, several snRNA-associated proteins have been identified through genetic techniques, including classical screens for temperature-sensitive or second-site-suppressor mutants. A number of pre-mRNA processing (PRP) mutants have been shown to be components of the spliceosomal snRNPs, based on immunoprecipitation studies with corresponding anti-PRP antisera. For example, two U4/U6 snRNP-associated proteins, PRP4 and PRP6, which are also present in the yeast tri-snRNP complex, have been shown to be required for tri-snRNP formation (44, 45). A transiently associated U6-specific protein, designated PRP24, which appears to mediate the association of U4 with U6 has also been found in *S. cerevisiae* (46). Lastly, two U5-specific proteins, PRP8 and PRP18, have been identified; both proteins are also present in the yeast tri-snRNP complex (47, 48). A direct association of PRP9 with yeast U2 snRNP could not be demonstrated by immunoprecipitation with anti-PRP9 serum. However, by analogy with its human homologue, the 60 kDa 17S U2-specific protein, PRP9 may be a loosely associated U2 snRNP protein, whose association does not withstand immunoprecipitation (49). A similar situation holds for PRP11 and PRP21 which are probable homologues of the 17S U2-specific 66 kDa and 110 kDa proteins (discussed in Chapter 7). Based on sequence comparisons, the yeast homologues of the metazoan U1-70K, U1-A, and the common D1 protein, denoted SNP1, MUD1, and SMD1, respectively, have also been identified (50–52). For more detailed discussion of yeast snRNP proteins see Chapter 7.

5. Structure of snRNP proteins

5.1 Structural motifs

Biochemical and molecular characterization of a number of snRNP proteins has revealed several interesting structural features. Only a small subset of mammalian snRNP proteins have been cloned. These include eight of the common proteins—E (53), F and G (194), D1 (54), D2 and D3 (55), and B/B′ (56, 57)—as well as C (58, 59), B″ (60), A (61), A′ (62), 70K (63–65), and the 60 kDa and 66 kDa U2-specific proteins (also called SAP 61 and SAP 62, respectively) (66, 67). 70K, A, and B″ possess one (70K) or two (A and B″) highly conserved RNA binding domains characteristic of a number of RNA-binding proteins. This domain, which has been called the RNA-recognition motif (RRM), the consensus sequence RNA-binding domain, or the RNP motif, consists of an 80 amino acid conserved sequence that contains two short, highly conserved regions denoted the RNP-1 and RNP-2 consensus (see 68 for review). The molecular structure of the N-terminal RNA-binding domain of the A protein has been investigated by X-ray crystallography and nuclear magnetic resonance studies. This functionally important region contains a four-stranded β-sheet and two α-helices which are packed against one face of the β-sheet (69, 70). The RNP-1 and RNP-2 consensus sequences are juxtaposed on the adjacent middle two β-strands (69, 70). In addition to the A and B″ protein, the U6-specific PRP24 protein from yeast also contains multiple (in this case, three) RNA-binding domains (46).

Several additional snRNP protein structural motifs have been identified. The metazoan U1-70K protein contains two mixed-charge amino acid clusters, particularly rich in arginine and serine, which are similar to those observed in a number of splicing factors including SF2/ASF (71, 72), U2AF (73), SC35 (74), and several alternative splicing factors in *Drosophila* (reviewed in 75, 76). Because of the highly basic nature of such RS-rich domains, they have been postulated to mediate binding to RNA through electrostatic interactions or interactions with other proteins (see also Section 8.1.3). The U1-C protein contains cysteine and histidine residues near the amino terminus, which can be arranged to form a zinc finger-like motif of the CC-HH type (77). The U2-specific 60 kDa and 66 kDa proteins and their probable yeast homologues, PRP6 and PRP9, also possess a similar zinc finger-like motif; mutation of some of the residues in this region has a critical effect on PRP6 and PRP9 function (78). It is not clear, however, whether zinc binds to these regions. The latter two proteins, as well as the U2-specific A′ protein, also possess leucine-rich motifs (62, 78). Leucine repeat motifs have been implicated in homo- and hetero-dimerization processes, so it is perhaps not surprising that the interaction between A′ and B″ (discussed in Section 7.3) involves the leucine-rich region of A′ (79). The Sm proteins, while lacking all of the above motifs, are particularly rich in proline and glycine, and, with the exception of the F protein, are basic in nature (41).

5.2 Post-translational modifications

At least one snRNP protein, namely 70K, is post-translationally modified. *In vivo* phosphorylation studies demonstrated that the U1-70K protein is phosphorylated at multiple serine residues in its RS domain (37). Interestingly, the 70K protein can be separated into 13 isoelectric variants, each of which is phosphorylated to a similar extent (37). Post-translational modification of this protein appears to be important for its function during splicing (see Section 8.1.3). Although the nature of their modification is unknown (phosphorylation has been ruled out by the above mentioned studies), evidence for the post-translational modification of other snRNP proteins, including C and A′, has also been reported (41, 80).

5.3 Structural similarities among snRNP proteins

Sequence comparisons have demonstrated structural similarities between some of the cloned mammalian snRNP proteins. For example, sequence comparison of the eight human Sm proteins (E, F, G, D1, D2, D3, B, and B′) has revealed two regions of homology (termed Sm boxes 1 and 2) which are shared by all the Sm proteins (194). These homology boxes are evolutionarily conserved and may be involved in protein–protein interactions which occur among the Sm proteins (126, 194, 195). Regions of extensive similarity are also found at the carboxy and amino termini of A and B″, suggesting that they may have a common evolutionary origin (60, 61).

In HeLa cells, B and B′ differ only at their carboxy termini, where B′ contains an additional proline-rich sequence of nine amino acids (57). Although both proteins are encoded by the same gene, an alternative splicing event generates the shorter B polypeptide (81). Interestingly, the isolation of B/B′ cDNAs from a variety of human and rodent cell types has revealed three closely related B/B′ isotypes (for a detailed discussion see 82). The functional significance of the tissue-specific expression of these B/B′ variants is not clear. On the basis of a strong correlation between the presence of one of these isoforms (referred to as the N protein) and the presence of an alternatively spliced product of the calcitonin gene-related peptide (CGRP), it has been proposed that this B/B′ isoform may play a role in the regulation of calcitonin/CGRP pre-mRNA alternative splicing (83–85).

5.4 Conservation of snRNP protein structure

As in the case of their snRNA counterparts, the structures of a number of snRNP proteins appear to be evolutionarily conserved, suggesting these proteins play an important role in snRNP morphogenesis and/or function. For example, a high level of amino acid sequence conservation is observed between the U1-70K genes of humans, *Xenopus laevis*, *Drosophila*, and plants (86; see 40 for detailed discussion). Significant sequence homology (30–40% amino acid identity) is observed between the human 70K, A, D1, and D3 proteins and their yeast equivalents (50–52, 55). Significantly, the human D1 protein can functionally replace its yeast homologue; a human D1 cDNA has been shown to complement an otherwise lethal D1 null allele in *S. cerevisiae* (87). More recently, conservation of both the structure and function of the human U2-specific 60 kDa, 66 kDa, and 114 kDa proteins and their yeast equivalents (PRP9, PRP11, and PRP21, respectively) has been demonstrated (66, 67, 88). Additional evidence for primary and/or secondary structure similarities can be inferred from the fact that various human antisera directed against snRNP proteins (for example, anti-Sm and anti-RNP) cross-react with both common and particle-specific proteins from diverse organisms, including yeast. Similarily, antibodies directed against yeast PRP8 and PRP9 proteins recognize HeLa snRNP proteins of comparable molecular weight and, in the case of PRP8, identical particle specificity (49, 89).

6. snRNP morphogenesis and transport

The spliceosomal snRNPs undergo a complex pathway of biogenesis which involves their shuttling between the nucleus and cytoplasm. The most probable snRNP assembly pathway, based primarily on studies with *Xenopus* oocytes and using the U1 snRNP as an example, is depicted schematically in Fig. 3. Subsequent to transcription, U1, U2, U4, and U5 snRNAs migrate to the cytoplasm where they are modified and assemble with the common snRNP proteins before returning to the nucleus to take part in the pre-mRNA splicing process. These newly synthesized snRNA molecules contain between one and 10 extra nucleotides at their 3′ end, which are trimmed in the cytoplasm (90–92). snRNA internal base modifications (primarily methylation), as well as hypermethylation of the 7-methylguanosine (m^7G) cap to a 2,2,7-trimethylguanosine (m_3G) cap are also cytoplasmic events (93, and reviewed in 2). The latter event is strictly dependent upon the association of the common snRNP proteins; disruption of Sm protein binding abolishes 5′ cap hypermethylation (93). Newly transcribed U6 snRNA, in contrast, remains in the nucleus where its 5′ end is converted to a γ-monomethylphosphate cap (3, 94). The 3′ end of U6 is also elongated by the non-templated addition of uridylates and converted, at least in part, to a 2′,3′ cyclic phosphate (95, 96). Since U6 is present solely in the nucleus, the formation of the U4/U6 snRNP particle, as well as that of the U4/U6·U5 tri-snRNP complex, must be nuclear events.

The common snRNP proteins are synthesized in excess over the snRNAs and remain in the cytoplasm until they associate with newly transcribed snRNA (97). The large cytoplasmic pools of E, F, G, D1, D2, D3, B, and B′ appear to contain preformed hetero-oligomeric protein complexes. A 6S complex containing E,F,G, and one or more of the D proteins has been detected (80, 98). This preformed complex is believed to associate with the snRNA, followed by B and B′ to form the core snRNP structure. Consistent with these *in vivo* data, recent *in vitro* assembly studies using *in vitro* translated proteins have demonstrated the formation of hetero-oligomer complexes containing E, F, G, D1, and D2; in addition, individual dimeric complexes between D1 and D2, and between D3 and B, as well as a heteromeric complex between E, F, and G were observed (55, 194, 196). The site of association of the particle-specific proteins is not clear. However, on the basis of evidence that several snRNP-specific proteins, such as A, C, and B″, migrate independently of core snRNP particles to the nucleus, it is believed that particle-specific protein assembly occurs in the nucleus (99–101).

Subsequent to the assembly of the core snRNP structure and cap hypermethylation, U1, U2, U4, and U5 RNP complexes are translocated to the nucleus. Nuclear transport of the snRNPs is an active, receptor-mediated process. The nuclear import pathway of U1, U2, U4, and U5 appears to be distinct from that of karyophilic proteins, whose transport is also receptor-mediated (99, 102–104). The selective interaction of karyophilic proteins with import receptors is determined by the presence of a nuclear localization signal (NLS), which may be a simple, short stretch of basic amino acids, or a more complex bipartite sequence. The snRNP NLS appears to be relatively complex. snRNPs U1, U2, U4, and U5 possess a bipartite NLS composed of the core snRNP structure and the m_3G cap [103–106]. However, not all snRNAs display the same m_3G cap requirements for nuclear import in *Xenopus* oocytes. Whereas U1 and U2 snRNAs absolutely require an intact m_3G cap structure, U4 and U5 snRNAs can enter the nucleus with non-physiological cap

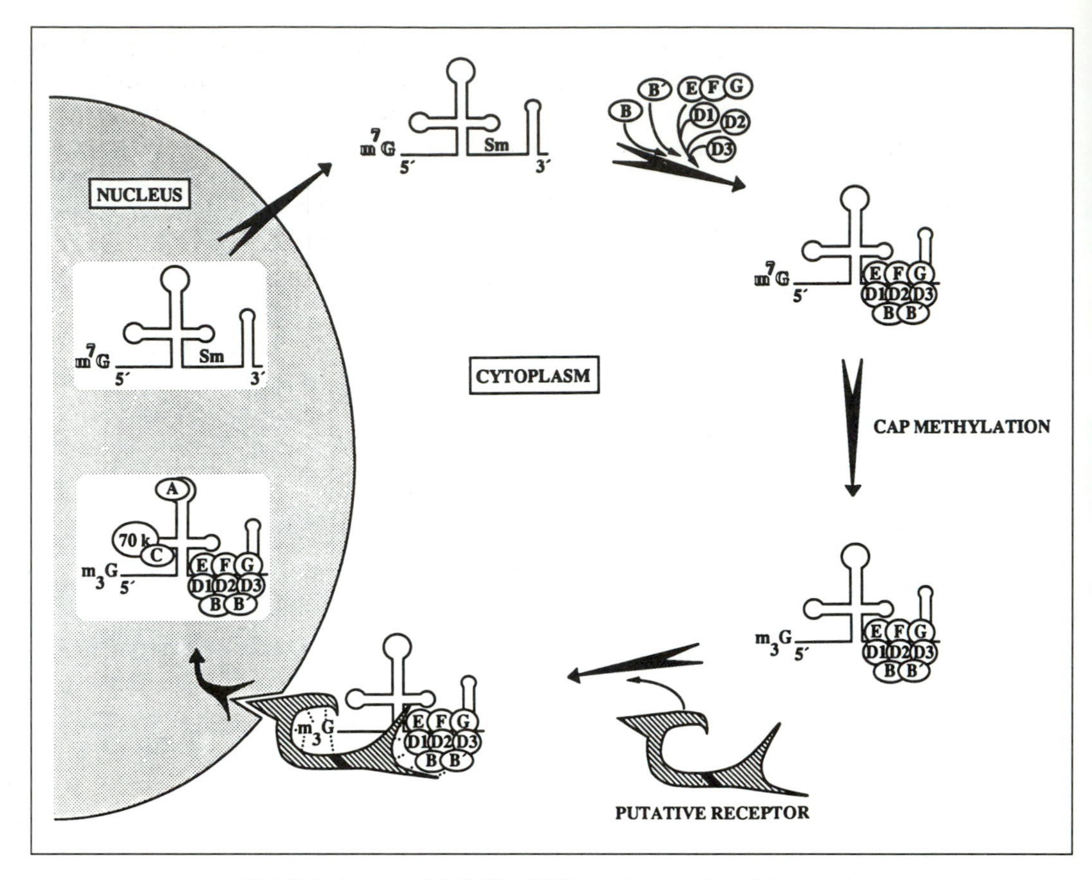

Fig. 3 Cartoon model of U1 snRNP morphogenesis and transport.

structures, although with reduced transport kinetics (102). More recently, using both micro-injected cells and an *in vitro* nuclear import system, it has been shown that the m_3G cap structure and the core snRNP also constitute a bipartite NLS in cultured mammalian cells (107, 197). Although the m_3G cap is not essential for the nuclear import of mammalian U1 and U2 snRNAs, it influences the rate of their import. The core snRNP, on the other hand, is necessary and sufficient to mediate snRNP nuclear targeting. snRNP nuclear import in mammalian cells occurs through nuclear pores, is ATP- and temperature-dependent, and requires soluble cytosolic factors. One or more of these cytosolic factors mediates the differential m_3G cap-dependence of U1 and U2 snRNA nuclear import in *Xenopus* oocytes and cultured mammalian cells (107).

7. snRNP higher order structure

The structure of a given snRNP particle arises from multiple RNA–protein and protein–protein interactions. For the spliceosomal snRNPs, only a limited number of these interactions, mostly RNA–protein in nature, have been elucidated. None the less, the study of snRNP structure has provided insight into general principles regarding the specificity of protein–RNA interactions. The investigation of snRNP structure relies on a number of methods, including:

(1) nuclease digestion or chemical modification of intact snRNPs (or those lacking a subset of their proteins);

(2) *in vitro* assembly in whole cell or nuclear extracts, or using *in vitro*-synthesized snRNAs and snRNP proteins;

(3) immunoelectron microscopy of snRNPs isolated under native conditions.

Information on the three-dimensional organization of snRNP components is based mainly on electron microscopy, through which a picture of the overall architecture of the spliceosomal snRNPs is slowly emerging. The current structural models of U1, U2, U5, and U4/U6 are based on information obtained primarily from studies carried out with *Xenopus* oocytes or HeLa cell extracts, or with snRNPs isolated from HeLa extracts. Information about the structure of yeast snRNPs is limited to the localization of the site of interaction of SNP1 and PRP4 with their cognate snRNAs and initial electron microscopy of isolated U1 and U4/U6·U5 tri-snRNP complexes.

7.1 Interactions of the common snRNP proteins

The association of the common snRNP proteins E, F, G, D1, D2, D3, B, and B′, with U1, U2, U4, and U5 snRNAs generates a common structural domain within the different RNP particles. As previously mentioned, the spliceosomal snRNAs (with the exception of U6) possess a common, evolutionarily conserved structural motif, referred to as domain A or, more commonly, the Sm site, which consists of a single-stranded region, PuA$(U)_n$GPu, often flanked by double-stranded stems (108). This motif serves as the association site for the eight common snRNP proteins. The presence of only one flanking stem-loop structure, either 3′ or 5′ to the single-stranded region, may suffice for stable interaction of the common proteins with the Sm site. In *S. cerevisiae*, for example, the Sm site of U1, U4, and the short form of U5 snRNA, lacks a 3′ hairpin structure (17). These snRNAs, none the less, support the binding of Sm proteins in yeast and are also capable of binding *Xenopus* Sm proteins when injected into *Xenopus* oocytes (109, 110). Because of the highly conserved nature of the single-stranded region of the Sm site, it was formerly assumed that protein–RNA interactions at the Sm site would involve the majority of these conserved nucleotides. However, saturation mutagenesis studies of this region of the yeast U5 snRNA revealed unexpected flexibility in the interaction between the Sm site and core proteins, insofar as mutation of only a few of the conserved nucleotides in this single-stranded region affected binding (111).

The single-stranded, uridine-rich region of the Sm site serves as the primary binding site for one or more of the common snRNP proteins. Despite the absence of a classical RNA-binding motif, a direct Sm site–protein interaction has been demonstrated for the G protein by cross-linking analyses; G protein could be UV cross-linked to the AAU stretch in the single-stranded region of the U1 snRNA Sm site (112). Whether G is the only protein that recognizes this site is not yet clear. As described in Section 6, prior to their association with U1, U2, U4, and U5 snRNAs, the common proteins interact with each other to form hetero-oligomeric complexes. Bearing this in mind, as well as the small size of the Sm site, it is most probable that in the snRNP core RNP domain protein–protein interactions dominate over protein–RNA ones. The existence of multiple contacts between several common proteins has been demonstrated by *in vitro* assembly studies (see Section 6) and is consistent with the highly stable nature of the core snRNP structure.

Although the snRNA structural elements required for the association of the common proteins appear to be quite similar, differences among snRNAs have been detected. For example, in *Xenopus laevis*, the single-stranded region of the U1 Sm site cannot functionally replace that of U5 (113). Mutational analyses of the U1 and U5 snRNA Sm site have also pointed to differences in the contributions of the surrounding stem–loop structures to common protein binding (113). These studies suggest that structural differences among the U1, U2, U4/U6, and U5 core domains may exist. Additional evidence in support of the proposed non-uniform nature of the core structure has recently been obtained by *in vitro* binding studies of the interaction of an N-terminal U1-70K fragment with purified U1, U2, and U5 snRNPs containing solely the common snRNP proteins. This N-terminal U1-70K fragment, which does not bind U1 snRNA, stably associates with U1, but not U2 or U5, core snRNP particles (114). Thus, although contributions from the snRNA cannot at present be ruled out, the observed specificity of this interaction is supportive of a unique core domain morphology for each snRNP species. Interestingly, whereas regions of the Sm site are not interchangeable among the snRNAs, the common snRNP proteins are functionally interchangeable as measured by *in vitro* splicing assays. For example, U2 or U5 snRNPs reconstituted *in vitro* with common proteins isolated from purified U1 snRNPs, restore splicing activity to extracts depleted of U2 or U5 snRNPs, respectively (198).

Investigation at the ultrastructural level has not revealed marked differences in the core structures of the spliceosomal snRNPs. High resolution electron microscopy of U1, U2, U5, and U4/U6 snRNPs containing only common proteins, demonstrates that these core snRNP particles share a similar, rather simple morphology—they are spherical, with a diameter of 8 nm (115–117).

7.2. Structure of the U1 snRNP

Considerable effort has been focused on elucidating the higher order structure of the U1 snRNP. Since this RNP particle is only of moderate complexity, a relatively detailed picture of its morphology (in comparison to other spliceosomal snRNPs) has emerged (Fig. 4). The U1-specific proteins are clustered at the 5′ end of the U1 snRNA molecule, covering stem–loops I and II (see Fig. 1 for U1 snRNA stem–loop designations). The U1-70K and A proteins interact specifically with stem–loops I

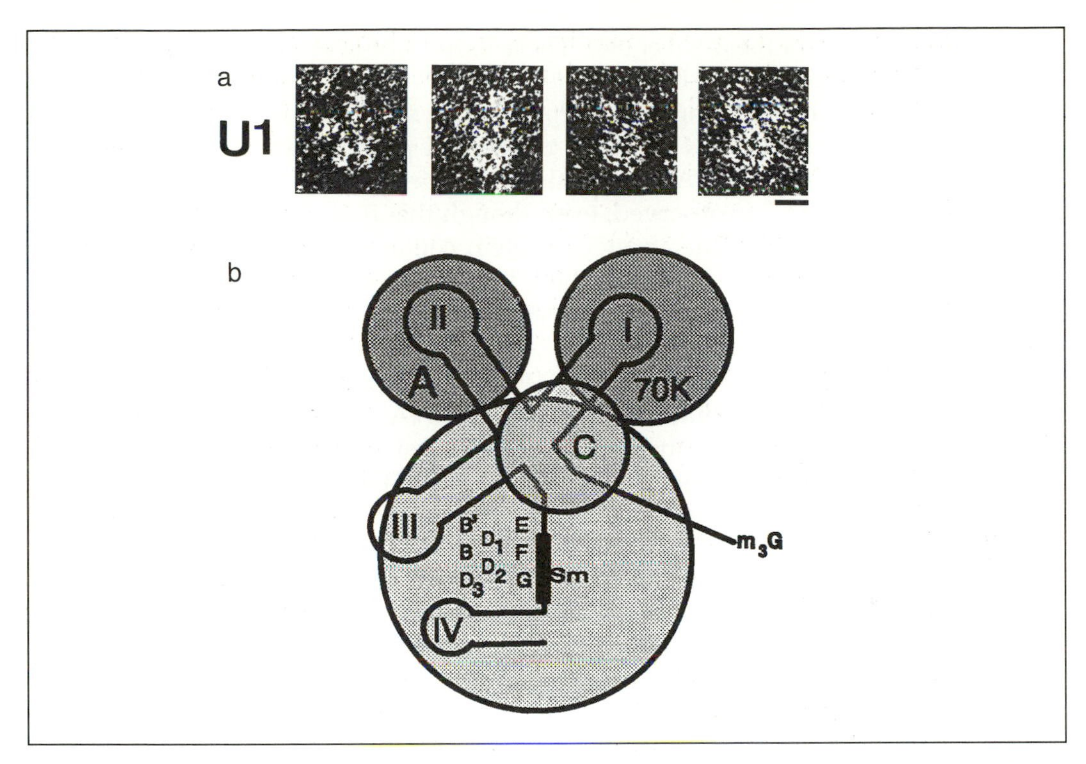

Fig. 4 Visualization of the human 12S U1 snRNP by electron microscopy and U1 snRNP structural model. (a) U1 snRNPs were isolated by immunoaffinity chromatography, followed by Mono Q chromatography (26) and negatively stained with uranyl formate before electron microscopy (115). The particles shown are oriented so that the protuberances point upward. Bar = 5 nm. (b) A cartoon structural model of the U1 snRNP, based on data gathered from a number of studies (see Section 7), is presented. The relative positions of the A and 70K proteins (dark circles), the C protein (small light circle), common snRNP proteins (large light circle) and U1 snRNA (line drawing) are shown.

and II, respectively, of the U1 snRNA (118–124). Deletion and mutation analyses of the U1 snRNA molecule have demonstrated that 70K and A protein binding occurs in a sequence-specific manner, which requires evolutionarily conserved bases in the single-stranded loops (120, 124). The association of 70K is believed to be stabilized by additional interactions with stem I (123). In *S. cerevisiae,* the yeast 70K homologue, SNP1, has been shown to bind specifically to the first 47 nucleotides of the yeast U1 snRNA, which include stem–loop I (125). Deletion and mutation analyses have also been performed in order to map those regions of the 70K and A protein required for specific U1 snRNA binding. In both instances, the N-terminal RRM, as well as several flanking amino acids, are required for a specific protein–RNA interaction (120, 124, 65). As discussed in Section 5.1, the tertiary structure of the N-terminal RRM of the A protein has been determined. Based on the position of several residues known to be essential for binding, the nucleotides of stem–loop II were proposed to interact with two regions of the A protein RRM: the four-stranded β-sheet, which contains the RNP-1 and RNP-2 consensus, and the loops on one end of the β-sheet (69). These loops, not RNP-1 or RNP-2, appear to mediate, at least in part, RNA-binding specificity.

Recent cross-linking and *in vitro* binding studies have also demonstrated an interaction between the N-terminus of the 70K protein and at least one of the common proteins (D2) (114). A stable interaction between U1-70K amino acids 1–97 and a core U1 snRNP is observed, even though this fragment does not exhibit U1 snRNA binding activity. Thus, protein–protein contacts appear to contribute to the association of the 70K protein with the U1 particle. These data additionally demonstrate a novel role for the common snRNP proteins in contributing to the binding sites for particle-specific proteins.

In contrast to 70K and A, the U1-specific C protein does not contain an RRM and probably does not bind directly to U1 snRNA. Association of the C protein with the U1 RNP is probably mediated, for the most part, by protein–protein interactions. Binding studies carried out with U1 snRNPs lacking C and various C protein deletion mutants have shown that the amino-terminal 45 amino acids suffice for U1 snRNP–C protein association (77). Interestingly, this C protein domain contains a zinc finger-like motif which appears to be essential for binding, since single point mutations in the cysteine and histidine residues of the zinc finger structure abolish C protein binding (77). More recent studies have demonstrated that the association of C with the U1 snRNP particle requires the presence of the N-terminal region of the 70K protein, as well as one or more common snRNP proteins (114). Thus, common snRNP proteins may be generally involved in mediating the interactions of particle-specific proteins with their cognate snRNP particles.

The structure of purified HeLa cell 12S U1 snRNPs, as well as the spatial arrangement of several U1 components, have been investigated by electron microscopy. Negatively stained 12S U1 snRNPs possess an almost circular main body (about 8 nm in diameter), with two protuberances 4–7 nm long and 3–4 nm wide (115) (Fig. 4). Through immunoelectron microscopy, the relative positions of the 70K and A proteins, and of the m_3G cap of the U1 snRNA, have been determined. Studies with antibodies specific for either the U1-70K or U1-A protein demonstrated that each of the U1 snRNP protuberances contains only one of these particle-specific proteins (127). Double antibody labelling experiments established that the 70K-containing protuberance is in closest proximity to the 5′ cap of the U1 snRNA (127). Since the binding site of 70K and A on the U1 snRNA is known, the positions of stem–loops I and II can be inferred to lie in, or at the base of, one of the two U1 snRNP protuberances (Fig. 4). The ultrastructure of the yeast U1 snRNP has also been recently elucidated by electron microscopy (43). In keeping with its more complex protein composition, U1 snRNPs isolated from *S. cerevisiae* exhibit a slightly larger and more elongated morphology than HeLa cell U1 snRNPs.

7.3 Structure of the U2 snRNP

Structural studies of the U2 snRNP have focused mainly on the interaction of the U2-specific proteins, A′ and B″, with U2 snRNA. Information regarding the association of these proteins with U2 snRNA has provided more general insight into how RNA binding is modulated by protein–protein interactions. Neither B″ nor A′, on its own, is able to interact specifically with the U2 snRNA. B″, which contains two RRMs, binds specifically to stem–loop IV of U2 only in the presence of the A′ protein (128); the former is able to bind non-specifically to other RNAs when A′ is absent (128). The binding specificity conferred on B″ by the A′ protein appears to be mediated by protein–protein interactions; in the absence of RNA, B″ and A′ associate with each other (129). This association requires the N-terminal region of both proteins (129). Interestingly, this region of A′ contains a leucine-rich segment which appears to mediate the A′–B″ intermolecular association (79). The N-terminal 88 amino acids of B″, which contain an RRM, also comprise the minimal U2 snRNA-binding domain (129). Analogous to the interactions of 70K and A with stem–loop structures of U1 snRNA, the single-stranded loop sequence of stem–loop IV is crucial for the specific binding of B″ and A′ to the U2 snRNA (128).

In comparison to A′ and B″, relatively little is known about the interaction of the more loosely associated 17S U2-specific proteins. However, whereas A′ and B″ associate with the 3′ half of the U2 snRNA molecule, several lines of evidence obtained from nuclease protection and electron microscopy studies (among others), suggest that most of the 17S U2-specific proteins are bound to the 5′ half of the U2 snRNA. Electron microscopy reveals that the 12S U2 snRNP possesses a main body 8 nm in diameter and an additional domain which is 4 nm long and 6 nm wide (Fig. 5a) (116). This additional domain was shown by immunoelectron microscopy to contain the B″ protein. The 17S U2 snRNP, in contrast, appears as a bipartite structure with two main globular domains connected by a short filamentous structure that is sensitive to RNase (Fig. 5b) (23). One globular domain probably contains the Sm proteins, A′, B″, and one or more additional 17S U2-specific proteins, whereas the other appears to contain several 17S U2-specific proteins which are clustered at the 5′ end of the U2 snRNA (Fig. 5) (23).

7.4 Structure of the U4/U6 snRNP

The U4/U6 snRNP appears to possess a rather simple morphology. Initial chemical modification studies comparing the accessibility of snRNA in isolated U4/U6 snRNPs containing solely the common snRNP proteins with those additionally containing the U4/U6-specific 60 kDa and 90 kDa proteins, have provided information about the site(s) of interaction of the U4/U6-specific proteins in HeLa cells. In particular, they indicate that the 60 kDa and 90 kDa proteins interact predominantly with the U4 central stem–loop and the majority of nucleotides in the single-stranded region just upstream of the U4 central stem–loop (Fig. 2) (Mougin *et al.*, unpublished). The site of interaction of the yeast U4/U6-specific protein, PRP4, on the other hand, has been mapped to the phylogenetically highly conserved 5′ stem–loop of the U4 snRNA (44). However, it is not yet clear whether PRP4 binds directly to the 5′ stem–loop and whether this region of U4 is sufficient for its association.

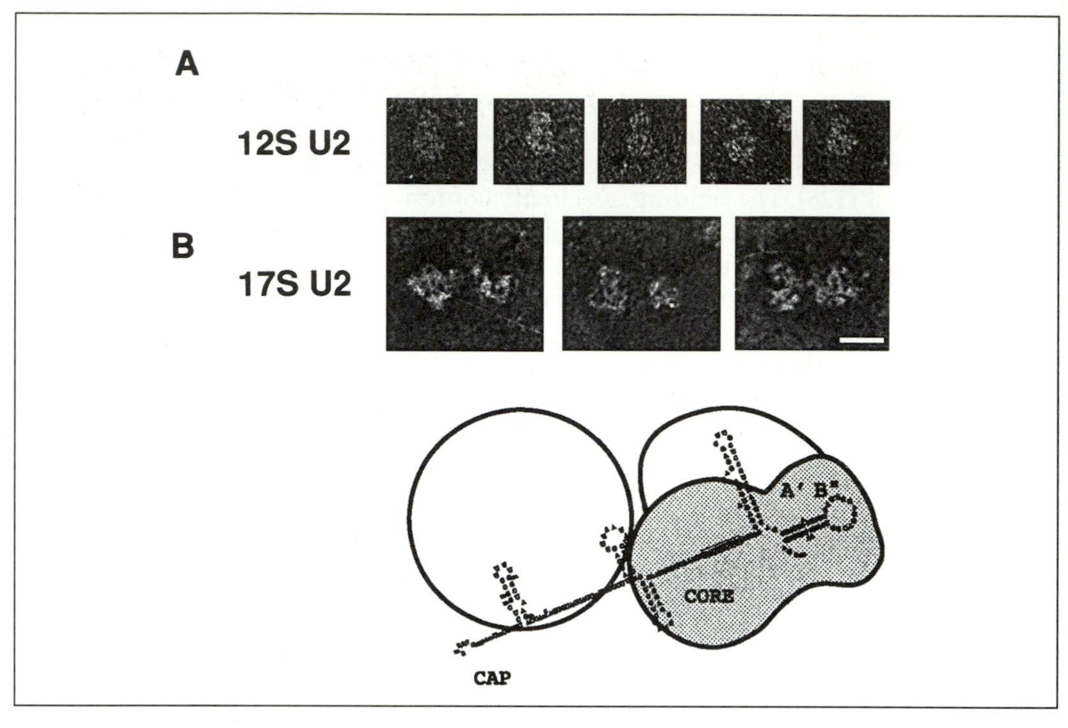

Fig. 5 Visualization of the 12S and 17S U2 snRNP by electron microscopy and structural model of 17S U2 snRNP. (a) 12S U2 snRNPs were isolated as described by Bach *et al.* (26) and visualized by electron microscopy after negative staining with uranyl formate (116). The particles are shown with the domain containing the A′ and B″ proteins pointing upwards. (b) 17S U2 snRNPs were isolated as described by Behrens *et al.* (23) and visualized by electron microscopy as described above. Bar = 7 nm for both 12S and 17S particles. Cartoon model of the 17S U2 snRNP, showing the U2 snRNA and probable positions of U2 proteins. The shaded oblong figure represents the A′, B″, and common (core) snRNP proteins (the 12S U2 particle). The additional circle and half-circle represent regions containing 17S U2-specific proteins.

7.5 Structure of the U5 snRNP

Despite the identification of multiple U5-specific proteins, the lack of U5-specific antibodies, coupled with the fact that no mammalian U5-specific proteins have been cloned, has hindered the identification of proteins that interact directly with the U5 snRNA molecule. Potential protein interaction sites have been identified in nuclear extracts by chemical and enzymatic modification studies of the U5 snRNA in the 20S U5 snRNP and U4/U6·U5 tri-snRNP complex (24). Only limited regions of the U5 snRNA are accessible to modification in both of these complexes and are therefore presumed to be free of protein. These include the evolutionarily conserved loop I of stem–loop I, which is involved in a base-pairing interaction with the pre-mRNA (see Section 8.3.1), and internal loop I (IL1) of stem–loop I (24, 130). Structure probing of U5 snRNA in purified 12S U5, which contains only the common snRNP proteins, as opposed to purified 20S U5 particles, indicated that stems Ib and Ic, as well as both sides of internal loop II, serve as the primary binding sites for one or

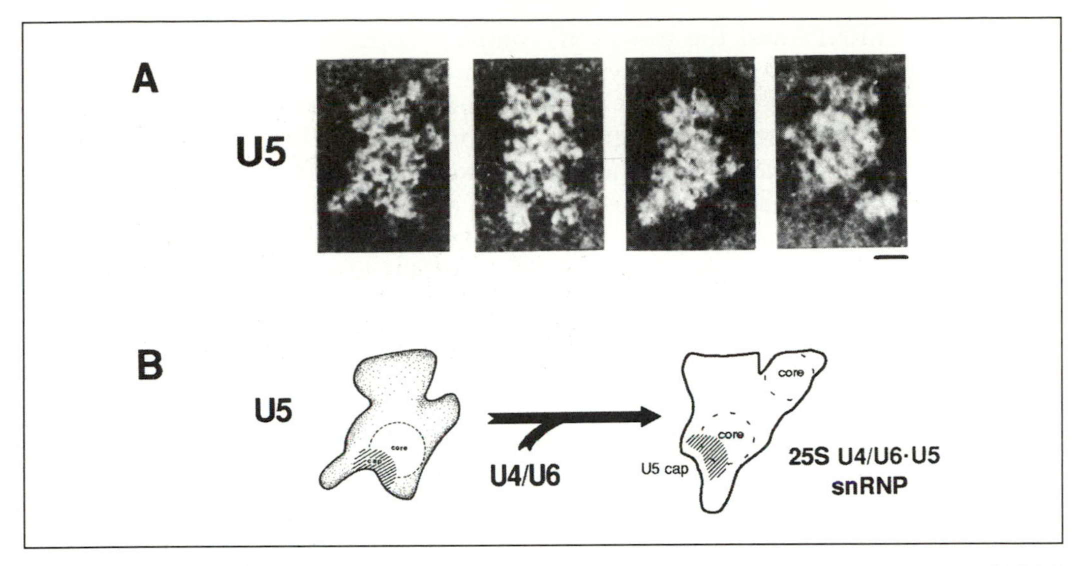

Fig. 6 Morphology of the 20S U5 snRNP and structural model of the U4/U6·U5 tri-snRNP particle. (a) 20S U5 snRNPs were isolated as previously described (26) and visualized by electron microscopy. Bar = 5 nm. (b) Cartoon models of the 20S U5 and 25S U4/U6·U5 snRNP. Dashed circles represent the approximate position of the U5 and U4/U6 common proteins (core) and shadowing the approximate position of the 5′ m_3G cap of the U5 snRNA.

more of the U5-specific proteins (130). In yeast, results consistent with those observed in HeLa cells have recently been obtained through complementation studies with U5 snRNA mutants. U5 mutants that contained the conserved Sm protein binding site, stem–loop I, internal loop II, and a stem closing internal loop II restored viability to *S. cerevisiae* cells that possess a lethal U5 gene disruption (131). The requirement for sequences encompassing internal loop II are consistent with the idea that it serves as the binding site for one or more functionally important U5-specific proteins. Due to the complex protein composition of the U5 snRNP, it is likely that many of the U5-specific proteins interact with the U5 particle via protein–protein interactions.

Consistent with its complex protein composition, investigation of the higher order structure of the U5 snRNP by electron microscopy has revealed a relatively complex morphology. The native 20S U5 snRNP has an elongated and highly indented shape, 20–23 nm long and 11–14 nm wide (116). A segmentation line divides the particle into a head and body region; the body, which is approximately twice as large as the head, also possesses three short protuberances (Fig. 6a).

7.6 Structure of the U4/U6·U5 tri-snRNP

Information regarding the structure of the 25S U4/U6·U5 particle is limited. As mentioned above, chemical and enzymatic modification studies with glycerol-gradient-fractionated nuclear extracts did not reveal differences between 20S

U5 snRNPs and the tri-snRNP complex, suggesting that the structure of the U5 snRNP is not significantly altered on association with U4/U6 (24). Comparable modification experiments have recently been carried out with purified HeLa U4/U6·U5 particles (Mougin *et al.*, unpublished). They demonstrate that all regions of the U5 snRNA, with the notable exception of the highly conserved loop I, are completely inaccessible to modifying agents in the tri-snRNP complex, whereas several regions of U4/U6 snRNA remain accessible. In addition, no alterations in the two-dimensional structure of the U4–U6 snRNA interaction domain were detected. *In vitro* splicing complementation studies using mutant U6 snRNAs have investigated whether particular regions of U6 are specifically required for tri-snRNP formation (132). Nucleotides in the 3′ terminal domain (nucleotides 74–76 of HeLa U6) as well as those in the central domain (nucleotides 33 and 37) were shown to be required for the efficient assembly of a functional tri-snRNP. It is not clear whether these regions are involved in intermolecular RNA–RNA or protein–RNA contacts that stabilize U4/U6 and U5 association, or whether they provide binding sites for proteins that mediate U4/U6 and U5 interaction.

The ultrastructure of purified HeLa cell and yeast tri-snRNP complexes has also been recently investigated by electron microscopy (Kastner *et al.*, unpublished). As depicted in Fig. 6b, 25S U4/U6·U5 complexes are typically triangular, with a maximal width and length of 20 nm and 30 nm, respectively. A characteristic indentation is also observed in the upper portion of the particle. The position of the U4/U6 and U5 core domains, as well as the 5′ m_3G cap of the U5 snRNA, have also been localized by immunoelectron microscopy. Interestingly, the general structure of the U4/U6·U5 particle appears to be evolutionarily conserved; tri-snRNPs isolated from the yeast *S. cerevisiae* possess an almost identical morphology under the electron microscope (43).

8. Spliceosomal snRNP function

The development of *in vitro* splicing systems has allowed biochemical investigation of the mechanism of nuclear pre-mRNA splicing in both metazoans and yeast. At least two temporally distinct splicing steps, which are identical in both yeast and mammals, can be detected (see 32 for review). In the first step of splicing, the pre-mRNA is cleaved at the 5′ splice site and the resultant free 5′ end of the intron is concomitantly linked to an adenosine residue (the branchpoint) approximately 18–30 nucleotides upstream from the 3′ splice site, forming a lariat intermediate. In the second step, the lariat intermediate, containing the intron and 3′ exon, is cleaved at the 3′ splice site and the 5′ and 3′ exons are ligated together. As previously mentioned, catalysis of the two-step pre-mRNA splicing reaction requires, in addition to non-snRNP protein factors, U1, U2, U5, and U4/U6 snRNPs. The spliceosomal snRNPs carry out a number of functions during the overall pre-mRNA splicing process, including the recognition and selection of splice sites and the formation of the multicomponent enzyme, the spliceosome, in which catalysis occurs. Spliceosome formation *in vitro* is an ordered process, with first U1 and then

U2 snRNPs binding to the 5′ splice site and branch site, respectively, followed by the interaction of U5 and U4/U6 snRNPs in the form of a U4/U6·U5 tri-snRNP complex (see Chapter 4 for a detailed discussion).

8.1 U1 snRNP function

8.1.1 Splice-site selection

The U1 snRNP carries out multiple functions during pre-mRNA splicing and its activity appears to be required mainly at points early in the splicing process. The primary role of U1 has long been considered to be the recognition of the 5′ splice site. The U1 snRNP has been shown to interact with this region of the pre-mRNA by means of a base-pairing interaction between the 5′ end of the U1 snRNA and the consensus splice junction sequence (Fig. 7) (133, 134). A stable U1–5′ splice-site complex may require the presence of non-snRNP protein factors; SF2/ASF has been shown to augment U1 binding *in vitro* (135, 136). In conjunction with 5′ splice-site recognition, U1 may play a regulatory role in 5′ splice-site selection. Through experiments using 5′ splice-site duplications or cryptic splice-site activation assays, it has been demonstrated that 5′ splice-site selection is often a function of the degree of complementarity between the 5′ end of the U1 snRNA and the 5′ splice site; sites with a higher degree of complementarity are preferentially used (137–139). However, determination of the location of the 5′ cleavage site may be influenced by other factors, such as the U5 snRNP (see below); base pairing between U1 and the 5′ splice site can be uncoupled from cleavage site selection in yeast (140, 141).

In addition to its interaction with the 5′ splice site, the U1 snRNP has also been shown to interact with the 3′ end of the intron. An association of U1 with the branchpoint region was suggested by studies investigating early spliceosome formation events (142, 143). Additionally, genetic suppression studies in the yeast *Schizosaccharomyces pombe* indicate that U1 snRNA can base pair with the conserved AG dinucleotide at the 3′ splice site; the 3′ splice site–U1 interaction also involves nucleotides at the 5′ end of the U1 snRNA molecule (Fig. 7) and has been shown to be required solely for the first step of splicing (144). However, a U1–3′ splice site interaction is not observed with all pre-mRNA splicing substrates and may even be a species-specific phenomenon (145). Not surprisingly, in light of its interaction with the branch site and/or 3′ splice site, the U1 snRNP also appears to influence the selection of 3′ splice sites. This effect has been observed for 3′ splice site duplications within a single intron containing pre-mRNA (146). The ability of U1 to interact with both the 5′ and 3′ regions of an intron led to a model in which U1 potentially contributes to the initial alignment of regions of the pre-mRNA which are recombined during splicing (Fig. 7b) (147).

8.1.2 Spliceosome assembly

The U1 snRNP is also considered to be an initiator of spliceosome formation. The association of the remaining spliceosomal snRNPs with the pre-mRNA is dependent, either directly or indirectly, on the presence of the U1 snRNP. For example, the

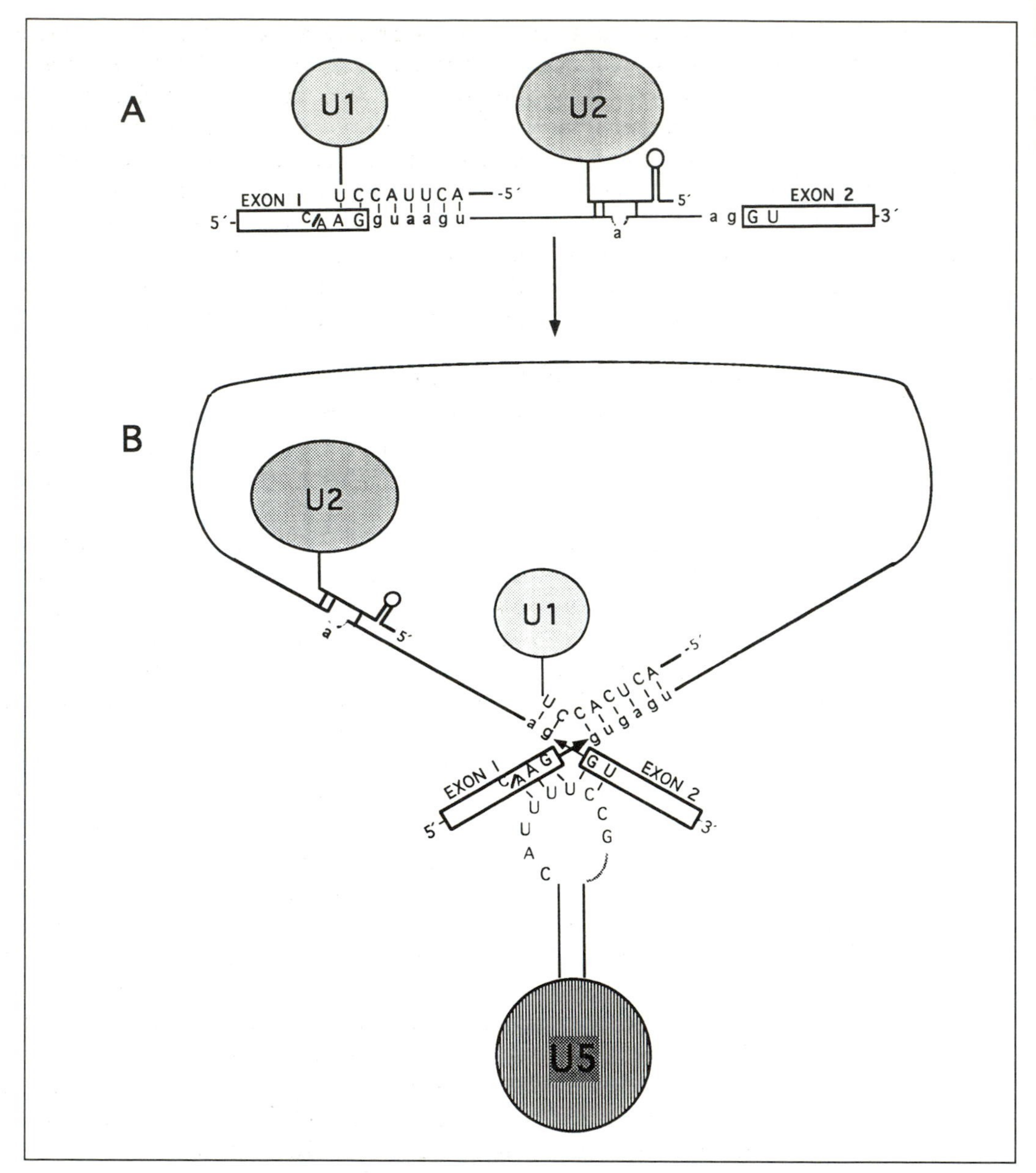

Fig. 7 The interactions of U1 and U5, and U2 snRNPs with the 5′ and 3′ splice sites, and branch site, respectively (adapted from 147). (a) Initial interactions between the 5′ end of U1 snRNA and the 5′ splice site and the 5′ end of U2 snRNA and the branchpoint. (b) Subsequent interactions of U5 snRNA with the 5′ and 3′ splice site and U1 with the 3′ splice site to form a Holliday-like structure, in which the 5′ and 3′ splice sites are juxtaposed. Note that the interactions of U1 and U5 with the pre-mRNA probably do not occur simultaneously, but rather successively. The conserved nucleotides at the 5′ end of U1 (positions 3–10) and in loop I of U5 are shown. Consensus nucleotides for vertebrate 5′ and 3′ splice sites and the branchpoint adenosine residue are shown. Upper case letters and lower case letters represent exon and intron sequences, respectively. The base pairing interaction between the fourth U of the U5 loop I and both exonic G residues at the 5′ and 3′ splice site are probably temporally distinct events, with the 5′ splice site interaction occurring prior to 5′ splice site cleavage and that at the 3′ splice site prior to 3′ splice site cleavage (see 170).

interaction of U2 with the branch site is greatly enhanced by the presence of U1 snRNPs both in mammalian cells (148) and in yeast (142, 149). In yeast, the interaction of U1 with the 5′ splice site gives rise to complexes which are committed to the splicing pathway (142); that is, pre-mRNA in these so-called commitment complexes can be chased into spliceosomes and ultimately into mature, spliced mRNA. In mammalian systems, pre-mRNA commitment involves the association of serine/arginine-rich protein factors, such as SC35, with the 5′ splice site (150) (see Chapter 6). Since such a factor has not been identified in yeast, and since yeast U1 snRNPs contain numerous additional proteins, it is tempting to speculate that one or more of the yeast U1-specific proteins possesses SC35-like activity.

8.1.3 Functions of the U1-specific proteins

While U1 snRNP function relies heavily upon base-pairing interactions of the U1 snRNA, protein components have also been shown to contribute to its activity. A general role for U1-specific proteins was initially suggested by splicing complementation experiments in *Xenopus* oocytes using mutant U1 snRNAs (151). Those mutants that did not support stable 70K, A, or C protein binding were unable to restore splicing in oocytes that had been depleted of their endogenous U1 snRNA molecules. A stable U1-snRNP–5′ splice site interaction appears to require snRNP proteins, since mild proteolysis of the U1 particle inhibits this association (152). Consistent with this observation, filter binding assays using purified HeLa cell U1 snRNPs that were gradually depleted of their specific proteins suggest that the U1-C protein augments the binding of U1 to the 5′ splice site (153). Direct evidence of an essential function for a U1-specific protein has recently been reported in mammals. In particular, *in vitro* splicing complementation studies have provided evidence that the phosphorylation state of the RS domain of the U1-70K protein is important for its activity in splicing (154). U1 particles containing normally phosphorylated 70K can restore splicing to extracts depleted of their endogenous U1, whereas those particles phosphorylated with γ-S-ATP (from which the phosphate groups cannot be removed) support spliceosome formation but not splicing. Interestingly, an interaction between the RS domain of the 70K protein and that of the splicing factor SF2/ASF has recently been demonstrated, indicating that the SF2/ASF-mediated interaction between U1 and the 5′ splice site involves the U1-70K protein (136, 155). The U1-70K protein also appears to play an essential role in yeast as deletion of the yeast 70K homologue (SNP1) abolishes cell viability (50). Curiously, the apparent functionally significant RS domain of the metazoan 70K protein is absent in yeast, suggesting a differential role for this protein in yeast versus higher eukaryotic pre-mRNA splicing.

8.2 U2 snRNP function

8.2.1 Branch site recognition

One role of the U2 snRNP is the recognition of the branch site. In mammals, the association of U2 snRNP with the branchpoint region requires ATP and protein factors such as SF1 and U2AF, the latter of which has been shown to bind to the polypyrimidine tract downstream from the branch site before U2 binding (156–158). In addition, RNA–RNA interactions also stabilize the association of U2 with the branch site and thereby contribute to branchpoint selection. In both yeast and mammals, compensatory base change studies have confirmed a base-pairing interaction between the branch site consensus sequence (which is highly conserved only in yeast) and a phylogenetically conserved single-stranded region (GUAGUA) of U2 snRNA located between stem–loops I and II (Figs 7 and 8) (159, 160). As mentioned above, a stable, functional U2–branch site interaction is also enhanced by the presence of the U1 snRNP.

8.2.2 Spliceosome assembly

U2 snRNPs have also been proposed to be involved in the deposition (assembly) of the U4/U6·U5 tri-snRNP into the spliceosome. Indeed, the formation of mature spliceosomes, but not that of pre-spliceosomes (that is, complexes containing U1 and U2 snRNPs and the pre-mRNA), is inhibited if the 5′ terminal 20 nucleotides of U2 snRNA are blocked with an antisense oligonucleotide (163). The dependence of spliceosome formation on this region of the U2 snRNA appears to be due to a base-pairing interaction between the 5′ end of U2 and the 3′ end of U6 snRNA (helix II, Fig. 8). This interaction was confirmed in mammalian cells by both biochemical and genetic studies, and has been shown to be essential for mammalian, but not yeast, pre-mRNA splicing (132, 161, 162, 164–166). An additional U2–U6 base-pairing interaction has recently been demonstrated in yeast (167) (see Fig. 8 and Section 8.3). This interaction has been proposed to contribute to the formation of the active centre of the spliceosome (see Section 8.3.3 for detailed discussion).

8.2.3 Functions of the U2-specific proteins

As with the U1 snRNP, the RNA component of the U2 snRNP clearly plays a critical role in its function. However, recent studies have also demonstrated an essential role for three of the 17S U2 snRNP-specific proteins. Direct evidence that the U2-specific 60 kDa protein is an essential splicing factor has been obtained through antibody inhibition experiments with anti-PRP9 serum. The latter serum specifically recognizes the U2–60 kDa protein, suggesting structural, as well as functional, homology to the yeast, splicing-essential PRP9 protein (49). The addition of this antiserum to HeLa cell splicing extracts inhibited splicing proper, as well as spliceosome formation; this antibody-mediated blockage could be reversed by the addition of an excess of the 17S, but not 12S, U2 snRNP (49). Since PRP9 mutants block splicing at an identical step in yeast (no pre-spliceosome formation is observed), these results demonstrate functional homology between a yeast and a human snRNP protein. The 60 kDa and 66 kDa U2-specific proteins have subsequently been cloned (66, 67) and sequence comparisons indicate they are homologues of the yeast PRP9 and PRP11 proteins which, in addition to PRP21, mediate the association of U2 with the pre-mRNA substrate in yeast (see Chapter 7). Interestingly, biochemical and functional comparisons between the mammalian splicing factor

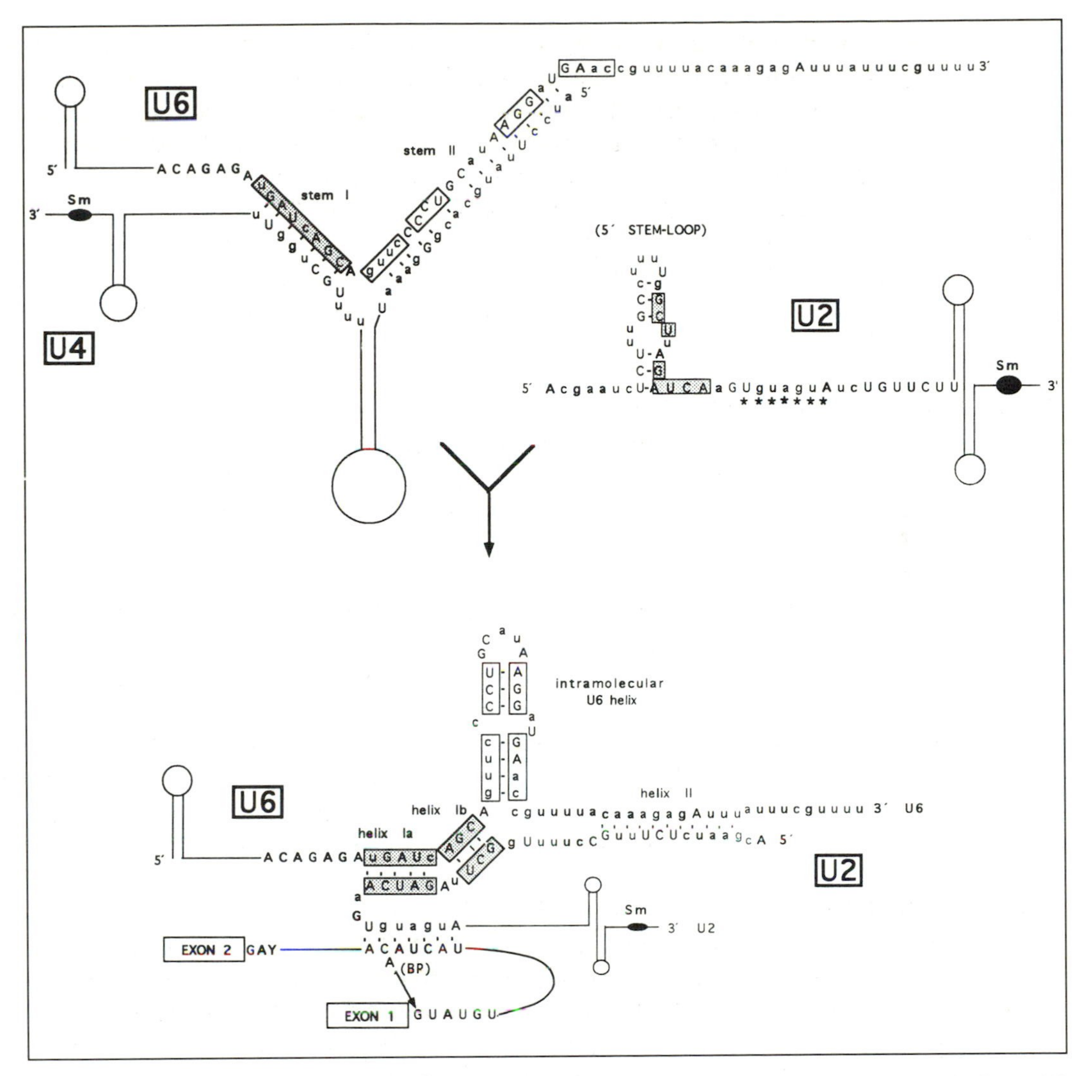

Fig. 8 Interactions between the U6 and U2 snRNAs within the spliceosome. The cartoon models of U6 and U2 snRNAs prior to and subsequent to their association are adapted from Madhani and Guthrie (167). The sequences of various regions of U4, U6, and U2 snRNAs shown here are from *S. cerevisiae*. Upper case letters represent phylogenetically conserved nucleotides. Nucleotides that base pair in helix I, and the U6 intramolecular helix of the U2–U6 structure, are indicated in the U2 and U4/U6 structures by boxes. Nucleotides in the U2 snRNA that base pair with the branch site are indicated with asterisks and the position of the U4 and U2 Sm sites by black ovals. The interaction between U2 and the yeast branchpoint region consensus (UACUAACA) is also shown. A representative yeast pre-mRNA with conserved intron sequences indicated is shown. (BP), branchpoint.

SF3a and several U2-specific proteins have demonstrated that the three subunits of SF3a, which include 60 kDa, 66 kDa, and 120 kDa proteins, are identical to the 17S U2-specific 60 kDa, 66 kDa, and 110 kDa proteins, respectively (168). Since integration of the U2 snRNP into the spliceosome requires both SF3a and SF3b, it is likely that additional, high molecular weight proteins of U2 are equivalent to SF3b and, thus, essential for U2 function (168). The apparent association of one or more of the 17S U2-specific proteins with the 5′ end of the U2 snRNA (see Section 7.3) suggests they may mediate, in part, the functionally important base-pairing interactions between U2 and the pre-mRNA branch site and/or U6 snRNA (Fig. 8).

8.3 U5 and U4/U6 snRNP function

8.3.1 Functions of U5 snRNA

Significant advances have been made in recent years in defining some of the roles of U5 and U4/U6 snRNPs in pre-mRNA splicing. These snRNP particles enter the spliceosome in the form of a U4/U6·U5 tri-snRNP complex subsequent to U1 and U2 snRNP binding (see Chapter 4). The interaction of the tri-snRNP complex was initially thought to arise mainly through contacts with the pre-bound U1 and U2 snRNPs rather than directly with the pre-mRNA molecule. Indeed, as discussed above, U2–U6 snRNA interactions have been detected. However, more recent studies have demonstrated a direct interaction of U5 snRNA with the 5′ and 3′ splice sites, as well as an interaction between U6 and the 5′ splice site. For example, genetic studies in yeast have demonstrated interactions between nucleotides in the invariant loop I sequence (GCCUUUUAC) of the U5 snRNA and exon sequences at both the 5′ and 3′ splice sites (Fig. 7) (169, 170). In addition, cross-linking analyses with HeLa cell splicing extracts revealed cross-links between U5 and both exon and intron sequences at the 5′ splice site, as well as exon sequences at the 3′ splice site (171, 172). Thus, the U5 snRNP, in addition to U1, appears to be involved in the recognition of 5′ and 3′ splice sites and, thereby, in the determination of the exact points of cleavage at these two sites. Due to its dual splice-site recognition capability, one role of U5 may be, in conjunction with U1, to bring 5′ and 3′ splice sites into close proximity before the first step of splicing (Fig. 7b) (170). The association of U1 and U5 with 5′ and 3′ splice sites appears be temporally distinct. The association of U1 with these regions of the pre-mRNA molecule appears to be disrupted before, or concomitant with, the first step of splicing, whereas the U5 snRNA–pre-mRNA interactions appear to persist after apparent dissociation of the U1 snRNP (173). The U5 snRNP is also required at stages subsequent to spliceosome formation; splicing complementation studies in HeLa cell nuclear extracts indicate that U5 snRNPs are required for both steps of the splicing reaction (174).

8.3.2 Functions of U4 snRNA

Interestingly, the functions of the individual snRNP components of the tri-snRNP complex appear to be temporally distinct. U4 snRNA, like U1 snRNA, appears to be required solely for spliceosome assembly and, therefore, not to be involved in subsequent catalytic events. For example, yeast spliceosomes that have lost their U4 snRNP are still capable of catalysing both cleavage–ligation steps of splicing (175). U4 snRNA has been proposed to act as an antisense negative regulator of U6 snRNA (17). That is, the base-pairing interaction between U4 and U6 is believed to maintain the U6 snRNA in an inactive conformation during spliceosome assembly.

There is good evidence that the association of U4 with U6 is disrupted before, or concomitant with, the first step of splicing (175, 176). This displacement of U4 snRNA is a prerequisite for the formation of a functionally important U2–U6 base pairing interaction (Fig. 8).

8.3.3 Functions of U6 snRNA

The proposal that U6 plays a central catalytic role in pre-mRNA splicing is based on a number of observations, including: the high degree of sequence conservation of the U6 snRNA (16); the discovery of introns in the U6 genes of some fungal species and the proposal that they arose by a reverse splicing process (177–179); and the observation that, at least in yeast, mutation of U6 snRNA at positions 51, 52, 58, and 59 specifically blocks the second step of splicing (180, 181). Recent studies using HeLa cell splicing extracts demonstrated that the intramolecular stem–loop region of the human U6 snRNA (positions 57–78, Fig. 2) not only is required for the base-pairing interaction with U4, but also functions during the first step of splicing (182). This region of the U6 snRNA appears to undergo a conformational change during splicing; it is initially sequestered in stem II of the U4–U6 interaction domain (Fig. 8), but after the dissociation of U4 may return to the intramolecular stem–loop conformation present in the free U6 snRNA.

Additional regions of U6 snRNA that may be directly involved in catalysis have recently been detected. Mutational analyses of the *S. cerevisiae* U6 and U2 snRNAs have revealed a base-pairing interaction between conserved nucleotides of U6 (positions 54 to 61) and nucleotides of U2 immediately upstream of the branch site recognition region (positions 21–23 and 26–30) (helix Ia and Ib, Fig. 8) (167). Formation of a U6–U2 helix in this region has been shown to be essential for cell viability and splicing, and requires the disruption of the U4–U6 base-pairing interaction (167). Mutation of the yeast U2 snRNA revealed that the absolutely conserved AGA trinucleotide in helix I (positions 25–27) is essential for the second step of splicing. The intermolecular structure created by the U2–U6 interaction in the vicinity of the branch point is similar to a structural domain required for group II intron self-splicing (Fig. 9). In particular, this U2–U6 helix mimics domain 5 of group II introns. Since domain 5 is required for 5′ splice-site cleavage, a similar role for this U2–U6 interaction domain (helix I) has been proposed. Consistent with this proposal, nucleotides of the U6 snRNA that interact with nucleotides just downstream of the 5′ splice site are located adjacent to helix I (see below; Fig. 10). Additionally, according to the model presented in Fig. 8, the proposed catalytically active U2–U6 domain is also in the vicinity of the branchpoint.

In addition to its interactions with U2, U6 snRNA has also been shown to base pair with conserved intron residues near the 5′ splice site. Cross-linking studies initially suggested that residues within the invariant ACAGAG sequence of the U6 snRNA (positions 47 to 52) were in close proximity to nucleotides at positions 4 to 6 of the intron (Fig. 10) (172, 183, 184). More recently, cross-linking studies using a pre-mRNA containing a photoactivatable 4-thiouridine group have identified a cross-link between the third adenosine of this conserved U6 snRNA sequence, and

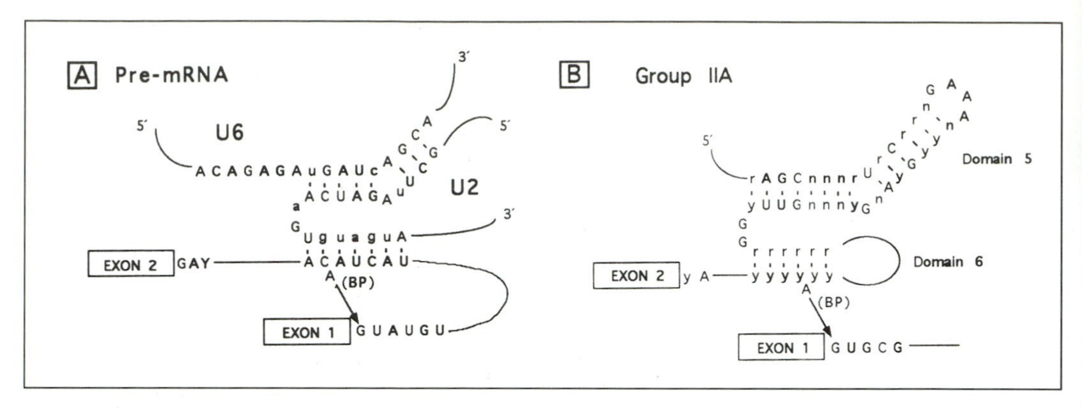

Fig. 9 Structural similarities between the U2–U6 snRNA interaction domain and conserved regions of group II self-splicing introns. (a) U2 and U6 snRNAs with a yeast consensus intron, as in Fig. 8. (b) The consensus sequence of domains 5 and 6 of group IIA self-splicing introns is shown. Upper case letters indicate highly conserved nucleotides; r and y indicate conserved purines and pyrimidines, respectively; and n represents variable nucleotides. From Madhani and Guthrie (167).

the second residue of the intron (Fig. 10) (173). Genetic suppression studies in yeast have subsequently confirmed a base-pairing interaction between the ACA trinucleotide of the conserved ACAGAG sequence of U6 and intron nucleotides 4–6 (Fig. 10) (185, 186). These studies additionally provided evidence that this U6 interaction specifies the precise site of nucleophilic attack at the 5′ splice junction. Thus, U6 has been proposed to align the 5′ splice site within the catalytic centre of the spliceosome in a manner conducive to the first step of splicing. The 5′ splice site is thus recognized initially by U1, but subsequently by both U5 and U6. These multiple recognition events have been proposed to decrease aberrant cleavage at the 5′ splice site (185). Aside from its interaction with the 5′ splice site, genetic data indicate that U6 snRNA may also contact nucleotides near the 3′ splice site (186). Thus, U6 snRNA could also influence the recognition/cleavage of the 3′ splice junction.

8.3.4 Functions of U5 and U4/U6-associated proteins

In yeast, a significant amount of evidence for the involvement of U5 and U4/U6 snRNP proteins in splicing is available. As previously discussed, several proteins essential for pre-mRNA splicing (PRP4, PRP6, PRP8, and PRP18) are components of U5 or U4/U6 snRNPs. Two of these snRNP proteins—the U4/U6-specific PRP4 and PRP6—are required for the formation of the U4/U6·U5 tri-snRNP complex and so are essential for spliceosome assembly (45). One or more of the mammalian tri-snRNP specific proteins appears to carry out an identical function. A splicing factor necessary for spliceosome formation is inactivated in heat-shocked cells (187). This essential splicing factor has been identified as one or more of the mammalian tri-snRNP-specific proteins, since splicing-inactive extracts prepared from heat-shocked cells could be complemented with 25S U4/U6·U5 complexes but not with a

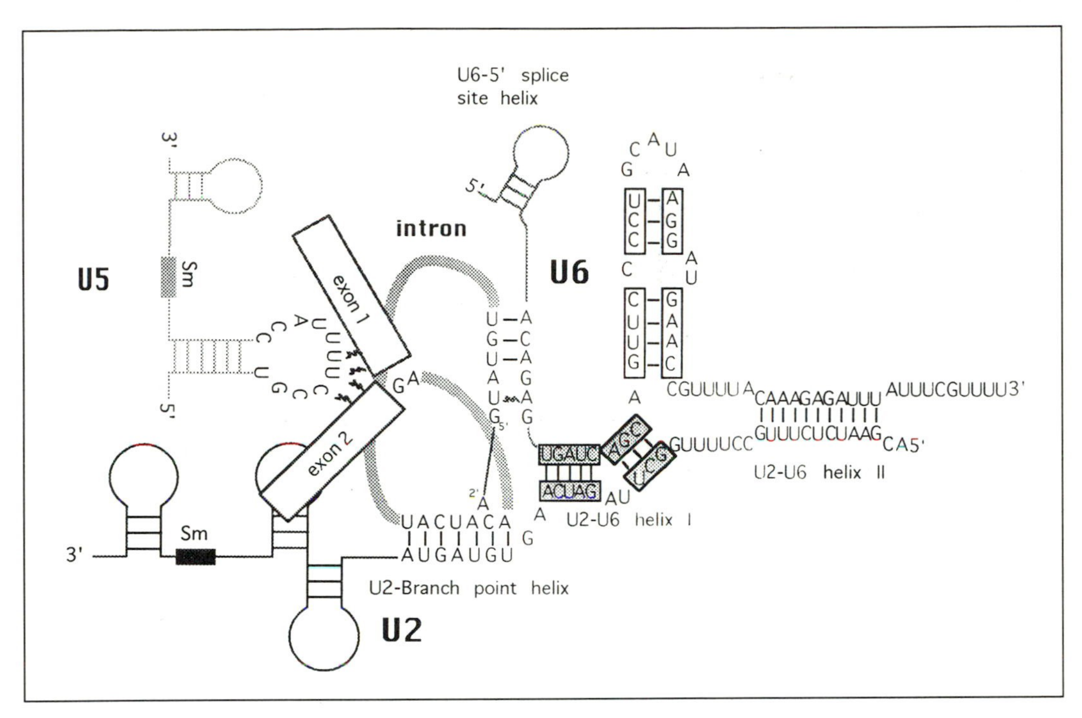

Fig. 10 Network of RNA interactions occurring among U2, U5, U6 and the pre-mRNA within the spliceosome. In this model, adapted from Wise (193), the pre-mRNA has undergone the first *trans*-esterification reaction. The excised exon 1 is held in place by interactions with nucleotides of the U5 conserved loop I sequence. Exon sequences are indicated by open boxes and intron sequences (other than conserved nucleotides at the 5′ splice site (GUAUGU), branch site (UACUAACA), and the 3′ splice site (AG)) are depicted by a thick shaded line. The 2′–5′-phosphodiester bond between the first intron nucleotide (G) and the branchpoint adenosine is depicted by a thin dark line. Watson–Crick base-pairing interactions are indicated by dashes and interactions confirmed by cross-linking are depicted by a wavy line. Cartoon structural models of U2, U5, and U6 are according to the metazoan consensus structure, whereas all nucleotide sequences are from the yeast *S. cerevisiae*.

combination of U5 and U4/U6 particles that lacked the tri-snRNP-specific proteins (188). The exact function of the U5-specific PRP8 and PRP18 proteins is not clear. PRP8, as well as its mammalian homologue (the U5-220 kDa protein), has been shown by cross-linking analyses to interact with the 5′ splice site during pre-mRNA splicing; thus, PRP8 could be envisaged to mediate the interaction of U5 snRNA with this region of the pre-mRNA (171, 189). PRP18, on the other hand, appears to play a role in the second step of splicing (48).

9. Summary and perspectives

The ribonucleoprotein complexes that serve as spliceosomal subunits, namely U1, U2, U4/U6, and U5 snRNPs, are both biochemically and structurally complex. Significant advances have recently been made in the biochemical characterization of these complexes. In HeLa cells, the vast majority of snRNP-associated proteins have already been identified, and a growing number are rapidly being identified in yeast. However, significantly less information about structural aspects of these particles, especially in the case of the 17S U2 and 25S U4/U6·U5 complexes, is currently available. Additional examination of the structure of individual snRNPs is clearly warranted, since it will undoubtedly contribute to the ultimate goal of elucidating the three-dimensional structure of the spliceosome.

The spliceosomal snRNPs play essential roles in the pre-mRNA splicing process; both the snRNA and protein constituents of U1, U2, U5, and U4/U6 snRNPs are required for their activity. The spliceosomal snRNAs play central roles in splice-site recognition and selection, and are likely to be directly involved in the catalysis of pre-mRNA splicing. During spliceosome formation and splicing catalysis, the snRNAs are involved in a number of dynamic RNA–RNA interactions, which appear to lead to the formation of one or more active sites within the spliceosome (Figs 7 and 10). The precise roles of the snRNP proteins are less well defined. A number of proteins have been shown to be essential for the association of a given snRNP particle with the pre-mRNA or with other snRNP particles and, thus, appear to be primarily required for spliceosome formation. snRNP proteins may also play important roles in regulating the formation and disruption of functionally important RNA–RNA interactions, or may shift the equilibrium between alternative snRNA conformations. They could also conceivably carry out enzymatic functions required for spliceosome formation or even splicing catalysis, functioning as helicases, isomerases, and/or kinases. None the less, the initial insights into the higher order structure of the spliceosome are consistent with the hypothesis that the active site, or sites, responsible for pre-mRNA cleavage–ligation events are composed of RNA and may be structurally similar to those identified in either group I or group II self-splicing introns (see Chapter 1). That is, the initial characterization of RNA–RNA contacts within the spliceosome has revealed a complex network of RNA–RNA interactions, some of which form structures paralleling those found in self-splicing introns (Figs 9 and 10). Although significant advances have been made in identifying the biochemical components of the spliceosome (snRNP components and splicing factors) as well as the structure of individual spliceosomal subunits (12S U1, 17S U2, and the 25S U4/U6·U5 tri-snRNP), relatively little is currently known about the three-dimensional architecture of the spliceosome. Additional information regarding RNA–RNA, RNA–protein, and protein–protein interactions in the spliceosome is needed before the catalytic mechanisms responsible for nuclear pre-mRNA splicing are clearly understood.

Acknowledgements

We thank Bertold Kastner and Patrizia Fabrizio for help in the preparation of figures and Christopher Marshallsay and Iain Mattaj for critical comments on the manuscript. The work from our laboratory was supported by grants from the Deutsche Forschungsgemeinschaft, the Bundesministerium für Forschung und

Technologie, and the Fonds der Chemischen Industrie. C.L.W. was supported, in part, by a fellowship from the Alexander von Humboldt Stiftung.

References

1. Mattaj, I. W., Tollervey, D., and Séraphin, B. (1993) Small nuclear RNAs in messenger RNA and ribosomal RNA processing. *FASEB J.*, **7**, 47.
2. Reddy, R. and Busch, H. (1988) Small nuclear RNAs: RNA sequences, structure, and modifications. In *Structure and function of major and minor small nuclear ribonucleoprotein particles* Birnstiel, M. L. (ed.). Springer-Verlag, Berlin, p. 1.
3. Singh, R. and Reddy, R. (1989) Gamma-monomethyl phosphate: A cap structure in spliceosomal U6 small nuclear RNA. *Proc. Natl Acad. Sci. USA*, **86**, 8280.
4. Kunkel, G. R., Maser, R. L., Calvet, J. P., and Pederson, T. (1986) U6 small nuclear RNA is transcribed by polymerase III. *Proc. Natl Acad. Sci. USA*, **83**, 8575.
5. Reddy, R., Henning, D., Das, G., Harless, M., and Wright, D. (1987) The capped U6 small nuclear RNA is transcribed by RNA polymerase III. *J. Biol. Chem.*, **262**, 75.
6. Moenne, A., Camier, S., Anderson, G., Margottin, F., Beggs, J., and Sentenac, A. (1990) The U6 gene of *Saccharomyces cerevisiae* is transcribed by RNA polymerase C (III) *in vivo* and *in vitro*. *EMBO J.*, **9**, 271.
7. Branlant, C., Krol, A., Lazar, E., Haendler, B., Jacob, M., Galego-Dias, L., *et al.* (1983) High evolutionary conservation of the secondary structure and of certain nucleotide sequences of U5 RNA. *Nucleic Acids Res.*, **11**, 8359.
8. Krol, A., Gallinaro, H., Lazar, E., Jacob, M., and Branlant, C. (1981) The nuclear 5S RNAs from chicken, rat, and man: U5 RNAs are encoded by multiple genes. *Nucleic Acids Res.*, **9**, 769.
9. Sontheimer, E. J. and Steitz, J. A. (1992) Three novel functional variants of human U5 small nuclear RNA. *Mol. Cell. Biol.*, **12**, 734.
10. Lund, E., Kahan, B., and Dahlberg, J. E. (1985) Differential control of U1 small nuclear RNA expression during mouse development. *Science*, **229**, 1271.
11. Mattaj, I. W. and Hamm, J. (1993) Regulated splicing in early development and stage-specific U snRNPs. *Development*, **105**, 183.
12. Siliciano, P. G., Kivens, W. J., and Guthrie, C. (1991) More than half of yeast U1 snRNA is dispensable for growth. *Nucleic Acids Res.*, **19**, 6367.
13. Igel, A. H. and Ares, M., Jr (1988) Internal sequences that distinguish yeast from metazoan U2 snRNA are unnecessary for pre-mRNA splicing. *Nature*, **334**, 450.
14. Shuster, E. O. and Guthrie, C. (1988) Two conserved domains of yeast U2 snRNA are separated by nonessential nucleotides. *Cell*, **55**, 41.
15. Shuster, E. O. and Guthrie, C. (1990) Human U2 snRNA can function in pre-mRNA splicing in yeast. *Nature*, **345**, 270.
16. Brow, D. and Guthrie, C. (1988) Spliceosomal RNA U6 is remarkably conserved from yeast to mammals. *Nature*, **334**, 213.
17. Guthrie, C. and Patterson, B. (1988) Spliceosomal snRNAs. *Annu. Rev. Genet.*, **22**, 387.
18. Raj, N. B. K., Ro-Choi, T. S., and Busch, H. (1975) Nuclear ribonucleoprotein complexes containing U1 and U2 RNA. *Biochemistry*, **14**, 4380.
19. Howard, E. F. (1978) Small nuclear RNA molecules in nuclear ribonucleoprotein complexes from mouse erythroleukemia cells. *Biochemistry*, **17**, 3228.
20. Lerner, M. R. and Steitz, J. A. (1979) Antibodies to small nuclear RNAs complexed with

proteins are produced by patients with systemic lupus erythematosus. *Proc. Natl Acad. Sci. USA*, **76**, 5495.
21. Brunel, C., Sri-Widada, J., and Jeanteur, P. (1985) snRNPs and scRNPs in eucaryotic cells. *Prog. Mol. Subcell.*, **9**, 1.
22. Lührmann, R. (1988) snRNP Proteins. In *Structure and function of major and minor small nuclear ribonucleoprotein particles*, Birnstiel, M. L. (ed.). Springer-Verlag, Berlin, p. 71.
23. Behrens, S.-E., Tyc, K., Kastner, B., Reichelt, J., and Lührmann, R. (1993) Small nuclear ribonucleoprotein (RNP) U2 contains numerous additional proteins and has a bipartite RNP structure under splicing conditions. *Mol. Cell. Biol.*, **13**, 307.
24. Black, D. L. and Pinto, A. L. (1989) U5 small nuclear ribonucleoprotein: RNA structure analysis and ATP-dependent interaction with U4/U6. *Mol. Cell. Biol.*, **9**, 3350.
25. Behrens, S.-E. and Lührmann, R. (1991) Immunoaffinity purification of a [U4/U6·U5] tri-snRNP from human cells. *Genes Dev.*, **5**, 1439.
26. Bach, M., Bringmann, P., and Lührmann, R. (1990) Purification of small nuclear ribonucleoprotein particles with antibodies against modified nucleotides of small nuclear RNAs. *Methods Enzymol.*, **181**, 232.
27. Tan, E. M. (1982) Autoantibodies to nuclear antigens (ANA): Their immunobiology and medicine. *Adv. Immunol.*, **33**, 167.
28. Lehmeier, T., Foulaki, K., and Lührmann, R. (1990) Evidence for three distinct D proteins, which react differentially with anti-Sm autoantibodies, in the cores of the major snRNPs U1, U2, U4/U6 and U5. *Nucleic Acids Res.*, **18**, 6475.
29. Hackl, W., Fischer, U., and Lührmann, R. (1993) A 69 kDa protein that associates reversibly with the Sm core domain of several spliceosomal snRNP species. *J. Cell Biol.*, **124**, 261.
30. Gozani, O., Patton, J. G., and Reed, R. (1994) A novel set of spliceosome-associated proteins and the essential splicing factor PSF bind stably to pre-mRNA prior to catalytic step II of the splicing reaction. *EMBO J.*, **13**, 3356.
31. Bach, M., Winkelmann, G., and Lührmann, R. (1989) 20S small nuclear ribonucleoprotein U5 shows a surprisingly complex protein composition. *Proc. Natl Acad. Sci. USA*, **86**, 6038.
32. Moore, M. J., Query, C. C., and Sharp, P. A. (1993) Splicing of precursors to messenger RNAs by the spliceosome. In *The RNA world*, Gesteland, R. F. and Atkins, J. F. (ed.). Cold Spring Harbor Laboratory Press, p. 303.
33. Hamm, J. and Mattaj, I. W. (1989) An abundant U6 snRNP found in germ cells and embryos of *Xenopus laevis*. *EMBO J.*, **8**, 4179.
34. Blencow, B. J., Carmo-Fonseca, M., Behrens, S.-E., Lührmann, R., and Lamond, A. I. (1993) Interaction of the human autoantigen p150 with splicing snRNPs. *J. Cell Sci.*, **105**, 685.
35. Gröning, K., Palfi, Z., Gupta, S., Cross, M., Wolff, T., and Bindereif, A. (1991) A new U6 small nuclear ribonucleoprotein-specific protein conserved between *cis*- and *trans*-splicing systems. *Mol. Cell. Biol.*, **11**, 2026.
36. Fujii, T., Mimori, T., Hama, N., Suwa, A., Akizuki, M., Tojo, T., *et al.* (1992) Characterization of autoantibodies that recognize U4/U6 small ribonucleoprotein particles in serum from a patient with primary Sjögren's syndrome. *J. Biol. Chem.*, **267**, 16412.
37. Woppmann, A., Will, C. L., Kornstädt, U., Zuo, P., Manley, J. L., and Lührmann, R. (1993) Identification of an snRNP-associated kinase activity that phosphorylates arginine/serine rich domains typical of splicing factors. *Nucleic Acids Res.*, **21**, 2815.
38. Paterson, T., Beggs, J. D., Finnegan, D. J., and Lührmann, R. (1991) Polypeptide com-

ponents of *Drosophila* small nuclear ribonucleoprotein particles. *Nucleic Acids Res.*, **19**, 5877.

39. Agabian, N. (1990) *Trans* splicing of nuclear pre-mRNAs. *Cell*, **61**, 1157.
40. Solymosy, F. and Pollák, T. (1993) Uridylate-rich small nuclear RNAs (UsnRNAs), their genes and pseudogenes, and UsnRNPs in plants: Structure and function. A comparative approach. *Crit. Rev. Plant Sci.*, **12**, 275.
41. Woppmann, A., Patschinsky, T., Bringmann, P., Godt, F., and Lührmann, R. (1990) Characterisation of human and murine snRNP proteins by two-dimensional gel electrophoresis and phosphopeptide analysis of U1-specific 70K protein variants. *Nucleic Acids Res.*, **18**, 4427.
42. Feeney, R. J., Sauterer, R. A., Feeney, J. L., and Zieve, G. W. (1989) Cytoplasmic assembly and nuclear accumulation of mature small nuclear ribonucleoprotein particles. *J. Biol. Chem.*, **264**, 5776.
43. Fabrizio, P., Esser, S., Kastner, B., and Lührmann, R. (1994) Isolation of *S. cerevisiae* snRNPs: Comparison of U1 and U4/U6·U5 to their human counterparts. *Science*, **264**, 261.
44. Bordonné, R., Banroques, J., Abelson, J., and Guthrie, C. (1990) Domains of yeast U4 spliceosomal RNA required for PRP4 protein binding, snRNP-snRNP interactions, and pre-mRNA splicing *in vivo*. *Genes Dev.*, **4**, 1185.
45. Gallisson, F. and Legrain, P. (1993) The biochemical defects of prp4-1 and prp6-1 yeast splicing mutants reveal that the PRP6 protein is required for the accumulation of the [U4/U6·U5] tri-snRNP. *Nucleic Acids Res.*, **21**, 1555.
46. Shannon, K. W. and Guthrie, C. (1991) Suppressors of a U4 snRNA mutation define a novel U6 snRNP protein with RNA-binding motifs. *Genes Dev.*, **5**, 773.
47. Whittaker, E., Lossky, M., and Beggs, J. D. (1990) Affinity purification of spliceosomes reveals that the precursor RNA processing protein PRP8, a protein in the U5 small nuclear ribonucleoprotein particle, is a component of yeast spliceosomes. *Proc. Natl Acad. Sci. USA*, **87**, 2216.
48. Horowitz, D. S. and Abelson, J. (1993) A U5 small nuclear ribonucleoprotein particle protein involved only in the second step of pre-mRNA splicing in *Saccharomyces cerevisiae*. *Mol. Cell. Biol.*, **13**, 2959.
49. Behrens, S.-E., Galisson, F., Legrain, P., and Lührmann, R. (1993) Evidence that the 60-kDa protein of 17S U2 small nuclear ribonucleoprotein is immunologically and functionally related to the yeast PRP9 splicing factor and is required for the efficient formation of prespliceosomes. *Proc. Natl Acad. Sci. USA*, **90**, 8229.
50. Smith, V. and Barrell, B. G. (1991) Cloning of a yeast U1 snRNP 70K protein homologue: Functional conservation of an RNA-binding domain between humans and yeast. *EMBO J.*, **10**, 2627.
51. Liao, X. C., Tang, J., and Rosbash, M. (1993) An enhancer screen identifies a gene that encodes the yeast U1 snRNP A protein: Implications for snRNP protein function in pre-mRNA splicing. *Genes Dev.*, **7**, 419.
52. Rymond, B. C. (1993) Convergent transcripts of the yeast *PRP38-SMD1* locus encode two essential splicing factors, including the D1 core polypeptide of small nuclear ribonucleoprotein particles. *Proc. Natl Acad. Sci. USA*, **90**, 848.
53. Stanford, D. R., Kehl, M., Perry, C. A., Holicky, E. L., Harvey, S. E., Rolhetter, N. M., *et al.* (1988) The complete primary structure of the human snRNP E protein. *Nucleic Acids Res.*, **16**, 10593.
54. Rokeach, L. A., Haselby, J. A., and Hoch, S. A. (1988) Molecular cloning of a cDNA encoding the human Sm-D autoantigen. *Proc. Natl Acad. Sci. USA*, **85**, 4832.

55. Lehmeier, T., Raker, V. A., Hermann, H., and Lührmann, R. (1994) cDNA cloning of the Sm proteins D2 and D3 from human snRNPs: Evidence for a direct D1-D2 interaction. *Proc. Natl Acad. Sci. USA*, **91**, 12317.
56. Ohosone, Y., Mimori, T., Griffith, A., Akizuki, M., Homma, M., Craft, J., and Hardin, J. A. (1989) Molecular cloning of cDNA encoding Sm autoantigen: Derivation of a cDNA for a B polypeptide of the U series of small nuclear ribonucleoprotein particles. *Proc. Natl Acad. Sci. USA*, **86**, 4249.
57. Van Dam, A., Winkel, I., Zijlstra-Baalbergen, J., Smeenk, R., and Cuypers, H. T. (1989) Cloned human snRNP proteins B and B′ differ only in their carboxy-terminal part. *EMBO J.*, **8**, 3853.
58. Sillekens, P. T. G., Beijer, R. P., Habets, W. J., and van Venrooij, W. J. (1988) Human U1 snRNP-specific C protein: complete cDNA and protein sequence and identification of a multigene family in mammals. *Nucleic Acids Res.*, **16**, 8307.
59. Yamamoto, K., Miura, H., Moroi, Y., Yoshinoya, S., Goto, M., Nishioka, K., *et al.* (1988) Isolation and characterization of a complementary DNA expressing human U1 small nuclear ribonucleoprotein C polypeptide. *J. Immunol.*, **140**, 311.
60. Habets, W. J., Sillekens, P. T. G., Hoet, M. H., Schalken, J. A., Roebroek, A. J. M., Leunissen, J. A. M., *et al.* (1987) Analysis of a cDNA clone expressing a human autoimmune antigen. Full-length sequence of the U2 small nuclear RNA-associated B″ antigen. *Proc. Natl Acad. Sci. USA*, **84**, 2421.
61. Sillekens, P. T. G., Habets, W. J., Beijer, R. P., and van Venrooij, W. J. (1987) cDNA cloning of the human U1 snRNP-associated A protein: extensive homology between U1 and U2 snRNP-specific proteins. *EMBO J.*, **6**, 3841.
62. Sillekens, P. T. G., Beijer, R. P., Habets, W. J., and van Venrooij, W. J. (1989) Molecular cloning of the cDNA for the human U2 snRNP-specific A′ protein. *Nucleic Acids Res.*, **17**, 1893.
63. Theissen, H., Etzerodt, M., Reuter, R., Schneider, C., Lottspeich, F., Argos, P., *et al.* (1986) Cloning of the human cDNA for the U1 RNA associated 70K protein. *EMBO J.*, **5**, 3209.
64. Spritz, R. A., Strunk, K., Surowy, C. S., Hoch, S. O., Barton, D. E., and Francke, U. (1987) The human U1-70K protein: cDNA cloning, chromosomal localization, expression, alternative splicing and RNA-binding. *Nucleic Acids Res.*, **15**, 10373.
65. Query, C. C., Bentley, R. C., and Keene, J. D. (1989) A common RNA recognition motif identified within a defined RNA binding domain of the 70K U1 snRNP protein. *Cell*, **57**, 89.
66. Bennett, M. and Reed, R. (1993) Correspondence between a mammalian spliceosome component and an essential yeast splicing factor. *Science*, **262**, 105.
67. Chiara, M. D., Champion-Arnaud, P., Buvoli, M., Nadal-Ginard, B., and Reed, R. (1994) Specific protein–protein interactions between the essential mammalian spliceosome-associated SAP61 and SAP114. *Proc. Natl Acad. Sci. USA*, **91**, 6403.
68. Burd, C. G. and Dreyfuss, G. (1994) Conserved structures and diversity of functions of RNA-binding proteins. *Science*, **265**, 615.
69. Nagai, K., Oubridge, C., Jessen, T. H., Li, J., and Evans, P. R. (1990) Crystal structure of the RNA-binding domain of the U1 small nuclear ribonucleoprotein A. *Nature*, **348**, 515.
70. Hoffman, D. W., Query, C. C., Golden, B. L., White, S. W., and Keene, J. D. (1991) RNA-binding domain of the A protein component of the U1 small nuclear ribonucleoprotein analyzed by NMR spectroscopy is structurally similar to ribosomal proteins. *Proc. Natl Acad. Sci. USA*, **88**, 2495.

71. Ge, H., Zuo, P., and Manley, J. L. (1991) Primary structure of the human splicing factor ASF reveals similarities with *Drosophila* regulators. *Cell*, **66**, 373.
72. Krainer, A. R., Mayeda, A., Kozak, D., and Binns, G. (1991) Functional expression of cloned human splicing factor SF2: Homology to RNA-binding proteins, U1 70K, and *Drosophila* splicing regulators. *Cell*, **66**, 383.
73. Zamore, P. D., Patton, J. G., and Green, M. R. (1992) Cloning and domain structure of the mammalian splicing factor U2AF. *Nature*, **355**, 609.
74. Fu, X.-D. and Maniatis, T. (1992) Isolation of a complementary DNA that encodes the mammalian splicing factor SC35. *Science*, **256**, 535.
75. Bingham, P. M., Chou, T.-B., Mims, I., and Zachar, Z. (1988) On/Off regulation of gene expression at the level of splicing. *Trends Genet.*, **4**, 134.
76. Mattox, W., Ryner, L., and Baker, B. S. (1992) Autoregulation and multifunctionality among *trans*-acting factors that regulate alternative pre-mRNA processing. *J. Biol. Chem.*, **267**, 19023.
77. Nelissen, R. L. H., Heinrichs, V., Habets, W. J., Simons, F., Lührmann, R., and van Venrooij, W. J. (1991) Zinc finger-like structure in U1-specific protein C is essential for specific binding to U1 snRNP. *Nucleic Acids Res.*, **19**, 449.
78. Legrain, P. and Choulika, A. (1990) The molecular characterization of PRP6 and PRP9 yeast genes reveals a new cysteine/histidine motif common to several splicing factors. *EMBO J.*, **9**, 2775.
79. Fresco, L. D., Harper, D. S., and Keene, J. D. (1991) Leucine periodicity of U2 small nuclear ribonucleoprotein particle (snRNP) A′ protein is implicated in snRNP assembly via protein–protein interactions. *Mol. Cell. Biol.*, **11**, 1578.
80. Fisher, D. E., Conner, G. E., Reeves, W. H., Wisniewolski, R., and Blobel, G. (1985) Small nuclear ribonucleoprotein particle assembly *in vivo*: demonstration of a 6S RNA-free core precursor and posttranslational modification. *Cell*, **42**, 751.
81. Chu, J.-L. and Elkon, K. B. (1991) The small nuclear ribonucleoproteins, SmB and B′, are products of a single gene. *Gene*, **97**, 311.
82. Rokeach, L. A. and Hoch, S. O. (1992) B-cell epitopes of Sm autoantigens. *Mol. Biol. Rep.*, **16**, 165.
83. McAllister, G., Roby-Shemkovitz, A., Amara, S. G., and Lerner, M. R. (1989) cDNA sequence of the rat U snRNP-associated protein N: Description of a potential Sm epitope. *EMBO J.*, **8**, 1177.
84. Li, S., Klein, E. S., Russo, A. F., Simmons, D. M., and Rosenfeld, M. G. (1989) Isolation of cDNA clones encoding small nuclear ribonucleoparticle-associated proteins with different tissue specificities. *Proc. Natl Acad. Sci. USA*, **86**, 9778.
85. Sharpe, N. G., Williams, D. G., Norton, P. M., and Latchman, D. S. (1989) Cell-type specific expression of the SmB′ splicing protein and alternative RNA splicing. *Biochem. Soc. Trans.*, **17**, 357.
86. Mancebo, R., Lo, P. C. H., and Mount, S. M. (1990) Structure and expression of the *Drosophila melanogaster* gene for the U1 small nuclear ribonucleoprotein particle 70K protein. *Mol. Cell. Biol.*, **10**, 2492.
87. Rymond, B. C., Rokeach, L. A., and Hoch, S. O. (1993) Human snRNP polypeptide D1 promotes pre-mRNA splicing in yeast and defines nonessential yeast Smd1p sequences. *Nucleic Acids Res.*, **21**, 3501.
88. Brosi, R., Gröning, K., Behrens, S.-E., Lührmann, R., and Krämer, A. (1993) Interaction of mammalian splicing factor SF3a with U2 snRNP and relationship of its 60-kDa subunit to yeast PRP9. *Science*, **262**, 102.

89. Anderson, G. J., Bach, M., Lührmann, R., and Beggs, J. D. (1989) Conservation between yeast and man of a protein associated with U5 small nuclear ribonucleoprotein. *Nature*, **342**, 819.
90. Eliceiri, G. L. and Sayavedra, M. S. (1976) Small RNAs in the nucleus and cytoplasm of HeLa cells. *Biochem. Biophys. Res. Commun.*, **72**, 507.
91. Neuman de Vegvar, H. E. and Dahlberg, J. E. (1990) Nucleocytoplasmic transport and processing of small nuclear RNA precursors. *Mol. Cell. Biol.*, **10**, 3365.
92. Yang, H., Moss, M. L., Lund, E., and Dahlberg, J. E. (1992) Nuclear processing of the 3′-terminal nucleotides of pre-U1 RNA in *Xenopus laevis* oocytes. *Mol. Cell. Biol.*, **12**, 1553.
93. Mattaj, I. (1986) Cap trimethylation of U snRNA is cytoplasmic and dependent on U snRNP protein binding. *Cell*, **46**, 905.
94. Vankan, P., McGuigan, C., and Mattaj, I. W. (1990) Domains of U4 and U6 snRNAs required for snRNP assembly and splicing complementation in *Xenopus* oocytes. *EMBO J.*, **9**, 3397.
95. Terns, M. P., Lund, E., and Dahlberg, J. E. (1992) 3′-End-dependent formation of U6 small nuclear ribonucleoprotein particles in *Xenopus laevis* oocyte nuclei. *Mol. Cell. Biol.*, **12**, 3032.
96. Lund, E. and Dahlberg, J. E. (1992) Cyclic 2′,3′-phosphates and nontemplated nucleotides at the 3′ end of spliceosomal U6 small nuclear RNA's. *Science*, **255**, 327.
97. Zieve, G. W. and Sauterer, R. A. (1990) Cell biology of the snRNP particles. *CRC Crit. Rev. Biochem. Mol. Biol.*, **25**, 1.
98. Sauterer, R. A., Goyal, A., and Zieve, G. W. (1990) Cytoplasmic assembly of small nuclear ribonucleoprotein particles from 6 S and 20 S RNA-free intermediates in L929 mouse fibroblasts. *J. Biol. Chem.*, **265**, 1048.
99. Kambach, C. and Mattaj, I. W. (1992) Intracellular distribution of the U1A protein depends on active transport and nuclear binding to U1 snRNA. *J. Cell Biol.*, **118**, 11.
100. Jantsch, M. F. and Gall, J. G. (1992) Assembly and localization of the U1-specific snRNP C protein in the amphibian oocyte. *J. Cell Biol.*, **119**, 1037.
101. Kambach, C. and Mattaj, I. W. (1994) Nuclear transport of the U2 snRNP-specific U2B″ protein is mediated by both direct and indirect signalling mechanisms. *J. Cell Sci.*, **107**, 1807.
102. Fischer, U., Darzynkiewicz, E., Tahara, S. M., Dathan, N. A., Lührmann, R., and Mattaj, I. W. (1991) Diversity in the signals required for nuclear accumulation of U snRNPs and variety in the pathways of nuclear transport. *J. Cell Biol.*, **113**, 705.
103. Michaud, N. and Goldfarb, D. (1992) Microinjected U snRNAs are imported to oocyte nuclei via the nuclear pore complex by three distinguishable targeting pathways. *J. Cell Biol.*, **116**, 851.
104. Fischer, U., Sumpter, V., Sekine, M., Satoh, T., and Lührmann, R. (1993) Nucleo-cytoplasmic transport of U snRNPs: Definition of a nuclear location signal in the Sm core domain that binds a transport receptor independently of the m3G cap. *EMBO J.*, **12**, 573.
105. Fischer, U. and Lührmann, R. (1990) An essential signaling role for the m3G cap in the transport of U1 snRNP to the nucleus. *Science*, **249**, 786.
106. Hamm, J., Darzynkiewicz, E., Tahara, S. M., and Mattaj, I. W. (1990) The trimethyl-guanosine cap structure of U1 snRNA is a component of a bipartite nuclear targeting signal. *Cell*, **62**, 569.
107. Marshallsay, C. and Lührmann, R. (1993) *In vitro* nuclear import of UsnRNPs: cytosolic factors from *Xenopus* oocytes mediate m3G cap dependence of U1 and U2 snRNP transport in somatic cells. *EMBO J.*, **13**, 222.

108. Branlant, C., Krol, A., Ebel, J. P., Lazar, E., Haendler, B., and Jacob, M. (1982) U2 RNA shares a structural domain with U1, U4 and U5 RNAs. *EMBO J.*, **1**, 1259.
109. Riedel, N., Wolin, S., and Guthrie, C. (1987) A subset of yeast snRNAs contains functional binding sites for the highly conserved Sm antigen. *Science*, **235**, 328.
110. Tollervey, D. and Mattaj, I. W. (1987) Fungal small nuclear ribonucleoproteins share properties with plant and vertebrate U snRNPs. *EMBO J.*, **6**, 469.
111. Jones, M. H. and Guthrie, C. (1990) Unexpected flexibility in an evolutionarily conserved protein-RNA interaction: Genetic analysis of the Sm binding site. *EMBO J.*, **9**, 2555.
112. Heinrichs, V., Hackl, W., and Lührmann, R. (1992) Direct binding of small nuclear ribonucleoprotein G to the Sm site of small nuclear RNA. Ultraviolet light cross-linking of protein G to the AAU stretch within the Sm site (AAUUUGUGG) of U1 small nuclear ribonucleoprotein reconstituted *in vitro*. *J. Mol. Biol.*, **227**, 15.
113. Jarmolowski, A. and Mattaj, I. W. (1993) The determinants for Sm protein binding to *Xenopus* U1 and U5 snRNAs are complex and non-identical. *EMBO J.*, **12**, 223.
114. Nelissen, R. L. H., Will, C. L., van Venrooij, W., and Lührmann, R. (1994) The association of the U1-specific 70K and C proteins with U1 snRNPs is mediated in part by common U snRNP proteins. *EMBO J.*, **13**, 4113.
115. Kastner, B. and Lührmann, R. (1989) Electron microscopy of U1 small nuclear ribonucleoprotein particles: Shape of the particle and position of the 5′ RNA terminus. *EMBO J.*, **8**, 277.
116. Kastner, B., Bach, M., and Lührmann, R. (1990) Electron microscopy of small nuclear ribonucleoprotein (snRNP) particles U2 and U5: Evidence for a common structure-determining principle in the major U snRNP family. *Proc. Natl Acad. Sci. USA*, **87**, 1710.
117. Kastner, B., Bach, M., and Lührmann, R. (1991) Electron microscopy of U4/U6 snRNP reveals a Y-shaped U4 and U6 RNA containing domain protruding from the U4 core RNP. *J. Cell Biol.*, **112**, 1065.
118. Patton, J. R. and Pederson, T. (1988) The M_r 70,000 protein of the U1 small ribonucleoprotein particle binds to the 5′stem–loop of U1 RNA and interacts with Sm domain proteins. *Proc. Natl Acad. Sci. USA*, **85**, 747.
119. Hamm, J., Kazmaier, M., and Mattaj, I. W. (1987) *In vitro* assembly of U1 snRNPs. *EMBO J.*, **6**, 3479.
120. Surowy, C. S., Van Santen, V. L., Scheib-Wixted, S. M., and Spritz, R. A. (1989) Direct, sequence-specific binding of the human U1-70K ribonucleoprotein antigen protein to loop I of U1 small nuclear RNA. *Mol. Cell. Biol.*, **9**, 4179.
121. Lutz-Freyermuth, C., and Keene, J. D. (1989) The U1 RNA-binding site of the U1 small nuclear ribonucleoprotein (snRNP)-associated A protein suggests a similarity with U2 snRNPs. *Mol. Cell. Biol.*, **9**, 2975.
122. Patton, J. R., Habets, W., van Venrooij, W. J., and Pederson, T. (1989) U1 small nuclear ribonucleoprotein particle-specific proteins interact with the first and second stem–loops of U1 RNA, with the A protein binding directly to the RNA independently of the 70K and Sm proteins. *Mol. Cell. Biol.*, **9**, 3360.
123. Bach, M., Krol, A., and Lührmann, R. (1990) Structure-probing of U1 snRNPs gradually depleted of the U1-specific proteins A, C and 70k. Evidence that A interacts differentially with developmentally regulated mouse U1 snRNA variants. *Nucleic Acids Res.*, **18**, 449.
124. Scherly, D., Boelens, W., van Venrooij, W. J., Dathan, N. A., Hamm, J., and Mattaj, I. W. (1989) Identification of the RNA binding segment of human U1 A protein and definition of its binding site on U1 snRNA. *EMBO J.*, **8**, 4163.

125. Kao, H.-Y. and Siliciano, P. G. (1992) The yeast homolog of the U1 snRNP protein 70K is encoded by the *SNP1* gene. *Nucleic Acids Res.*, **20**, 4009.
126. Cooper, M., Johnston, L. H., and Beggs, J. D. (1995) Identification and characterization of Uss1p (Sdb23p): a novel U6 snRNA-associated protein with significant similarity to core proteins of small nuclear ribonucleoproteins. *EMBO J.*, **14**, 2066.
127. Kastner, B., Kornstädt, U., Bach, M., and Lührmann, R. (1992) Structure of the small nuclear RNP particle U1: Identification of the two structural protuberances with RNP-antigens A and 70K. *J. Cell Biol.*, **116**, 839.
128. Scherly, D., Boelens, W., Dathan, N. A., van Venrooij, W. J., and Mattaj, I. W. (1990) Major determinants of the specificity of interaction between small nuclear ribonucleoproteins U1A and U2B″ and their cognate RNAs. *Nature*, **345**, 502.
129. Scherly, D., Dathan, N. A., Boelens, W., van Venrooij, W. J., and Mattaj, I. W. (1990) The U2B″ RNP motif as a site of protein–protein interaction. *EMBO J.*, **9**, 3675.
130. Bach, M. and Lührmann, R. (1991) Protein–RNA interactions in 20S U5 snRNPs. *Biochim. Biophys. Acta*, **1088**, 139.
131. Frank, D. N., Roiha, H., and Guthrie, C. (1994) Architecture of the U5 small nuclear RNA. *Mol. Cell. Biol.*, **14**, 2180.
132. Wolff, T. and Bindereif, A. (1992) Reconstituted mammalian U4/U6 snRNP complements splicing: A mutational analysis. *EMBO J.*, **11**, 345.
133. Zhuang, Y. and Weiner, A. (1986) A compensatory base change in U1 snRNA suppresses a 5′ splice site mutation. *Cell*, **46**, 827.
134. Siliciano, P. G. and Guthrie, C. (1988) 5′ splice site selection in yeast: genetic alterations in base-pairing with U1 reveal additional requirements. *Genes Dev.*, **2**, 1258.
135. Eperon, I. C., Ireland, D. C., Smith, R. A., Mayeda, A., and Krainer, A. R. (1993) Pathways for selection of 5′ splice sites by U1 snRNPs and SF2/ASF. *EMBO J.*, **12**, 3607.
136. Kohtz, J. D., Jamison, S. F., Will, C. L., Zuo, P., Lührmann, R., Garcia-Blanco, M. A., and Manley, J. L. (1994) Protein–protein interactions and 5′-splice-site recognition in mammalian mRNA precursors. *Nature*, **368**, 119.
137. Eperon, L. P., Estibeiro, J. P., and Eperon, I. C. (1986) The role of nucleotide sequences in splice site selection in eukaryotic pre-messenger RNA. *Nature*, **324**, 280.
138. Nelson, K. K. and Green, M. R. (1990) Mechanism for cryptic splice site activation during pre-mRNA splicing. *Proc. Natl Acad. Sci. USA*, **87**, 6253.
139. Nelson, K. and Green, M. R. (1988) Splice site selection and ribonucleoprotein complex assembly during *in vitro* pre-mRNA splicing. *Genes Dev.*, **2**, 319.
140. Séraphin, B., Kretzner, L., and Rosbash, M. (1988) A U1 snRNA: pre-mRNA base pairing interaction is required early in yeast spliceosome assembly but does not uniquely define the 5′ cleavage site. *EMBO J.*, **7**, 2533.
141. Séraphin, B. and Rosbash, M. (1990) Exon mutations uncouple 5′ splice site selection from U1 snRNA pairing. *Cell*, **63**, 619.
142. Séraphin, B. and Rosbash, M. (1989) Identification of functional U1 snRNA–pre-mRNA complexes committed to spliceosome assembly and splicing. *Cell*, **59**, 349.
143. Michaud, S. and Reed, R. (1993) A functional association between the 5′ and 3′ splice sites is established in the earliest prespliceosome complex (E) in mammals. *Genes Dev.*, **7**, 1008.
144. Reich, C. I., VanHoy, R. W., Porter, G. L., and Wise, J. A. (1992) Mutations at the 3′ splice site can be suppressed by compensatory base changes in U1 snRNA in fission yeast. *Cell*, **69**, 1159.
145. Séraphin, B. and Kandels-Lewis, S. (1993) 3′ splice site recognition in *S. cerevisiae* does not require base pairing with U1 snRNA. *Cell*, **73**, 803.

146. Goguel, V., Liao, X., Rymond, B. C., and Rosbash, M. (1991) U1 snRNP can influence 3′-splice site selection as well as 5′-splice site selection. *Genes Dev.*, **5**, 1430.
147. Steitz, J. A. (1992) Splicing takes a holliday. *Science*, **257**, 888.
148. Barabino, S. M. L., Blencowe, B. J., Ryder, U., Sproat, B. S., and Lamond, A. I. (1990) Targeted snRNP depletion reveals an additional role for mammalian U1 snRNP in spliceosome assembly. *Cell*, **63**, 293.
149. Ruby, S. W. and Abelson, J. (1988) An early hierarchic role of U1 small nuclear ribonucleoprotein in spliceosome assembly. *Science*, **242**, 1028.
150. Fu, X.-D. (1993) Specific commitment of different pre-mRNAs to splicing by single SR proteins. *Nature*, **365**, 82.
151. Hamm, J., Dathan, N. A., Scherly, D., and Mattaj, I. W. (1990) Multiple domains of U1 snRNA, including U1 specific protein binding sites, are required for splicing. *EMBO J.*, **9**, 1237.
152. Mount, S. M., Pettersson, I., Hinterberger, M., Karmas, A., and Steitz, J. A. (1983) The U1 small nuclear RNA-protein complex selectively binds a 5′ splice site *in vitro*. *Cell*, **33**, 509.
153. Heinrichs, V., Bach, M., Winkelmann, G., and Lührmann, R. (1990) U1-specific protein C needed for efficient complex formation of U1 snRNP with a 5′ splice site. *Science*, **247**, 69.
154. Tazi, J., Kornstädt, U., Rossi, F., Jeanteur, P., Cathala, G., Brunel, C., and Lührmann, R. (1993) Thiophosphorylation of U1-70K protein inhibits pre-mRNA splicing. *Nature*, **363**, 283.
155. Wu, J. Y. and Maniatis, T. (1993) Specific interactions between proteins implicated in splice site selection and regulated alternative splicing. *Cell*, **75**, 1061.
156. Krämer, A. and Utans, U. (1991) Three protein factors (SF1, SF3 and U2AF) function in pre-splicing complex formation in addition to snRNPs. *EMBO J.*, **10**, 1503.
157. Ruskin, B., Zamore, P. D., and Green, M. R. (1988) A factor, U2AF, is required for U2 snRNP binding and splicing complex assembly. *Cell*, **52**, 207.
158. Krämer, A. (1988) Presplicing complex formation requires two proteins and U2 snRNP. *Genes Dev.*, **2**, 1155.
159. Parker, R., Siliciano, P. G., and Guthrie, C. (1987) Recognition of the TACTAAC box during mRNA splicing in yeast involves base pairing to the U2-like snRNP. *Cell*, **49**, 229.
160. Wu, J. and Manley, J. L. (1989) Mammalian pre-mRNA branch site selection by U2 snRNP involves base pairing. *Genes Dev.*, **3**, 1553.
161. Wu, J. and Manley, J. L. (1991) Base pairing between U2 and U6 snRNAs is necessary for splicing of a mammalian pre-mRNA. *Nature*, **352**, 818.
162. Datta, B. and Weiner, A. M. (1991) Genetic evidence for base pairing between U2 and U6 snRNA in mammalian mRNA splicing. *Nature*, **352**, 821.
163. Lamond, A. I., Sproat, B., Ryder, U., and Hamm, J. (1989) Probing the structure and function of U2 snRNP with antisense oligonucleotides made of 2′-OMe RNA. *Cell*, **58**, 383.
164. Hausner, T.-P., Giglio, L. M., and Weiner, A. M. (1990) Evidence for base-pairing between mammalian U2 and U6 small nuclear ribonucleoprotein particles. *Genes Dev.*, **4**, 2146.
165. McPheeters, D. S. and Abelson, J. (1992) Mutational analysis of the yeast U2 snRNA suggests a structural similarity to the catalytic core of group I introns. *Cell*, **71**, 819.
166. Bordonné, R. and Guthrie, C. (1992) Human and human–yeast chimeric U6 snRNA

genes identify structural elements required for expression in yeast. *Nucleic Acids Res.*, **20**, 479

167. Madhani, H. D. and Guthrie, C. (1992) A novel base-pairing interaction between U2 and U6 snRNAs suggests a mechanism for the catalytic activation of the spliceosome. *Cell*, **71**, 803.
168. Brosi, R., Gröning, K., Behrens, S.-E., Lührmann, R., and Krämer, A. (1993) Interaction of mammalian splicing factor SF3a with U2 snRNP and relation of its 60-kD subunit to yeast PRP9. *Science*, **262**, 102.
169. Newman, A. and Norman, C. (1991) Mutations in yeast U5 snRNA alter the specificity of 5′ splice-site cleavage. *Cell*, **65**, 115.
170. Newman, A. J. and Norman, C. (1992) U5 snRNA interacts with exon sequences at 5′ and 3′ splice sites. *Cell*, **68**, 1.
171. Wyatt, J. R., Sontheimer, E. J., and Steitz, J. A. (1992) Site-specific cross-linking of mammalian U5 snRNP to the 5′ splice site before the first step of pre-mRNA splicing. *Genes Dev.*, **6**, 2542.
172. Wassarman, D. A. and Steitz, J. A. (1992) Interactions of small nuclear RNA's with precursor messenger RNA during *in vitro* splicing. *Science*, **257**, 1918.
173. Sontheimer, E. J. and Steitz, J. A. (1993) The U5 and U6 small nuclear RNAs as active site components of the spliceosome. *Science*, **262**, 1989.
174. Winkelmann, G., Bach, M., and Lührmann, R. (1989) Evidence from complementation assays *in vitro* that U5 snRNP is required for both steps of mRNA splicing. *EMBO J.*, **8**, 3105.
175. Yean, S.-L. and Lin, R.-J. (1991) U4 small nuclear RNA dissociates from a yeast spliceosome and does not participate in the subsequent splicing reaction. *Mol. Cell. Biol.*, **11**, 5571.
176. Blencowe, B. J., Sproat, B. S., Ryder, U., Barabino, S., and Lamond, A. I. (1989) Antisense probing of the human U4/U6 snRNP with biotinylated 2′-OMe RNA oligonucleotides. *Cell*, **59**, 531.
177. Brow, D. A. and Guthrie, C. (1989) RNA processing: Splicing a spliceosomal RNA. *Nature*, **337**, 14.
178. Tani, T. and Ohshima, Y. (1989) The gene for the U6 small nuclear RNA in fission yeast has an intron. *Nature*, **337**, 87.
179. Tani, T. and Ohshima, Y. (1991) mRNA-type introns in U6 small nuclear RNA genes: Implications for the catalysis in pre-mRNA splicing. *Genes Dev.*, **5**, 1022.
180. Fabrizio, P. and Abelson, J. (1990) Two domains of yeast U6 small nuclear RNA required for both steps of nuclear precursor messenger RNA splicing. *Science*, **250**, 404.
181. Madhani, H. D., Bordonné, R., and Guthrie, C. (1990) Multiple roles for U6 snRNA in the splicing pathway. *Genes Dev.*, **4**, 2264.
182. Wolff, T. and Bindereif, A. (1993) Conformational changes of U6 RNA during the spliceosome cycle: An intramolecular helix is essential both for initiating the U4–U6 interaction and for the first step of splicing. *Genes Dev.*, **7**, 1377.
183. Sawa, H. and Abelson, J. (1992) Evidence for a base-pairing interaction between U6 small nuclear RNA and the 5′ splice site during the splicing reaction in yeast. *Proc. Natl Acad. Sci. USA*, **89**, 11269.
184. Sawa, H. and Shimura, Y. (1992) Association of U6 snRNA with the 5′ splice site region of pre-mRNA in the spliceosome. *Genes Dev.*, **6**, 244.
185. Kandels-Lewis, S. and Séraphin, B. (1993) Role of U6 snRNA in 5′ splice site selection. *Science*, **262**, 2035.

186. Lesser, C. F. and Guthrie, C. (1993) Mutations in U6 snRNA that alter splice site specificity: Implications for the active site. *Science*, **262**, 1982.
187. Bond, U. (1988) Heat shock but not other stress inducers leads to the disruption of a subset of snRNPs and inhibition of *in vitro* splicing in HeLa cells. *EMBO J.*, **7**, 3509.
188. Utans, U., Behrens, S.-E., Lührmann, R., Kole, R., and Krämer, A. (1992) A splicing factor which is inactivated during *in vivo* heat shock is functionally equivalent to the [U4/U6·U5] triple-snRNP-specific proteins. *Genes Dev.*, **5**, 1439.
189. Whittaker, E. and Beggs, J. D. (1991) The yeast PRP8 protein interacts directly with pre-mRNA. *Nucleic Acids Res.*, **19**, 5483.
190. Ares, M., Jr and Igel, A. H. (1990) Lethal and temperature-sensitive mutations and their suppressors identify an essential structural element in U2 small nuclear RNA. *Genes Dev.*, **4**, 2132.
191. Myslinski, E. and Branlant, C. (1991) A phylogenetic study of U4 snRNA reveals the existence of an evolutionarily conserved secondary structure corresponding to 'free' U4 snRNA. *Biochimie*, **73**, 17.
192. Rinke, J., Appel, B., Digweed, M., and Lührmann, R. (1985) Localization of a base paired interaction between small nuclear RNAs U4 and U6 in intact U4/U6 ribonucleoprotein particles by psoralen cross-linking. *J. Mol. Biol.*, **185**, 721.
193. Wise, J. A. (1993) Guides to the heart of the spliceosome. *Science*, **262**, 1978.
194. Hermann, H., Fabrizio, P., Raker, V. A., Foulaki, K., Hornig, H., Brahms, H., and Lührmann, R. (1995) snRNP Sm proteins share two evolutionarily conserved sequence motifs which are involved in Sm protein-protein interactions. *EMBO J.*, **14**, 2076.
195. Séraphin, B. (1995) Sm and Sm-like proteins belong to a large family: identification of proteins of the U6 as well as the U1, U2, U4 and U5 snRNPs. *EMBO J.*, **14**, 2089.
196. Raker, V. A., Plessel, G., and Lührmann, R. (1996) The snRNP core assembly pathway: identification of stable core protein heteromeric complexes and an snRNP subcore particle *in vitro*. *EMBO J.*, **15**, 2256.
197. Fischer, U., Heinrich, J., Van Zee, K., Fanning, E., and Lührmann, R. (1994) Nuclear transport of U1 snRNP in somatic cells: Differences in signal requirement compared with *Xenopus laevis* oocytes. *J. Cell Biol.*, **125**, 971.
198. Ségault, V., Will, C. L., Sproat, B. S., and Lührmann, R. (1995) *In vitro* reconstitution of mammalian U2 and U5 snRNPs active in splicing: Sm proteins are functionally interchangeable and are essential for the formation of functional U2 and U5 snRNPs. *EMBO J.*, **14**, 4010.

6 Mammalian pre-mRNA splicing factors

JAVIER F. CÁCERES and ADRIAN R. KRAINER

1. Introduction

Since the discovery of pre-mRNA splicing in 1977, considerable progress has been made in dissecting the *cis*-elements and *trans*-acting factors involved in carrying out this complex reaction. Multiple protein and nucleoprotein components are necessary to catalyse pre-mRNA splicing, which occurs within a particle termed the spliceosome. The components of the splicing machinery, and/or factors involved in the assembly of the spliceosome, include the small nuclear ribonucleoprotein particles (snRNPs) U1, U2, U4/U6, and U5; the polypeptides that associate with hnRNA to form hnRNP particles (hnRNP proteins); and a large number of non-snRNP essential splicing factors (for earlier reviews see 1–6).

Whereas the components of the splicing apparatus in yeast have been largely identified and analysed using powerful genetic methods (Chapter 7), the study of mammalian splicing factors has relied primarily on biochemical methods. The first *in vitro* systems to study splicing of exogenous pre-mRNA were developed in 1984, and soon thereafter it became possible to fractionate splicing extracts to identify and characterize individual constituents of the mammalian splicing machinery. The best characterized components are the snRNPs, and in particular their snRNA moieties, in large measure because they can be studied with nucleic acid techniques and also because of the availability of autoimmune antibodies (Chapter 5).

The protein components of the mammalian splicing machinery have been studied in a number of ways:

(1) the polypeptide constituents of snRNP particles have been characterized primarily from the structural point of view and in terms of their snRNA-binding properties (Chapter 5);
(2) the hnRNP proteins were studied for many years as hnRNA-packaging proteins, which were of course suspected of involvement in RNA-processing reactions (Chapter 3);
(3) these and other proteins have been characterized more recently, as active or passive components of the spliceosome during its assembly (Chapter 4);
(4) several proteins have been identified on the basis of their specific binding to pre-mRNA sequence elements important for splicing;

(5) finally, several protein factors have been identified by biochemical fractionation based on functional complementation assays.

As expected, there is considerable overlap between components identified by different strategies, and this overlap may increase as each factor is characterized further.

In this chapter we will focus on the description of mammalian protein factors which have already been shown by biochemical assays to be required for one or both catalytic steps of general pre-mRNA splicing. Some of these components also appear to be involved in the regulation of alternative pre-mRNA splicing (Chapter 8), and these activities will also be described here. Selected factors that appear to function in alternative splicing, but are not known to be required for constitutive splicing, will be described insofar as their mechanism of action involves interactions with the constitutive splicing machinery. We will emphasize components for which the primary sequence has already been determined.

2. The SR protein family

Among the most extensively characterized metazoan splicing factors are a group of proteins related to each other by the arrangement of characteristic domains, by extensive sequence similarity, and by related biochemical activities. These proteins are known collectively as the SR protein family (7) and can be subdivided on the basis of structure into two subgroups, whose prototypes are the splicing factors SF2/ASF and SC35. These factors, which were identified and functionally characterized as splicing factors before the existence of a larger family became clear, will be described first.

2.1 Pre-mRNA splicing factor SF2/ASF

SF2/ASF is a 28 kDa protein factor that has an activity essential for constitutive pre-mRNA splicing *in vitro* (8), and in addition strongly influences splice-site selection in a concentration-dependent manner. The addition of SF2/ASF to nuclear extracts switches splice-site selection of pre-mRNAs bearing competing 5′ splice sites towards the most proximal 5′ splice site (9, 10).

SF2/ASF was identified in two separate ways. Using selective inactivation and biochemical complementation of mammalian splicing extracts, an essential splicing activity was shown to be required to complement an otherwise inactive cytosolic HeLa S100 extract (11). A single factor possessing complementing activity, termed SF2 (splicing factor 2), was purified from HeLa cell nuclear extracts, and was subsequently shown to modulate alternative splicing *in vitro* in a concentration-dependent manner. Increasing concentrations of SF2 were shown to switch splice-site selection towards the downstream-most 5′ splice site of a β-thalassemia pre-mRNA, in which a single-base change at position 1 of the 5′ splice site prevents normal splicing and activates three nearby cryptic 5′ splice sites. A similar effect was observed using β-globin model pre-mRNAs with duplicated 5′ splice sites (8, 9). A protein factor termed ASF (alternative splicing factor) was independently identified and purified from 293 cells, on the basis of an alternative splicing assay (10). In this assay, ASF was added to HeLa cell nuclear extracts and caused a change in 5′ splice-site selection of an SV40 early pre-mRNA, favouring the downstream small t-antigen 5′ splice site. Subsequently, sequence analysis of cloned cDNAs showed that SF2 and ASF were identical factors (12, 13). This protein of 248 amino acids has two copies of the RNA-recognition motif (RRM), separated by a glycine-rich hinge, and a C-terminal arginine- and serine-rich (RS or SR) domain.

The RRM is a conserved region of around 80 amino acids, which is present in one or several copies in a large family of RNA-binding proteins, many of which are involved in pre-mRNA and pre-rRNA processing, such as snRNP polypeptides, hnRNP proteins, and poly-(A) binding protein (reviewed in 14–16). Several of the most conserved residues cluster into two short submotifs, the RNP-1 octamer and the RNP-2 hexamer. A structural model for the N-terminal RRM of the U1-A polypeptide of U1 snRNP (Chapter 5) has been derived on the basis of X-ray diffraction and NMR studies (17, 18). It consists of a four-stranded antiparallel β-sheet and two α-helices, in the sequence $\beta_1\alpha_1\beta_2\beta_3\alpha_2\beta_4$. The RRM of hnRNP C1/C2 has a very similar folded structure to that of U1-A (19). In the folded structures, the two α-helices are on one side of the β-sheet, with the RNP-2 and RNP-1 segments lying adjacent on the central β_1 and β_3 strands. A hydrophobic core composed of conserved residues is probably involved in assuming and maintaining the correct tertiary structure, whereas several solvent-exposed aromatic residues within RNP-2 and RNP-1 are thought to contact bound RNA through ring-stacking interactions with single-stranded bases. These predicted stacking interactions have been recently confirmed in a co-crystal structure of the U1-A N-terminal RRM with bound RNA (20). Substitution of conserved residues within the RNP-1 and RNP-2 submotifs affects binding of the U1-A and U1-70K proteins to U1 snRNA (21, 22). Phylogenetic analyses and sequence alignments suggest that the RRM is an ancient conserved region, whose current diversity arose through duplication of genes and intragenic domains, with retention of a common tertiary structural fold (15, 16).

The second RRM of SF2/ASF is atypical, in that it shows poor conservation of the usually conserved RNP-2 and RNP-1 submotifs. However, the conservation of other typical RRM residues strongly suggests that its tertiary fold is similar to that of canonical RRMs (15). The other characteristic domain of SF2/ASF is the C-terminal RS domain, which is also found in several mammalian and *Drosophila* protein splicing factors. The RS domain of SF2/ASF is rich in consecutive RS dipeptides, in contrast to other proteins, in which Arg and Ser residues are dispersed and do not show the same type of periodicity. It is unclear what constitutes a minimal RS domain, both from the statistical relevance and protein structural standpoints (reviewed in 15).

Detailed structural and functional analyses of SF2/ASF allowed the identification of distinct domains of the protein that are involved in the constitutive and in the alternative splicing activities of this protein (23, 24). Whereas both an intact

N-terminal RRM and an RS domain are essential for the constitutive splicing activity, the RS domain is surprisingly dispensable for the *in vitro* alternative splicing activity. The requirement for an intact RS domain for complementation of inactive S100 extracts seems to be very stringent, since SF2/ASF mutant derivatives with an RT or KS domain are inactive, demonstrating that the presence of positive charges interspersed with polar residues that may be phosphorylated is not sufficient for constitutive splicing (23). In addition, it has been shown that RRM1 and RRM2 synergize for binding to RNA (23, 24).

Analysis of additional cDNA clones has shown that SF2/ASF pre-mRNA can itself be alternatively spliced by utilization of different 3′ splice sites (13). This alternative splicing generates the standard SF2/ASF mRNA when the distal 3′ splice site is chosen, and ASF-2 when the proximal 3′ splice site is selected. In addition, retention of an ASF pre-mRNA intron generates a third isoform, ASF-3. The two alternatively spliced isoforms of SF2/ASF, ASF-2, and ASF-3, lack the RS domain. Whether they encode stable proteins *in vivo*, which may have a regulatory role, is unknown. Even if these mRNAs do not encode stable proteins, the alternative splicing from which they originate may serve to down-regulate the levels of authentic SF2/ASF. Interestingly, ASF alternative transcripts are highly conserved between mice and humans (25).

2.2 Pre-mRNA splicing factor SC35

A monoclonal antibody raised against size-fractionated mammalian spliceosomes allowed the identification of another splicing factor, known as SC35 (spliceosomal component 35) (26). This protein is required for mammalian spliceosome assembly and was localized to the speckled region of the nucleoplasm (26, 27; see Chapter 2). A cDNA encoding SC35 was initially cloned and shown to encode an open reading frame (designated PR264), corresponding to the opposite strand of a reportedly *trans*-spliced c-*myb* exon (28). The SC35/PR264 gene is extremely well conserved in chicken and humans (28, 29) and codes for three major mRNA isoforms that are developmentally regulated and may represent alternative spliced forms of SC35/PR264. SC35 has a single RRM at the N-terminus, and a C-terminal RS domain, but lacks the second degenerate RRM that is present in SF2/ASF. Although SF2/ASF and SC35 only display 31% overall amino acid sequence identity, the two proteins have very similar activities *in vitro*. Thus, both proteins complement inactive S100 extracts and promote the selection of proximal alternative 5′ splice sites (30). In addition, SF2/ASF can complement a HeLa cell nuclear extract that has been depleted with SC35 monoclonal antibody (probably co-depleting SC35 and SF2/ASF), demonstrating that these proteins are interchangeable at least in these *in vitro* assays. It was also shown that both proteins modulate alternative 3′ splice-site selection *in vitro* (30).

The notion that these were not unique proteins, but rather members of a larger family of splicing regulators, became apparent when several proteins closely related in sequence were identified in different species, such as a 55 kDa protein in

Drosophila, termed SRp55 or B52 (31, 32), and a mouse protein, termed X16 (33), both of which display high homology and share structural features with SF2/ASF and SC35.

2.3 Other SR proteins

A family of nuclear phosphoproteins was identified by their common reactivity with a monoclonal antibody, termed mAb104, which was initially screened for its reactivity against active sites of RNA polymerase II transcription on lampbrush amphibian chromosomes (7, 34). The first suggestion that these proteins might be involved in mRNA processing came from studies showing that a similar pattern of nuclear immunostaining was obtained with the mAb104 and with anti-snRNP antibodies (32). This family of proteins is characterized by the presence of one or two RRMs, and a C-terminal domain rich in serine (S) and arginine (R) residues (Figs 1–3). The amino acid composition of the latter domain provided the basis for naming members of this family as SR proteins. Domains with repeated Arg-Ser dipeptides are often found in general splicing factors, such as SR proteins and U2AF, in gene-specific alternative splicing factors such as *Drosophila* SWAP, Tra, and Tra2 (see Chapter 8), and in the U1-70K snRNP protein (Chapter 5). This domain is a target for phosphorylation and while the highly basic RS domain can be expected to interact with the phosphate backbone of RNA, the state of serine phosphorylation should influence these putative electrostatic interactions. It is presently unclear whether phosphorylated or dephosphorylated RS domains are the active form of this domain (see Section 5.3). A role for this domain in nuclear localization of two *Drosophila* splicing regulators, SWAP and Tra has been described (35; Chapter 2). However, this is not the sole function of this domain, since the RS domains of U2AF^{65} and SF2/ASF have been shown to be essential for constitutive splicing *in vitro*, and in the case of SF2/ASF, both Arg and Ser residues are required (discussed in Sections 2.1 and 3).

cDNAs that encode the most abundant SR proteins have been isolated and nucleotide or partial amino acid sequence analysis showed that some of these proteins correspond to previously identified splicing regulators (7) (Fig. 1). For instance, SRp30a corresponds to SF2/ASF, SRp30b to SC35, and SRp20 to the mouse protein X16. Other cDNAs represent novel SR proteins, such as SRp75, 9G8, and SRp30c (36–38).

All mAb104-immunoreactive proteins can be purified from a variety of cell lines and animal tissues by a simple two-step purification procedure that consists of ammonium sulphate fractionation, followed by selective precipitation in the presence of millimolar concentrations of $MgCl_2$ (7, 32). Whereas the first step is a

---

Fig. 1 Primary structure of the known human SR proteins. The amino acid sequence of the eight known human SR proteins is shown in the one letter code, with the proteins listed in decreasing order of molecular mass. Arginine–serine (RS) or serine–arginine (SR) dipeptides are highlighted as white letters in black boxes. This clustering of dipeptides constitutes an RS domain, which in these proteins is found towards the C-terminus.

HUMAN SR PROTEINS

SRp75

MPRVYIGRLSYQARERDVERFFKGYGKILEVDLKNGYGFVEFDDLRDADDAVYELNGKDLCGERVI
VEHARGPRRDGSYGSGRSGYGYRRSGRDKYGPPTRTEYRLIVENLSSRCSWQDLKDYMRQAGEVTY
ADAHKGRKNEGVIEFVSYSDMKRALEKLDGTEVNGRKIRLVEDKPGSRRRRSYSRSRSHSRSRSRS
RHSRKSRSRSGSSKSSHSKSRSRSRSGSRSRSKSRSRSQSRSRSKKEKSRSPSKDKSRSRSHSAGK
SRSKSKDQAEEKIQNNDNVGKPKSRSPSRHKSKSKSRSRSQERRVEEEKRGSVEQGQEQEKSLRQS
RSRSRSKAGSRSRSRSRSKSKDKRKSRKRSREESRSRSRSRSKSERSRKRGSKRDSKAGSSKKKKK
EDTDRSQSRSPSRSVSKEREHAKSESSQREGRGESENAGRNEETRSRSRSNSKSKPNLPSESRSRS
KSASKTRSRSKSRSRSASRSPSRSRSRSHSRS

SRp55

MPRVYIGRLSYNVREKDIQRFFSGYGRLLEVDLKNGYGFVEFEDSRDADDAVYELNGKELCGEHVI
VEHARGPRRDRDGYSYGSRSGGGGYSSRRTSGRDKYGPPVRTEYRLIVENLSSRCSWQDLKDFMRQ
AGEVTYADAHKERTNEGVIEFRSYSDMKRALDKLDGTEINGRNIRLIEDKPRTSHRRSYSGSRSRS
RSRRRSRSRSRRSSRSRSRSISKSRSRSRSRSKGRSRSRSKGRKSRSKSKSKPKSDRGSHSHSRSR
SKDEYEKSRSRSRSRSPKENGKGDIKSKSRSRSQSRSNSPLPVPPSKARSVSPPPKRATSRSRSRS
RSKSRSRSRSSSRD

SRp40

MSGCRVFIGRLNPAAREKDVERFFKGYGRIRDIDLKRGFGFVEFEDPRDADDAVYELDGKELCSER
VTIEHARARSRGGRGRGRYSDRFSSRRPRNDRRNAPPVRTENRLIVENLSSRVSWQDLKDFMRQAG
EVTFADAHRPKLNEGVVEFASYGDLKNAIEKLSGKEINGRKIKLIEGSKRHSRSRSRSRSRTRSSS
RSRSRSRSRSRKSYSRSRSRSRSRSRSKSRSVSRSPVPEKSQKRGSSSRSKSPASVDRQRSRSRSR
SRSVDSGN

SF2/ASF

MSGGGVIRGPAGNNDCRIYVGNLPPDIRTKDIEDVFYKYGAIRDIDLKNRRGGPPFAFVEFEDPRD
AEDAVYGRDGYDYDGYRLRVEFPRSGRGTGRGGGGGGGGAPRGRYGPPSRRSENRVVVSGLPPSG
SWQDLKDHMREAGDVCYADVYRDGTGVVEFVRKEDMTYAVRKLDNTKFRSHEGETAYIRVKVDGPR
SPSYGRSRSRSRSRSRSRSRSNSRSRSYSPRRSRGSPRYSPRHSRSRSRT

SC35

MSYGRPPPDVEGMTSLKVDNLTYRTSPDTLRRVFEKYRRVGDVYIPRDRYTKESRGFAFVRFHDKR
DAEDAMDAMDGAVLDGRELRVQMARYGRPPDSHHSRRGPPPRRYGGGYGRRSRSPRRRRRSRSRS
RSRSRSRSRSRYSRSKSRSRTRSRSRSTSKSRSARRSKSKSSSVSRSRSRSRSRSRSRSPPPVSKR
ESKSRSRSKSPPKSPEEEGAVSS

SRp30c

MSGWADERGGEGDGRIYVGNLPTDVREKDLEDLFYKYGRIREIELKNRHGLVPFAFVRFEDPRDAE
DAIYGRNGYDYGQCRLRVEFPRTYGGRGGWPRGGRNGPPTRRSDFRVLVSGLPPSGSWQDLKDHMR
EAGDVCYADVQKDGVGMVEYLRKEDMEYALRKLDDTKFRSHEGETSYIRVYPERSTSYGYSRSRSG
SRGRDSPYQSRGSPHYFSPFRPY

9g8

MSRYGRYGGETKVYVGNLGTGAGKGELERAFSYYGPLRTVWIARNPPGFAFVEFEDPRDAEDAVRG
LDGKVICGSRVRVELSTGMPRRSRFDRPPARRPFDPNDRCYECGEKGHYAYDCHRYSRRRRSRSRS
RSHSRSRGRRYSRSRSRSRGRRSRSASPRRSRSISLRRSRSASLRRSRSGSIKGSRYFQSPSRSRS
RSRSISRPRSSRSKSRSPSPKRSRSPSGSPRRSASPERMD

SRp20

MHRDSCPLDCKVYVGNLGNNGNKTELERAFGYYGPLRSVWVARNPPGFAFVEFEDPRDAADAVREL
DGRTLCGCRVRVELSNGEKRSRNRGPPPSWGRRPRDDYRRRSPPPRRRSPRRRSFSRSRSRSLSRD
RRRERSLSRERNHKPSRSFSRSRSRSRSNERK

Fig. 2 Domain organization of human SR proteins. The eight known human SR proteins are drawn to scale in decreasing order of molecular mass. The grey boxes denote the RNA-recognition motifs (RRMs). The vertical black lines indicate the location of RS or SR dipeptides, with the thickness being proportional to the number of consecutive dipeptides; RSR or SRS was scored as a single dipeptide repeat.

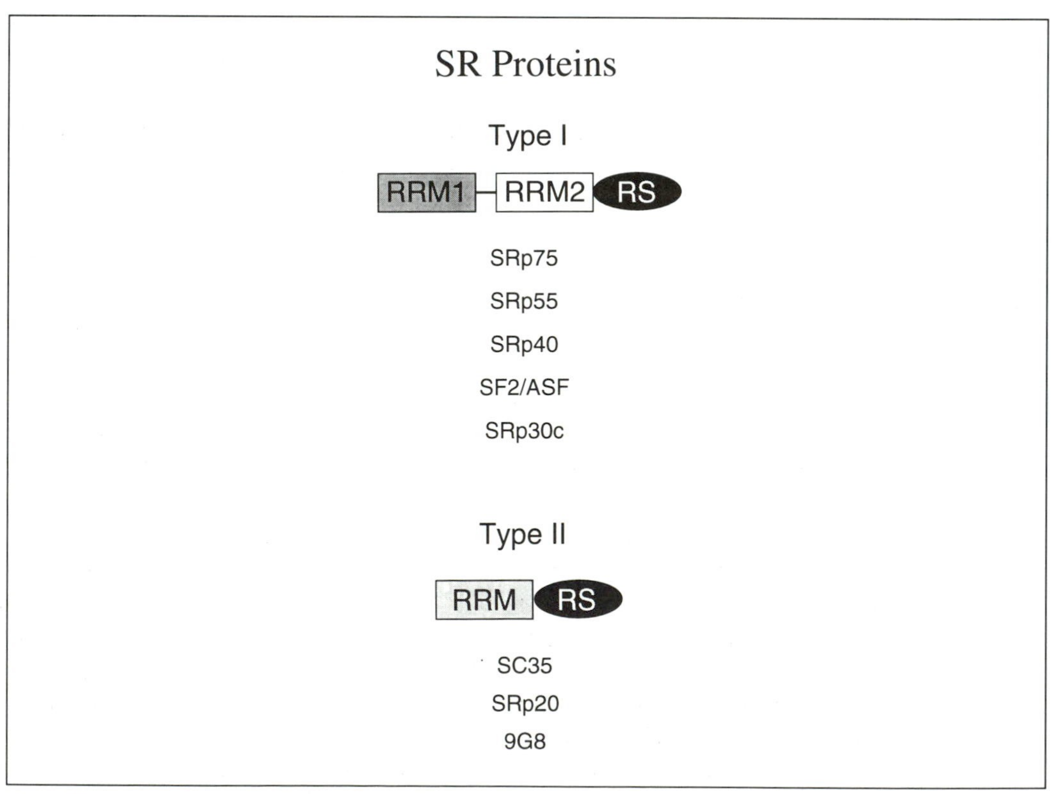

Fig. 3 Two subfamilies of SR proteins. SR proteins can be distinguished on the basis of the presence or absence of a second RRM. The two diagrams show the respective domain structures, with the known SR proteins of each type listed below each diagram. In SR proteins with two RRMs, the second one is atypical, in that it lacks conserved RNP-2 and RNP-1 submotif residues. The two RRMs are typically separated by a glycine-rich flexible linker, shown by a thin line. The first RRM of these proteins, and the single RRM of the remaining proteins, resemble conventional RRMs.

standard protein fractionation technique and had also been used as the initial step in the purification of SF2/ASF (8), the basis for the unusual insolubility in the presence of magnesium remains unclear, though extremely useful. Using this protocol, SR proteins with approximate molecular masses of 75, 55, 40, 30, and 20 kDa were purified from several sources, such as human HeLa cells, *Drosophila* Kc cells, mouse 10T 1/2 cells, and calf thymus. Direct staining and/or immunoblotting showed that the apparent sizes of SR proteins are highly conserved among different species, from human to *C. elegans* (7, 32).

Several, though not all, SR proteins contain in addition to the conserved RRM at the N-terminus, a second atypical central RRM, which includes an invariant heptapeptide (SWQDLKD) that constitutes a characteristic signature for this class of SR proteins (15). The difference in size among different members of the SR family is due to the presence or absence of the second RRM, and to the variable length of the C-terminal RS domain (Figs 2 and 3). SRp75 is the largest SR protein and contains two RRMs, followed by the longest RS domain present in proteins belonging to this family (315 amino acids) (36).

B52, a *Drosophila* protein that is a nearly identical variant of dSRp55, was identified by means of a monoclonal antibody that recognizes transcriptionally active chromatin on *Drosophila* polytene chromosomes. B52 brackets the sites of RNA polymerase II localization on chromatin on both sides, and similar changes in the distribution of B52 and RNA polymerase II occur upon induction of the heat shock genes (31). This is not inconsistent with a role in pre-mRNA splicing, taking into consideration the co-transcriptional nature of pre-mRNA splicing *in vivo* (see Chapter 2). B52 was also found to be associated with active chromatin in non-polytene chromosomes in *Drosophila* cell lines. A B52 null mutant is lethal at the embryonic or larval stages (39, 40), demonstrating that an individual SR protein is essential for viability and development. Moreover, overexpression of B52 in specific cell-types or developmental stages leads to various developmental abnormalities (41). Therefore, the concentration of B52 is important for normal development, and the observed phenotype may be due to inappropriate alternative splicing of one or more unidentified cellular targets.

A murine cDNA, termed X16, was cloned as a gene preferentially expressed in pre-B cell lines, relative to mature B cell lines (33). The X16 protein sequence is identical to human SRp20 (7), and is also highly homologous to a *Drosophila* protein termed RBP1 (42). Mouse X16, HeLa SRp20, and *Drosophila* RBP1 contain a single N-terminal RRM repeat and a C-terminal RS domain. Interestingly, the tissue-specific distribution of this factor, as analysed by northern blotting, changes substantially among different mouse tissues; for instance, expression is high in thymus, spleen, and testes, and low or undetectable in liver, kidney, brain, and heart (33). The *Drosophila* RBP1 protein was cloned by means of a polymerase chain reaction (PCR) assay with degenerate primers that recognize conserved features in the RNP-2 and RNP-1 submotifs of the RRM (43). This nuclear protein, which is expressed in all tissues, has a very similar distribution to the *Drosophila* protein B52. The rbpl pre-mRNA can be alternatively spliced to generate, in addition to the

RBP1-1 protein, another isoform that lacks the RS domain. The multiple coding capacity of this splicing factor gene resembles that of SF2/ASF, which also encodes three different proteins. The RBP1-1 protein was assayed for constitutive and alternative splicing in heterologous HeLa cell extracts, showing similar, albeit not identical, properties to those of SF2/ASF (42).

A mAb originally raised against heterogeneous nuclear ribonucleoproteins (hnRNPs) allowed the identification of another member of the SR family of proteins, a novel essential splicing factor of 35 kDa that was termed 9G8 (37). This protein factor has a single RRM at the N-terminus and a C-terminal RS domain, separated by an atypical region of approximately 40 amino acids, which includes a CCHC motif similar to the zinc knuckle motif found in the SLU7 splicing factor in yeast (44; see Chapter 7). The RRM is highly homologous to those of SRp20 and RBP1, but the homology is less pronounced when compared with SC35 and SF2/ASF. Another distinctive feature of this protein is the presence within the RS domain of a RRSRSXSX consensus sequence repeated six times, which is not found in any other RS domain. A 9G8-depleted nuclear extract can be complemented by SC35 but not by SF2/ASF for splicing of adenovirus E1A and β-globin pre-mRNAs, suggesting that SC35 and 9G8 are functionally related (37).

HRS, a rat homologue of human SRp40, was identified as an immediate-early expressed gene by substraction and differential screening of cDNA libraries from an insulin-induced hepatoma cell line (45). Different forms of HRS mRNA are temporally regulated during the growth response. A long mRNA isoform that arises via alternative splicing is generated at late times, suggesting that HRS autoregulates processing of its own pre-mRNA.

In summary, different members of the SR family of protein regulators, which share structural and functional features, were identified using a variety of different approaches. These approaches include biochemical complementation of inactive mammalian splicing extracts, *in vitro* alternative splicing assays, the use of monoclonal antibodies raised against spliceosomal components or active chromatin fractions, degenerate polymerase chain reaction cDNA cloning, simple biochemical purification procedures, or differential growth-related gene expression.

2.4 Role of SR proteins in constitutive and alternative splicing

The SR protein family can be divided into two different subgroups, according to the number of RRMs that are present (Fig. 3). While members of both groups have C-terminal RS domains that differ in size, members of the first group, which comprises SRp30a (SF2/ASF), SRp40, SRp55, and SRp75, have two RRMs (one canonical RRM and one atypical RRM); whereas members of the second group, which includes SRp20 (X16, RBP1), 9G8, and SRp30b (SC35), have only the first RRM. Although it was first speculated that the presence of one or both copies of the RRM could determine the specificity of these splicing regulators, SF2/ASF and SC35 behave very similarly in constitutive and alternative splicing assays *in vitro* (7, 30, 37). The high degree of phylogenetic conservation of SR proteins of the same size, in species ranging from vertebrates to invertebrates, suggests that each individual protein has a specific function in the regulation of alternative splicing.

Every member of this family of proteins has been tested for its function in constitutive and alternative splicing. When SRp20, SRp30a, SRp30b, SRp40, SRp55, and SRp75 gel-purified from either HeLa cells or calf thymus were incubated with HeLa S100 cytosolic extracts that have all the components required for splicing activity except for the entire set of SR proteins, each individual protein was able to complement the inactive cytosolic extracts for splicing of β-globin and *Drosophila fushi tarazu* (*ftz*) pre-mRNAs (7). Although no differences among SR proteins were noted in this constitutive splicing assay, several differences have been reported in the ability of these proteins to regulate alternative splicing. For example, SRp30b (SC35) promotes exclusively the use of a proximal 5′ splice site in an E1A pre-mRNA *in vitro*, while SRp40, SRp55, and SRp75 showed preference for the proximal 5′ splice site but also selected the distal 5′ splice site (7). When an SV40 pre-mRNA was tested with the same protein preparations, SRp30b selected preferentially the proximal 5′ splice site, while SRp40, SRp55, and SRp75 showed a strong preference for the distal 5′ splice site. These experiments suggest that different sets of SR proteins have different substrate- and site-specificities (or specific activities) in the regulation of alternative splicing. Substrate-specific differences between SR proteins were also observed using an *in vivo* co-transfection assay for alternative splicing (38; see Section 2.7.1)

Additional evidence for a differential role of individual SR proteins in the regulation of splicing came from experiments in which different SR proteins were examined for their role in committing different pre-mRNAs to the splicing pathway. The commitment complex is the first functional pre-spliceosomal complex assembled on pre-mRNA that cannot be competed by excess cold pre-mRNA (Chapter 4). Individual SR proteins were preincubated with different pre-mRNAs and the resulting complexes were challenged by addition of splicing extracts in the presence of excess competitor RNA (46). When β-globin pre-mRNA was used, a competitor RNA containing a 5′ splice site inhibited splicing, except when the pre-mRNA had been pre-incubated with SC35. Thus, SC35 was able to commit β-globin pre-mRNA to the splicing pathway. SC35 displayed the highest activity, followed by SF2/ASF and SRp55, while SRp20 had no detectable activity. In contrast, when an HIV tat pre-mRNA was analysed, only SF2/ASF was able to commit this pre-mRNA to splicing, whereas other SR proteins had no effect. These experiments showed that different SR proteins are able to commit different pre-mRNAs to the splicing pathway, and this may represent a critical step at which alternative and tissue-specific splicing is regulated. Although it is unclear whether these stable complexes between pre-mRNA and purified SR proteins reflect the natural order of spliceosome assembly, as presumably observed during commitment experiments with nuclear extracts (Chapter 4), the results do show that individual SR proteins have different off rates when bound to different pre-mRNAs. These experiments suggest that different pre-mRNAs may require different sets of SR proteins for splicing *in vivo*, and also that SR proteins are required in the early steps of the splicing reaction. The latter implication is consistent with the observation that the earliest stages of spliceosome assembly are indeed blocked in extracts lacking SR proteins (9, 26).

Several lines of evidence suggest that SR proteins may associate with each other to form a protein or RNP complex in splicing extracts (see Section 2.6.1). For example, immunodepletion with antibodies against SC35 or 9G8 co-precipitates other SR proteins, although this may be in part due to cross-reactivity (30, 37). If such SR protein complexes indeed exist, it is possible that the addition of a single SR protein to an inactive S100 extract, which has limiting but detectable amounts of SR proteins, drives assembly of the complexes so that they are no longer limiting for splicing. Thus, the possibility that a single SR protein is not active unless some or all the others are also present, has not been ruled out.

2.5 hnRNP A/B proteins antagonize SR proteins to modulate alternative splicing

An activity that antagonizes SF2/ASF in the selection of 5′ splice sites is present in HeLa cell nuclear extracts, and upon biochemical purification of this activity, a single, active polypeptide was found to be identical to hnRNP A1 (47). An excess of hnRNP A1 favours the use of distal 5′ splice sites when two competing 5′ splice sites are present, whereas high levels of SF2/ASF promote the use of proximal 5′ splice sites (relative to the 3′ splice site), and these effects are position-dependent, rather than sequence-dependent. Therefore, the balanced activities of these two factors seem to regulate alternative splicing *in vitro*. hnRNP A1, which contains two copies of the RRM and a C-terminal domain rich in glycine residues, does not inhibit the activity of the SR proteins in constitutive splicing when present in excess (47). Moreover, although hnRNP A1, like other hnRNP proteins, appears to be associated with most or all transcripts *in vivo* (Chapter 3), and also binds to most or all pre-mRNAs to form a non-specific H-complex *in vitro* (Chapter 4), it appears not to be essential for splicing, since a cell line that has an inactive hnRNP A1 locus is still viable (48).

The relative amounts of hnRNP A1 and SF2/ASF have also been shown to affect alternative splicing by exon inclusion or skipping (Chapter 8). Thus, it has been shown that SF2/ASF can promote exon inclusion of alternatively spliced exons while hnRNP A1 favours exon skipping, although this effect of hnRNP A1 is dependent on several parameters, such as the size of the internal alternative exon and the strength of the polypyrimidine tract in the preceding intron (49). Significantly, an excess of hnRNP A1 fails to promote inappropriate exon skipping of natural constitutively spliced pre-mRNAs. In the case of SF2/ASF, the ability to promote alternative exon inclusion may well be mechanistically related to its 5′ and/or 3′ splice-site switching activities, since exon inclusion is formally the same as selection of proximal 5′ and 3′ splice sites. The antagonistic effect of hnRNP A1 is not unique to this protein, as shown by the fact that other members of the hnRNP family of proteins—hnRNP A2 and hnRNP B1—also antagonize SF2/ASF, with even stronger activity (50). hnRNP A1^B, an alternatively spliced isoform of hnRNP A1 (51), is much less active in alternative splicing, raising the possibility that alternative splicing of the A1/A1^B pre-mRNA is a regulatory switch for hnRNP A1 activity (50, 52).

2.6 Mechanism of action of SR proteins

In vitro studies with individual SR proteins have shown that these proteins are required for catalytic step I of pre-mRNA splicing and that, in their absence, spliceosome assembly is blocked at an early stage (7, 8, 26). In addition, SR proteins can modulate alternative splice-site selection in a concentration-dependent manner with many different pre-mRNAs (7, 9, 30, 37, 38, 42, 53), and more recently they have been implicated in the recognition of purine-rich splicing enhancer elements (see Section 2.6.3). An important, not fully resolved question, is whether the same molecular interactions involving each of the domains of SR proteins are involved in each of these splicing-related activities. Conversely, an important issue is whether a particular interaction with protein or RNA is relevant to a single or to all of these splicing-related activities.

2.6.1 Protein–protein interactions mediated by RS domains

Functional interactions between the 5′ and 3′ splice sites during the earliest steps of spliceosome assembly have been demonstrated in yeast and mammals, strongly suggesting that components bound to these splice sites may interact directly or through unknown bridging factors (54; see Chapter 4). The functional association between the splice sites occurs during the assembly of the pre-spliceosome E complex, in which splicing commitment is established (55–57). The fact that the SR proteins, which modulate 5′ splice-site selection, are also required for this early step suggests that they are good candidates to mediate these interactions. The first experimental evidence that SR proteins are involved in the 5′/3′ splice-site interaction was obtained by RNase T1 protection/immunoprecipitation experiments. These experiments showed that SC35 mediates direct or indirect interactions in the pre-spliceosome complex with both U1 snRNP and U2 snRNP particles at the 3′ splice site (54).

Several techniques, such as far-western blotting and co-immunoprecipitation, have been used to study protein–protein interactions between factors bound at the 5′ and 3′ splice sites. Far-western experiments showed that a U1 snRNP-specific protein, U1-70K (Chapter 5), can interact with a subset of the SR proteins (58–60). These interactions are mediated by the RS domain present in both the U1-70K protein and in the SR proteins. In addition, SC35 interacts with itself, with SF2/ASF, and with the *Drosophila* Tra and Tra-2 splicing regulators. All the players involved in these interactions share a structural feature—the RS domain—and when the RS domain of one of the interacting proteins is deleted, the proteins no longer interact. Because RS domains are highly phosphorylated (at a large but still undetermined number of serine residues) and because unphoshorylated RS domains would be expected to electrostatically repel each other because of the large number of arginine residues, it seems likely that the phosphorylation state of each RS domain has a profound effect on the protein–protein interactions it mediates.

The use of the yeast two-hybrid system to detect protein–protein interactions (59, 60) extended these observations and demonstrated specific interactions between SC35 or SF2/ASF and both the U1-70K polypeptide and $U2AF^{35}$. Additional experiments in which a third protein was expressed in the same cell, in addition to the DNA binding and activation domain fusion proteins, showed that either SC35 or SF2/ASF can interact simultaneously with $U2AF^{35}$ and U1-70K, whereas $U2AF^{35}$ can interact simultaneously with SF2/ASF and $U2AF^{65}$ (59). These observations led to a model involving a network of protein–protein interactions, in which SC35 (and/or other SR proteins) interacts on one hand with the U1 snRNP at the 5′ splice site via the U1-70K protein, and on the other hand with $U2AF^{65}$ at the polypyrimidine tract near the 3′ splice site via its associated $U2AF^{35}$ subunit (Fig. 4). Thus, SR proteins may act as bridging factors to facilitate functional interactions between the 5′ and 3′ splice sites in the pre-spliceosome. The fact that individual SR proteins are able to interact with themselves or with other members of this family in the two-hybrid and far-western assays, is consistent with the possibility that SC35, SF2/ASF and probably other SR proteins are capable of forming homotypic and/or heterotypic higher order complexes that may be important for splicing *in vivo*. Although the RS domain is required to mediate these interactions, not all RS-containing proteins interact, as illustrated by the lack of interactions between U1-70K and either subunit of U2AF. What structural features of this type of domain, which has low compositional complexity, are responsible for specificity in RS domain–RS domain interactions, remains unknown. It is worth noting, however, that the RS domain sequences of particular SR protein homologues are highly conserved phylogenetically (for example, they are 100% conserved between mouse and human SF2/ASF), implying that at least some of the interactions that these domains participate in are highly specific.

2.6.2 Cooperative binding of the U1 snRNP and SF2/ASF to 5′ splice sites

Purified U1 snRNP can interact with pre-mRNA containing a 5′ splice site, but this interaction requires additional factors, probably reflecting the fact that mammalian 5′ splice sites are highly divergent compared to those found in budding yeast (reviewed in 61). Therefore, factors that stabilize the interactions between U1 snRNP and 5′ splice sites are required. The formation of stable complexes between a pre-mRNA containing a 5′ splice site and purified U1 snRNP particles was studied in the presence or absence of SF2/ASF (58). Only low levels of complex formation were detected in the absence of SF2/ASF, and the RS domain of SF2/ASF is required for this cooperative binding. The order of addition is also important: when SF2/ASF is pre-incubated with the template and the U1 snRNP is subsequently added, efficient complex formation is observed. In contrast, when U1 snRNP was added first, only low levels of complex formation were observed. These

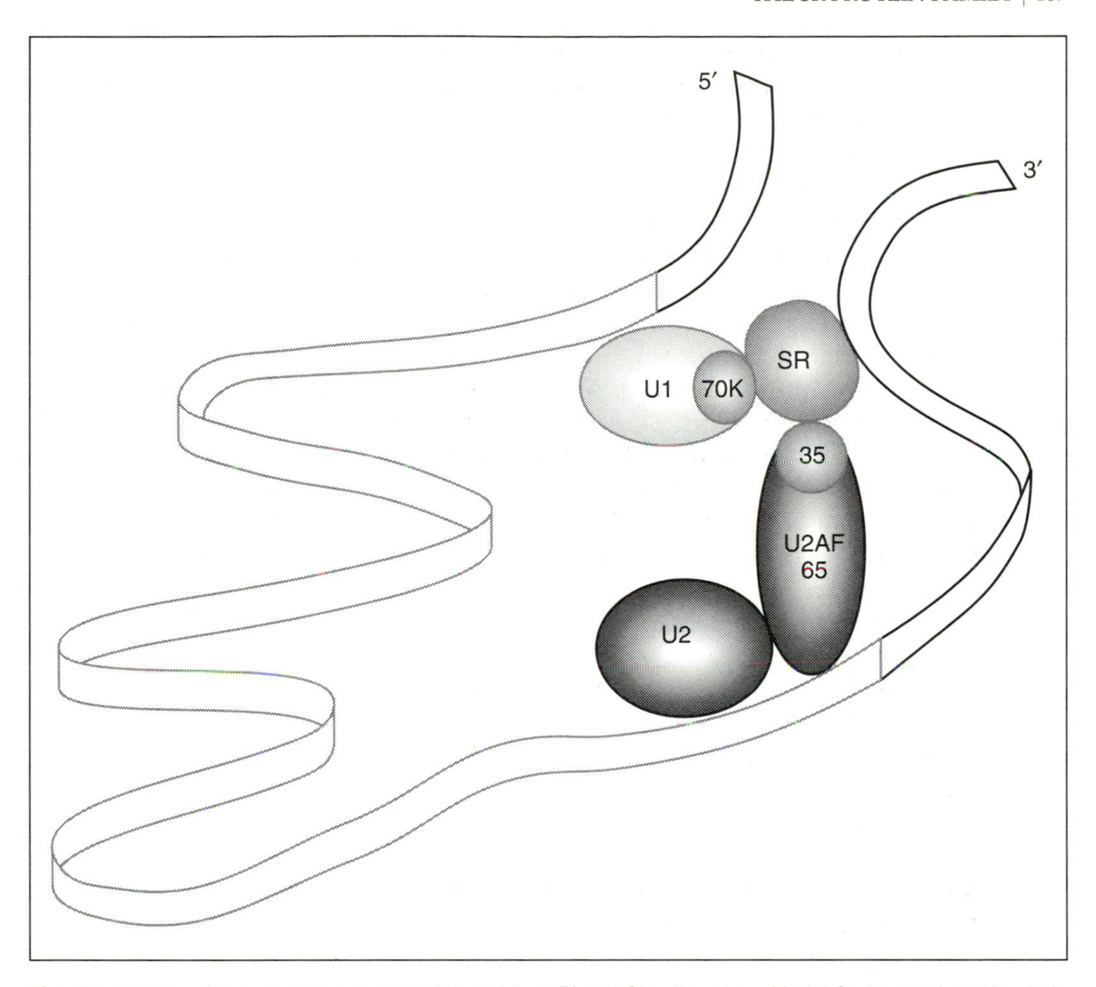

Fig. 4 A network of protein–protein interactions bridges 5′ and 3′ splice sites. Model for interactions that help define the splice sites, based primarily on two-hybrid and far-western data for each of the pair-wise interactions. The ribbon represents a pre-mRNA with two exons (dark edges) and one intron (light edges). Proteins and snRNPs are represented by labeled circles or ellipses, with the relevant subunits indicated. The SR protein is shown bound to the downstream exon, based on examples of enhancer-dependent splicing. In pre-mRNAs lacking this type of enhancer, at least one SR protein is required for splicing, but the binding site is not known. SR proteins make contacts with the 70K subunit of U1 snRNP and promote base pairing of U1 snRNA to the 5′ splice site. SR proteins also contact the small subunit of U2AF ($U2AF^{35}$); the large subunit of U2AF ($U2AF^{65}$) contacts the polypyrimidine tract portion of the 3′ splice site element and promotes base-pairing of U2 snRNA to the branch site.

data suggest that SF2/ASF interacts first with the pre-mRNA and then recruits the U1 snRNP particle to the 5′ splice site.

RNA binding and competition experiments with purified SF2/ASF lacking the RS domain uncovered a binding preference for the 5′ splice sites of two different pre-mRNAs (62). Other experiments suggest that SR proteins act at or near 5′ splice sites, and in some cases may even bypass the requirement for U1 snRNP

(63, 64). Thus, when U1 snRNP was made limiting by depletion with immobilized 2′O-methyl oligonucleotides complementary to U1 snRNA, or debilitated by binding of similar oligonucleotides, splicing was inhibited, as expected; surprisingly, addition of excess SR proteins restored normal splicing, although in one study a cryptic 5′ splice site was also activated (64). Whether the excess of SR proteins is able to carry out the U1 functions, allows debilitated or extremely limiting U1 to function at some level, and/or enables another component to take over the function of U1, remains to be determined.

RNase H protection experiments in splicing extracts also showed that SF2/ASF enhanced the binding of U1 snRNP to 5′ splice sites (65). In the case of pre-mRNAs containing multiple 5′ splice sites, increasing SF2/ASF led to multiple U1 occupancy. This observation suggested a model to explain the mechanism of action of SF2/ASF in alternative splicing (Chapter 8). The model proposes that SF2/ASF enhances the interaction of U1 snRNP with all competing 5′ splice sites indiscriminately; subsequently, the selection of the proximal 5′ splice site reflects an integral aspect of the splicing mechanism (with an unknown molecular basis). Thus, a relatively weak proximal 5′ splice site can only be selected in the presence of excess SF2/ASF, which essentially acts to equalize the relative strengths of all alternative 5′ splice sites. Interestingly, hnRNP A1, which antagonizes SF2/ASF in alternative splicing and promotes use of distal alternative 5′ splice sites in certain cases (47), was found to decrease the affinity of U1 snRNP for 5′ splice sites (Ian Eperon, personal communication). In this case, excess hnRNP A1 should result in use of the site with highest affinity for U1 snRNP.

2.6.3 SR proteins and exonic splicing enhancers

In certain metazoan pre-mRNAs, exonic sequences have been shown to affect the efficiency of splicing of the adjacent upstream or downstream intron (66, 67). Both splicing enhancers and silencers, which respectively activate or repress adjacent splice sites, have been described.

In the *Drosophila* P element transposase pre-mRNA, two pseudo 5′ splice sites in the exon preceding the regulated intron form non-productive complexes with U1 snRNP, apparently resulting in steric hindrance for binding of U1 snRNP to the authentic 5′ splice site, and thus abolishing splicing of intron 3 in somatic cells (68). In contrast, splicing enhancers have been defined as exonic elements that bind *trans*-acting factors in a sequence-specific manner and activate splicing of introns located some distance away from the protein binding sites. Among the splicing enhancers that have been best characterized in terms of their bound *trans*-acting factors are those present in specific exons of *Drosophila dsx*, human IgM, growth hormone, human fibronectin, and chicken cardiac troponin T pre-mRNAs (69–74). In this section we will describe a few examples of these splicing enhancers, emphasizing the role of *trans*-acting factors. Additional discussion of these elements in the context of spliceosome assembly and of alternative splicing can be found in Chapters 4 and 8.

The *Drosophila dsx* pre-mRNA enhancer was the first enhancer for which both *cis*-

acting sequences and *trans*-acting factors were well defined, thanks to the power of genetics. The information derived from this system has been very useful to understand mammalian enhancers as well. Sex-specific alternative splicing of the *dsx* pre-mRNA involves the activation of a female-specific 3′ splice site by the products of the transformer and transformer-2 genes (reviewed in 75; see Chapter 8). Both Tra and Tra-2 contain RS domains, and the latter also contains an RRM. The female-specific acceptor site has a suboptimal polypyrimidine tract interrupted by purines and is poorly recognized in males. When these purines are mutated to the consensus sequence, the *dsx* female 3′ splice site is constitutively used both *in vivo* and *in vitro*. Positive control of this splicing event is exerted by an exonic element located about 300 nucleotides downstream of the female-specific 3′ splice site, which contains six copies of a 13-nuclear repeat (76). Tra and Tra-2 are thought to act by recruiting SR proteins to this splicing enhancer element, and this multi-protein complex assembled on the regulatory sequence may facilitate recognition of the adjacent 3′ splice site by the general splicing machinery (69). A commitment complex formation assay was developed, in which labeled *dsx* RNA containing the enhancer sequence was pre-incubated with Tra, Tra-2, and different human SR proteins, followed by addition of HeLa cell splicing extracts in the presence of unlabeled RNA competitor. These experiments showed that individual SR proteins differ in their ability to activate this splice site. Recombinant SC35 and SRp55 proteins, as well as human SRp40, SRp75, and SRp55, complemented Tra and Tra-2 very efficiently in this commitment complex formation assay; in contrast SRp20 and SF2/ASF were inactive (69).

Vertebrate exonic enhancers were first identified in the mouse immunoglobulin μ heavy chain pre-mRNA (70). Secreted or membrane-bound forms of IgM are specified by mRNAs that differ at their 3′ ends as a result of alternative 3′ processing and splicing (Chapters 8 and 9). Exon M2 (specific for the membrane form) contains a purine-rich enhancer, designated ERS (exon recognition sequence), which was shown to be required for efficient splicing. Detailed mutational analyses showed that a number of different alternating purine sequences were functional, whereas oligo(A) or oligo(G) were inactive, and interruption of alternating oligo-purine elements by U residues markedly reduced the level of stimulation (77). In addition, these exonic elements are functionally interchangeable among different pre-mRNAs. Thus, the IgM ERS and similar sequences present in an avian retrovirus (ASLV) and in chicken cardiac troponin T (cTNT), when placed downstream of the regulated female-specific *Drosophila dsx* 3′ splice site, allow its use in HeLa cell extracts, in the absence of the Tra and Tra-2 proteins or the 13-nucleotide repeats (70).

The last intron of the bovine growth hormone (bGH) gene pre-mRNA is spliced inefficiently due to a weak 5′ splice site. Splicing of this intron is dependent on binding of *trans*-acting factors to a purine-rich exonic splicing enhancer (ESE) within the last exon, which is required for activation of splicing of the preceding bGH intron (71, 78, 79). This element is required only in the context of a weak 5′ splice site, since mutation of this site into a mammalian consensus 5′ splice site circumvents the requirement for the ESE. RNA-protein crosslinking assays with the exonic element (termed FP element) identified an RNA-binding protein of approximately 30 kDa that was present in HeLa nuclear extracts but absent from S100 cytosolic extracts (71). UV cross-linking followed by immunoprecipitation with a monoclonal antibody identified the cross-linked protein as SF2/ASF (72). Using the FP RNA or a control RNA as competitors in the cross-linking assay, it was shown that SF2/ASF binds specifically to the FP RNA, both in crude extracts and with purified protein. In addition, purified or recombinant human SF2/ASF protein stimulated splicing of bGH pre-mRNA with the FP sequence, but had no effect when that sequence was deleted. The stimulatory effect of SF2/ASF on bGH splicing was counteracted by hnRNP A1, which was also shown to bind to the FP element, albeit with limited specificity. In contrast to SF2/ASF, another SR protein, SC35, did not interact specifically with the bGH ESE, and accordingly failed to activate splicing of the bGH last intron.

In the case of the human fibronectin pre-mRNA, an 81-nucleotide sequence located in the central region of the alternatively spliced exon ED1 (also known as EIIIA) is required for alternative splicing of this exon (73, 74). The presence of this exonic enhancer, which also consists of a purine-rich sequence, is required for use of the upstream 3′ splice site, resulting in inclusion of this alternatively spliced exon. The enhancer fails to work when moved farther than 291 nucleotides from the 3′ splice site (74). In this case, as with bGH, SR proteins are thought to mediate functional recognition of the enhancer, since total SR proteins were shown to form a specific complex (using a gel-mobility shift assay) with fibronectin RNA containing the enhancer element. Whether this is due to specific interactions with a single SR protein, or with a subset of, or all, SR proteins remains to be established.

Purine-rich elements associated with the stimulation of 3′ splice site use were also identified in the chicken cardiac troponin T exon 5, and in the calcitonin/CGRP exon 4 (Chapter 8). The cTNT splicing enhancer is not required for developmental regulation of exon 5 alternative splicing in primary skeletal muscle cultures, as shown by the fact that the ratio of exon inclusion to skipping remains the same in exons having or lacking the element. Therefore, this element appears to be involved in determining the efficiency of constitutive splicing (80).

To compensate for the presence of a weak 5′ splice site, inclusion of the rat or human calcitonin-specific exon 4 in mature mRNA in non-neural cells requires initial binding of factors to specific exon 4 sequences (81–84). Exon skipping leading to CGRP-1 mRNA in neural cells may be accomplished either by low levels of these putative factors in neural cells, or by the presence of a specific dominant neural *trans*-acting factor to inhibit CT-specific splicing. Two exonic elements (A and B), which have an additive effect, have been identified. There is some sequence similarity between element A and the 13-nucleotide repeat of the *Drosophila dsx* gene, while the B element is purine-rich. Partially purified rat brain factors of 43 kDa and 41 kDa, which bind specifically to sequences surrounding the CT-specific 3′ splice site, inhibit CT splicing (81).

Splicing enhancers are required to activate splicing of weak introns that have suboptimal splicing signals, and the presence of a purine-rich stretch in the downstream exon may be an integral component of the recognition process for introns with weak 3′ splice sites. *In vitro* splicing of the HIV *tat/rev* intron between the first and second coding exons is also activated by SF2/ASF, but not by SC35 (8, 46). Purine-rich sequences in the downstream exon function as an exonic enhancer (85, 86). An additional adjacent element functions as a splicing silencer, and its removal results in increased splicing efficiency. In this and other retroviruses, it it important to modulate the relative amounts of unspliced pre-mRNA and spliced mRNA (including pre-mRNA that is exported to the cytoplasm), since both are necessary for the synthesis of viral proteins and/or for packaging into virions (87; Chapter 8). It appears that enhancers that modulate splicing efficiency, together with SR proteins that are responsible for, or participate in, their functional recognition, are an important element of how the necessary balance of cytoplasmic pre-mRNA and mRNA is achieved.

It seems likely that the specificity of SR proteins in regulating the efficiency or pattern of splicing of different sets of genes may be imparted by the specific recognition of different exonic enhancer sequences. Although tissue-specific or developmentally modulated vertebrate splicing enhancers or silencers have not been identified yet, binding of splicing enhancers by individual SR proteins may be a general mechanism for regulating alternative splicing of pre-mRNAs that contain these elements, whether of the purine-rich class or of a different class. The SR proteins may mediate interactions between the splicing enhancer and the 3′ splice site through protein–protein interactions with the small subnunit of U2AF ($U2AF^{35}$) (59). The identity of the ESE should determine which *trans*-acting factors regulate the pattern of alternative splicing of a particular pre-mRNA. A downstream consensus 5′ splice site can also function as an ESE (88). For example, experiments with the preprotachykinin pre-mRNA showed that binding of U1 snRNP to the 5′ splice site downstream of an alternatively spliced exon influences the binding of $U2AF^{65}$ to the upstream 3′ splice site (89). These observations are consistent with a protein–protein interaction bridge across the exon, involving U1 snRNP (presumably U1–70K) and $U2AF^{65}$, which serves to define the exon–intron boundaries. These interactions were predicted by the exon-definition model, according to which, splice sites are defined primarily by the interactions between bound factors across exons, rather than across the usually much longer introns (90, 91). An analysis of spliceosome assembly (Chapter 4) showed that similar complexes involving U1 snRNP and SR proteins are assembled on RNAs containing 5′ or 3′ splice sites, as on enhancer-containing RNAs (92). The same network of interactions involving SR proteins, U1 snRNP, and U2AF was proposed to be involved in spliceosome assembly of enhancer-dependent and –independent pre-mRNAs, as well as pre-mRNAs in which exon-bridging interactions are essential. In one case the interactions occur between 5′ and 3′ splice sites across the intron, in another case they occur across an exon, that is between an upstream 3′ splice site and a downstream 5′ splice site, and in the third case, analogous interactions would take place between an exonic enhancer and the 3′ splice site immediately upstream (Figs 4 and 5).

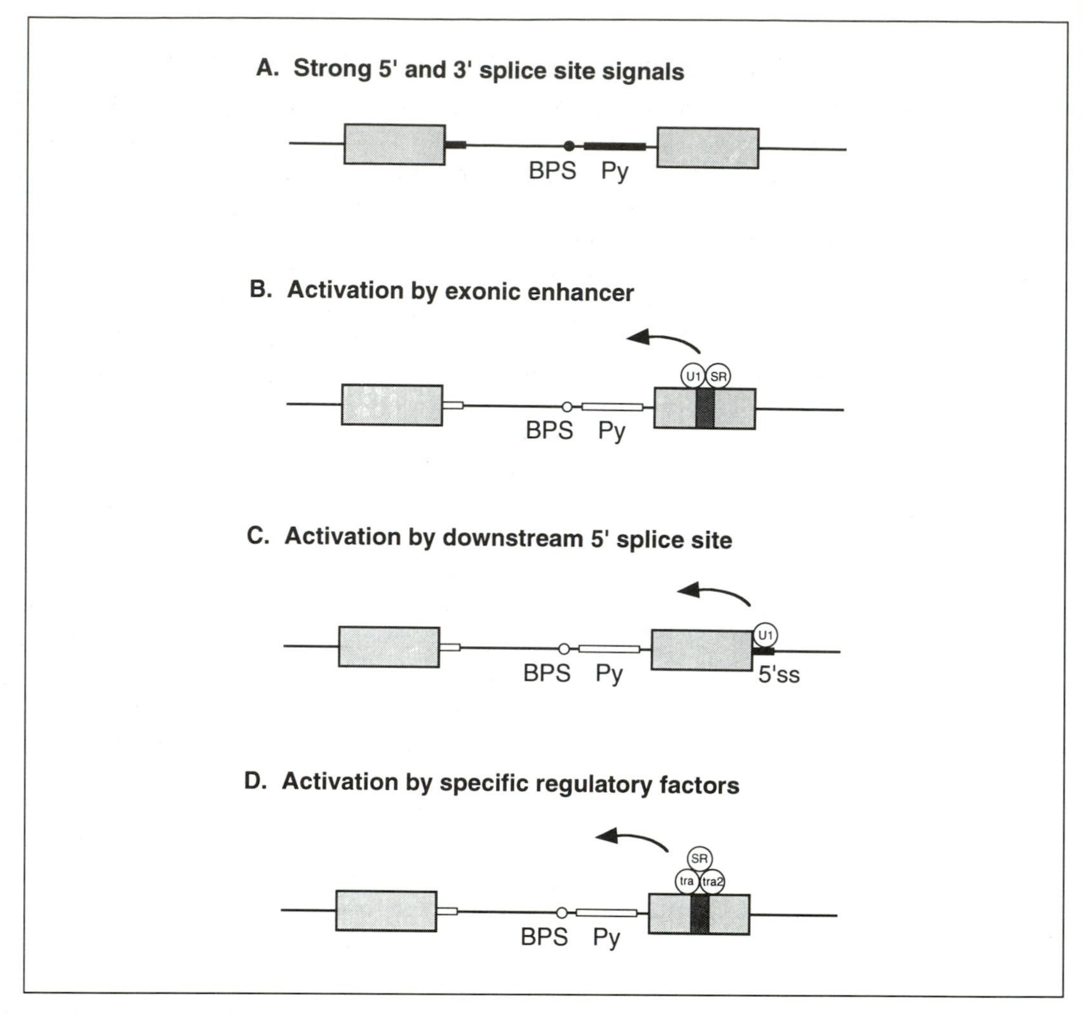

Fig. 5 Definition of splice sites. The diagrams show different ways in which exon–intron boundaries are defined. For pre-mRNAs with strong 5′ and 3′ splice sites (filled thin boxes and dot), which resemble the consensus elements, most of the required information appears to be present at the splice sites (a). When the 5′ and/or 3′ splice site is weak (open thin boxes and dot), other elements or interactions can have a pronounced effect on recognition of the intron. These include a downstream exonic enhancer element, which is often purine rich, and to which U1 snRNA and SR proteins bind (b); the 5′ splice site of the following intron, which binds U1 snRNP (c); and a specific enhancer complex, such as the *Drosophila* doublesex enhancer, which is recognized by transformer and transformer-2 proteins, together with SR proteins (d). Interactions between factors bound to adjacent introns (C) constitute an example of exon definition. Similar networks of protein–protein interactions (Fig. 4), whether across an intron or an exon, are thought to mediate recognition of exon–intron boundaries. The branchpoint sequence (BPS) and polypyrimidine tract (Py) upstream of the 3′ splice site are indicated. Exons are represented by shaded boxes, exonic enhancers by darker shading, introns by thin lines, and conserved intron elements by open or filled thin boxes (5′ splice site and polypyrimidine tract portion of the 3′ splice site) or dots (branch site).

2.7 Regulation of SR protein expression and activity

Antagonistic effects between SR proteins and hnRNP A/B proteins (or combinations thereof) may function to regulate alternative splicing events in living cells. Thus, unique combinations of these proteins, in some cases in conjunction with gene-specific regulators, may regulate alternative splicing choices in different tissues, developmental situations, or in response to the metabolic state of the cell (Chapter 8). While some of the required specificity may arise from interactions with specific sequences present only on certain pre-mRNAs, additional specificity may be achieved through the regulated spatial and temporal co-expression of individual SR or hnRNP proteins and pre-mRNAs with suitable target sequences.

2.7.1 Variable ratios of SR/hnRNP proteins

The above model for regulation of alternative splicing by SR proteins and their antagonists predicts that individual SR and hnRNP A/B proteins have unique tissue distributions and abundances, which result in corresponding changes in the ratio of individual SR and hnRNP A/B proteins. This is indeed the case, as shown by tissue-specific variations in the total and relative amounts of SR proteins (93), as well as by changes in the molar ratio of SF2/ASF to hnRNP A1 that naturally occur over a wide range in different rat tissues (Hanamura and Krainer, unpublished). For instance, in the case of the mouse protein SRp20/X16, it has been shown that its expression at the mRNA level is much higher in pre-B cells than in mature B cells (33).

Evidence for the role of SR protein abundance in modulating alternative splice-site selection *in vivo* comes from experiments in which an SF2/ASF cDNA was transiently over-expressed in HeLa cells and changes in the pattern of alternative splicing of cotransfected reporter genes were analysed (94). Over-expression of SF2/ASF in HeLa cells caused selection of the most proximal 5′ splice site in β-thalassemia and adenovirus E1A pre-mRNAs, prevented abnormal exon skipping (due to over-expression) of a β-tropomyosin pre-mRNA, and promoted inclusion of a clathrin light chain B neurone-specific exon. In contrast, over-expression of hnRNP A1 caused activation of a distal 5′ splice site in the E1A pre-mRNA (52, 94). These results showed that changes in the intracellular abundance of SF2/ASF or hnRNP A1 (and presumably in their relative ratio) can influence different modes of alternative splicing. Similar experiments have been carried out with other members of the SR family, several of which could be shown to be active in this assay, although striking substrate-specific differences could be observed in some cases (38). Together, the *in vitro* and transient transfection results suggest that different patterns of alternative splicing can be controlled in a tissue-specific manner or developmentally regulated fashion by changes in the expression of a few master genes. Thus, changes in the nuclear concentration of a limited number of general splicing factors, rather than expression of numerous gene-specific regulators, may control the expression of a wide variety of genes.

An interesting variation on the theme of modulation of the ratio between SR proteins and hnRNP A/B proteins involves the sequestration of factors, rather than changes in their expression. Stévenin and co-workers have obtained evidence consistent with the notion that major late transcripts of adenovirus sequester SR proteins, presumably by binding to high affinity binding sites (95, 96). As a result, SR proteins become limiting, and hence during the early-to-late switch of the viral life cycle, alternative splicing of the E1A transcripts changes towards the distal 5′ splice site, consistent with an excess of hnRNP A/B proteins over SR proteins.

2.7.2 Regulation of SR protein gene expression

Unique ratios of SR and hnRNP proteins can be established in a tissue-specific or developmentally regulated manner by regulating the expression of their genes transcriptionally or post-transcriptionally.

The SF2/ASF, 9G8, SC35/PR264, and SRp40/HRS pre-mRNAs can be alternatively spliced to give rise to mRNAs encoding protein isoforms that lack the RS domain (13, 38, 45, 97, 98). Although the existence of these alternatively spliced isoforms as stable proteins *in vivo* has not been demonstrated, it has been shown that recombinant ASF-2 and ASF-3 (two isoforms of SF2/ASF that lack the RS domain) are inactive in 5′ splice-site switching and in S100 complementation assays *in vitro* (24). Moreover, recombinant ASF-2 and ASF-3 were reported to act as dominant inhibitors of authentic SF2/ASF to repress splicing. Even if these isoforms are not stably expressed *in vivo*, it seems likely that alternative splicing of SR protein pre-mRNAs, which is a phylogenetically conserved phenomenon, plays a role in regulating the expression levels of individual SR proteins. Regulation of SR protein alternative splicing might be accomplished by SR proteins themselves or by other splicing regulators.

Analysis of the promoter sequences of the splicing regulator SC35/PR264 gene revealed the presence of several myb-recognition elements, which interact *in vitro* with the c-myb DNA-binding domain. Thus, c-myb is able to *trans*-activate SC35 expression, providing the first example of a splicing regulator whose expression is modulated by a nuclear proto-oncogene (99). c-myb is preferentially expressed in inmature hematopoietic cells, but upon differentiation there is a reduction in its levels, concomitantly with a reduction in SC35/PR264. The expression of mouse X16/SRp20 has been reported to be induced by serum (33), whereas gene expression of another SR protein, rat HRS/SRp40, is induced by insulin (45). The fact that genes coding for splicing factors are up-regulated by mitogens may reflect the requirement for increased RNA processing due to the large increase in gene expression during late G1 phase. In addition, quantitative differences in up-regulation of individual SR protein genes, or the genes encoding their antagonists, is expected to result in changes in the alternative splicing patterns of many pre-mRNAs. Many such changes have been reported in proliferating versus quiescent cells.

The intracellular level of hnRNP A1 protein changes as a function of the proliferative state in certain cell types, such as in stimulated lymphocytes and in exponentially growing fibroblasts (100, 101; Chapter 3). Likewise, hnRNP A1 gene expression is much higher in transformed cell lines than in differentiated tissues, and its expression is modulated by serum and by epidermal growth factor in Rat-1 cells (102). Therefore, there is a correlation between cell proliferation and increased hnRNP A1 expression that is limited to certain cell types. Analysis of the hnRNP A1 promoter sequences by DNase I footprinting revealed the presence of multiple binding sites for transcriptional regulators, including GC-rich elements, two Sp1-binding sites, a CAAT box, and a putative binding site for the USF/MLTF and myc gene products (101). The complexity of this promoter is consistent with its capacity to modulate its response to different cellular and physiological stimuli. Changes induced by proliferation in the level of hnRNP A1 (and/or of the antagonizing SR proteins) might modify the splicing pathways involved in the expression of numerous genes.

There are many examples of changes in the patterns of alternative splicing of cellular pre-mRNAs as a result of oncogenic transformation. These examples include the fibronectin gene (103, 104), the tropomyosin gene family (105), the CD44 family of glycoproteins (106), and the oestrogen receptor (107), among many others. Thus, in malignant cells, as well as in mitogenically stimulated cells, the mechanisms that regulate alternative splicing appear to be altered. The reprogramming of the machinery that regulates alternative splicing during malignant transformation may be due to changes in the expression or post-translational modification of general splicing factors, such as SF2/ASF and hnRNP A1. If alteration of the SF2/hnRNP A1 ratio proves to be a general mechanism for determining a specific splicing pattern, it may affect other cellular genes whose expression is associated with, and perhaps responsible for, the establishment or maintenance of the transformed phenotype.

2.7.3 Phosphorylation of SR proteins

All SR proteins are phosphorylated *in vivo*, as shown for example by the fact that mAb104, which reacts with all the family members, recognizes a shared phosphoepitope (32). The U1–70K protein also exists as a phosphoprotein *in vivo* (108; Chapter 5). It is presently unclear whether phosphorylated and/or dephosphorylated RS domains are the active form of this domain in SR proteins. Whereas the highly basic RS domain can be expected to interact with the phosphate backbone of RNA, the state of serine phosphorylation may influence these electrostatic interactions. Cycles of phosphorylation–dephosphorylation may therefore modulate charge-based interactions of the SR proteins with other components of the splicing machinery.

A role for reversible protein phosphorylation in the regulation of both constitutive and alternative splicing has been postulated. Recent studies showed that while phosphorylation is critical for spliceosome assembly, serine/threonine protein phosphatases are required for both catalytic steps of pre-mRNA splicing but do not affect the assembly of the spliceosome (109–111). Addition of specific protein phosphatase inhibitors (such as okadaic acid, tautomycin, and mycrocystin-LR, which are specific inhibitors of PP1 and PP2A protein serine/threonine phosphatases) block both catalytic steps of the splicing reaction in HeLa cell nuclear extracts, without affecting the assembly of splicing complexes. This inhibition is relieved upon addition of purified mammalian PP1 or PP2A protein phosphatases. Moreover, similar experiments showed that changing the levels of active PP1 in nuclear extracts results in changes in alternative 5′ splice-site selection (112).

A U1 snRNP-associated kinase activity that phosphorylates serine residues in the RS domains of both U1 snRNP 70K protein and the splicing factor SF2/ASF has been described (113). In the case of SF2/ASF, the intact protein is efficiently phosphorylated, whereas an alternatively spliced variant that lacks the RS domain (ASF-3) is not. The presence of this associated kinase activity in purified U1 snRNP made it possible to generate U1 snRNP particles containing either phosphorylated or thiophosphorylated U1–70K protein (114). Complementation studies with both forms of U1 snRNP showed that the thiophosphorylated form failed to restore splicing activity to a U1-depleted extract; one interpretation of this result is that dephosphorylation of this U1-specific protein is critical for splicing activity.

A novel serine kinase, SRPK1, specifically phosphorylates SR proteins on RS domain serine residues *in vitro* (115, 116). This protein kinase, which is highly related to a *C. elegans* kinase and to the fission yeast kinase dsk1 (117), is cell-cycle regulated and induces disassembly of nuclear speckles upon addition to permeabilized cells. This disassembly is detected as a redistribution of SC35 and snRNPs, which are present in speckles (Chapter 2), whereas several other nuclear constituents are unaffected. It was proposed that SRPK1 controls the intranuclear distribution of splicing factors in interphase cells and the reorganization of the speckled nuclear domain during mitosis (115). This kinase phosphorylates different members of the SR family of proteins, as well as $U2AF^{65}$, at least *in vitro*. It is possible that this kinase activity is the same as the U1 snRNP-associated kinase activity described above.

3. U2AF

Partially purified U2 snRNP particles cannot interact stably with the pre-mRNA branch site, which is complementary to a region of U2 snRNA (Chapter 5), unless an activity termed U2AF (U2-auxiliary factor) is present (118). U2AF binds to the polypyrimidine tract that is part of the 3′ splice site and facilitates or stabilizes the interaction between the adjacent upstream branch site and U2 snRNP. Thus, U2AF binding triggers spliceosome assembly by facilitating the ATP-dependent binding of U2 snRNP. A branch site RNase protection assay in the presence of partially purified U2 snRNP and fractions derived from HeLa cell nuclear extract allowed the identification, purification, and cloning of U2AF (119–121). This factor is composed of two subunits of 65 kDa and 35 kDa, which are associated, as shown by co-sedimentation on glycerol gradients. Chromatography of HeLa cell nuclear extracts over poly(U)-Sepharose in high salt resulted in the selective depletion of U2AF activity, and the depleted extracts were inactive in splicing. Purified or recombinant $U2AF^{65}$ is sufficient to restore splicing activity, demonstrating that the

65 kDa polypeptide is an essential splicing factor. The role of the small U2AF^{35} subunit remains unknown.

cDNAs encoding the large subunit of U2AF have been isolated in humans, mouse, and *Drosophila* (121–123). Related proteins or open reading frames have also been described in budding and fission yeast (124, 125; Chapter 7). The hU2AF^{65} subunit contains a short N-terminal RS domain, followed by three RRMs (Fig. 6). Structural and functional studies showed that whereas both the RRMs and the RS domain are essential for biochemical complementation of splicing in depleted extracts, the RS domain is not required for sequence-specific binding to pre-mRNAs (121). The RS domain of U2AF^{65} appears to perform a different function from that of the RS domain present in SR proteins. First, the position of this domain is different in the two kinds of protein (C-terminal versus N-terminal); second, the RS domain of SR proteins is considerably longer; third, the RS repeats of the two proteins differ in periodicity and in the nature of other interspersed amino acids; fourth, mutational analyses of the RS domains of SF2/ASF and U2AF^{65} give very different results in their respective constitutive splicing assays. Although the RS domain is essential for the general splicing activity of both factors, the RS domain of U2AF^{65} is very tolerant of substitutions, whereas that of SF2/ASF exhibits a stringent requirement for both Arg and Ser residues. Thus, in SF2/ASF, replacement of the Arg residues for Lys or Gly, or of the Ser residues for Thr or Gly, results in loss of general splicing activity (23). In contrast, in U2AF^{65} only positive charge seems to be required, and a short tract of Arg or Lys residues in place of the natural RS domain results in protein with significant levels of activity (J. Válcarcel, personal communication).

U2AF^{65} has RNA-annealing activity with complementary single-stranded RNA or single-stranded DNA substrates, and the RS domain is essential for this activity (126). An annealing activity has also been described for SF2/ASF (8) and hnRNP A1 (127), and this type of activity may be essential to facilitate base-pairing interactions between snRNAs and pre-mRNA, or to facilitate alternative RNA conformations that are conducive to splicing.

The notion that the RS domain of U2AF^{65} acts as an effector domain was suggested from experiments with the *Drosophila* transformer pre-mRNA (Chapter 8). The sex-lethal protein (Sxl) inhibits the use of a tra pre-mRNA proximal, non-sex specific, default 3′ splice site. It does so by binding to its polypyrimidine tract, thus apparently diverting U2AF to the distal female-specific 3′ site. In the absence of the

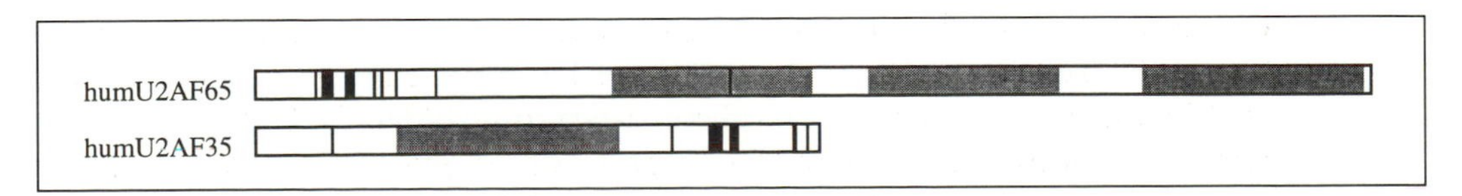

Fig. 6 Domain organization of human U2AF subunits. The large and small subunits of human U2AF are drawn to scale. The grey boxes denote the RNA-recognition motifs (RRMs). The vertical black lines indicate the location of RS or SR dipeptides, with the thickness being proportional to the number of consecutive dipeptides; RSR or SRS was scored as a single dipeptide repeat.

Sxl protein, the default 3′ splice site is selected. However, a chimeric Sxl-U2AF^{65} protein that contains the RS domain from U2AF^{65} fused to Sxl, stimulated the use of the default 3′ splice site instead of repressing it, demonstrating that the presence of an RS domain can convert a splicing repressor into a splicing activator (128).

The 35 kDa subunit of U2AF contains an RS domain but lacks canonical RRM motifs, which is consistent with the observation that this polypeptide does not bind to pre-mRNA (129). There is, however, a region reminiscent of an RRM (124) (Fig. 6). The region of human U2AF^{65} required for interaction with U2AF^{65} has been mapped and is distinct from both the RRMs and the RS domain (129). Homologues of both U2AF subunits have been identified in *Drosophila* with apparent molecular weights of 50 kDa and 37 kDa, respectively (123). The 50 kDa polypeptide is a true homologue as shown by the fact that the recombinant protein restores splicing activity to a heterologous U2AF-depleted HeLa cell nuclear extract. In addition, the large subunit of *Drosophila* U2AF is required for viability, since a lethal mutation in this chromosomal locus can be rescued by germline transformation with the dU2AF^{50} subunit cDNA.

Homologues or putative homologues for the large subunit have also been identified in *S. pombe* and in *S. cerevisiae*. The *S. pombe* homologue, encoded by the prp2 gene (no relation to the *S. cerevisiae* splicing factor of the same name), is required for splicing, as shown by the analysis of temperature-sensitive mutations (125; Chapter 7). Not only does the 59 kDa *S. pombe* PRP2 protein have sequence similarity to hU2AF^{65}, but also the human U2AF^{65} cDNA rescues growth of a prp2 temperature-sensitive mutant strain at the restrictive temperature. This observation demonstrates that PRP2 is a true functional homologue of human U2AF. In *S. cerevisiae*, an open reading frame of unknown function has similar sequence and domain organization as human U2AF^{65} (124). In addition, the MUD2 gene product, which is involved in splicing (Chapter 7), also shares some primary structure features with hU2AF^{65}, and parallels between the functional properties of the two proteins have been noted (130).

4. Other purified factors required for catalytic step I

Chromatographic fractionation of HeLa cell extracts, followed by *in vitro* splicing with combinations of column fractions has been used by several groups to attempt to reconstitute splicing with partially purified or highly purified components. This approach is technically very challenging, because of the very large number of required protein components, the fact that at least five multi-subunit snRNP particles are also required, and the fact that both the snRNPs and several multi-subunit proteins display heterogeneous chromatographic behaviour, which is due to loosely associated subunits, heterogeneous patterns of phosphorylation, and so on. The main functional assay is reconstitution of splicing activity with a purified fraction and with other complementing fractions that may be fairly crude, but which are limiting for the factor being assayed. Because in some cases reconstitution of the full splicing activity is inefficient, which renders the assay somewhat insensitive, assembly of specific pre-spliceosomal complexes has also been successfully used as a purification assay. Although several partially purified fractions have been described, and their activities partially characterized, we will focus on components that have been fully purified, and for which sequence information is available.

4.1 SF1 and SF3 requirement for pre-spliceosome assembly

Three protein fractions were reported to be required, in addition to a fraction containing snRNPs, for formation of the ATP-dependent pre-spliceosomal complex A (Chapter 4), as well as for the catalytic steps of splicing (131). One of these fractions was shown to be replaceable by purified U2AF, and the other two fractions were termed SF1 and SF3. At least one SR protein is known to be required for A complex assembly as well, and this activity was probably present in one of the fractions, such as the snRNP fraction. In earlier work, a fraction coincidentally designated SF2 (distinct from the factor SF2/ASF) was also required for spliceosome assembly (132). This activity co-purified with creatine phosphokinase (133), which is required for ATP regeneration from ADP and creatine phosphate in *in vitro* splicing reactions.

SF1 has been purified to homogeneity and migrates as a single 75 kDa polypeptide (134). Its activity survives boiling, as well as renaturation after SDS-PAGE. Other than its requirement for splicing and for A complex assembly, its mechanism of action is currently unknown. The purified protein was reported to be inactive in several RNA-binding assays, as well as in RNA annealing and ATPase assays. The fact that SF1 does not bind RNA makes it unlikely that it corresponds to SRp75, despite the apparent molecular weight similarity and involvement in A complex assembly.

A fraction containing SF3 activity was subsequently fractionated into two required components, designated SF3a and SF3b (135). Both of these are required for assembly of the ATP-dependent pre-spliceosomal complex A, together with additional protein fractions and U1 and U2 snRNPs. The three subunits of SF3a are stable constituents of the 17S form of U2 snRNP (Chapter 5), and in fact SF3a is required (together with SF3b) to convert the 12S into the 17S U2 snRNP particle (136). It is the latter form of U2 snRNP that is thought to be the active form during assembly of the pre-spliceosomal complex A (Chapter 4). Whereas SF3b remains to be further purified and characterized, SF3a has been extensively studied.

4.1.1 Primary structure of SF3a

SF3a consists of three subunits of 120 kDa, 66 kDa and 60 kDa, all of which are present in a fraction purified to apparent homogeneity (137). Monoclonal antibodies specific for the 66 kDa subunit immunoprecipitate all three polypeptides, confirming the subunit relationship. The sequence of cDNAs encoding the 120 kDa subunit reveals significant homology to the yeast splicing factor PRP21 (Chapter 7). Both proteins contain a repeated motif known as a SURP module, which was first identified in the *Drosophila* alternative splicing regulator SWAP (138; Chapter 8). The C-terminus of SF3a120 has a proline-rich region and a ubiquitin-like domain. Deletion analysis showed that the SURP module region is involved in interactions with the 60 kDa subunit, while a region downstream is required for binding to the 66 kDa subunit (137).

The 60 kDa subunit of SF3a is homologous to the yeast PRP9 splicing factor, with approximately 30% sequence identity between the two proteins. Both proteins contain a putative zinc-finger motif near the C-terminus. The regions spanning the homologous zinc-finger motif have been swapped between the human and yeast proteins, and the chimeric yeast protein containing the human zinc-finger region is functional *in vivo*: it allows growth at the restrictive temperature of a yeast temperature-sensitive strain with a mutation in PRP9 (139). This experiment shows that the domain from human SF3a60 can function correctly within the framework of yeast PRP9, presumably in the context of splicing. Several proteins involved in splicing contain putative zinc-finger motifs with characteristic spacings of cysteine and histidine residues, besides SF3a60. These include the SR protein 9G8 (37), the U1-C snRNP polypeptide (Chapter 5), and the yeast proteins SLU7, PRP6, PRP9, and PRP11 (Chapter 7). The putative zinc-finger domains of PRP9 and U1-C have been implicated in dimerization interactions, and mutation of the second of two motifs in PRP9 results in a dominant lethal phenotype (140).

Experiments from several laboratories demonstrated that each of the subunits of the heterotrimeric SF3a are highly conserved between yeast and mammals, as are the subunit interactions of the complex (136, 139, 141–143; see Chapters 4, 5, and 7). There is a one-to-one correspondence between SF3a120 (SAP114) and PRP21, SF3a60 (SAP61) and PRP9, and SF3a66 (SAP62) and PRP11.

5. PSF and related proteins

Human PSF (PTB-associated splicing factor) was initially identified as a 100-kDa polypeptide that forms a complex with PTB (polypyrimidine tract-binding protein) (144). PTB was purified on the basis of its binding to an adenovirus intron 3′ splice site, and was subsequently shown to be identical to hnRNP I (145–148; see also Chapters 3, 4, and 8). Depletion of poly(U)-binding proteins from nuclear extracts caused loss of splicing of an α-tropomyosin substrate, and activity could be restored by addition of U2AF together with a fraction containing primarily PTB and PSF. Subsequently, PTB proved to be dispensable in this complementation assay, whereas PSF was found to be required. For example, antibodies to PSF were used to immunodeplete this protein, which blocked splicing, and recombinant protein was added back to restore splicing. On the basis of these results, PSF was proposed to be a splicing factor involved in catalytic step I of the reaction (144).

Recently, PSF was shown to be stably associated with isolated spliceosomal complex C, in which the pre-mRNA has already undergone the first catalytic step of splicing (149; Chapter 4). Within complex C, PSF is bound to the pre-mRNA, as shown by UV cross-linking. In the same study, immunodepletion and add-back experiments showed that PSF is required for catalytic step II during splicing of an α-tropomyosin pre-mRNA intron. In the earlier immunodepletion study the inhibition of splicing was less extensive, and the procedure was such that the first step also appears to have been partially inhibited, perhaps non-specifically. Thus, the most recent evidence implicates this factor in catalytic step II, rather than in catalytic step I. It remains to be determined whether the requirement for PSF is general, or whether it is a specialized factor involved in processing of certain pre-mRNAs with extensive upstream polypyrimidine tracts.

The sequence of PSF cDNAs reveals the presence of two RRMs preceded by an N-terminal domain rich in Pro and Gln residues, reminiscent of the activation domains of certain transcription factors (144). The recombinant protein binds to the polypyrimidine tracts of introns, even in the absence of PTB. The protein is 707 amino acids long but migrates more slowly than expected on SDS-PAGE.

A 54 kDa human protein of unknown function, termed $p54^{nrb}$, has extensive homology to human PSF (150). The two proteins share a 320-amino acid region with greater than 70% sequence identity. The first half of this region of ungapped identity spans two RRMs, which are separated by a single alanine. A *Drosophila* protein, variously known as NONA, BJ6, or $NONA^{diss}$, has a highly homologous 320 amino acid domain. Although its molecular function is unknown, this gene product is ubiquitously expressed and has highly pleiotropic effects in the central nervous system (151). Thus, it appears that PSF belongs to a family of proteins, although it remains to be demonstrated whether the other family members are also involved in splicing, and if so, whether individual proteins have unique functions, such as distinct substrate specificities. $p54^{nrb}$ was also shown to be an RNA-binding protein, as expected from the sequence, but unlike PSF, it was not found to bind preferentially to polypyrimidine tracts.

6. Other mammalian and viral splicing factors

6.1 HRH1

Recently, a human splicing factor designated HRH1 (human RNA helicase 1) was identified and shown to have sequence homology to yeast PRP22 and to other putative RNA helicases of the DEAH motif class, including the yeast splicing factors PRP2 and PRP16 (Chapter 7). Whereas these and several other putative RNA helicases of the DEAD, motif class are involved in splicing in *S. cerevisiae*, no homologous proteins had been implicated in mammalian splicing before the identification of HRH1. Human cDNAs encoding HRH1 were identified by RT-PCR using degenerate oligonucleotides corresponding to conserved motifs of DEAH proteins (152). Of the yeast DEAH splicing factors, human HRH1 appears to be most closely related in sequence to PRP22, which is involved in releasing mature mRNA from the spliceosome (Chapter 7). Moreover, the human cDNA allows a PRP22 temperature-sensitive yeast strain to grow at the restrictive temperature, indicating that the yeast and human protein have similar functional properties (152). An interesting feature of the 139 kDa HRH1 protein is the presence of an RS domain near the N-terminus, whereas yeast PRP22 lacks this type of domain. Like other proteins with RS domains, HRH1 was shown to interact, via its RS domain, with the RS domains of other splicing factors, by methods including the two-hybrid assay.

6.2 SF4 and SF53/4

Upon assembly of pre-spliceosomal complexes with partially purified fractions containing snRNPs, U2AF, SF1, SF3, and presumably other required components, it was found that the resulting complexes did not carry out catalytic step I (cleavage of the pre-mRNA at the 5′ splice site and lariat formation). Addition of another partially purified fraction, termed SF4, allowed this catalytic step to occur (153). It was shown that the SF4 component(s) is heat labile and *N*-ethyl maleimide sensitive. When further purification and characterization is reported, it will be of interest to see if SF4 is related to, for example, the yeast protein PRP2, an RNA-dependent ATPase required for the first transesterification subsequent to spliceosome assembly (Chapter 7).

SF53/4 was identified as an 88 kDa polypeptide required for an early step in spliceosome assembly (154). Monoclonal antibodies were raised against large 200S RNP particles isolated from mammalian nuclei (Chapter 3). One of these antibodies, 53/4, recognized an 88 kDa polypeptide, and could be used to immunodeplete splicing extracts of the protein, and consequently of splicing activity. Activity was restored by addition of the protein eluted from the immunoaffinity column. It should be possible to use the same antibody to clone cDNAs encoding SF53/4.

6.3 Adenovirus E4 ORF3 and ORF6

The adenovirus early region E4 encodes two proteins, E4 ORF3 and ORF6, which have opposite effects on the regulation of alternatively spliced mRNAs from the viral major late transcription unit. While the ORF3 protein facilitates exon inclusion of an alternative exon (the i-leader) of the tripartite leader region, the ORF6 protein promotes exon skipping (155). Furthermore, these adenovirus-encoded proteins can modulate alternative splicing of a chimeric β-globin transcript containing multiple exons. Thus, the levels of E4 ORF3 and ORF6 during virus infection may have important consequences for alternative splicing in an adenovirus-infected cell. These antagonistic effects on 5′ splice-site selection resemble the properties of hnRNP A1 versus SF2/ASF in alternative splicing, although no sequence homology between these adenovirus proteins and either hnRNP or SR proteins has been detected.

7. Summary and perspectives

Although numerous metazoan splicing proteins have already been identified through biochemical and other approaches, it is clear by reference to the presumably less complex splicing apparatus of budding yeast (Chapter 7) and to the size and complexity of the spliceosome (Chapter 4), that numerous splicing factors remain to be identified. This task may be facilitated by the ongoing genome projects and characterization of expressed sequence tags, together with sequence comparisons with known yeast splicing factors. In addition, a complementary approach is the sequencing of spliceosomal constituents identified by 2-dimensional gel analysis (Chapter 4). In these cases, biochemical approaches will be required to establish functional assays to test the involvement of these proteins in splicing. As more required factors are identified, it is anticipated that purification of novel splicing factors on the basis of biochemical complementation will also be facilitated, provided that the known factors are available in large quantities in a biochemically active state. Much work remains to be done not only in identifying splicing proteins, but also in determining their precise mechanisms of action in assembling a functional spliceosome, driving conformational rearrangements, hydrolysing ATP, modulating alternative splicing, and—in conjunction with the snRNPs—helping to select the correct splice sites and perhaps contribute to the formation of active sites for one or both catalytic steps.

Acknowledgements

We thank members of our laboratory for many helpful discussions. Our work is supported by grant GM42699 from the NIH, grant CA13106 from the NCI, and by the Pew Charitable Trusts.

References

1. Krainer, A. R. and Maniatis, T. (1988) RNA splicing. In *Transcription and splicing*, Hames, B. D. and Glover, D. M. (eds.). IRL Press, Oxford, p. 131.
2. Green, M. R. (1991) Biochemical mechanisms of constitutive and regulated pre-mRNA splicing. *Annu. Rev. Cell. Biol.*, **7**, 559.
3. Moore, M. J., Query, C. C., and Sharp, P. A. (1993) Splicing of precursors to messenger RNAs by the spliceosome. In *The RNA world*, Gesteland, R. F. and Atkins, J. F. (eds). Cold Spring Harbor Laboratory Press, Cold Spring Harbor, p. 303.
4. Rio, D. (1993) Splicing of pre-mRNA: mechanism, regulation and role in development. *Curr. Opin. Genet. Dev.*, **3**, 574.
5. Krämer, A. (1993) Mammalian protein factors involved in nuclear pre-mRNA splicing. *Mol. Biol. Rep.*, **18**, 93.
6. Lamm, G. and Lamond, A. I. (1993) Non-snRNP protein splicing factors. *Biochim. Biophys. Acta*, **1173**, 247.
7. Zahler, A. M., Lane, W. S., Stolk, J. A. and Roth, M. B. (1992) SR proteins: a conserved family of pre-mRNA splicing factors. *Genes Dev.*, **6**, 837.
8. Krainer, A. R., Conway, G. C., and Kozak, D. (1990) Purification and characterization of SF2, a human pre-mRNA splicing factor. *Genes Dev.*, **4**, 1158.
9. Krainer, A. R., Conway, G. C., and Kozak, D. (1990) The essential pre-mRNA splicing factor SF2 influences 5′ splice site selection by activating proximal sites. *Cell*, **62**, 35.

10. Ge, H. and Manley, J. L. (1990) A protein factor, ASF, controls alternative splicing of SV40 early pre-mRNA *in vitro*. *Cell*, **62**, 25.
11. Krainer, A. R. and Maniatis, T. (1985) Multiple factors including the small nuclear ribonucleoproteins U1 and U2 are necessary for pre-mRNA splicing *in vitro*. *Cell*, **42**, 725.
12. Krainer, A. R., Mayeda, A., Kozak, D., and Binns, G. (1991) Functional expression of cloned human splicing factor SF2: homology to RNA-binding proteins, U1 70K and *Drosophila* splicing regulators. *Cell*, **66**, 383.
13. Ge, H., Zuo, P., and Manley, J. L. (1991) Primary structure of the human splicing factor ASF reveals similarities with *Drosophila* regulators. *Cell*, **66**, 373.
14. Bandziulis, R. J., Swanson, M. S., and Dreyfuss, G. (1989) RNA-binding proteins as developmental regulators. *Genes Dev.*, **3**, 431.
15. Birney, E., Kumar, S., and Krainer, A. R. (1993) Analysis of the RNA-recognition motif and RS and RGG domains: conservation in metazoan pre-mRNA splicing factors. *Nucleic Acids Res.*, **21**, 5803.
16. Kenan, D. J., Query, C. C., and Keene, J. D. (1991) RNA recognition: towards identifying determinants of specificity. *Trends Biochem. Sci.*, **16**, 214.
17. Nagai, K., Oubridge, C., Jessen, T. H., Li, J., and Evans, P. R. (1990) Crystal structure of the RNA-binding domain of the U1 small nuclear ribonucleoprotein A. *Nature*, **348**, 515.
18. Hoffman, D. W., Query, C. C., Golden, B. L., White, S. W., and Keene, J. D. (1990) RNA-binding domain of the A protein component of the U1 small nuclear ribonucleoprotein analyzed by NMR spectroscopy is structurally similar to ribosomal proteins. *Proc. Natl Acad. Sci. USA*, **88**, 2495.
19. Wittekind, M., Görlach, M., Friedrichs, M., Dreyfuss, G., and Mueller, L. (1992) ^{13}C and ^{15}N NMR assignments and global folding pattern of the RNA-binding domain of the human hnRNP C proteins. *Biochemistry*, **31**, 6254.
20. Oubridge, C., Ito, N., Evans, P. R., Teo, C.-H., and Nagai, K. (1994) Crystal structure at 1.92 Å resolution of the RNA-binding domain of the U1A spliceosomal protein complexed with an RNA hairpin. *Nature*, **372**, 432.
21. Scherly, D., Boelens, W., van Venrooij, W. J., Dathan, N., Hamm, J., and Mattaj, I. W. (1989) Identification of the RNA binding segment of human U1A protein and definition of its binding site on U1 snRNA. *EMBO J.*, **8**, 4163.
22. Surowy, C. S., Van Santen, V. L., Scheib-Wixted, S. M., and Spritz, R. A. (1989) Direct, sequence-specific binding of the human U1-70K ribonucleoprotein antigen protein to loop I of U1 small nuclear RNA. *Mol. Cell. Biol.*, **9**, 4179.
23. Cáceres, J. F. and Krainer, A. R. (1993) Functional analysis of pre-mRNA splicing factor SF2/ASF structural domains. *EMBO J.*, **12**, 4715.
24. Zuo, P. and Manley, J. L. (1993) Functional domains of the human splicing factor ASF/SF2. *EMBO J.*, **12**, 4727.
25. Tacke, R., Boned, A., and Goridis, C. (1992) ASF alternative transcripts are highly conserved between mouse and man. *Nucleic Acids Res.*, **20**, 5482.
26. Fu, X.-D. and Maniatis, T. (1990) Factor required for mammalian spliceosome assembly is localized to discrete regions in the nucleus. *Nature*, **343**, 437.
27. Spector, D. L., Fu, X.-D., and Maniatis, T. (1991) Associations between distinct pre-mRNA splicing components and the cell nucleus. *EMBO J.*, **10**, 3467.
28. Vellard, M., Sureau, A., Soret, J., Martinerie, C., and Perbal, B. (1992) A potential splicing factor is encoded by the opposite strand of the trans-spliced c-myb exon. *Proc. Natl Acad. Sci. USA*, **89**, 2511.

29. Fu, X.-D. and Maniatis, T. (1992) Isolation of a complementary DNA that encodes the mammalian splicing factor SC35. *Science*, **256**, 535.
30. Fu, X.-D., Mayeda, A., Maniatis, T., and Krainer, A. R. (1992) General splicing factors SF2 and SC35 have equivalent activities in vitro and both affect alternative 5′ and 3′ splice site selection. *Proc. Natl Acad. Sci. USA*, **89**, 11224.
31. Champlin, D. T., Frasch, M., Saumweber, H., and Lis, J. T. (1991) Characterization of a *Drosophila* protein associated with boundaries of transcriptionally active chromatin. *Genes Dev.*, **5**, 1611.
32. Roth, M. B., Zahler, A. M., and Stolk, J. A. (1991) A conserved family of nuclear phosphoproteins localized to sites of polymerase II transcription. *J. Cell Biol.*, **115**, 587.
33. Ayane, M., Preuss, U., Köhler, G., and Nielsen, P. J. (1991) A differentially expressed murine RNA encoding a protein with similarities to two types of nucleic acid binding motifs. *Nucleic Acids Res.*, **19**, 1273.
34. Roth, M. B., Murphy, C., and Gall, J. G. (1990) A monoclonal antibody that recognizes a phosphorylated epitope stains lampbrush chromosome loops and small granules in the amphibian germinal vesicle. *J. Cell. Biol.*, **111**, 2217.
35. Li, H. and Bingham, P. M. (1991) Arginine/Serine rich domains of the su(wa) and tra RNA processing regulators target proteins to a subnuclear compartment implicated in splicing. *Cell*, **67**, 335.
36. Zahler, A. M., Neugebauer, K. M., Stolk, J. A., and Roth, M. B. (1993) Human SR proteins and isolation of a cDNA encoding SRp75. *Mol. Cell. Biol.*, **13**, 4023.
37. Cavaloc, Y., Popielarz, M., Fuchs, J.-P., Gattoni, R., and Stévenin, J. (1994) Characterization and cloning of the human splicing factor 9G8: a novel 35 kDa factor of the serine/arginine protein family. *EMBO J.*, **13**, 2639.
38. Screaton, G. R., Cáceres, J. F., Mayeda, A., Bell, M. V., Plebanski, M., Jackson, D. G., *et al.* (1995) Identification and characterization of three members of the human SR family of pre-mRNA splicing factors. *EMBO J.*, **14**, 4336.
39. Ring, H. Z. and Lis, J. T. (1994) The SR protein B52/SRp55 is essential for *Drosophila* development. *Mol. Cell. Biol.*, **14**, 7499.
40. Peng, X. and Mount, S. M. (1995) Genetic enhancement of RNA-processing defects by a dominant mutation in B52, the *Drosophila* gene for an SR protein splicing factor. *Mol. Cell. Biol.*, **15**, 6273.
41. Kraus, M. E. and Lis, J. T. (1994) The concentration of B52, an essential splicing factor and regulator of splice site choice *in vitro*, is critical for *Drosophila* development. *Mol. Cell. Biol.*, **14**, 5360.
42. Kim, Y.-J., Zuo, P., Manley, J. L., and Baker, B. S. (1992) A *Drosophila* RNA binding protein RBP1 is localized to transcriptionally active sites of chromosomes and shows a functional similarity to human splicing factor ASF/SF2. *Genes Dev.*, **6**, 2569.
43. Kim, Y.-J. and Baker, B. S. (1993) Isolation of RRM-type RNA-binding protein genes and the analysis of their relatedness by using a numerical approach. *Mol. Cell. Biol.*, **13**, 174.
44. Green, L. M. and Berg, J. M. (1989) A retroviral Cys-Xaa2-Cys-Xaa4-His-Xaa4-Cys peptide binds metal ions: spectroscopic studies and a proposed three-dimensional structure. *Proc. Natl Acad. Sci. USA*, **86**, 4047.
45. Diamond, R. H., Du, K., Lee, V. M., Mohn, K. L., Haber, B. A., Tewari, D. S. *et al.* (1993) Novel delayed-early and highly insulin-induced growth response genes: identification of HRS, a potential regulator of alternative pre-mRNA splicing. *J. Biol. Chem.*, **268**, 15185.

46. Fu, X.-D. (1993) Specific commitment of different pre-mRNAs to splicing by single SR proteins. *Nature*, **365**, 82.
47. Mayeda, A. and Krainer, A. R. (1992) Regulation of alternative pre-mRNA splicing by hnRNP A1 and splicing factor SF2. *Cell*, **68**, 365.
48. Ben-David, Y., Boni, M. R., Chabot, B., DeKoven, A., and Bernstein, A. (1992) Retroviral insertions downstream of the heterogeneous nuclear ribonucleoprotein A1 gene in erythroleukemia cells: evidence that A1 is not essential for cell growth. *Mol. Cell. Biol.*, **12**, 4449.
49. Mayeda, A., Helfman, D. M., and Krainer, A. R. (1993) Modulation of exon skipping and inclusion by heterogenous nuclear ribonucleoprotein A1 and pre-mRNA splicing factor SF2/ASF. *Mol. Cell. Biol.*, **13**, 2993.
50. Mayeda, A., Munroe, S. H., Cáceres, J. F., and Krainer, A. R. (1994) Function of conserved domains of hnRNP A1 and other hnRNP A/B proteins. *EMBO J.*, **13**, 5483.
51. Buvoli, M., Cobianchi, F., Bestagno, M. G., Mangiarotti, A., Bassi, M. T., Biamonti, G., *et al.* (1990) Alternative splicing in the human gene for the core protein A1 generates another hnRNP protein. *EMBO J.*, **9**, 1229.
52. Yang, X., Bani, M. R., Lu, S.-J., Rowan, S., Ben-David, Y., and Chabot, B. (1994) The A1 and $A1^B$ proteins of heterogeneous nuclear ribonucleoparticles modulate 5′ splice site selection *in vivo*. *Proc. Natl Acad. Sci. USA*, **91**, 6924.
53. Mayeda, A., Zahler, A. M., Krainer, A. R., and Roth, M. B. (1992) Two members of a conserved family of nuclear phosphoproteins are involved in pre-mRNA splicing. *Proc. Natl Acad. Sci. USA*, **89**, 1301.
54. Fu, X.-D. and Maniatis, T. (1992) The 35-kDa mammalian splicing factor SC35 mediates specific interactions between U1 and U2 small nuclear ribonucleoprotein particles at the 3′ splice site. *Proc. Natl Acad. Sci. USA*, **89**, 1725.
55. Michaud, S. and Reed, R. (1991) An ATP-independent complex commits pre-mRNA to the mammalian spliceosome assembly pathway. *Genes Dev.*, **5**, 2534.
56. Michaud, S. and Reed, R. (1993) A functional association between the 5′ and 3′ splice sites is established in the earliest prespliceosome complex (E) in mammals. *Genes Dev.*, **7**, 1008.
57. Jamison, S. F., Crow, A., and García-Blanco, M. A. (1992) The spliceosome assembly pathway in mammalian extracts. *Mol. Cell. Biol.*, **12**, 4279.
58. Kohtz, J. D., Jamison, S. F., Will, C. L., Zuo, P., Lührmann, R., García-Blanco, M., *et al.* (1994) Protein–protein interactions and 5′ splice-site recognition in mammalian mRNA precursors. *Nature*, **368**, 119.
59. Wu, J. Y. and Maniatis, T. (1993) Specific interactions between proteins implicated in splice site selection and regulated alternative splicing. *Cell*, **75**, 1061.
60. Amrein, H., Hedley, M. L., and Maniatis, T. (1994) The role of specific protein–RNA and protein–protein interactions in positive and negative control of pre-mRNA splicing by transformer2. *Cell*, **76**, 735.
61. Horowitz, D. S. and Krainer, A. R. (1994) Mechanisms for selecting 5′ splice sites in mammalian pre-mRNA splicing. *Trends Genet.*, **10**, 100.
62. Zuo, P. and Manley, J. L. (1994) The human splicing factor ASF/SF2 can specifically recognize pre-mRNA 5′ splice sites. *Proc. Natl Acad. Sci. USA*, **91**, 3363.
63. Crispino, J. D., Blencowe, B. J., and Sharp, P. A. (1994) Complementation by SR proteins of pre-mRNA splicing reactions depleted of U1 snRNP. *Science*, **265**, 1866.
64. Tarn, W.-Y. and Steitz, J. A. (1994) SR proteins can compensate for the loss of U1 snRNP functions *in vitro*. *Genes Dev.*, **8**, 2704.

65. Eperon, I. C., Ireland, D. C., Smith, R. A., Mayeda, A., and Krainer, A. R. (1993) Pathways for selection of 5′ splice sites by U1 snRNPs and SF2/ASF. *EMBO J.*, **12**, 3607.
66. Reed, R. and Maniatis, T. (1986) A role for exon sequences and splice-site proximity in splice-site selection. *Cell*, **46**, 681.
67. Furdon, P. J. and Kole, R. (1988) The length of the downstream exon and the substitution of specific sequences affect pre-mRNA splicing *in vitro*. *Mol. Cell. Biol.*, **8**, 860.
68. Siebel, C. W., Fresco, L. D., and Rio, D. C. (1992) The mechanism of somatic inhibition of *Drosophila* P-element pre-mRNA splicing: multiprotein complexes at an exon pseudo-5′ splice site control U1 snRNP binding. *Genes Dev.*, **6**, 1386.
69. Tian, M. and Maniatis, T. (1993) A splicing enhancer complex controls alternative splicing of doublesex pre-mRNA. *Cell*, **74**, 105.
70. Watakabe, A., Tanaka, K., and Shimura, Y. (1993) The role of exon sequences in splice site selection. *Genes Dev.*, **7**, 407.
71. Sun, Q., Hampson, R. K., and Rottman, F. M. (1993) *In vitro* analysis of bovine growth hormone pre-mRNA alternative splicing: exon sequences and trans-acting factor(s). *J. Biol. Chem.*, **268**, 15659.
72. Sun, Q., Mayeda, A., Hampson, R. K., Krainer, A. R., and Rottman, F. M. (1993) General splicing factor SF2/ASF promotes alternative splicing by binding to an exonic splicing enhancer. *Genes Dev.*, **7**, 2598.
73. Mardon, H. J., Sebastio, G., and Baralle, F. E. (1987) A role for exon sequences in alternative splicing of the human fibronectin gene. *Nucleic Acids Res.*, **15**, 7725.
74. Lavigueur, A., La Branche, H., Kornblihtt, A. R., and Chabot, B. (1993) A splicing enhancer in the human fibronectin alternate ED1 exon interacts with SR proteins and stimulates U2 snRNP binding. *Genes Dev.*, **7**, 2405.
75. Mattox, W., Ryner, L., and Baker, B. S. (1992) Autoregulation and multifunctionality among *trans*-acting factors that regulate alternative pre-mRNA processing. *J. Biol. Chem.*, **267**, 19023.
76. Nagoshi, R. N. and Baker, B. S. (1990) Regulation of sex-specific RNA splicing at the *Drosophila* doublesex gene: cis-acting mutations in exon sequences alter specific RNA splicing patterns. *Genes Dev.*, **4**, 89.
77. Tanaka, K., Watakabe, A., and Shimura, Y. (1994) Polypurine sequences within a downstream exon function as a splicing enhancer. *Mol. Cell. Biol.*, **14**, 1347.
78. Hampson, R. K., LaFollette, L., and Rottman, F. M. (1989) Alternative processing of bovine growth hormone mRNA is influenced by downstream exon sequences. *Mol. Cell. Biol.*, **9**, 1604.
79. Dirksen, W. P., Hampson, R. K., Sun, Q., and Rottman, F. M. (1994) A purine-rich exon sequence influences alternate splicing of bovine growth hormone pre-mRNA. *J. Biol. Chem.*, **269**, 6431.
80. Xu, R., Teng, J., and Cooper, T. A. (1993) The cardiac troponin T alternative exon contains a novel purine-rich positive splicing element. *Mol. Cell. Biol.*, **13**, 3660.
81. Roesser, J. R., Liittschwager, K., and Leff, S. E. (1993) Regulation of tissue-specific splicing of the calcitonin/calcitonin gene-related peptide gene by RNA-binding proteins. *J. Biol. Chem.*, **268**, 8366.
82. Cote, G. J., Stolow, D. T., Peleg, S., Berget, S. M., and Gagel, R. F. (1992) Identification of exon sequences and an exon binding protein involved in alternative RNA splicing of calcitonin/CGRP. *Nucleic Acids Res.*, **20**, 2361.
83. Van Oers, C. C. M., Adema, G. J., Zandberg, H., Moen, T. C., and Baas, P. D. (1994) Two different sequence elements within exon 4 are necessary for calcitonin-specific splicing

of the human calcitonin/calcitonin gene-related peptide I pre-mRNA. *Mol. Cell. Biol.*, **14**, 951.

84. Yeakley, J. M., Hedjran, F., Morfin, J.-P., Merillat, N., Rosenfeld, M. G., and Emeson, R. B. (1993) Control of calcitonin/calcitonin gene-related peptide pre-mRNA processing by constitutive intron and exon elements. *Mol. Cell. Biol.*, **13**, 5999.
85. Amendt, B. A., Si, Z. H., and Stoltzfus, C. M. (1995) Presence of exon splicing silencers within human immunodeficiency virus type 1 tat exon 2 and tat-rev exon 3: evidence for inhibition mediated by cellular factors. *Mol. Cell. Biol.*, **15**, 4606.
86. Staffa, A. and Cochrane, A. (1995) Identification of positive and negative splicing regulatory elements within the terminal tat-rev exon of human immunodeficiency virus type 1. *Mol. Cell. Biol.*, **15**, 4597.
87. Varmus, H. (1988) Retroviruses. *Science*, **240**, 1427.
88. Kuo, H. C., Nasim, F. H., and Grabowski, P. J. (1991) Control of alternative splicing by the differential binding of U1 small nuclear ribonucleoprotein particle. *Science*, **251**, 1045.
89. Hoffman, B. E. and Grabowski, P. J. (1992) U1 snRNP targets an essential splicing factor, $U2AF^{65}$, to the 3′ splice site by a network of interactions spanning the exon. *Genes Dev.*, **6**, 2554.
90. Robberson, B. L., Cote, G. J., and Berget, S. M. (1990) Exon definition may facilitate splice site selection in RNAs with multiple exons. *Mol. Cell. Biol.*, **10**, 84.
91. Talerico, M. and Berget, S. M. (1990) Effect of 5′ splice site mutation on splicing of the preceding intron. *Mol. Biol. Cell*, **10**, 6299.
92. Staknis, D. and Reed, R. (1994) SR proteins promote the first specific recognition of pre-mRNA and are present together with the U1 small nuclear ribonucleoprotein particle in a general splicing enhancer complex. *Mol. Cell. Biol.*, **14**, 7670.
93. Zahler, A. M., Neugebauer, K. M., Lane, W. S., and Roth, M. B. (1993) Distinct functions of SR proteins in alternative pre-mRNA splicing. *Science*, **260**, 219.
94. Cáceres, J. F., Stamm, S., Helfman, D. M., and Krainer, A. R. (1994) Regulation of alternative splicing *in vivo* by overexpression of antagonistic splicing factors. *Science*, **265**, 1706.
95. Gattoni, R., Chebli, K., Himmelspach, M., and Stévenin, J. (1991) Modulation of alternative splicing of adenoviral E1A transcripts: factors involved in the early-to-late transition. *Genes Dev.*, **5**, 1847.
96. Himmelspach, M., Cavaloc, Y., Chebli, K., Stévenin, J., and Gattoni, R. (1995) Titration of serine/arginine (SR) splicing factors during adenoviral infection modulates E1A pre-mRNA alternative splicing. *RNA*, **1**, 794.
97. Popielarz, M., Cavaloc, Y., Mattei, M. G., Gattoni, R., and Stévenin, J. (1995) The gene encoding human splicing factor 9G8. Structure, chromosomal localization, and expression of alternatively processed transcripts. *J. Biol. Chem.*, **270**, 17830.
98. Sureau, A. and Perbal, B. (1994) Several mRNAs with variable 3′ untranslated regions and different stability encode the human PR264/SC35 splicing factor. *Proc. Natl Acad. Sci. USA*, **91**, 932.
99. Sureau, A., Soret, J., Vellard, M., Crochet, J., and Perbal, B. (1992) The PR264/c-myb connection: expression of a splicing factor modulated by a nuclear protooncogene. *Proc. Natl Acad. Sci. USA*, **89**, 11683.
100. Celis, J. E., Bravo, R., Arenstorf, H. P., and LeStourgeon, W. M. (1986) Identification of proliferation-sensitive human proteins amongst components of the 40 S hnRNP particles: identity of hnRNP core proteins in the HeLa protein catalogue. *FEBS Lett.*, **194**, 101.

101. Biamonti, G., Bassi, M. T., Cartegni, L., Mechta, F., Buvoli, M., Cobianchi, F., *et al.* (1993) Human hnRNP protein A1 gene expression: structural and functional characterization of the promoter. *J. Mol. Biol.*, **230**, 77.
102. Planck, S. R., Listerud, M. D., and Buckley, S. D. (1988) Modulation of hnRNP A1 protein by epidermal growth factor in Rat-1 cells. *Nucleic Acids Res.*, **16**, 11663.
103. Magnuson, V. L., Young, M., Schattenberg, D. G., Mancini, M. A., Chen, D., Steffensen, B., *et al.* (1991) The alternative splicing of fibronectin pre-mRNA is altered during aging and in response to growth factors. *J. Biol. Chem.*, **266**, 14654.
104. Zardi, L., Carnemolla, B., Siri, A., Petersen, T. E., Paolella, G., Sebastio, G., *et al.* (1987) Transformed human cells produce a new fibronectin isoform by preferential alternative splicing of a previously unobserved exon. *EMBO J.*, **6**, 2337.
105. Matsumura, F., Lin, J. J.-C., Yamashiro-Matsumura, S., Thomas, G. P., and Topp, W. C. (1983) Differential expression of tropomyosin forms in the microfilaments isolated from normal and transformed rat cultured cells. *J. Biol. Chem.*, **258**, 13954.
106. Gunthert, U., Hofmann, M., Rudy, W., Reber, S., Zoller, M., Haussmann, J., *et al.* (1991) A new variant of glycoprotein CD44 confers metastatic potential to rat carcinoma cells. *Cell*, **65**, 13.
107. Miksicek, R. J., Lei, Y., and Wang, Y. (1993) Exon skipping gives rise to alternatively spliced forms of the estrogen receptor in breast tumor cells. *Breast Cancer Res. Treat.*, **26**, 163.
108. Woppmann, A., Patschinsky, T., Bringmann, P., Godt, F., and Lührmann, R. (1990) Characterization of human and murine snRNP proteins by two-dimensional gel electrophoresis and phosphopeptide analysis of U1-specific 70K protein variants. *Nucleic Acid Res.*, **18**, 4427.
109. Tazi, J., Daugeron, M.-C., Cathala, G., Brunel, C., and Jeanteur, P. (1992) Adenosine phosphorothioates (ATPαS and ATPγS) differentially affect the two steps of mammalian pre-mRNA. *J. Biol. Chem.*, **267**, 4322.
110. Mermoud, J. E., Cohen, P., and Lamond, A. I. (1992) Ser/Thr specific protein phosphatases are required for both catalytic steps of pre-mRNA splicing. *Nucleic Acids Res.*, **20**, 5263.
111. Mermoud, J. E., Cohen, P. T. W., and Lamond, A. I. (1994) Regulation of mammalian spliceosome assembly by a protein phosphorylation mechanism. *EMBO J.*, **13**, 5679.
112. Cardinali, B., Cohen, P. T. W., and Lamond, A. I. (1994) Protein phosphatase I can modulate alternative 5′ splice site selection in a HeLa splicing extract. *FEBS Lett.*, **352**, 276.
113. Woppmann, A., Will, C. L., Kornstadt, U., Zuo, P., Manley, J. L., and Lührmann, R. (1993) Identification of an snRNP-associated kinase activity that phosphorylates arginine/serine rich domains typical of splicing factors. *Nucleic Acid Res.*, **21**, 2815.
114. Tazi, J., Kornstadt, U., Rossi, F., Jeanteur, P., Cathala, G., Brunel, C., *et al.* (1993) Thiophosphorylation of U1-70K protein inhibits pre-mRNA splicing. *Nature*, **363**, 283.
115. Gui, J.-F., Lane, W. S., and Fu, X.-D. (1994) A serine kinase regulates intracellular localization of splicing factors in the cell cycle. *Nature*, **369**, 678.
116. Gui, J-F., Tronchère, H., Chandler, S. D., and Fu, X.-D. (1994) Purification and characterization of a kinase specific for the serine and arginine-rich pre-mRNA splicing factors. *Proc. Natl Acad. Sci. USA*, **91**, 10824.
117. Takeuchi, M. and Yanagida, M. (1993) A mitotic role for a novel fission yeast protein kinase dsk1 with cell cycle stage dependent phosphorylation and localization. *Mol. Biol. Cell*, **4**, 247.

118. Ruskin, B., Zamore, P. D., and Green, M. R. (1988) A factor, U2AF, is required for U2 snRNP binding and splicing complex assembly. *Cell*, **52**, 207.
119. Zamore, P. D. and Green, M. R. (1989) Identification, purification and biochemical characterization of U2 small nuclear ribonucleoprotein auxiliary factor. *Proc. Natl Acad. Sci. USA*, **86**, 9243.
120. Zamore, P. D. and Green, M. R. (1991) Biochemical characterization of U2 small nuclear ribonucleoprotein auxiliary factor: an essential pre-mRNA splicing factor with a novel intranuclear distribution. *EMBO J.*, **10**, 207.
121. Zamore, P. D., Patton, J. G., and Green, M. R. (1992) Cloning and domain structure of the mammalian splicing factor U2AF. *Nature*, **355**, 609.
122. Sailer, A., MacDonald, N. J., and Weissmann, C. (1992) Cloning and sequencing of the murine homologue of the human splicing factor $U2AF^{65}$. *Nucleic Acids Res.*, **20**, 2374.
123. Kanaar, R., Roche, S. E., Beall, E. L., Green, M. R., and Rio, D. C. (1993) The conserved pre-mRNA splicing factor U2AF from *Drosophila*: requirement for viability. *Science*, **262**, 569.
124. Birney, E., Kumar, S., and Krainer, A. R. (1993) A putative homolog of $U2AF^{65}$ in *S. cerevisiae. Nucleic Acids Res.*, **20**, 4663.
125. Potashkin, J., Naik, K., and Wentz-Hunter, K. (1993) U2AF homolog required for splicing *in vivo. Science*, **262**, 573.
126. Lee, C.-G., Zamore, P. D., Green, M. R., and Hurwitz, J. (1993) RNA annealing activity is intrinsically associated with U2AF. *J. Biol. Chem.*, **268**, 13472.
127. Munroe, S. H. and Dong, X. F. (1992) Heterogeneous nuclear ribonucleoprotein A1 catalyzes RNA.RNA annealing. *Proc. Natl Acad. Sci. USA*, **89**, 895.
128. Valcárcel, J., Singh, R., Zamore, P. D., and Green, M. R. (1993) The protein Sex-lethal antagonizes the splicing factor U2AF to regulate alternative splicing of transformer pre-mRNA. *Nature*, **362**, 171.
129. Zhang, M., Zamore, P. D., Carmo-Fonseca, M., Lamond, A. I., and Green, M. R. (1992) Cloning and intracellular localization of the U2 small nuclear ribonucleoprotein auxiliary factor small subunit. *Proc. Natl Acad. Sci. USA*, **89**, 8769.
130. Abovich, N., Liao, X. C., and Rosbash, M. (1994) The yeast MUD2 protein: an interaction with PRP11 defines a bridge between commitment complexes and U2 snRNP addition. *Genes Dev.*, **8**, 843.
131. Krämer, A. and Utans, U. (1991) Three protein factors (SF1, SF3 and U2AF) function in pre-splicing complex formation in addition to snRNPs. *EMBO J.*, **10**, 1503.
132. Krämer, A., Frick, M., and Keller, W. (1987) Separation of multiple components of HeLa cell nuclear extracts required for pre-messenger RNA splicing. *J. Biol. Chem.*, **262**, 17630.
133. Krämer, A. and Keller, W. (1990) Preparation and fractionation of mammalian extracts active in pre-mRNA splicing. *Methods Enzymol.*, **181**, 3.
134. Krämer, A. (1992) Purification of splicing factor SF1, a heat-stable protein that functions in the assembly of presplicing complex. *Mol. Cell. Biol.*, **12**, 4545.
135. Brosi, R., Hauri, H.-P., and Krämer, A. (1993) Separation of splicing factor SF3 into two components and purification of SF3a activity. *J. Biol. Chem.*, **268**, 17640.
136. Brosi, R., Gröning, K., Behrens, S.-E., Lührmann, R., and Krämer, A. (1993) Interaction of mammalian splicing factor SF3a with U2 snRNP and relation of its 60-kD subunit to yeast PRP9. *Science*, **262**, 102.
137. Krämer, A., Mulhauser, F., Wersig, C., Groning, K., and Bilbe, G. (1995) Mammalian splicing factor SF3a120 represents a new member of the SURP family of proteins and is

homologous to the essential splicing factor PRP21p of *Saccharomyces cerevisiae*. *RNA*, **1**, 260.
138. Spikes, D. A., Kramer, J., Bingham, P. M., and Van Doren, K. (1994) SWAP pre-mRNA splicing regulators are a novel, ancient protein family sharing a highly conserved sequence motif with the prp21 family of constitutive splicing proteins. *Nucleic Acids Res.*, **22**, 4510.
139. Krämer, A., Legrain, P., Mulhauser, F., Groning, K., Brosi, R., and Bilbe, G. (1994) Splicing factor SF3a60 is the mammalian homologue of PRP9 of *S. cerevisiae*: the conserved zinc finger-like motif is functionally exchangeable *in vivo*. *Nucleic Acids Res.*, **22**, 5223.
140. Legrain, P., Chapon, C., and Galisson, F. (1993) Interactions between PRP9 and SPP91 splicing factors identify a protein complex required in prespliceosome assembly. *Genes Dev.*, **7**, 1390.
141. Chiara, M. D., Champion-Arnaud, P., Buvoli, M., Nadal-Ginard, B., and Reed, R. (1994) Specific protein–protein interactions between the essential mammalian spliceosome-associated proteins SAP 61 and SAP 114. *Proc. Natl Acad. Sci. USA*, **91**, 6403.
142. Behrens, S.-E., Galisson, F., Legrain, P., and Lührmann, R. (1993) Evidence that the 60-kDa protein of 17S U2 small nuclear ribonucleoprotein is immunologically and functionally related to the yeast PRP9 splicing factor and is required for the efficient formation of prespliceosomes. *Proc. Natl Acad. Sci. USA*, **90**, 8229.
143. Bennet, M. and Reed, R. (1993) Correspondence between a mammalian spliceosome component and an essential yeast splicing factor. *Science*, **262**, 105.
144. Patton, J. G., Porro, E. B., Galceran, J., Tempst, P., and Nadal-Ginard, B. (1993) Cloning and characterization of PSF, a novel pre-mRNA splicing factor. *Genes Dev.*, **7**, 393.
145. García-Blanco, M. A., Jamison, S. F., and Sharp, P. A. (1989) Identification and purification of a 62,000-dalton protein that binds specifically to the polypyrimidine tract of introns. *Genes Dev.*, **3**, 1874.
146. Patton, J. G., Mayer, S. A., Tempst, P., and Nadal-Ginard, B. (1991) Characterization and molecular cloning of a polypyrimidine tract-binding protein: a component of a complex necessary for pre-mRNA splicing. *Genes Dev.*, **5**, 1237.
147. Gil, A., Sharp, P. A., Jamison, S. F., and García-Blanco, M. (1991) Characterization of cDNAs encoding the polypyrimidine tract-binding protein. *Genes Dev.*, **5**, 1224.
148. Ghetti, A., Piñol-Roma, S., Michael, W. M., Morandi, C., and Dreyfuss, G. (1992) hnRNP I, the polypyrimidine tract-binding protein: distinct nuclear localization and association with hnRNAs. *Nucleic Acids Res.*, **20**, 3671.
149. Gozani, O., Patton, J. G., and Reed, R. (1994) A novel set of spliceosome-associated proteins and the essential splicing factor PSF bind stably to pre-mRNA prior to catalytic step II of the splicing reaction. *EMBO J.*, **13**, 3356.
150. Dong, B., Horowitz, D. S., Kobayashi R., and Krainer, A. R. (1993) Purification and cDNA cloning of HeLa cell $p54^{nrb}$, a nuclear protein with two RNA recognition motifs and extensive homology to human splicing factor PSF and *Drosophila* NONA/BJ6. *Nucleic Acids Res.*, **21**, 4085.
151. Hall, J. C. (1994) The mating of a fly. *Science*, **264**, 1702.
152. Ono, Y., Ohno, M., and Shimura, Y. (1994) Identification of a putative RNA helicase (HRH1), a human homolog of yeast Prp22. *Mol. Cell. Biol.*, **14**, 7611.
153. Utans, U. and Krämer, A. (1990) Splicing factor SF4 is dispensable for the assembly of a functional splicing complex and participates in the subsequent steps of the splicing reaction. *EMBO J.*, **9**, 4119.

154. Ast, G., Goldblatt, D., Offen, D., Sperling, J., and Sperling, R. (1991) A novel splicing factor is an integral component of 200S large nuclear ribonucleoprotein (lnRNP) particles. *EMBO J.*, **10**, 425.
155. Nordqvist, K., Öhman, K., and Akusjärvi, G. (1994) Human adenovirus encodes two proteins which have opposite effects on accumulation of alternatively spliced mRNAs. *Mol. Cell. Biol.*, **14**, 437.

7 Pre-mRNA splicing factors in the yeast *Saccharomyces cerevisiae*

PETER E. HODGES, MARY PLUMPTON, and JEAN D. BEGGS

1. Introduction

The budding yeast *Saccharomyces cerevisiae* provides a model system to study pre-mRNA splicing using a combination of classical genetics, molecular biology, and biochemistry (see 1–4 for other recent reviews). Attention must be paid to those aspects of yeast splicing that are not general to all eukaryotes. Initial discoveries left some doubt about the universality of the yeast mechanism. For instance, the yeast splicing machinery has more stringent requirements for branchpoint and splice site sequences (4), yeast U1 and U2 snRNAs are unusually large (5–7), and initial genetic screens for yeast splicing mutants failed to identify genes encoding any of the snRNP proteins known from mammals (see below). However, the universal nature of spliceosome function is now accepted; many splicing factor homologues are recognized as conserved between organisms, and the mechanisms of spliceosome assembly, deduced from mammalian and yeast splicing systems, are identical. This review summarizes the discoveries found in yeast splicing, compares splicing in yeast and in other eukaryotes, and enumerates the advantages of working with yeast as a model system.

2. Splicing in yeast

2.1 Yeast introns

Yeast introns differ from those of higher eukaryotes in size, prevalence, and degree of conservation of primary sequence elements. Only 2–5% of nuclear genes in *S. cerevisiae* contain an intron (4), whereas in mammals genes that do not contain an intron are the exception. Even in the fission yeast *Schizosaccharomyces pombe* at least half of the genes are interrupted (8). Of those introns known in *S. cerevisiae*, most lie in the extreme 5′ end of the pre-mRNA. It has been proposed that gene conversion from reverse—transcribed cDNAs has eliminated most introns in *S. cerevisiae* (9). Due to incomplete reverse transcripts and the requirement for flanking homologies for gene conversion, those introns that lie at the 5′ extremes of the genes are least likely to be removed. A disproportionate number of ribosomal protein genes are interrupted by introns (30 of 49 genes reported in reference 10). These introns could allow coordination in ribosomal biogenesis by regulation of pre-mRNA splicing (11), but this has not been demonstrated.

Introns in *S. cerevisiae* are generally much smaller than those of higher eukaryotes. They show a bimodal size distribution with about a third of introns between 80 and 120 bases and the remaining larger introns typically 200–600 bases long (4). Some larger introns require gene-specific intron sequences for efficient splicing; base pairing between intron sequences may help bring the intron ends together (12, 13). There are no examples in yeast genes of the extremely large introns commonly seen in metazoan genes; the largest observed intron is 1001 bases in *DBP2*, which may be the target of regulated splicing (see below) (14). There are no known examples in the yeast nuclear genome of nested genes contained within introns. While higher eukaryotic genes often contain multiple introns, in general the intron-containing genes in *S. cerevisiae* contain only one. The *MATal* gene, which contains two small introns (15), is an exception.

Cis-acting intron sequences at the 5′ splice site and, in particular, at the branchpoint, are more stringently conserved in *S. cerevisiae* than in other eukaryotes (4). In higher eukaryotes, sequence degeneracy may allow a more flexible system, permitting the use of alternative splice sites that can be selected in, for example, a tissue- or sex-specific manner (see Chapter 8). No example of alternative splice site usage in yeast is known. The flexibility of the mammalian splicing system has been demonstrated *in vitro*; mammalian splicing extracts will remove introns from yeast and plant transcripts (16, 17). *S. cerevisiae*, however, will not accurately splice transcripts derived from several other eukaryotes tested (18–21). This greater sequence specificity of intron recognition in *S. cerevisiae* must be imposed by the sequence-specific binding of *trans*-acting factors in the splicing machinery. Presumably protein splicing factors mediate this effect either directly by binding to the pre-mRNA, or indirectly by influencing the stringency of RNA–RNA interactions within the spliceosome.

2.2 Regulation of splicing in yeast

It has been proposed that the evolutionary advantages of maintaining pre-mRNA introns are the production of alternative proteins from a single gene by alternative splicing, the coordinate regulation of dispersed genes by regulated splicing, and the potential to create new genes by shuffling exons through recombination between introns. Higher eukaryotes have provided many examples of alternative splicing, on–off regulation of gene expression by splicing, and the evolution of genes in exon modules. In yeast, however, none of these phenomena commonly occurs. In fact, it must be kept in mind that most yeast genes do not contain introns at all. Despite this, there are several examples in yeast of regulation of gene expression by splicing.

2.2.1 Regulation of *MER2* pre-mRNA splicing by MER1 protein

Expression of the *MER2* gene is essential for meiosis (22). Although the gene is transcribed during both mitosis and meiosis, the single intron in the pre-mRNA is only spliced during meiosis and production of a functional open reading frame requires splicing. Meiosis-specific splicing of *MER2* transcripts requires the activity of the *MER1* gene, which is transcribed only in meiosis. The *MER2* intron is unusual in three respects: it does not lie at the immediate 5′ end of the mRNA; the 5′ splice site (GU**U**CGU) deviates from the strict consensus in yeast (GU**A**YGU); and the distance between the intron branchpoint and the 3′ splice site (nine nucleotides) is remarkably short. The deviant 5′ splice site plays a role in regulated splicing. Changing this splice site to a consensus sequence alleviates the need for the MER1 protein for *MER2* pre-mRNA splicing (23). Compensatory mutations in the U1 snRNA can also overcome the requirement for MER1 protein. The interaction with U1 snRNA is more extensive than expected; base pairing between the eighth nucleotide of the *MER2* intron and the first nucleotide of U1 influences splicing efficiency. Presumably, the MER1 protein acts as a positive regulator of splicing by promoting the association of U1 with the *MER2* 5′ splice site, although other scenarios are possible (23). *MER1* may control the splicing of additional meiotic transcripts, since expression of an intron-less *MER2* gene in a *merl*-deficient strain does not correct defective sporulation, defective reciprocal crossing over, or spore inviability (22). The predicted MER1 protein is 31 kDa (24) and contains a 45 amino acid motif shared with the hnRNP K protein—the KH motif—that is proposed to bind RNA (25).

Meiosis-specific splicing has also been observed in *S. pombe* (26). The *mes1*$^{+}$ gene, which is essential for the second meiotic division in fission yeast, has a 75 nucleotide intron with splice site sequences that deviate from the *S. pombe* consensus sequences. Splicing of the *mes1* transcript requires the function of the *mei2*$^{+}$ gene which encodes a factor required for commitment of cells to meiosis. The mechanism by which the Mei2 protein directly or indirectly influences splicing of the *mes1* transcript is not known. However, it is interesting that the *mes1* 5′ splice site (GU**U**AGU) resembles that of the *MER2* intron of *S. cerevisiae*, suggesting the possibility that this may constitute a meiosis-specific splicing signal (26).

2.2.2 Autoregulation of ribosomal protein L32 pre-mRNA splicing

The yeast ribosomal protein L32 has been shown to auto-regulate the splicing of its pre-mRNA (11, 27, 28). In the presence of excess unassembled L32 protein (due to excess L32 translation or inhibited rRNA synthesis) splicing of *RPL32* pre-mRNA is inhibited. The regulation requires base pairing between the 5′ splice site and a sequence in the 5′ untranslated region of exon 1. It has been demonstrated that the L32 protein can bind to this RNA structure, and inhibit splicing of its pre-mRNA *in vitro*. It might be presumed that L32 binding to this RNA region prevents U1 snRNA association with the 5′ splice site, but Vilardell and Warner (11) showed that U1 snRNP is bound to the stalled splicing complexes formed with L32 protein.

More work will be necessary to understand exactly how L32 binding aborts spliceosome assembly. Like the *MER2* pre-mRNA, the 5′ splice site of *RPL32* (GUC**A**GU) deviates from the consensus and contains extended complementarity to U1 snRNA over the first eight intron bases.

2.2.3 Potential regulation of *MATa1* and *DBP2* pre-mRNA splicing

The splicing of *MATa1* and *DBP2* transcripts is suspected to be regulated as their introns are atypical. The intron interrupting *DBP2* is positioned in the 3′ half of the open reading frame and is 1001 bases long—the longest yet discovered in *S. cerevisiae* and about twice as long as the next longest (14). Furthermore, an intron interrupts the analogous position of the human and *S. pombe* genes (where again the intron is abnormally long), implying evolutionary conservation of a mechanism of regulated splicing. *DBP2* encodes a DEAD-box protein, that is, it contains amino acid motifs, including the motif D-E-A-D, characteristic of RNA helicases, although only a few of these proteins have demonstrated RNA unwinding activity (see Section 4.5). DBP2 protein is the homologue of human p68, which has proven RNA-helicase activity (29). This accommodates the hypothesis that the helicase may autoregulate its own splicing, although there is currently no evidence for the regulation of this intron's splicing.

In contrast, the *MATa1* locus contains two introns, which are the shortest known in yeast (15). These introns are inefficiently spliced under normal conditions due to their short length from 5′ splice site to branchpoint (30). Splicing is drastically reduced by mutation of the *AAR2* gene; *aar2* mutants accumulate unspliced *MATa1* transcripts but not actin pre-mRNA (31). Functionally, AAR2 resembles the mammalian SF2/ASF splicing factor in promoting the splicing of short introns (see Chapters 6 and 8). However, the AAR2 protein sequence does not resemble any other splicing factor. It contains a putative leucine zipper and an acidic carboxyl terminus. The *aar2* mutation results in defective mitotic growth as well, implying that the splicing of other pre-mRNAs may depend on AAR2 protein.

2.2.4 Regulation of splicing by metabolic signals

Finally, there are hints throughout the yeast splicing literature of metabolic regulation of splicing. The *RNA1* gene affects not only the splicing of pre-mRNA (32), but multiple steps in the nuclear processing of mRNA, tRNA, and rRNA. However, the RNA1 protein appears to be cytoplasmic, which could reflect a role in signal transduction, nuclear-cytoplasmic transport, or maintaining the integrity of the nucleus (33, 34). The *rna1-1* mutation can be suppressed by mutation or deletion of the *SRN1/HEX2/REG1* gene, which functions in the glucose-repression pathway (34). A mutation in the *SRN1* gene can also partially suppress the heat sensitivity of several pre-mRNA processing mutations—*prp2, prp3, prp4, prp5, prp6*, and *prp8* (35). This is consistent with the observations that the splicing mutants *prp2, prp3, prp4, prp7, prp11*, and *snp1* can grow at higher temperatures on non-glucose carbon sources than on glucose (34; P. Siliciano, unpublished data). Therefore, in *S. cerevisiae*, some aspect of nuclear splicing efficiency is regulated according to the metabolic state of the cell, such that glucose repression exacerbates inefficient splicing.

Maddock *et al.* (36) have independently isolated an *srn1* allele (*spp43*) that suppresses *prp2, prp3, prp4,* and *prp11* mutations, as well as three additional suppressors of *prp3* and *prp4*. One gene, *SPP41*, may control the transcription of multiple *PRP* genes, since *spp41* mutant strains show reduced expression of reporter genes fused to the *PRP3* or *PRP4* upstream regions, and SPP41 protein binds to the RAP1 transcription factor. Thus one mechanism for the global regulation of splicing may be co-ordinated transcription of several *PRP* genes. Another regulation could operate through the phosphorylation of splicing factors. At present, no yeast splicing factors are known to be phosphorylated. However, in *S. pombe*, a mutation affecting pre-mRNA splicing has allowed the first cloning of a putative protein kinase implicated in splicing (37, 38).

3. Biochemical and genetic methods to study the spliceosomal cycle in yeast

3.1 Biochemistry of the *in vitro* splicing reaction

Lin *et al.* (39) developed an *in vitro* assay for yeast pre-mRNA splicing using whole-cell extracts, similar to that developed for mammalian cell nuclear extracts. The yeast and mammalian assays have similar biochemical requirements. While both systems require ATP for spliceosome assembly and splicing, yeast splicing reactions are not supplemented with creatine phosphate, as ATP is not depleted in these extracts in phosphate buffers. Splicing in yeast extract is rapid, commencing without the 10–30 minute lag period typical of mammalian extract assays.

An advantage to *in vitro* studies is afforded by yeast genetics through the *p*recursor *R*NA *p*rocessing (*prp*) mutants, defective in pre-mRNA splicing. Many of these strains allow the preparation of cell extracts lacking a specific protein function, by heat-inactivating extracts from temperature-sensitive *prp* mutants (40). Splicing is a strictly ordered pathway (see Fig. 1); reactions in heat-inactivated *prp* mutant extracts are stalled at the stage at which the missing protein function is first required. This is a valuable method of synchronizing a splicing reaction. A novel variation on this approach is provided by dominant negative mutant forms of PRP proteins (41, 42). For example, mutant PRP2 protein, added in excess to a wild-type cell extract, is capable of stalling the splicing reaction specifically at the step requiring PRP2 action. Similar dominant negative versions of mammalian proteins could overcome the lack of genetic manipulation of splicing extracts.

3.2 Genetic identification of splicing factors

Unlike the majority of vertebrate splicing factors which have been identified by biochemical studies (see Chapter 6), the yeast splicing factors have been identified mainly through genetic approaches.

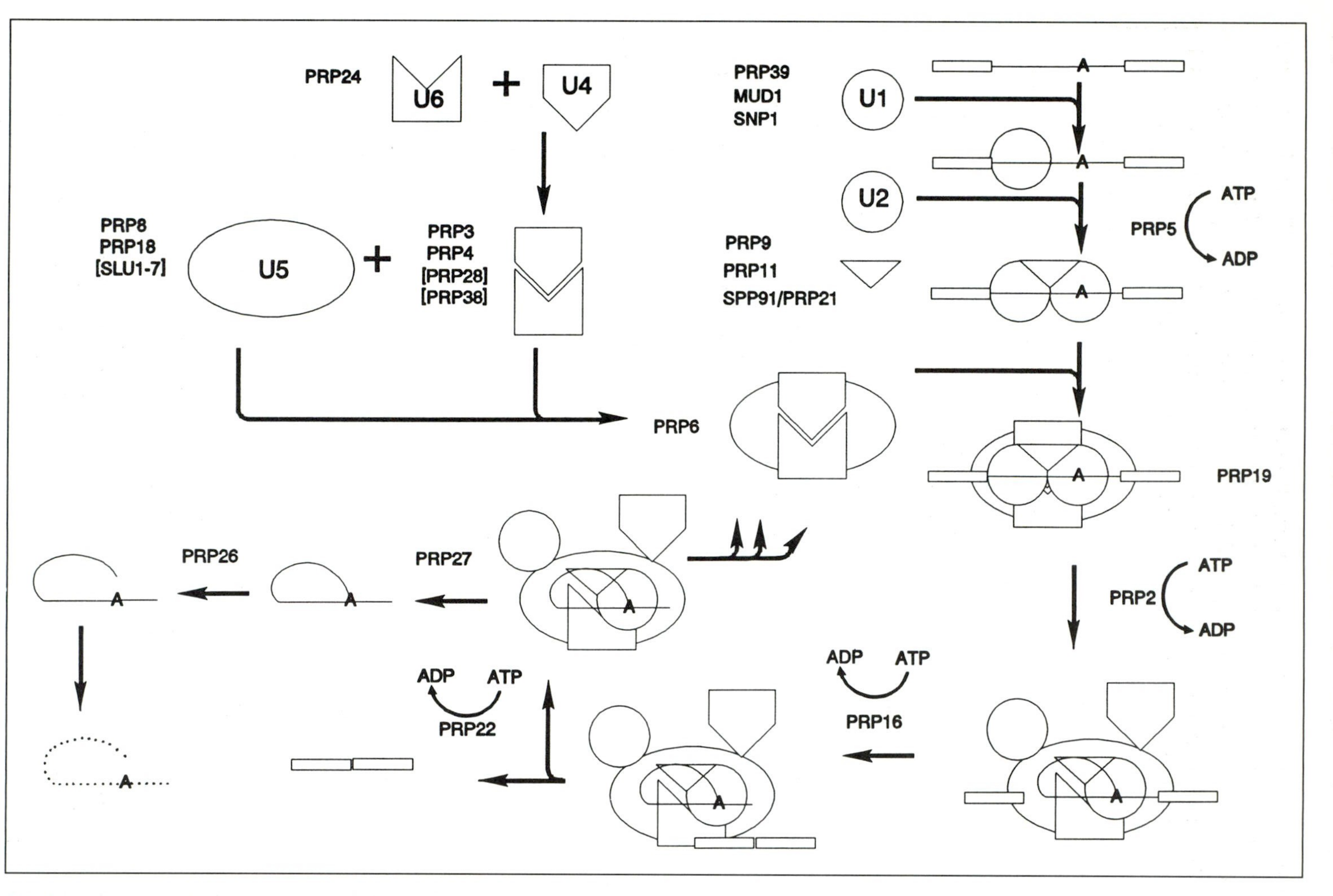

Fig. 1 Yeast splicing factors in the pre-mRNA splicing cycle. In the substrate pre-mRNA the open boxes represent exons, and A indicates the branchpoint nucleotide in the intron. The individual snRNPs are represented by different geometrical shapes. Each protein splicing factor is named alongside the snRNP or the earliest stage in complex assembly or splicing with which a physical, functional, or genetic association is detected. Square brackets indicate factors for which the indicated interaction is based on genetic data that has not yet been supported by evidence for a physical or functional association. Steps which are known to require ATP hydrolysis are indicated, and the DEAD/H protein acting at each step is assumed to be at least partially responsible for that ATP requirement. Details of commitment complex formation are not shown (see Fig. 2).

Over 30 splicing factors have already been identified (see Table 1). A critical advantage of yeast genetics has been the isolation of splicing mutants that arrest the splicing cycle at distinct stages. This allows the complex steps of spliceosome assembly to be ordered into a pathway and allows the biochemical analysis of what are normally transient steps. Splicing factors in yeast are most often identified by

Table 1 Yeast splicing factors

Gene	Accession number	MV (kDa)	Step blocked	Association/ activity	Sequence motif	References
PRP2	X55936	100	1	Spliceosome	DEAH; zinc finger-like	40, 43–47
PRP3	–	56	1	U4/U6 snRNP		48, 49
PRP4	M26597	52	1	U4/U6 snRNP	β-subunit of G protein	50–56
PRP5	M33191	96	1	U2 addition	DEAD	44, 48, 57, 58
PRP6	X53465	104	1	U4/U6·U5	TPR/PW, zinc finger-like	55, 59–62
PRP7	–		1			40, 48
PRP8	Z24732	280	1	U5	Proline-rich, acidic N-end	63–69
PRP9	X53466	63	1	U2 addition	zinc finger-like	58–61, 70, 71
PRP11	M21316	30	1	U2 addition	zinc finger-like	60, 70, 72, 73
PRP16	M3152	120	2	Spliceosome	DEAH	74–80
PRP17/ SLU4	–	52	2		β-subunit of G protein	48, 81–83
PRP18	L03536	28	2	U5 snRNP		81, 84–86
PRP19	L09721	57	1	Spliceosome		81, 87–89
PRP21/ SPP91	X67564	33	1	U2 addition	'SURP' modules of SWAP	58, 70, 71, 81 90, 91
PRP22	X58681	130	mRNA release	mRNA release	DEAH	81, 92
PRP24	Z49260	51	1	U6 snRNP	RRMs	48, 81, 93
PRP25	–		1			81
PRP26	M62813	48	Intron turnover	Debranching activity		81, 94
PRP27	–		Intron turnover			81
PRP28	X56934	67	1		DEAD	95
PRP38	L04669	28	1		Acidic, serine-rich C-end	96, 97
PRP39	L29224	75	1	U1 snRNP		98
SPP2	Z75056	21	1			48, 99
SLU1	–		1			82
SLU2	–		1			82
SLU7	X67810	44	2		Zinc knuckle-like	82, 83
SNP1	X59986	34	–	U1 snRNP	RRM	100–102
MUD1	S56634	33	–	U1 snRNP	RRMs	102
MUD2	–	58	–	U2 addition	U2AF-like, RRMs	103
SMD1	L04669	16	1	core snRNP		97, 104
MER1	M31304	31	–	U1 addition	KH motif	22–25
ARR2	D90455	41	–		Leucine zipper, acidic C-end	31
*prp2**	L22577	59	1		U2AF-like, RRMs	105
*prp4**	L10739	43	1		Serine/threonine kinase	37, 38

* *S. pombe.*

their function (or more accurately, the deficiency caused by their mutation), which gives the study of yeast an appearance that is very different from that in other organisms. For instance, although the mammalian core snRNP proteins, snRNP-specific proteins, and spliceosome-associated factors are well catalogued, the functions of many of these proteins are still obscure.

As splicing is an essential process, conditional mutations (conferring heat- or cold-sensitivity) have been sought most commonly, allowing propagation of the mutants under permissive conditions. The affected genes can be cloned with relative ease in yeast, and efficient homologous recombination can be used to introduce designed mutations onto the genome. The cloned genes, in turn, are often used as the basis for further genetic screens to identify components which physically or functionally interact.

The first conditional pre-mRNA splicing mutants were fortuitously isolated 10 years before the discovery of splicing itself. In a genetic screen to isolate temperature-sensitive mutants thought to be defective in RNA metabolism, 10 complementation groups (*rna2* to *rna11*) were shown to specifically block ribosome biosynthesis (106, 107). Later it was revealed that this phenotype was in fact a secondary effect of a defect in pre-mRNA splicing as a result of the disproportionally large number of intron-containing ribosomal protein genes in *S. cerevisiae* (4, 32). These mutants were renamed *prp* for *p*recursor *R*NA *p*rocessing.

Vijayraghavan *et al.* (81) identified another set of genes whose products are involved in pre-mRNA splicing. They screened a pool of 750 temperature-sensitive yeast strains by northern blot analysis using an intron-containing probe, and isolated a further 11 complementation groups (*prp17* to *rna27*) that affect various stages of the splicing pathway. Similarly, the temperature-sensitive *prp38* mutation was identified as causing accumulation of unspliced pre-mRNA (96).

In an attempt to identify novel splicing factors, Strauss and Guthrie (95) sought cold-sensitive mutants. Mutational defects affecting macromolecular assembly are often enhanced at low temperature, possibly because assembly processes involve largely hydrophobic interactions and therefore are particularly sensitive to cold (108). From a pool of 18 cold-sensitive mutants, one accumulated pre-mRNA at the restrictive temperature of 16°C, identifying the *PRP28* gene.

The interaction of splicing factors with intron RNA sequences can be used as a basis for mutant selection. Couto *et al.* (74) introduced a mutation of the branch-point (UACUA**A**C to UACUA**CC**) in the intron of a selectable marker gene. They isolated a mutation of the *PRP16* gene that functions as a *trans*-acting suppressor, allowing increased usage of the aberrant lariat that is formed.

Several groups have screened for mutations that are lethal in combination with viable mutations (so-called synergistic, synthetic lethal, or enhancer mutations). The conserved loop I of U5 snRNA has been shown to influence splice site choice (109, 110), and several mutations in this loop cause a temperature-sensitive phenotype. Frank *et al.* (82) identified mutations that are lethal in combination with these U5 snRNA mutations. Two strategies were used to provide wild-type U5 snRNA that could be removed to test the mutants, a galactose-regulated U5 gene or a U5 gene on a plasmid that could be counter-selected. Screening for mutants that could grow with the wild-type U5 snRNA but not with the mutant snRNA, 13 *slu* (*s*ynergistic *l*ethal with *U*5 snRNA) mutants were isolated. Seven of these confer a temperature-sensitive phenotype independent of the U5 mutation, which facilitates their study and cloning of the genes.

Since few of the *prp* mutations affect the early steps of splicing or the U1 snRNP, Liao *et al.* (102) designed a screen to identify genes that interact with U1 snRNA. Expression of a U1 snRNA deleted for most of the long, yeast-specific internal regions and containing a point mutation in the conserved A loop causes a strict temperature sensitivity. Sixteen *mud* (*m*utant-*u*-*d*ie) genes have been identified from strains that grow when expressing a wild-type U1 snRNA but die when expressing only the mutant U1 snRNA. The success of the strategy was proven when the first characterized *MUD* gene, *MUD1*, was shown to encode a homologue of the mammalian U1 snRNP protein, U1A.

4. Yeast splicing factors

Many of the *PRP* genes have been cloned by complementing the temperature sensitivity of the mutant strains with yeast genomic plasmid DNA libraries. All are single-copy genes that encode proteins (for references and current list see Table 1). Most of the genes are essential, with the exceptions of *PRP18, PRP26, SNP1, MUD1,* and *MUD2*. (*MER1* is essential for meiosis only.) An additional source of potential splicing factors comes from the initiative to sequence the yeast genome. Genes with homology to splicing factors identified in other organisms can be tested for their role in splicing.

4.1 snRNAs and core snRNP proteins

Yeast spliceosomal snRNAs evaded identification for a number of years owing to their low abundance (only 10–500 copies per cell; 111) and the relatively large number of distinct snRNA species (many of which are non-essential) that exist in yeast compared to mammalian cells (reviewed in 112; see Chapter 5). Eventually, the essential, single-copy genes encoding homologues of the mammalian U1, U2, U4, U5, and U6 snRNAs were cloned and identified by virtue of their limited primary sequence and secondary structural similarity to the corresponding metazoan snRNAs. Surprisingly, both U1 and U2 snRNAs in *S. cerevisiae* are much larger than their counterparts in all other species, although large deletions show that much of the yeast-specific sequences are not essential for growth. Yeast U5 snRNA also contains an additional stem–loop, whereas U4 and U6 snRNAs more closely resemble their mammalian homologues in size. Like their mammalian counterparts, the yeast U1, U2, U4, and U5 snRNAs were shown to be trimethylguanosine-capped (111), and to possess an Sm-binding site consensus motif. Yeast snRNPs are precipitable with human Sm antisera, demonstrating their association with proteins in the form of snRNPs that contain cross-reactive polypeptides (113, 114). This implies that at least some of the core proteins are conserved in yeast; however, only one yeast gene encoding a core snRNP protein has been identified to date. The yeast *SMD1* gene, cloned by virtue of its close proximity to the *PRP38* locus, encodes a protein predicted to be 40% identical to the mammalian D1 protein. *In vivo* depletion of the yeast protein abolishes the first step of splicing and results in reduced levels of U1, U2, U4, U5, but not U6 snRNAs (97). The functional homology of yeast SMD1 and human snRNP core D1 protein is indicated by the ability of the human cDNA to complement an *smd1* null allele (104).

The biochemical purification of snRNP particles from yeast (115) will allow more direct comparison of these particles with their mammalian counterparts. Six proteins, ranging from 10 kDa to 20 kDa, are identified as core proteins of the yeast spliceosomal snRNPs. Antiserum against the human F protein strongly cross-reacts with one of these proteins. At least superficially, the yeast U4/U6·U5 tri-snRNP appears to have a similar protein composition and morphology to the human particle, but the yeast U1 snRNP is larger and more complex, consistent with the much greater size of the U1 RNA in yeast.

4.2 Factors required for formation of a commitment complex containing U1 snRNP and pre-mRNA

The genes encoding three yeast U1-specific snRNP proteins have been found; the homologues of the mammalian U1-70K and U1A proteins, and PRP39. During sequencing of chromosome IX of *S. cerevisiae* the *SNP1* gene was identified as encoding a protein with 30% overall identity to the human U1-70K protein (100). Regions of greatest identity between SNP1 and U1-70K include glycine-rich stretches and an RNA recognition motif (RRM). Interestingly, the yeast homologue does not contain a domain similar to the mammalian RS and RD/RE dipeptide-rich carboxy terminus (see Chapter 6). The product of the yeast *SNP1* gene produced as a fusion protein in *E. coli* was demonstrated to bind specifically to a 47 nucleotide fragment of yeast U1 snRNA (101) that corresponds to the binding site of U1-70K on the metazoan U1 snRNA. Gene disruption data indicate that although the *SNP1* gene is essential under certain conditions (100), the requirement for SNP1 function is apparently carbon source- and strain-dependent (P. Siliciano, unpublished data).

The *MUD1* gene, encoding a U1A-like protein, was identified in a screen for mutants that enhance the adverse effects of mild U1 snRNA mutations (102). The MUD1 and U1A proteins each contain two RRMs separated by a spacer region of similar length, and their functional similarity was revealed by the demonstration that U1 snRNA was co-immunoprecipitated using a monoclonal antibody directed against epitope-tagged MUD1 protein. Surprisingly, the *MUD1* gene is non-essential, and is not required for splicing *in vitro*, which could explain why it has not been identified in screens for conditional lethal mutants.

The *PRP39* gene encodes a U1 snRNP-associated 75 kDa protein. Antibodies recognizing PRP39 protein immunoprecipitate U1 snRNPs and pre-mRNA-containing complexes. Extracts depleted of PRP39 contain U1 snRNPs with increased electrophoretic mobility, and are unable to form either of the earliest detectable U1 snRNP–pre-mRNA associations of splicing, commitment complexes CC1 or CC2 (see Chapter 4). It was proposed that PRP39 facilitates or stabilizes the snRNP–5′ splice site interaction (98).

As mentioned above (Section 2.2), both the MER1 and the AAR2 proteins act at the level of targeting U1 snRNA to the pre-mRNA and regulate splicing by promoting the use of specific suboptimal 5′ splice sites.

4.3 Factors required for the addition of U2 snRNP to the pre-spliceosome

The power of the yeast system to define genetic and physical interactions between splicing factors is best demonstrated in the intricate pathway of U2 snRNP addition to the pre-spliceosome (Fig. 2). Two commitment complexes form on pre-mRNA without the hydrolysis of ATP, as predicted by competition experiments and visualized by native gel electrophoresis (116, 119, 120). Formation of either complex requires U1 snRNP and a pre-mRNA containing a 5′ splice site. Commitment complex 2 requires a functional branchpoint, although it does not contain U2 snRNP, leading to the proposal that this complex contains a protein factor (X in Fig. 2) that recognizes the branchpoint (116, 119, 120). The addition of U2 snRNP to this assembly requires a multiprotein complex of PRP9, PRP11, and PRP21/SPP91, as well as ATP hydrolysis. At this point, there are likely to be rearrangements of the RNA interactions in the pre-spliceosome that may involve PRP5, the putative RNA helicase that interacts genetically with the proteins in this complex. These interactions are detailed below (reviewed in 17).

Two possible homologues of the mammalian splicing factor U2AF^{65} (see Chapter 6) have been reported from *S. cerevisiae*, and one from *S. pombe*. The *S. pombe* gene *prp2*$^{+}$ encodes a 59 kDa polypeptide with extensive overall similarity to, and structural domains characteristic of, U2AF^{65}, and is essential for the first step of pre-mRNA splicing (105). The *S. cerevisiae YCL11c* gene was identified based on the sequence similarity of a predicted open reading frame in chromosome III, found in a database search (121, 122). The function of this protein has not been determined, but it is not essential for yeast splicing (103). It has an arginine-rich amino terminus containing four RS dipeptides and four RGG tripeptides. Both of these motifs are common in mammalian splicing factors (123) but have not yet been found in any other yeast protein.

The *MUD2* gene was identified by its interaction with a mutant U1 snRNA gene, and shows some resemblance to U2AF^{65} (102, 123). MUD2 is present in commitment complex 2, before any ATP-dependent step and before U2 snRNP addition, and in that way behaves like U2AF (see Chapters 4 and 6). If MUD2 interacts with the 3′ end of the intron, as does U2AF^{65}, and interacts with the U1 snRNP, as suggested by the genetic interaction with U1 snRNA, then this could be the first bridge that links U1 snRNP with the branchpoint, the factor X proposed by Séraphin and Rosbash (116, 117, 119, 120). Curiously, it is not essential to wild-type

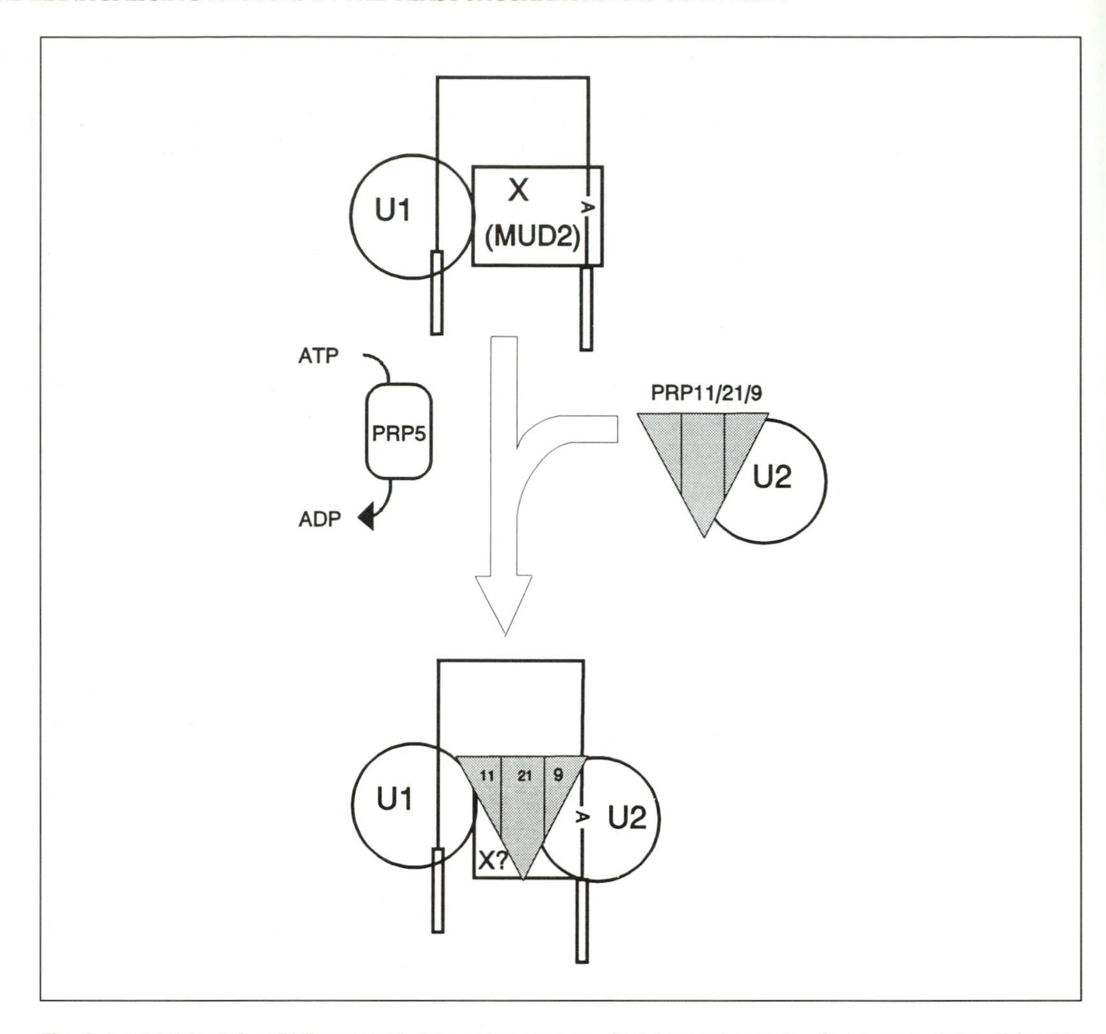

Fig. 2 A model for U2 snRNP assembly into spliceosomes. Commitment complex 2, as proposed by Séraphin and Rosbash (116), contains factor(s) X that would function like the mammalian U2AF to promote the interaction of U1 snRNP with the branchpoint region (A) of the pre-mRNA. MUD2 is a component of factor X in *S. cerevisiae*. It is not known whether factor X/MUD2 persists in the spliceosome after snRNP association. The PRP5, PRP9, PRP11, and PRP21 proteins are required for the association of U2 snRNP with commitment complex 2 to form the prespliceosome complex (58, 59, 91). Interaction of the DEAD protein, PRP5, with U2 snRNP is suggested by genetic data (58, 161); however, the nature of the molecular interactions involved are not known. By analogy with PRP2 and PRP16, PRP5 may interact only transiently with the splicing complex. A complex (shaded), in which PRP21 apparently links PRP9 and PRP11, is suggested by genetic interactions (see Section 4.3) and by two-hybrid interactions (70, 71). The association of this PRP9–PRP21–PRP11 complex with U2 snRNP is suggested by genetic interactions (see text), and by the apparent association of PRP9 with U2 snRNP indicated by co-immunoprecipitation (59). It is also suggested by the structural and functional homology of PRP9, PRP11, and PRP21 with the proteins in the SF3a complex that is a component of human 17S U2 snRNPs (see text; reviewed in 117). It is proposed (70) that the PRP9–PRP21–PRP11 complex may serve as a bridge between the U1 and U2 snRNPs to form the prespliceosome and to commit pre-mRNAs to the splicing pathway. The molecular contacts in this prespliceosome complex remain to be identified. The mammalian homologues of PRP9, PRP11, and PRP21 (SAP60, SAP61, and SAP114) contact the 3′ region of the intron (117, 118); however, to date no such data exist for the yeast proteins.

yeast, implying that this putative bridging factor is not obligatory for the addition of U2 snRNP to the pre-spliceosome.

The products of the *PRP9*, *PRP11*, and *PRP21/SPP91* genes form a complex that is proposed to connect the U2 snRNP to the U1 snRNP–pre-mRNA complex (58, 59, 70, 71, 90, 91). In HeLa cells, a 17S U2 snRNP particle contains several proteins not seen in the smaller 12S particle (124). Three of these proteins form a stable interaction, and have been purified as a splicing factor (SF3a) required for the association of the 12S U2 snRNP with the pre-spliceosome (see Chapters 4–6). Immunological cross-reaction and similarities in their deduced amino acid sequences indicate that PRP9 corresponds to the 60 kDa subunit (SAP 61), PRP11 corresponds to the 66 kDa subunit (SAP 62), and SPP91/PRP21 corresponds to the 120 kDa subunit (SAP 114) of the HeLa 17S U2 snRNP (118, 125–127). Molecular biology experiments in yeast have clarified how these proteins interact.

The products of *PRP5, PRP9, PRP11,* and *PRP21* are required for U2 snRNP binding to the U1-containing complex, as determined by the behaviour of heat-inactivated extracts from the *prp* mutant strains (58, 59, 70, 71, 90, 91). Genetic interactions have been demonstrated between temperature-sensitive alleles of *PRP5, PRP9, PRP11,* and *PRP21,* and between these genes and mutant U2 alleles (58; 161), indicating functional association between these *PRP* gene products and U2 snRNA. The *PRP21* gene was independently isolated (and called *SPP91*) through suppression of the temperature-sensitive *prp9-1* phenotype (90), and was shown to be equivalent to *PRP21* (91). The SPP91/PRP21 protein binds directly to PRP9 and PRP11 proteins, acting as a bridge, to form a multimolecular complex (70, 71). PRP9 protein itself forms homodimers, which requires two zinc finger-like regions of the protein (71). Interaction with PRP21/SPP91 is lost by two mis-sense mutations in the PRP9 protein and one in PRP11, identifying possible interaction domains in these proteins (70, 71). PRP21/SPP21 contains a protein motif called the 'surp' module after its discovery in the *Drosophila* SWAP splicing regulator and in PRP21 (D. Spikes *et al.*, unpublished). This module contains basic and aromatic residues, reminiscent of RNA-binding domains, but its function is uncertain.

The *PRP5* gene shows genetic interactions with *PRP9*, *PRP21/SPP91*, and *PRP11*, as well as with U2 snRNA. PRP5 is a member of the DEAD-box family of proteins (57), and so is proposed to be an ATP-dependent RNA helicase. The PRP5 protein associates with splicing commitment complexes without ATP (58), so the activity of PRP5 is likely to be one of the (if not the only) ATP-requiring steps in U2 snRNP binding to the pre-mRNA.

If PRP5 hydrolyses ATP to cause a conformational shift in the pre-spliceosomal RNAs, there are a few RNA structures that could be substrates. The U2 snRNA is a candidate, since the *prp5-1* temperature-sensitive mutation is lethal in combination with either of two mutations in the U2 snRNA (58). An alternative RNA target for PRP5 activity is the U1 snRNA–5′-splice-site helix. It has been observed that the interaction between U1 snRNP and the spliceosome is weakened during the course of splicing, perhaps following an ATP-dependent RNA conformational change disrupting the bonds between U1 snRNA and the 5′ splice site region. Liao *et al.* (128)

have demonstrated that a mutation in the 5′ end of the U1 snRNA which disrupts its base pairing with the pre-mRNA 5′ splice site allows some U2 snRNP addition to bypass the ATP requirement. The simplest interpretation of this phenomenon is that complexes containing the mutant U1 snRNA have a reduced requirement for the ATP-dependent helicase activity. Thus the U1 mutation may permit a conformational change, perhaps catalysed by PRP5, that normally requires ATP (128).

4.4 Factors required for formation and function of the U4/U6·U5 particle

The PRP8 protein is exceptionally large (280 kDa), and immunologically cross-reacting proteins of 200–300 kDa have been detected in humans (65, 129, 130), rats (131), mice (P. E. H., unpublished), *Drosophila* (132), *Caenorhabditis elegans* (P. E. H., unpublished), tobacco, and pea (69). PRP8 protein is found associated with U5 snRNPs, with U4/U6·U5 snRNP particles, and with spliceosomes in *S. cerevisiae* and in HeLa cell nuclear extracts (63, 65, 66, 129). It is the only yeast splicing factor yet identified that clearly associates with the U5 snRNP before the assembly of the U4/U6·U5 tri-snRNP. It is required for assembly of the tri-snRNP and for the association of any of these snRNPs with the pre-spliceosome, stabilizing these snRNAs (68). In the absence of PRP8, levels of the U4, U5, and U6 snRNAs decline dramatically, suggesting a role for PRP8 in stabilizing these snRNAs within the tri-snRNP. Genetic data also implicate PRP8 in stabilizing RNA structures, since in two circumstances PRP8 activity is counterbalanced against the activity of a putative RNA helicase. First, the *prp8-1* mutation can be suppressed by mutation of an abundant DEAD-box protein, SPP81/DED1 (133). Second, another *prp8* allele can suppress the effects of a *prp28* mutation (95). The PRP28 protein (see below) is proposed to unwind the U4/U6 helices during spliceosome assembly. Beyond its role in spliceosome assembly, the PRP8 protein is likely to play a direct role in spliceosome function. PRP8 is in direct contact with the pre-mRNA in the spliceosome, as has been demonstrated by the UV-cross-linking of substrate RNA to the yeast PRP8 protein (67) and to the human homologue (130, 134). Contact with the RNA substrate is initiated before the first step of splicing, is maintained with the lariat intermediate RNA before step 2, and continues with the excised intron product (S. Teigelkamp and J. D. B., unpublished). The interactions of PRP8 with RNA, its role in spliceosome assembly, and its large size support the proposal that PRP8 forms a scaffold on which the RNA interactions of splicing take place (66).

The PRP8 protein sequence contains no obvious similarities to motifs found in other splicing factors, or to other RNA-binding proteins. Comparison to the sequence databases has identified a homologous *C. elegans* gene, encoding a 272 kDa protein with 61% overall identity to yeast PRP8 protein as well as partial sequences of homologues from humans, rice, *Arabidopsis*, and *Plasmodium* (162). Alignment of the two proteins may help identify functional domains; for instance, two regions over 60 amino acids in length are greater than 90% identical. The most striking feature of the yeast protein is an amino-terminal domain consisting of four polyproline stretches spaced by strongly acidic peptides. However, the *C. elegans* gene does not contain a similar sequence, so the function of this domain may be specific to yeast.

The *PRP18* gene product has been shown by immunoprecipitation to be associated with the U4/U6·U5 tri-snRNP (85). It is primarily U5 snRNP-associated, since depletion of U6 snRNA leaves the PRP18 protein in association with U4 and U5 snRNAs but U5 depletion prevents its snRNP association. Yeast cells with a deleted *PRP18* gene are viable, but grow poorly and accumulate splicing intermediates. *In vitro*, the PRP18 protein is only required for the second cleavage–ligation reaction (84–86). Extracts immunodepleted of PRP18 accumulate splicing intermediates. Following ATP-depletion of these PRP18Δ extracts, splicing intermediates can be chased into reaction products upon addition of purified PRP18 protein alone, without addition of ATP (86). This supports the belief that the RNA cleavage–ligation reactions are, in themselves, ATP-independent transesterifications (135). The protein sequence of PRP18 does not contain any known motifs, but allows the identification of partial sequences of rice and *C. elegans* homologues in the database.

Seven other genes, *SLU1* to *SLU7*, have been identified through a genetic interaction with U5 snRNA, and so encode proteins that may associate with the U5 snRNP, the U4/U6·U5 tri-snRNP, or with U5 snRNA in the spliceosome (82). It is unclear what splicing defect is conferred by the *slu3*, *slu5*, and *slu6* mutations. *SLU1* and *SLU2* are essential for the first cleavage–ligation step of splicing, and have yet to be characterized in any detail.

SLU4 (allelic with *PRP17*) and *SLU7* are required for the second step of splicing. As *slu4-1* and *slu7-1* are lethal in combination with mutations in *PRP16* and *PRP18*, it has been proposed that the SLU4/PRP17, SLU7, PRP16, and PRP18 proteins and U5 snRNA interact functionally at the second catalytic step (82). The *slu7-1* mutant allele produces a protein that inhibits the use of 3′ splice sites distant from the branchpoint, but has no effect on the use of closely positioned 3′ sites (83). This effect is not seen for all mutants of the gene. The protein sequence includes three closely spaced cysteines and a histidine, proposed to form a zinc knuckle. Mutation of two of the cysteines to serines inhibits the use of both proximal and distal 3′ splice sites equally.

The U6 snRNA is unstable following heat-inactivation of *prp3*, *prp4*, *prp6*, *prp19*, *prp24*, and *prp38* temperature-sensitive strains (55, 96), implying that the proteins encoded by these genes are likely to be components of U6-containing particles. The *PRP6* gene product was initially reported to be U4/U6-associated (59). It was later shown that its association with U4/U6 was dependent on U5 snRNA, that is, specific to the U4/U6·U5 particle (55). Analysis of complexes formed in prp6–1 heat-inactivated extracts demonstrated that PRP6 protein is required for the accumulation of stable U4/U6·U5 tri-snRNP (55). The observation that the PRP6 protein is localized to discrete subnuclear compartments may indicate the accumulation of U4/U6·U5 triple snRNPs at specific nuclear sites (55, 136).

PRP4 protein associates with the U4/U6 particle through interaction with the 5′

stem–loop of U4 snRNA, although no direct RNA binding has been demonstrated (50, 52–54, 56). The PRP4 protein does not contain any characterized RNA-binding motif. The amino-terminal 108 residues (including an exceptionally acidic region) are dispensable for PRP4 function. A central region is essential, but does not show similarity to other known proteins. The carboxy-terminal half of the protein contains five copies of a G_β repeat (51, 137). This motif, named for the prototype seen in the signal transduction G protein β-subunit, consists of a peptide of about 40 amino acids repeated in five to eight copies. Members of this gene family function in pathways as diverse as signal transduction, transcriptional regulation, cell cycle control, cytoskeletal assembly, and RNA processing (137). The repeat domain is seen also in PRP17/SLU4, a protein of similar size that is required for step 2 of splicing. A yeast U3 snoRNP protein, involved in rRNA processing in the nucleolus, contains a G_β repeat domain (138), as does the 50 kDa subunit of CstF, the cleavage stimulatory factor for the mammalian polyadenylation complex (139; see Chapter 9).

The exact function of the G_β repeat is unknown, but a common feature of all members of the family is assembly into multi-protein complexes. Further, it is proposed that at least some of the G_β-like proteins interact with proteins containing another motif, the tetratricopeptide (34 amino acid) repeat (TPR) (140). These repeats are proposed to form intramolecular α-helices (snap helices). The TPR motif is seen in an equally diverse range of proteins, including the splicing factor PRP6, so an association is suggested between PRP4 or PRP17/SLU4 and PRP6. The prototype for this interaction is an instructive model. The yeast transcriptional-repressor proteins TUP1 (G_β-like) and SSN6/CYC8 (with TPR repeats) associate in high-molecular-weight complexes (141). Neither protein has DNA-binding ability, but they can bind to several DNA-binding proteins that target specific genes for repression. If a chimeric protein is expressed that fuses the SSN6/CYC8 protein to a DNA-binding protein, it can repress the targeted gene directly, but only in the presence of TUP1 (142). The exact interactions of DNA-binding proteins, SSN6/CYC8, and TUP1 are still unclear, but similar genetic and biochemical dissection of complexes containing PRP4, PRP17/SLU4, and PRP6 is possible, following the example of the PRP9–PRP11–PRP21/SPP91 complex (see above).

PRP3 interacts genetically with *PRP4*, in that *prp3-1* and *prp4-1* are synthetic lethal and *PRP3* in high copy number suppresses the *prp4-1* defect. These *PRP3* and *PRP4* interactions and their requirement for U6 snRNA stability are compatible with the association of both PRP3 and PRP4 proteins with U4/U6 particles (99; J. Anthony and J. L. Woolford, unpublished).

The *prp24-1* allele was isolated as a temperature-sensitive suppressor of a cold-sensitive mutation in U4 snRNA, and immunoprecipitation experiments demonstrated an association of PRP24 protein with U6 snRNA. The PRP24 protein contains three RNA-binding motifs, leading to a model in which PRP24 may be a U6-binding protein that plays a role in moderating the interaction of the U4 and U6 snRNAs (93).

Genetic evidence implicates the activity of PRP28, a DEAD-box protein, in the destabilization of the U4/U6 base pairing, proposed to occur just before the initial cleavage–ligation step. Yeast strains containing cold-sensitive *prp28* alleles are defective for the first step of splicing *in vivo* (95). *prp28* mutations and mutant alleles of the U6-binding protein PRP24 are lethal in combination. In addition, suppression analysis has detected a genetic link between PRP28 and the U5 snRNP protein PRP8. These data suggest a model in which PRP28 acts to unwind the U4/U6 helix, with PRP24 acting to stabilize the unwound form, and PRP8 acts in opposition, stabilizing the base-paired form of U4/U6 in the context of the U4/U6·U5 tri-snRNP (68, 95). Thus the three factors may act in concert during the stabilization–destabilization cycle, the balance between these activities determining the extent of unwinding.

PRP19 has been shown to associate with splicing complexes after or together with U6 snRNA entry, but before the first step of splicing (87, 88). The PRP19 protein is not associated with splicing complexes in extracts depleted of either U2 or U6 snRNAs, however it is present in the complete spliceosomes formed in *prp2–1* heat-inactivated extracts (88). Functional PRP19 protein fractionates from yeast extracts in a high molecular weight complex with a number of other polypeptides, at least one of which is another essential splicing factor (89).

4.5 The DEAH proteins

It has been proposed that three similar proteins (PRP2, PRP16, and PRP22) regulate the activity of the spliceosome at three consecutive points (step 1, step 2, and the release of the mature mRNA from the spliceosome, respectively). These proteins form a subfamily within the DEAD-box superfamily of proteins, and contain a change of DEAD to DEAH in one of the characteristic motifs (45, 75, 92). They are proposed to catalyse ATP-dependent conformational changes of RNA substrates within the spliceosome.

4.5.1 PRP2

PRP2 is an extrinsic factor that interacts transiently with the spliceosome before and during step 1. In the absence of PRP2 (in heat-inactivated *prp2* extracts), pre-mRNA-containing complexes accumulate (44). Immunoprecipitation with anti-PRP2 antibodies showed that PRP2 associates transiently with these complete spliceosomes; a low level of immunoprecipitation of normal spliceosomes was observed, which increased upon use of mutant pre-mRNA unable to undergo step 1 (46). Two methods have been developed to stall the association of PRP2 in spliceosomes, allowing biochemical dissection of this transient step. First, using heat-inactivated *prp2* extracts, Kim and Lin (143) isolated spliceosomes that require only PRP2 and ATP to perform step 1. Second, a mis-sense mutation introduced into PRP2 at a conserved sequence motif results in dominant negative effects; the mutant protein can enter spliceosomes but is unable to perform its function and is stalled (41). The initial interaction with the spliceosome may be with spliceosomal components, but PRP2 then binds directly to the pre-mRNA (42). It has been proposed that PRP2 binds to a site of pre-mRNA/snRNA interaction and catalyses an ATP-dependent change in this interaction. The resulting rearrangement of the spliceosome creates a catalytic 'ribocentre' that performs the first splicing reaction. The hydrolysis of ATP allows PRP2 to be released from the spliceosome (42).

Multiple copies of the *SPP2* gene can suppress a temperature-sensitive *prp2* mutation, but not a null allele of *PRP2* (99). This dosage-dependent suppression is reciprocal; extra copies of *PRP2* suppress temperature-sensitive alleles of *spp2* (K. Kim and J. L. Woolford, unpublished). Temperature-sensitive mutants of *spp2* are defective in splicing both *in vivo* and *in vitro*, identifying SPP2 as a splicing factor that interacts with PRP2.

4.5.2 PRP16

A transitory association of PRP16 with splicing complexes at step 2 has been demonstrated, similar to that observed with PRP2 at step 1. ATP hydrolysis by PRP16 is required for the second step and for PRP16 release from complexes (77). Mutations of *PRP16* have been isolated as suppressors that allow use of introns with mutations at the branchpoint (74, 79). All of these mutant proteins demonstrate reduced ATPase activity *in vitro* and cause an accumulation of splicing intermediates *in vivo*. This has led to the proposal that, in spliceosomes containing an aberrant branch, the slowed rate of ATP hydrolysis by the mutant PRP16 protein prevents a normal discard pathway that degrades mis-spliced RNAs (75, 79, 80).

4.5.3 PRP22

PRP22 encodes a third member of the DEAH family, and is proposed to regulate the release of the spliced exons from the post-splicing complex. Mutant *prp22* cells incubated at the non-permissive temperature accumulate large amounts of excised lariat intron, suggesting that intron turnover is an active process. In extracts derived from mutant *prp22* cells, mRNA is synthesized but it is not released from post-splicing complexes (92). Thus, PRP22 may act on one or more RNAs to potentiate a conformational change culminating in message release. If spliceosome disassembly is essentially a reversal of the earlier events in the cycle, the function of PRP22 protein may closely resemble that of the DEAD/H proteins that act earlier in splicing, but affecting a different directionality of events.

4.5.4 Models for the function of DEAH proteins

The DEAD superfamily of proteins was discovered by sequence comparison of the eukaryotic translation initiation factor eIF-4A and the human p68 protein (144). Large blocks of absolute amino acid identity unveiled a core region containing seven sequence motifs, which are now the hallmark of more than 30 members involved in a wide variety of cellular processes (reviewed in 145–149). Several members of this family, including eukaryotic translation initiation factor eIF-4A (150), have been shown to possess ATP-dependent RNA helicase activity *in vitro*. Considering the strong conservation of sequence motifs required for this activity, it is likely that other family members, including the PRP proteins, will use ATP hydrolysis to alter the structure of RNA substrates *in vitro*. Both PRP2 and PRP16 have been shown to possess RNA-dependent ATPase activity (47, 76). The failure to demonstrate RNA unwinding for PRP2 (47) and PRP16 (76) may be due to stringent substrate specificity or the requirement for additional factors (eIF-4A requires the auxiliary factor eIF-4B, for example; 150).

Detailed functional analysis of the sequence motifs remains limited to the prototype of the group, eIF-4A (151, 152). The two N-terminal motifs, GKT and DEAD/H, are special forms of the previously defined ATPase motifs A and B, respectively (153–155). Mutations within these motifs of eIF-4A affect ATP binding and/or hydrolysis. The DEAD motif is also required to couple ATP hydrolysis with RNA helicase activity in eIF-4A. The HRIGRXXR motif, changed in the DEAH proteins to QRIGRXXR, is involved in RNA binding, as well as influencing ATP hydrolysis. The SAT motif is essential for RNA unwinding. It remains to be seen whether these functional assignments hold for other members of the DEAD/H family.

Currently, there are two working hypotheses which might explain the function of PRP2 and PRP16 in splicing. The first is based on the 'reduced fidelity' phenotype of the prp16–1 protein, allowing this factor to overcome the splicing defect caused by a branchpoint mutation. As suggested by Couto *et al.* (74) and elaborated by Burgess and Guthrie (75, 79, 80), PRP16 might act to proofread the branch site residue. By analogy to translation elongation factors (156), the rate of ATP hydrolysis by PRP16 is proposed to act as a clock, discriminating between correct and aberrant splicing intermediates and targeting mis-spliced products for abortive degradation. Thus PRP2, PRP16, and PRP22 might effect sequential fidelity checks accompanied by ATP-dependent conformational changes that ensure the directionality of the splicing process.

The second hypothesis involves an analogy with the prototype of the helicase family, eIF-4A, which uses the energy derived from ATP hydrolysis to unwind messenger RNA secondary structure to facilitate ribosome binding (157). The analogy is suggested by the isolation of a dominant negative *PRP2* mutation (41) that alters the conserved SAT motif, which in eIF-4A is specifically required for the RNA unwinding activity (151). In PRP2 this mutation results in a prolonged interaction of this protein with spliceosomes, where it remains in contact with the pre-mRNA. This suggests that the pre-mRNA may be part of the target of an RNA conformational change mediated by PRP2 protein which the mutant protein is unable to perform.

Perhaps the RNA helicase-like splicing factors combine characteristics of both models, using ATP hydrolysis to effect RNA conformational changes that are required to permit catalysis, and also to ensure the correct directionality of reactions. Rigid substrate specificity might account for their role in determining the fidelity of splicing. Biochemical analysis of the effects of certain mutations within these splicing factors holds the answer.

4.6 Factors required for mRNA release, intron degradation, and snRNP recycling

Following the splicing reaction, the mature mRNA is released from the spliceosome while the excised intron is retained within a post-splicing complex, at least *in vitro*

(158, 159). Three mutations have been identified that prevent spliceosome disassembly and intron degradation: *prp22* (discussed above), *prp26*, and *prp27* (81). It is likely that the conditional lethal phenotypes of *prp22* and *prp27* mutants are due to a block in the recycling of spliceosomal components that depletes the pool of free splicing factors and prevents further rounds of processing at the non-permissive temperature. The function of the *prp27* gene and the nature of the PRP27 protein are unknown.

The *PRP26* gene encodes a 2′–5′-phosphodiesterase, the intron-debranching enzyme (81, 94). *prp26* mutants accumulate excised lariat introns, suggesting a role for the debranching activity in intron turnover. The gene is not essential, and no substantial accumulation of branched ribonucleotides occurs in yeast cells lacking debranching enzyme. A *prp26* mutation was isolated as a suppressor of transposition of the endogenous retroposon Tyl, although it is not clear how debranching is involved in retrotransposition. A debranching enzyme has also been identified in mammalian extracts (160), and the yeast gene sequence has allowed the identification of a partial cDNA homologue from *C. elegans*.

5. Prospects

The goal of studying pre-mRNA splicing in yeast is an ambitious one: to understand the roles of all of the components in the biochemical reaction of splicing. Understanding splicing in this model system will provide a foundation for understanding the complexities of splice-site discrimination and alternative splicing in higher eukaryotes. The side-benefits have been and will be generous, such as setting general principles for RNA–protein interactions and understanding how human diseases can be caused by mutations affecting the process.

Currently, there are many gaps in the picture of yeast splicing. There are still many splicing factors to be identified; the genetic screens have been far from exhaustive. Future genetic screens are likely to be increasingly sophisticated, seeking factors that interact with a specific component at a specific time in the cycle. Genetic approaches will be complemented by cross-species comparisons. An expected benefit from the determination of the complete sequence of the yeast genome is the identification of homologues of the splicing factors identified in mammals and *Drosophila*. Finally, biochemical characterization of the splicing reaction, aided by the tricks of yeast reverse genetics, should allow identification and purification of the full cast of factors in spliceosomes stalled at any point in the cycle. The challenge inherent in all of these approaches is to convert an ‘interaction’, be it genetic or molecular, into a meaningful understanding of what binds where, when, with what effect. This is where the value of yeast genetics will be most appreciated.

Acknowledgements

We are very grateful to the following for communicating results in advance of publication: Jaime Arenas, Manuel Ares, Soo-Chen Cheng, Angela Krämer, Pierre

Legrain, Melissa Moore, Shirleen Roeder, Michael Rosbash, Stephany Ruby, Brian Rymond, Bertrand Séraphin, Paul Siliciano, and John Woolford. M.P. was supported by a Science and Engineering Research Council Studentship and by the Cancer Research Campaign. P.E.H. was supported by The Royal Society. J.D.B. holds a Royal Society E.P.A. Cephalosporin Fund Senior Research Fellowship.

References

1. Beggs, J. D. and Plumpton, M. (1992) Genetic studies of pre-mRNA splicing in yeast. In *Nucleic acids and molecular biology* (Vol. 6), Eckstein, F. and Lilley, D. M. J. (ed.). Springer-Verlag, Berlin, p. 187.
2. Brown, J. D., Plumpton, M., and Beggs, J. D. (1992) The genetics of nuclear pre-mRNA splicing: a complex story. *Antonie van Leeuwenhoek*, **62**, 35.
3. Woolford, J. L., Jr and Peebles, C. L. (1992) RNA splicing in lower eukaryotes. *Curr. Biol.*, **2**, 712.
4. Rymond, B. C. and Rosbash, M. (1993) Yeast pre-mRNA splicing. In *The molecular and cellular biology of the yeast* Saccharomyces: *gene expression*. Jones, E. W., Pringle, J. R., and Broach, J. R. (ed.). Cold Spring Harbor Laboratory Press, Cold Spring Harbor, p. 143.
5. Ares, M. (1986) U2 RNA from yeast is unexpectedly large and contains homology to U4, U5 and U6 small nuclear RNAs. *Cell*, **47**, 49.
6. Kretzner, L., Rymond, B. C., and Rosbash, M. (1987) *S. cerevisiae* U1 RNA is large and has limited primary sequence homology to metazoan U1 snRNA. *Cell*, **50**, 593.
7. Siliciano, P., Jones, M. H., and Guthrie, C. (1987) *Saccharomyces cerevisiae* has a U1-like small nuclear RNA with unexpected properties. *Science*, **237**, 1484.
8. Potashkin, J., Li, R., and Frendeway, D. (1989) Pre-mRNA splicing mutants of *Schizosaccharomyces pombe*. *EMBO J.*, **8**, 551.
9. Fink, G. R. (1987) Pseudogenes in yeast? *Cell*, **49**, 5.
10. Woolford, J. L., Jr and Wagner, J. R. (1991) The ribosome and its synthesis. In *The molecular and cellular biology of the yeast* Saccharomyces: *genome dynamics, protein synthesis, and energetics*. Broach, J. R., Pringle, J. R., and Jones, E. W. (ed.). Cold Spring Harbor Laboratory Press, Cold Spring Harbor, p. 587.
11. Vilardell, J. and Warner, J. R. (1994) Regulation of splicing at an intermediate step in the formation of the spliceosome. *Genes Dev.*, **8**, 211.
12. Newman, A. (1987) Specific accessory sequences in *Saccharomyces cerevisiae* introns control assembly of pre-mRNAs into spliceosomes. *EMBO J.*, **6**, 3833.
13. Parker, R. and Patterson, B. (1987) Architecture of fungal introns: implications for spliceosome assembly. In *Molecular biology of RNA: new perspectives*. Inouye, M. and Dudock, B. S. (ed.). Academic Press, San Diego, p. 133.
14. Iggo, R. D., Jamieson, D. J., MacNeill, S. A., Southgate, J., McPheat, J., and Lane, D. P. (1991) p68 RNA helicase: identification of a nucleolar form and cloning of related genes containing a conserved intron in yeasts. *Mol. Cell. Biol.*, **11**, 1326.
15. Miller, A. M. (1984) The yeast MATa1 gene contains two introns. *EMBO J.*, **3**, 1061.
16. Ruskin, B., Pikielny, C. W., Rosbash, M., and Green, M. R. (1986) Alternative branchpoints are selected during splicing of a yeast pre-mRNA in mammalian and yeast extracts. *Proc. Natl Acad. Sci. USA*, **83**, 2022.
17. Brown, J. W. S., Feix, G., and Frendeway, D. (1986) Accurate *in vitro* splicing of two pre-mRNA plant introns in a HeLa cell nuclear extract. *EMBO J.*, **5**, 2749.

18. Beggs, J. D., Van den Berg, J., Van Ooyen, A., and Weissmann, C. (1980) Abnormal expression of chromosomal rabbit β-globin gene in *Saccharomyces cerevisiae*. *Nature*, **283**, 835.
19. Langford, C., Nellen, W., Niesing, J., and Gallwitz, D. (1983) Yeast is unable to excise foreign intervening sequences from hybrid gene transcripts. *Proc. Natl Acad. Sci. USA*, **80**, 1486.
20. Watts, F., Castle, C., and Beggs, J. (1983) Aberrant splicing of *Drosophila* alcohol dehydrogenase transcripts in *Saccharomyces cerevisae*. *EMBO J.*, **2**, 2085.
21. Watts, F. Z., Simanis, V., Castle, C., and Beggs, J. D. (1983) Splicing of the transcripts of heterologous genes cloned in *Saccharomyces cerevisiae*. In *Gene expression in yeast. (Proceedings of the Alko Symposium, Helsinki 1983)*. Korhola, M. and Väisänen, E. (ed.). Foundation for Biotechnical and Industrial Fermentation Research, Helsinki, p. 43.
22. Engebrecht, J., Voelkel-Meiman, K., and Roeder, G. S. (1991) Meiosis-specific RNA splicing in yeast. *Cell*, **66**, 1257.
23. Nandabalan, K., Price, L., and Roeder, G. S. (1993) Mutations in U1 snRNA bypass the requirement for a cell type-specific RNA splicing factor. *Cell*, **73**, 407.
24. Engebrecht, J. and Roeder, G. S. (1990) *Mer1*, a yeast gene required for chromosome pairing and genetic recombination, is induced in meiosis. *Mol. Cell. Biol.*, **10**, 2379.
25. Siomi, H., Matunis, M. J., Michael, W. M., and Dreyfuss, G. (1993) The pre-mRNA binding K protein contains a novel evolutionarily conserved motif. *Nucleic Acids Res.*, **21**, 1193.
26. Kishida, M., Nagai, T., Nakaseko, Y., and Shimoda, C. (1994) Meiosis-dependent messenger-RNA splicing of the fission yeast *Schizosaccharomyces pombe mes1*$^{+}$ gene. *Curr. Genet.*, **25**, 497.
27. Dabeva, M. D., Post-Beittenmiller, M. A., and Warner, J. R. (1986) Autogenous regulation of splicing of the transcript of a yeast ribosomal protein gene. *Proc. Natl Acad. Sci. USA*, **83**, 5854.
28. Eng, F. J. and Warner, J. R. (1991) Structural basis for the regulation of splicing of a yeast messenger RNA. *Cell*, **65**, 797.
29. Hirling, H., Scheffner, M., Restle, T., and Stahl, H. (1989) RNA helicase activity associated with the human p68 protein. *Nature*, **339**, 562.
30. Köhler, K. and Domdey, H. (1988) Splicing and spliceosome formation of the yeast *MATa1* transcript require a minimum distance from the 5′ splice site to the internal branch acceptor site. *Nucleic Acids Res.*, **16**, 9457.
31. Nakazawa, N., Harashima, S., and Oshima, Y. (1991) *AAR2*, a gene for splicing premRNA of the *MATa1* cistron in cell type control of *Saccharomyces cerevisiae*. *Mol. Cell. Biol.*, **11**, 5693.
32. Rosbash, M., Harris, P. K. W., Woolford, J. L, Jr., and Teem, J. L. (1981) The effect of temperature-sensitive RNA mutants on the transcription products from cloned ribosomal protein genes of yeast. *Cell*, **24**, 679.
33. Hopper, A. K., Traglia, H. M., and Dunst, R. W. (1990) The yeast *RNA1* gene product necessary for RNA processing is located in the cytosol and apparently excluded from the nucleus. *J. Cell Biol.*, **111**, 309.
34. Tung, K.-S., Norbeck, L. L., Nolan, S. L., Atkinson, N. S., and Hopper, A. K. (1992) *SRN1*, a yeast gene involved in RNA processing, is identical to *HEX2/REG1*, a negative regulator in glucose repression. *Mol. Cell. Biol.*, **12**, 2673.
35. Pearson, N. J., Thornburn, P. C., and Haber, J. E. (1982) A suppressor of temperature-sensitive *rna* mutations that affect mRNA metabolism in *Saccharomyces cerevisiae*. *Mol. Cell. Biol.*, **2**, 571.

36. Maddock, J. R., Weidenhammer, E. M., Adams, C. C., Lunz, R. L. and Woolford, J. L., Jr (1994) Extragenic suppressors of *Saccharomyces cerevisiae prp4* mutations identify a negative regulator of *PRP* genes. *Genetics*, **136**, 833.
37. Rosenberg, G. H., Alahari, S. K., and Käufer, N. F. (1991) *prp4* from *Schizosaccharomyces pombe*, a mutant deficient in pre-mRNA splicing isolated using genes containing artificial introns. *Mol. Gen. Genet.*, **226**, 305.
38. Alahari, S. K., Schmidt, H., and Käufer, N. F. (1993) The fission yeast *prp4*$^{+}$ gene involved in pre-mRNA splicing codes for a predicted serine/threonine kinase and is essential for growth. *Nucleic Acids Res.*, **17**, 4079.
39. Lin, R.-J., Newman, A. J., and Abelson, J. (1985) Yeast mRNA splicing *in vitro*. *J. Biol. Chem.*, **260**, 14780.
40. Lustig, A. J., Lin, R.-J., and Abelson, J. (1986) The yeast RNA gene products are essential for mRNA splicing *in vitro*. *Cell*, **47**, 953.
41. Plumpton, M., McGarvey, M., and Beggs, J. D. (1993) A dominant negative mutation in the conserved RNA helicase motif 'SAT' causes splicing factor PRP2 to stall in spliceosomes. *EMBO J.*, **13**, 879.
42. Teigelkamp, S., McGarvey, M., Plumpton, M., and Beggs, J. D. (1993) The splicing factor PRP2, a putative RNA helicase, interacts directly with pre-mRNA. *EMBO J.*, **13**, 888.
43. Lee, M. G., Lane, D. P., and Beggs, J. D. (1986) Identification of the *RNA2* protein of *Saccharomyces cerevisiae*. *Yeast*, **2**, 59.
44. Lin, R.-J., Lustig, A. J., and Abelson, J. (1987) Splicing of yeast nuclear pre-mRNA *in vitro* requires a functional 40S spliceosome and several extrinsic factors. *Genes Dev.*, **1**, 7.
45. Chen, J.-H. and Lin, R.-J. (1990) The yeast PRP2 protein, a putative RNA-dependent ATPase, shares extensive homology with two other pre-mRNA splicing factors. *Nucleic Acids Res.*, **18**, 6447.
46. King, D. S. and Beggs, J. D. (1990) Interactions of PRP2 protein with pre-mRNA splicing complexes in *Saccharomyces cerevisiae*. *Nucleic Acids Res.*, **18**, 6559.
47. Kim, S.-H., Smith, J., Claude, A., and Lin, R.-J. (1992) The purified yeast pre-mRNA splicing factor PRP2 is an RNA-dependent NTPase. *EMBO J.*, **11**, 2319.
48. Ruby, S. W. and Abelson, J. (1991) Pre-mRNA splicing in yeast. *Trends Genet.*, **7**, 79.
49. Last, R. L. and Woolford, J. L., Jr (1986) Identification and nuclear localization of the yeast pre-messenger RNA processing components: *PRP2* and *PRP3* proteins. *J. Cell Biol.*, **103**, 2104.
50. Petersen-Bjørn, S., Soltyk, A., Beggs, J. D., and Friesen, J. D. (1989) *PRP4 (RNA4)* from *Saccharomyces cerevisiae*: its gene product is associated with the U4/U6 small nuclear ribonucleoprotein particle. *Mol. Cell. Biol.*, **9**, 3698.
51. Dalrymple, M. A., Peterson-Bjørn, S., Friesen, J. D., and Beggs, J. D. (1989) The product of the *PRP4* gene of *S. cerevisiae* shows homology to β subunits of G proteins. *Cell*, **58**, 811.
52. Banroques, J. and Abelson, J. N. (1989) PRP4: a protein of the yeast U4/U6 small nuclear ribonucleoprotein particle. *Mol. Cell. Biol.*, **9**, 3710.
53. Bordonné, R., Banroques, J., Abelson, J., and Guthrie, C. (1990) Domains of yeast U4 spliceosomal RNA required for PRP4 protein binding, snRNP-snRNP interactions, and pre-mRNA splicing *in vivo*. *Genes Dev.*, **4**, 1185.
54. Xu, Y., Peterson-Bjørn, S., and Friesen, J. D. (1990) The PRP4 (RNA4) protein of *Saccharomyces cerevisiae* is associated with the 5′ portion of the U4 small nuclear RNA. *Mol. Cell. Biol.*, **10**, 1217.

55. Galisson, F. and Legrain, P. (1993) The biochemical defects of *prp4-1* and *prp6-1* yeast splicing mutants reveal that the PRP6 protein is required for the accumulation of the [U4/U6·U5] tri-snRNP. *Nucleic Acids Res.*, **21**, 1555.
56. Hu, J., Xu, Y., Schappert, K., Harrington, T., Wang, A., Braga, R., Mogridge, J., and Friesen, J. D. (1994) Mutational analysis of the PRP4 protein of *Saccharomyces cerevisiae* suggests domain structure and snRNP interactions. *Nucleic Acids Res.*, **22**, 1724.
57. Dalbadie-McFarland, G. and Abelson, J. (1990) PRP5: A helicase-like protein required for mRNA splicing in yeast. *Proc. Natl Acad. Sci. USA*, **87**, 4236.
58. Ruby, S. W., Chang, T.-H., and Abelson, J. (1993) Four yeast spliceosomal proteins (PRP5, PRP9, PRP11, and PRP21) interact to promote U2 snRNP binding to pre-mRNA. *Genes Dev.*, **7**, 1909.
59. Abovich, N., Legrain, P., and Rosbash, M. (1990) The yeast *PRP6* gene encodes a U4/U6 small nuclear ribonucleoprotein particle (snRNP) protein, and the *PRP9* gene encodes a protein required for U2 snRNP binding. *Mol. Cell. Biol.*, **10**, 6417.
60. Legrain, P. and Choulika, A. (1990) The molecular characterization of *PRP6* and *PRP9* yeast genes reveals a new cysteine/histidine motif common to several splicing factors. *EMBO J.*, **9**, 2775.
61. Legrain, P., Chapon, C., Schwob, E., Martin, R., Rosbash, M., and Dujon, B. (1991) Cloning of the two essential yeast genes, *PRP6* and *PRP9*, and their rapid mapping, disruption and partial sequencing using a linker insertion strategy. *Mol. Gen. Genet.*, **225**, 199.
62. Legrain, P., Chapon, C., and Galisson, F. (1991) Proteins involved in mitosis, RNA synthesis and pre-mRNA splicing share a common repeating motif. *Nucleic Acids Res.*, **19**, 2509.
63. Lossky, M., Anderson, G. J., Jackson, S. P., and Beggs, J. D. (1987) Identification of a yeast snRNP protein, and detection of snRNP–snRNP interactions. *Cell*, **51**, 1019.
64. Jackson, S. P., Lossky, M., and Beggs, J. D. (1988) Cloning of the *RNA8* gene of *Saccharomyces cerevisiae*, detection of the RNA8 protein, and determination that it is essential for nuclear pre-mRNA splicing. *Mol. Cell. Biol.*, **8**, 1067.
65. Anderson, G. J., Bach, M., Lührmann, R., and Beggs, J. D. (1989) Conservation between yeast and man of a protein associated with U5 small ribonucleoprotein. *Nature*, **342**, 819.
66. Whittaker, E., Lossky, M., and Beggs, J. D. (1990) Affinity purification of spliceosomes reveals that the precursor RNA processing protein PRP8, a protein in the U5 small nuclear ribonucleoprotein particle, is a component of yeast spliceosomes. *Proc. Natl Acad. Sci. USA*, **87**, 2216.
67. Whittaker, E. and Beggs, J. D. (1991) The yeast PRP8 protein interacts directly with pre-mRNA. *Nucleic Acids Res.*, **19**, 5483.
68. Brown, J. D. and Beggs, J. D. (1992) Roles of PRP8 protein in the assembly of splicing complexes. *EMBO J.*, **11**, 3721.
69. Kulesza, H., Simpson, G. G., Waugh, R., Beggs, J. D., and Brown, J. W. S. (1993) Detection of a plant protein analogous to the yeast spliceosomal protein, PRP8. *FEBS Lett.*, **318**, 4.
70. Legrain, P. and Chapon, C. (1993) Identification of the yeast PRP9/PRP11/SPP91 splicing factor complex required for prespliceosome assembly. *Science*, **262**, 108.
71. Legrain, P., Chapon, C., and Galisson, F. (1993) Interactions between PRP9 and SPP91 splicing factors identify a protein complex required in prespliceosome assembly. *Genes Dev.*, **7**, 1390.

72. Chang, T.-H., Clark, M. W., Abelson, J., Cusick, M. E., and Lustig, A. J. (1988) RNA11 protein is associated with the yeast spliceosome and is localized in the periphery of the nucleus. *Mol. Cell. Biol.*, **8**, 2379.
73. Schappert, K. and Friesen, J. D. (1991) Genetic studies of the *PRP11* gene of *Saccharomyces cerevisiae*. *Mol. Gen. Genet.*, **226**, 277.
74. Couto, J. R., Tamm, J., Parker, R., and Guthrie, C. (1987) A *trans*-acting suppressor restores splicing of a yeast intron with a branchpoint mutation. *Genes Dev.*, **1**, 445.
75. Burgess, S., Couto, J. R., and Guthrie, C. (1990) A putative ATP binding protein influences the fidelity of branchpoint recognition in yeast splicing. *Cell*, **60**, 705.
76. Schwer, B. and Guthrie, C. (1991) PRP16 is an RNA-dependent ATPase that interacts transiently with the spliceosome. *Nature*, **349**, 494.
77. Schwer, B. and Guthrie, C. (1992) A conformational rearrangement in the spliceosome is dependent on PRP16 and ATP hydrolysis. *EMBO J.*, **11**, 5033.
78. Schwer, B. and Guthrie, C. (1992) A dominant negative mutation in a spliceosomal ATPase affects ATP hydrolysis but not binding to the spliceosome. *Mol. Cell. Biol.*, **12**, 3540.
79. Burgess, S. M. and Guthrie, C. (1993) A mechanism to enhance mRNA splicing fidelity: the RNA-dependent ATPase Prp16 governs usage of a discard pathway for aberrant lariat intermediates. *Cell*, **73**, 1377.
80. Burgess, S. M. and Guthrie, C. (1993) Beat the clock: paradigms for NTPases in the maintenance of biological fidelity. *Trends Genet.*, **18**, 381.
81. Vijayraghavan, U., Company, M., and Abelson, J. (1989) Isolation and characterization of pre-mRNA splicing mutants of *Saccharomyces cerevisiae*. *Genes Dev.*, **3**, 1206.
82. Frank, D., Patterson, B., and Guthrie, C. (1992) Synthetic lethal mutations suggest interactions between U5 small nuclear RNA and four proteins required for the second step of splicing. *Mol. Cell. Biol.*, **12**, 5197.
83. Frank, D. and Guthrie, C. (1992) An essential splicing factor, *SLU7*, mediates 3′ splice site choice in yeast. *Genes Dev.*, **6**, 2112.
84. Vijayraghavan, U. and Abelson, J. (1990) PRP18, a protein required for the second reaction in pre-mRNA splicing. *Mol. Cell. Biol.*, **10**, 324.
85. Horowitz, D. S. and Abelson, J. (1993) A U5 small nuclear ribonucleoprotein particle protein involved only in the second step of pre-mRNA splicing in *Saccharomyces cerevisiae*. *Mol. Cell. Biol.*, **13**, 2959.
86. Horowitz, D. S. and Abelson, J. (1993) Stages in the second reaction of pre-mRNA splicing: the final step is ATP independent. *Genes Dev.*, **7**, 320.
87. Cheng, S.-C., Tarn, W.-Y., Tsao, T. Y., and Abelson, J. (1993) PRP19: A novel spliceosomal component. *Mol. Cell. Biol.*, **13**, 1876.
88. Tarn, W.-Y., Lee, K. R., and Cheng, S.-C. (1993) The yeast PRP19 protein is not tightly associated with small nuclear RNAs, but appears to associate with the spliceosome after binding of U2 to the pre-mRNA and prior to formation of the functional spliceosome. *Mol. Cell. Biol.*, **13**, 1883.
89. Tarn, W.-Y., Hsu, C.-H., Huang, K.-T., Chen, H.-R., Kao, H. Y., Lee, K.-R., *et al.* (1994) Functional association of essential splicing factor(s) with PRP19 in a protein complex. *EMBO J.*, **13**, 2421.
90. Chapon, C. and Legrain, P. (1992) A novel gene, *spp91–1*, suppresses the splicing defect and the pre-mRNA nuclear export in the *prp91* mutant. *EMBO J.*, **11**, 3279.
91. Arenas, J. E. and Abelson, J. N. (1993) The *Saccharomyces cerevisiae PRP21* gene product is an integral component of the prespliceosome. *Proc. Natl Acad. Sci. USA*, **90**, 6771.

92. Company, M., Arenas, J., and Abelson, J. (1991) Requirement of the RNA helicase-like protein PRP22 for release of messenger RNA from spliceosomes. *Nature*, **349**, 487.
93. Shannon, K. W. and Guthrie, C. (1991) Suppressors of a U4 snRNA mutation define a novel U6 snRNP protein with RNA-binding motifs. *Genes Dev.*, **5**, 773.
94. Chapman, K. B. and Boeke, J. D. (1991) Isolation and characterization of the gene encoding yeast debranching enzyme. *Cell*, **65**, 483.
95. Strauss, E. J. and Guthrie, C. (1991) A cold-sensitive mRNA splicing mutant is a member of the RNA helicase family. *Genes Dev.*, **5**, 629.
96. Blanton, S., Srinivasan, A., and Rymond, B. C. (1992) *PRP38* encodes a yeast protein required for pre-mRNA splicing and maintenance of stable U6 small nuclear RNA levels. *Mol. Cell. Biol.*, **12**, 3939.
97. Rymond, B. C. (1993) Convergent transcripts of the yeast *PRP38-SMD1* locus encode two essential splicing factors, including the D1 core polypeptide of small nuclear ribonucleoprotein particles. *Proc. Natl Acad. Sci. USA*, **90**, 848.
98. Lockhart, S. R. and Rymond, B. C. (1994) Commitment of yeast pre-mRNA to the splicing pathway requires a novel U1 small nuclear ribonucleoprotein polypeptide, Prp39p. *Mol. Cell. Biol.*, **14**, 3623.
99. Last, R. L., Maddock, J. R., and Woolford, J. L., Jr (1987) Evidence for related functions of the RNA genes of *Saccharomyces cerevisiae*. *Genetics*, **117**, 619.
100. Smith, V. and Barrel, B. G. (1991) Cloning of a yeast U1 snRNP 70K protein homolog: Functional conservation of an RNA-binding domain between humans and yeast. *EMBO J.*, **10**, 2627.
101. Kao, H. Y. and Siliciano, P. G. (1992) The yeast homolog of the U1 snRNP protein 70K is encoded by the *SNP1* gene. *Nucleic Acids Res.*, **20**, 4009.
102. Liao, X. C., Tang, J., and Rosbash, M. (1993) An enhancer screen identifies a gene that encodes the yeast U1 snRNP A protein: implications for snRNP protein function in pre-mRNA splicing. *Genes Dev.*, **7**, 419.
103. Abovich, N., Liao, X. C., and Rosbash, M. (1994) The yeast MUD2 protein: an interaction with PRP11 defines a bridge between commitment complexes and U2 snRNP addition. *Genes Dev.*, **8**, 843.
104. Rymond, B. C., Rokeach, L. A., and Hoch, S. O. (1993) Human snRNP polypeptide-D1 promotes pre-messenger RNA splicing in yeast and defines nonessential yeast Smd1p sequences. *Nucleic Acids Res.*, **21**, 3501.
105. Potashkin, J., Naik, K., and Wentz-Hunter, K. (1993) U2AF homolog required for splicing *in vivo*. *Science*, **262**, 573.
106. Hartwell, L. H. (1967) Macromolecule synthesis in temperature-sensitive mutants of yeast. *J. Bacteriol.*, **93**, 1662.
107. Hartwell, L. H., McLaughlin, C. S., and Warner, J. R. (1970) Identification of ten genes that control ribosome formation in yeast. *Mol. Gen. Genet.*, **109**, 42.
108. Cantor, C. R. and Schimmel, P. R. (ed.) (1980) *Biophysical Chemistry*. W. H. Freeman, San Francisco.
109. Newman, A. J. and Norman, C. (1991) Mutations in yeast U5 snRNA alter the specificity of 5′ splice-site cleavage. *Cell*, **65**, 115.
110. Newman, A. J. and Norman, C. (1992) U5 snRNA interacts with exon sequences at 5′ and 3′ splice sites. *Cell*, **68**, 743.
111. Wise, J. A., Tollervey, D., Maloney, D., Swerdlow, H., Dunn, E. J., and Guthrie, C. (1983) Yeast contains small nuclear RNAs encoded by single copy genes. *Cell*, **35**, 743.

112. Guthrie, C. (1986) Finding functions for small nuclear RNAs in yeast. *Trends Biochem. Sci.*, **11**, 430.
113. Siliciano, P. G., Brow, D. A., Roiha, H., and Guthrie, C. (1987) An essential snRNA from *S. cerevisiae* has properties predicted for U4, including interaction with a U6-like snRNA. *Cell*, **50**, 585.
114. Tollervey, D. and Mattaj, I. W. (1987) Fungal small nuclear ribonucleoproteins share properties with plant and vertebrate U-snRNPs. *EMBO J.*, **6**, 469.
115. Fabrizio, P., Esser, S., Kastner, B., and Lührmann, R. (1994) Isolation of *S. cerevisiae* snRNPs: comparison of U1 and U4/U6·U5 to their human counterparts. *Science*, **264**, 261.
116. Séraphin, B. and Rosbash, M. (1991) The yeast branchpoint sequence is not required for the formation of a stable U1 snRNA-pre-mRNA complex and is recognized in the absence of U2 snRNA. *EMBO J.*, **10**, 1209.
117. Hodges, P. E. and Beggs, J. D. (1995) U2 fulfils a commitment. *Current Biol.*, **4**, 264.
118. Bennet, M. and Reed, R. (1993) Correspondence between a mammalian spliceosome component and an essential yeast splicing factor. *Science*, **262**, 105.
119. Séraphin, B. and Rosbash, M. (1989) Identification of functional U1 snRNA–pre-mRNA complexes committed to spliceosome assembly and splicing. *Cell*, **59**, 349.
120. Rosbash, M. and Séraphin, B. (1991) Who's on first? The U1 snRNP-5′ splice site interaction and splicing. *Trends Genet.*, **16**, 187.
121. Birney, E., Kumar, S., and Krainer, A. R. (1992) A putative homolog of U2AF^{65} in *S. cerevisiae*. *Nucleic Acids Res.*, **20**, 4663.
122. Birney, E., Kumar, S., and Krainer, A. R. (1993) A putative homolog of U2AF^{65} in *S. cerevisiae*. *Nucleic Acids Res.*, **21**, 1333.
123. Mattaj, I. W. (1993) RNA recognition: a family matter? *Cell*, **73**, 837.
124. Behrens, S.-E., Tyc, K., Kastner, B, Reichelt, J., and Lührmann, R. (1993) Small nuclear ribonucleoprotein (RNP) U2 contains numerous additional proteins and has a bipartite RNP structure under splicing conditions. *Mol. Cell. Biol.*, **13**, 307.
125. Behrens, S.-E., Galisson, F., Legrain, P., and Lührmann, R. (1993) Evidence that the 60-kDa protein of 17S U2 small nuclear ribonucleoprotein is immunologically and functionally related to the yeast PRP9 splicing factor and is required for the efficient formation of prespliceosomes. *Proc. Natl Acad. Sci. USA*, **90**, 8229.
126. Brosi, R., Gröning, K., Behrens, S. E., Lührmann, R., and Krämer, A. (1993) Interaction of mammalian splicing factor SF3a with U2 snRNP and relation of its 60-kD subunit to yeast PRP9. *Science*, **262**, 102.
127. Krämer, A. (1993) Mammalian protein factors involved in nuclear pre-mRNA splicing. *Mol. Biol. Rep.*, **18**, 93.
128. Liao, X. C., Colot, H. V., Wang, Y., and Rosbash, M. (1992) Requirements for U2 snRNP addition to yeast pre-mRNA. *Nucleic Acids Res.*, **16**, 4237.
129. Pinto, A. L. and Steitz, J. A. (1989) The mammalian analog of the yeast PRP8 splicing protein is present in the U4/U5/U6 small nuclear ribonucleoprotein particle and the spliceosome. *Proc. Natl Acad. Sci. USA*, **86**, 8742.
130. García-Blanco, M. A., Anderson, G. J., Beggs, J. D., and Sharp, P. A. (1990) A mammalian protein of 220-kDa binds pre-mRNAs in the spliceosome: A potential homologue of the yeast PRP8 protein. *Proc. Natl Acad. Sci. USA*, **87**, 3082.
131. Guialis, A., Moraitou, M., Patrinou-Georgoula, M., and Dangli, A. (1991) A novel 40S multi-snRNP complex isolated from rat liver nuclei. *Nucleic Acids Res.*, **19**, 287.
132. Paterson, T., Beggs, J. D., Finnegan, D. J., and Lührmann, R. (1991) Polypeptide components of *Drosophila* small nuclear ribonucleoprotein particles. *Nucleic Acids Res.*, **19**, 5877.

133. Jamieson, D. J., Rahe, B., Pringle, J., and Beggs, J. D. (1991) A suppressor of a yeast splicing mutation (*prp8-1*) encodes a putative ATP-dependent RNA helicase. *Nature*, **349**, 715.
134. Wyatt, J. R., Sontheimer, E. J., and Steitz, J. A. (1992) Site-specific cross-linking of mammalian U5 snRNP to the 5′ splice site before the first step of pre-mRNA splicing. *Genes Dev.*, **6**, 2542.
135. Moore, M. J. and Sharp, P. A. (1993) Evidence for two active sites in the spliceosome provided by stereochemistry of pre-mRNA splicing. *Nature*, **365**, 364.
136. Elliot, D. J., Bowman, D. S., Abovich, N., Fay, F. S., and Rosbash, M. (1992) A yeast splicing factor is localized in discrete subnuclear domains. *EMBO J.*, **11**, 3731.
137. Neer, E. J., Schmidt, C. J., and Smith, T. (1993) LIS is more. *Nature Genet.*, **5**, 3.
138. Jansen, R., Tollervey, D., and Hurt, E. C. (1993) A U3 snoRNP protein with homology to splicing factor PRP4 and Gβ domains is required for ribosomal RNA processing. *EMBO J.*, **12**, 2549.
139. Takagaki, Y. and Manley, J. L. (1992) A human polyadenylation factor is a G protein β-subunit homologue. *J. Biol. Chem.*, **267**, 23471.
140. Goebl, M. and Yanagida, M. (1991) The TPR snap helix: a novel protein repeat motif from mitosis to transcription. *Trends Biochem. Sci.*, **16**, 173.
141. Williams, F. E., Varanasi, U., and Trumbly, R. J. (1991) The CYC8 and TUP1 proteins involved in glucose repression in *Saccharomyces cerevisiae* are associated in a protein complex. *Mol. Cell. Biol.*, **11**, 3307.
142. Keleher, C. A., Redd, M. J., Schultz, J., Carlson, M., and Johnson, A. D. (1992) Ssn6-Tup1 is a general repressor of transcription in yeast. *Cell*, **68**, 709.
143. Kim, S.-H. and Lin, R.-J. (1993) Pre-mRNA splicing within an assembled yeast spliceosome requires an RNA-dependent ATPase and ATP hydrolysis. *Proc. Natl Acad. Sci. USA*, **90**, 888.
144. Ford, M. J., Anton, I. A., and Lane, D. P. (1988) Nuclear protein with sequence homology to translation factor eIF-4A. *Nature*, **332**, 736.
145. Linder, P., Lasko, P. F., Leroy, P., Nielsen, P. J., Nishi, K., Schnier, J., *et al.* (1989) Birth of the D-E-A-D box. *Nature*, **340**, 246.
146. Fuller-Pace, F. V. and Lane, D. P. (1992) RNA helicases. In *Nucleic acids and molecular biology* (Vol. 6), Eckstein, F. and Lilley, D. M. J. (ed.). Springer-Verlag, Berlin, p. 159.
147. Schmid, S. R. and Linder, P. (1992) D-E-A-D protein family of putative RNA helicases. *Mol. Microbiol.*, **6**, 283.
148. Pause, A. and Sonenberg, N. (1993) Helicases and RNA unwinding in translation. *Curr. Opin. Struct. Biol.*, **3**, 953.
149. Fuller-Pace, F. V. (1994) RNA helicases: modulators of RNA structure. *Trends Cell. Biol.*, **4**, 271.
150. Rozen, F., Edery, I., Meerovitch, K., Dever, T. E., Merrick, W. C., and Sonenberg, N. (1990) Bidirectional RNA helicase activity of eukaryotic translation initiation factors 4A and 4F. *Mol. Cell. Biol.*, **10**, 1134.
151. Pause, A. and Sonenberg, N. (1992) Mutational analysis of a DEAD box RNA helicase: the mammalian translation initiation factor eIF-4A. *EMBO J.*, **11**, 2643.
152. Pause, A., Méthot, N., and Sonenberg, N. (1993). The HRIGRXXR region of the DEAD box RNA helicase eukaryotic translation initiation factor 4A is required for RNA binding and ATP hydrolysis. *Mol. Cell. Biol.*, **13**, 6789.
153. Walker, J. E., Saraste, M., Runswick, M. J., and Gay, N. J. (1982) Distantly related sequences in the alpha-subunits and beta-subunits of ATP synthase, myosin, kinases,

and other ATP-requiring enzymes and a common nucleotide binding fold. *EMBO J.*, **1**, 945.

154. Linder, P., Lasko, P. F., Ashburner, M., Leroy, P., Nielsen, P. J., Nishi, K., *et al.* (1989) Birth of the D-E-A-D box. *Nature*, **337**, 121.
155. Gorbalenya, A. E., Koonin, E. V., Donchenko, A. P., and Blinov, V. M. (1989) Two related superfamilies of putative helicases involved in replication, recombination, repair and expression of DNA and RNA genomes. *Nucleic Acids Res.*, **17**, 4713.
156. Weissbach, H. (1980) In *Ribosomes: structure, function and genetics*, Chamblis, G. (ed.). University Park Press, Baltimore, p. 377.
157. Sonenberg, N. (1988) Cap-binding proteins of eukaryotic messenger RNA: functions in initiation and control of translation. *Prog. Nucleic Acid Res. Mol. Biol.*, **35**, 173.
158. Konarska, M. M. and Sharp, P. A. (1987) Interactions between small nuclear ribonucleoprotein particles in formation of spliceosomes. *Cell*, **49**, 763.
159. Cheng, S.-C. and Abelson, J. (1987) Spliceosome assembly in yeast. *Genes Dev.*, **1**, 1014.
160. Ruskin, B. and Green, M. R. (1985) Role of the 3′ splice site consensus sequence in mammalian pre-mRNA splicing. *Nature*, **317**, 732.
161. Wells, S. E. and Ares, M. (1994) Interactions between highly conserved U2 small nuclear RNA structures and Prp5p, Prp9p, Prp11p and Prp21p proteins. *Mol. Cell. Biol.*, **14**, 6337.
162. Hodges, P. E., Jackson, S. P., Brown, J. D., and Beggs, J. D. (1995) Extraordinary sequence conservation of the PRP8 splicing factor. *Yeast*, **11**, 337.

8 Alternative pre-mRNA splicing

YUNG-CHIH WANG, MEENAKSHI SELVAKUMAR, and
DAVID M. HELFMAN

1. General overview

The formation of a mature mRNA from a primary transcript generally requires excision of intervening sequences (introns) with the subsequent joining together (splicing) of exons. The general splicing reaction of metazoan introns has been well characterized and involves a two-step process, which occurs in a large ribonuclear protein complex, termed the spliceosome (see Chapter 4; reviewed in 1, 2). In the first step, a 5′ exon is cleaved and two intermediates are formed: a 5′ exon with a 3′ hydroxyl end, and an RNA molecule containing the downstream intron sequence and 3′ exon, in which the 5′ terminal guanosine of the intron is covalently linked via a 2′–5′-phosphodiester bond to a residue termed the branchpoint, usually an adenosine, located 18–40 nucleotides upstream of the 3′ splice site. In the second step, cleavage at the 3′ splice site results in the release of a lariat intron and concomitantly the two exons are ligated together.

1.1 Splice-site selection

Despite great advances in our understanding of the general splicing reaction, a number of important problems remain unresolved. How splicing is carried out with such fidelity is still not understood, either at the level of cleavage and ligation at a specific 5′ or 3′ splice site or at the level of joining exons together without inadvertently failing to include an exon ('exon skipping'). Most genes contain multiple exons; for example, the collagen gene has more than 50 exons, and the dystrophin gene has more than 70 exons. Clearly if the splicing machinery were to make an error at the level of a single nucleotide, or miss including an exon, the result would probably be an mRNA that encoded a nonfunctional protein. In addition, some introns are very large and how the individual 5′ and 3′ splice sites are brought together over great distances is not known. For example, the dystrophin gene is over two million base pairs long and contains a number of introns that are larger than 200,000 nucleotides (3). Whether the 5′ and 3′ splice sites of these introns are brought together by free diffusion or require an active mechanism remains to be determined. Although consensus sequences are known for 5′ and 3′ splice sites, it is still not possible to predict with 100% accuracy the position of authentic splice sites based solely on sequence analysis. Finally, the identities of the cellular factors that mediate pre-mRNA splicing are just beginning to emerge. They include a group of small nuclear ribonucleoprotein particles (snRNPs) including U1, U2, U4, U5, and U6, as well as other proteins (see Chapters 5 and 6; reviewed in 1, 2). In the case of alternative splicing, two additional questions need to be considered: (1) the nature of the splicing signals in alternatively spliced genes that allows them to be regulated in a tissue- and cell type-specific manner; and (2) what are the cellular factors that are responsible for alternative splicing.

1.2 Alternative splicing

Alternative splicing of primary RNA transcripts is a widespread mechanism that has been identified in the genes of a variety of organisms including *Drosophila*, *Caenorhabditis elegans*, and vertebrates. Different patterns of alternative RNA splicing have been observed (Fig. 1). Alternative RNA splicing functions predominantly in two ways. The first is in the on/off control of gene expression; that is, in regulating the production of a functional or non-functional protein product usually by inclusion of a stop codon; examples include genes that encode the proteins involved in sex determination, suppressor-of-white-apricot, and P-element transposase in *Drosophila*. The second function of alternative splicing is in the formation of multiple protein isoforms. A growing number of cellular and viral genes have been characterized, which encode multiple protein isoforms via the use of alternatively spliced exons (for reviews, see 2, 4–(8)). In many cases alternative RNA splicing contributes to developmentally regulated and tissue-specific patterns of gene expression. In the case of viruses, the use of alternative splicing may facilitate the expression of a larger number of proteins from a relatively small genome. The organization of genes that are subject to alternative processing is diverse, and can be due to the use of alternative promoters and alternative exons located in internal regions, as well as at the 3′ end of genes (reviewed in 4).

Considerable progress has been made in recent years in identifying the *cis*-acting elements and cellular factors that mediate alternative splicing. A variety of mechanisms are involved in alternative splice site selection, including inhibition via factors that specifically block 5′ or 3′ splice sites and activation via factors that enhance the use of specific splice sites. The first part of this review presents a more general overview of the *cis*-acting elements and cellular factors that relate to mechanisms of alternative splice site selection. The second part discusses specific model systems that have been studied in detail, which illustrate various positive and negative mechanisms of regulation.

2. *Cis*-acting elements

A number of features in the pre-mRNA can contribute to alternative splice-site selection, including intron size, the relative strength of 5′ splice sites, the pyrimidine

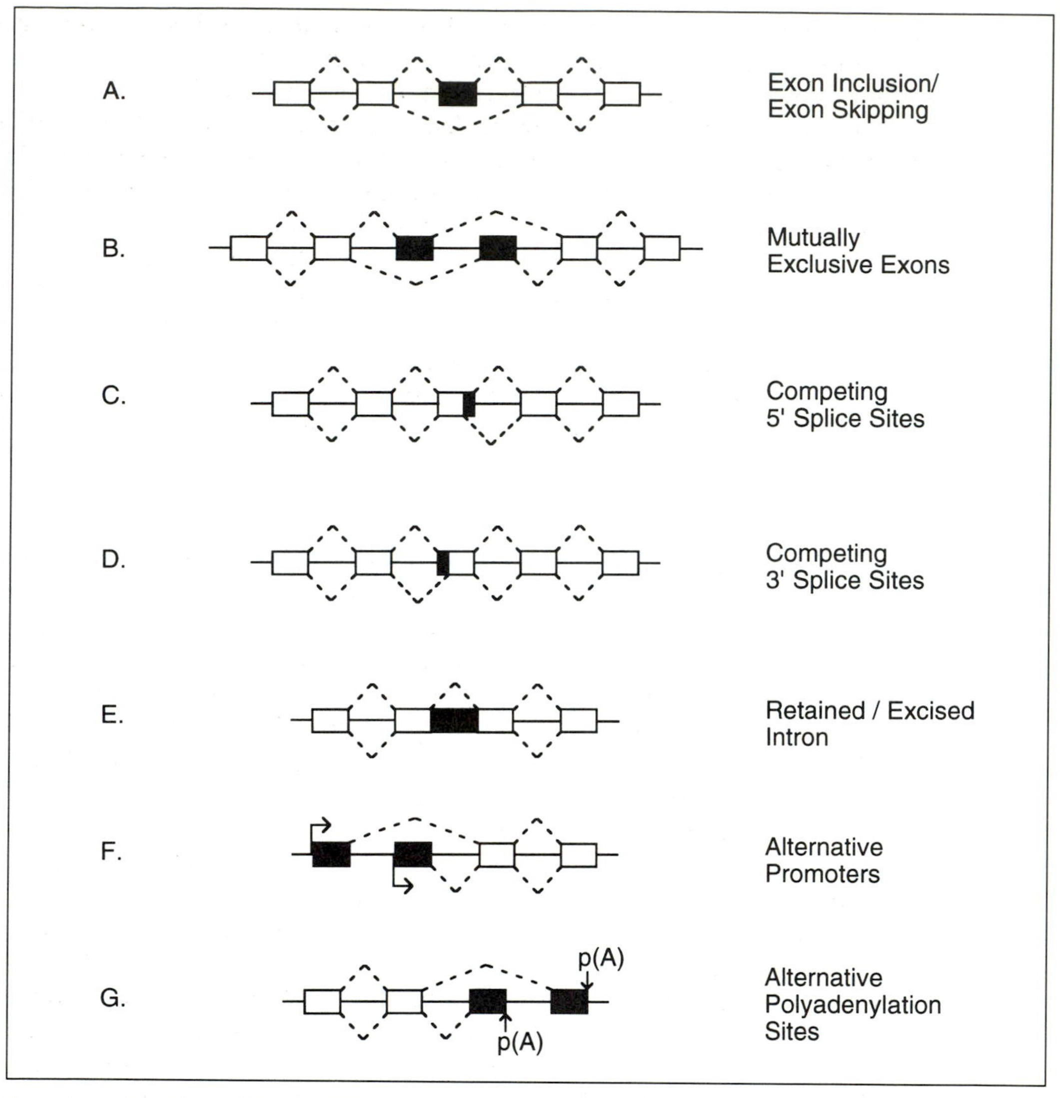

Fig. 1 The alternative splicing patterns that can be used by multiexon-containing pre-mRNAs. Boxes represent exons and lines represent introns. Open boxes represent the constitutively expressed exons, whereas solid boxes represent the alternatively spliced exons. Dash lines represent splicing pathways. In f, the arrows indicate the alternative transcription initiation sites, which define the 5′ boundaries of the alternative first exons. In g, the arrows indicate the alternative sites of 3′ cleavage and polyadenylation (see Chapter 9), which define the 3′ boundaries of the alternative last exons. More complex combinations of patterns a–g are found in complex transcription units.

content of the 3′ splice sites, the location of branchpoints, the presence of multiple alternative branchpoints, and specific exon sequences. In addition, exon and intron sequences may play a role in alternative splicing by regulating the accessibility of different exons to the splicing machinery through the formation of secondary structures. In many cases the use of alternative splice sites is due to a balance between the strength of competing splice sites. Thus, an emerging theme is the importance of 5′ and 3′ splice-site strengths as defined by their match to the consensus sequences and the complementarity of the 5′ splice site and branch site to the corresponding regions of U1 and U2 snRNAs, respectively, as well as the presence and content of the polypyrimidine tract associated with a given 3′ splice site. Sequences within some exons, as well as intronic sequences other than those associated with the 5′ and 3′ splice sites, can play a role in splice-site selection. In some cases weak 5′ and 3′ splice sites can be compensated for or activated via factors that promote the binding of U1 snRNP or U2 snRNP. Studies of alternatively spliced exons have demonstrated that a number of elements can influence the use of an upstream 3′ splice site. These include sequences within the 3′ splice site, the adjacent downstream exon (9, 10), a 5′ splice site downstream of the 3′ splice site (11), and, for 3′ terminal exons, the polyadenylation site (poly(A) site) (12).

2.1 Relative strength and cooperation between signals

Early studies concentrated on defining the *cis*-acting elements required for efficient intron removal and demonstrated the importance of sequences at the 5′ and 3′ splice sites, including the branchpoint sequence and polypyrimidine tract located near the 3′ splice site (13–17). The polypyrimidine tract likely functions as a binding site for U2AF, a factor required for stable binding of the U2 snRNP particle to the branch site (18). U2AF appears to have a preference for poly U-rich sequences over poly C- or purine-rich sequences (19). The binding of U2AF is also influenced by other factors that can compete or enhance its binding to the intron (20–22). More recently, studies of some alternatively spliced pre-mRNAs revealed various levels of recognition required for use of certain 3′ splice sites, which probably contribute to splice-site selection. For instance, in β-tropomyosin and immunoglobulin μ heavy chain pre-mRNAs, activation of a 3′ splice site requires that the adjoining exon first be spliced to a downstream exon. The sequences responsible for this effect appear to reside in the purine-rich motifs located within the downstream activating exon (9, 23, 24). Purine-rich motifs also function within an exon to promote the use of its own 3′ splice site (25–28). Another example is in preprotachykinin pre-mRNA, in which the 5′ splice site of a downstream common exon contains the critical *cis*-acting information that helps activate the use of an upstream exon (11). Similarly, sequences within the 5′ splice site of an exon can promote the use of its own 3′ splice site (11). On the other hand, it has also been shown that the sequences in the upstream 3′ splice site of an exon can facilitate the use of a downstream 5′ splice site (24). These observations nicely fit the exon definition model proposed by Berget and colleagues which postulates that exons are recognized and defined as units during early spliceosome assembly by factors that recognize the 3′ and 5′ splice sites bordering an exon (29).

Support for the exon definition model comes from studies demonstrating that sequences at the 5′ splice site of an exon facilitate the use of the upstream 3′ splice site (11, 29–31). In this model, the U1 snRNP positioned at the 5′ splice site facilitates the binding of the U2 snRNP to the upstream 3′ splice site. Whether sequences within the exon contribute to, or are required for, the cooperative interaction of splicing factors across an exon remains to be determined. Finally, sequences at the 3′ end of a gene, including the poly(A) site, can contribute to the use of the upstream 3′ splice site, and thus can be a target for regulation (32). Collectively, these studies demonstrate that multiple *cis*-acting elements can act cooperatively to stimulate the use of a given 3′ or 5′ splice site. As described below, a number of alternatively spliced exons are regulated by factors that may either promote or antagonize this cooperation.

2.2 Distant branchpoints

Several studies have demonstrated the use of branchpoints a relatively long distance upstream of the 3′ splice sites of some alternatively spliced viral and cellular genes. Splicing of adenovirus E1A RNA was found to use multiple branchpoints located 51–59 nucleotides from the 3′ splice site (33). The use of distant branch sites, located greater than 100 nucleotides upstream of the AG dinucleotide of the 3′ splice junction, has been reported for both β- and α-tropomyosin genes (34, 35). In the α-tropomyosin gene, the use of a distant branch site plays a direct role in preventing the two mutually exclusive exons 2 and 3 from being spliced together (36). The molecular basis for this is the proximity of the 5′ splice site of exon 2 relative to the branchpoint used for exon 3. The branchpoint upstream of exon 3 is located 177 nucleotides from the 3′ splice site, and is only 42 nucleotides from the 5′ splice site of exon 2. Presumably the proximity of the 5′ splice site and branchpoint results in steric hindrance, thereby preventing appropriate interactions of splicing factors with these splice sites (36). In contrast, this mechanism cannot explain the mutually exclusive use of exons 6 and 7 in the rat and chicken β-tropomyosin genes, because the 5′ splice site of exon 6 in these genes is located at least 136 nucleotides upstream of the branchpoint sequences used for exon 7 (34, 35). However, the distant branchpoints of β-tropomyosin genes play a different role in alternative splicing. Point mutations or deletions in sequences between the distant branchpoints and 3′ splice site result in activation of the skeletal muscle-specific exon in a non-muscle cellular environment (37–39). The mechanism by which these mutually exclusive exons are prevented from being joined together is still poorly understood. Studies of the fibronectin gene revealed the use of a distant branchpoint (40), but whether or how this is involved in alternative splicing has not been determined. Studies of the *Drosophila suppressor-of-white-apricot* (*SWAP*) gene demonstrated that a remote branchpoint approximately 100 nucleotides from the 3′ splice site is used in the regulated first intron (41). Collectively, these studies raise the possibility that the use of branchpoints located a relatively long distance from a 3′ splice site may be an essential feature of regulation of some alternatively spliced exons. Whether or not such distant branchpoints are found in introns that are not associated with alternative splice-site selection is unknown.

2.3 RNA secondary structure

RNA secondary structure has been suggested to play a role in alternative RNA splicing by regulating the accessibility of different splice sites to the splicing machinery (42–48). In principle, RNA secondary structure could function in two ways to regulate splice-site selection: by sequestering specific splice sites, or by providing a binding site for cellular factors. Early studies of both chicken and rat β-tropomyosin genes using computer algorithms found that the intron sequences upstream and downstream of a skeletal muscle-specific exon could form a secondary structure, and thereby sequester the exon (37, 49). Thus, based on computer analysis there is phylogenetic conservation of this putative secondary structure. Subsequent *in vitro* analyses demonstrated the existence of such a structure in the chicken β-tropomyosin pre-mRNA, and provided experimental evidence for its role in splice-site regulation (48). However, further experimental results in both the chicken and rat systems are inconsistent with the hypothesis that these intron sequences function in the regulation of splicing via RNA secondary structures (39, 50), although there is evidence in support of a role for secondary structure confined to the exon in the avian gene (47). Work by Eperon *et al.* (45) using model pre-mRNAs has suggested that there is a critical time period in which an RNA can fold after transcription, which could contribute to the regulation of tissue-specific splicing if the rate of transcription along a gene is subject to regulatory control. For example, a protein that has homology to a family of presumed RNA helicases was found to be required for splicing of the yeast mitochondrial cytochrome b and cytochrome c oxidase subunit genes (51). Interestingly, this protein was not required for splicing of all introns in these genes but only specific group I and group II introns (see Chapter 1). Such an intron-specific RNA helicase could, in principle, lead to tissue-specific alternative splice-site selection. In addition, an activity that destabilizes RNA helices has been found in a number of hnRNP preparations (52–54). Thus it is possible that a factor simply binds to a pre-mRNA and regulates splicing by preventing the formation of a secondary structure.

3. Cellular factors involved in alternative splicing

The cellular factors that regulate alternative RNA splicing can be divided broadly into general splicing factors and cell type-specific factors, which will be discussed in their relevant sections below.

3.1 General splicing factors

These proteins, except hnRNPs, have been shown to be essential in the process of constitutive RNA splicing (see Chapter 6), as well as have a role in alternative splice-site selection, acting via their substrate specificity, concentration, or competition with other factors. Many snRNP and non-snRNP splicing factors are characterized by the presence of an RS domain consisting of arginine and serine residues, whose length and location are variable. These proteins include general splicing factors such as U1 snRNP-70K (see Chapter 5), U2AF, and the SR protein family (see Chapter 6), as well as cell type-specific factors such as Tra and Tra-2 (see 4.1.1), and SWAP (see 4.1.3). Although the serine residues of the RS domains can be phosphorylated, it is not clear how the cycle of phosphorylation/dephosphorylation contributes to splicing (reviewed in 55). A study by Cardinali *et al.* (56) suggests that the phosphorylation state of proteins in HeLa nuclear extracts can affect alternative 5′ splice-site selection *in vitro*. It is not known at present how the RS domain contributes to alternative splicing although it has been shown to be necessary for nuclear localization *in vivo* (for example, in SWAP) and for constitutive splicing *in vitro* (for example, in SR proteins) but not for RNA binding (57).

3.1.1 SR proteins

SR proteins are a family of highly conserved splicing factors that have been identified in *Drosophila* and various vertebrates (58–65). The SR family of proteins contains members of 20, 30, 40, 55, and 75 kDa (62). The complexity of this group of proteins has not been fully characterized and no doubt other members of this family will be identified. Some members exhibit similar electrophoretic mobilities, making their identification impossible by one-dimensional analysis. For example, SF2/ASF and SC35 (also known as SRp30a and SRp30b, respectively) co-migrate at approximately 30 kDa. Primary sequence analysis of SR proteins demonstrates that each family member contains one or two copies of a characteristic RNA-recognition motif (RRM) at the amino-terminal end of the protein and an RS domain at the carboxy-terminal region (62).

SR proteins function in both general and alternative pre-mRNA splicing. Their role as essential pre-mRNA splicing factors was first demonstrated biochemically when one member of the family (referred to as SF2, 33 kDa) was purified to apparent homogeneity based on its ability to complement a HeLa cell S100 cytosolic fraction for functional splicing activity (58). These authors also demonstrated that SF2 was able to influence the choice of alternative 5′ splice-site selection using model pre-mRNAs derived from β-globin pre-mRNA (66). In an independent study, a factor termed ASF (alternative splicing factor, 30–35 kDa) was found to affect the choice of alternative 5′ splice sites in *in vitro* experiments using the early region of simian virus (SV40) involving the large T and small t splice choices (59). Structural analysis of cDNAs encoding SF2 and ASF revealed that these proteins were identical (67, 68). Collectively, these studies demonstrated that high concentrations of SF2/ASF promote the use of proximal 5′ splice sites, whereas low concentrations favour the use of distal 5′ splice sites (59, 66). Thus, although this factor is also required for general splicing, differences in the relative concentration or activity of this factor in different cell types could, in principle, play a role in regulated alternative splicing. Selection of proximal 5′ splice sites has also been shown for SC35, another SR protein (61, 69), unlike SRp40, SRp55, and SRp75 which in some cases select both proximal and distal splice sites (62). Another mode of regulation by SR proteins is through substrate specificity for different pre-mRNAs

(70), suggesting that complex patterns of control could be achieved by these factors. Additionally, it has been shown that differences exist in the relative levels of the different SR proteins in various tissues, thereby contributing to another level of control (70). Also, the actions of SR proteins can be antagonized by various hnRNP proteins including A1 and other hnRNP A/B proteins *in vitro* (71, 72) and *in vivo* (73). These results demonstrated that alternative RNA splicing can be regulated by interactions between SR proteins and certain hnRNP proteins.

The mechanism by which SR proteins alter splice-site selection and their target sequences in the pre-mRNA are just beginning to be elucidated. Purine-rich motifs in pre-mRNAs including human growth hormone and fibronectin appear to be targets for binding SR proteins. Purine-rich motifs have also been described in a number of genes including immunoglobulin, troponin T, and β-tropomyosin (23–25, 27, 28, 38, 74–77) although it remains to be determined if SR proteins function through these sequences and if so, whether their sequence requirements are different. In *Drosophila*, studies of alternative splicing of the doublesex pre-mRNA have demonstrated that the alternative splicing factors Tra and Tra-2 function to activate the female-specific exon by stabilizing the interaction of SR proteins (78, 79). Thus, protein–protein interactions via SR proteins can play a role in regulating alternative splice-site selection. It is also possible that SR proteins act as bridging factors between specific 5′ and 3′ splice sites, and thus may also regulate alternative splicing.

3.1.2 U2AF

U2 snRNP auxiliary factor (U2AF) is an essential splicing factor that is required for the binding of U2 snRNP to the pre-mRNA branch site (18). U2AF is a heterodimer composed of a 65 kDa ($U2AF^{65}$) and a 35 kDa ($U2AF^{35}$) subunit (80). $U2AF^{65}$ contains two functional domains consisting of a sequence-specific RNA-binding region containing three RNA-recognition motifs, and an N-terminal short arginine/serine-rich motif which is essential for splicing but not for binding to RNA (81). The 35 kDa subunit was found to contain an arginine/serine-rich motif (RS) and is not essential for constitutive splicing in extracts from which U2AF was depleted (82). Recent studies have demonstrated an indirect role for U2AF in alternative splicing. In studies involving the sex-determination pathway in *Drosophila melanogaster* it has been shown that the protein Sex-lethal competes with U2AF in binding to the polypyrimidine tract of a regulated 3′ splice site, thereby preventing the use of the non-sex-specific splice site (21). As a result, U2AF binds to a female-specific splice site for which it has lower affinity. Furthermore, regulated splicing of another gene (*doublesex*) in the sex determination cascade occurs by the interaction of U2AF with members of the SR protein family such as SC35, and splicing regulators such as Tra and Tra-2 (79). Thus protein–protein as well as protein–RNA interactions can play an important role in regulating alternative splicing.

3.1.3 hnRNPs

Following transcription, the nascent transcript becomes associated with a distinct set of proteins known as heterogeneous nuclear ribonucleoproteins (hnRNPs) (reviewed in 83; see Chapter 3). The hnRNP complexes that form immediately after transcription are thought to represent the first step in pre-mRNA processing. The hnRNPs comprise a large family of proteins. For example, human cells contain at least 20 major proteins that are resolved by two-dimensional gel electrophoresis. Some studies have demonstrated that specific hnRNP proteins exhibit sequence-specific binding properties (84, 85), preferentially assemble on different pre-mRNAs (86), and can affect the patterns of alternatively spliced pre-mRNAs *in vitro* (71) and *in vivo* (73, 87, 88). Although they are associated with most nascent transcripts, a requirement for hnRNA proteins in general splicing has not been unambiguously shown.

Experiments analysing the proteins associated with pre-mRNA before spliceosome assembly *in vitro* demonstrate that unique combinations of hnRNPs assemble on specific pre-mRNAs (86). The observation that different introns can associate with distinct combinations of hnRNPs before spliceosome assembly suggests that the hnRNP complexes that form immediately after transcription might play a critical role in alternative splice-site selection. Recently, experiments using engineered and natural pre-mRNAs have demonstrated that different hnRNPs including A1, A2, and B1 can antagonize the effects of SR proteins in regulating alternative RNA splicing (71–73, 89). Given the complexity of hnRNPs it will not be surprising if other hnRNPs also are found to play a role in splice-site selection. In addition, these data raise the possibility that changes in the levels of various hnRNPs could contribute to regulated tissue-specific splicing patterns.

Another hnRNP that has been implicated in splicing regulation is the polypyrimidine tract binding protein (PTB), which is identical to hnRNP I (90). PTB was first identified in HeLa cell nuclear extracts on the basis of its ability to UV cross-link to the polypyrimidine tract associated with the 3′ splice sites of pre-mRNAs (91). PTB is ubiquitously expressed in all cell types and tissues examined, and multiple forms of PTB are expressed via alternative RNA splicing (92–94). To date there is no evidence that these mRNAs are produced in a tissue-specific manner, or that different PTB isoforms bind to different RNA sequences, but this needs to be further investigated. UV cross-linking studies in nuclear extracts under splicing conditions have been used to study the binding of PTB to the β-globin intron 1 (95), adenovirus 2 major late pre-mRNA, HIV *tat* pre-mRNA, a fibronectin intron (91), and α-tropomyosin introns 1 and 2 (17, 94). The binding of PTB to RNA was found to correlate with efficient splicing of pre-mRNAs, and led to the hypothesis that PTB is involved in splicing (91, 94). Subsequent biochemical complementation studies indicated that nuclear extracts depleted by poly(U)-affinity chromatography are splicing incompetent, but upon addition of a fraction containing PTB, as well as an associated 100 kDa protein (PTB-associated splicing factor, or PSF), splicing can be restored (94, 96). It has been hypothesized that PTB functions in a complex to recognize the pyrimidine stretch in the 3′ splice site region, and through a U1 snRNP interaction forms the earliest commitment complex (92). Studies in the rat β-tropomyosin gene have demonstrated that PTB binds to sequences upstream of a skeletal muscle-specific exon, which are involved in blockage of this exon in non-muscle cells (97), but thus far no functional assays have demonstrated a direct role for these interactions in splice-site regulation.

3.2 Cell type-specific splicing factors

In *Drosophila*, genetic approaches have facilitated the identification and subsequent biochemical characterization of tissue-specific proteins that alter splicing patterns (see below). However, tissue-specific regulators (factors) of splicing, similar to those identified in *Drosophila* systems, have not yet been identified in vertebrates. The lack of functional systems to reproduce tissue-specific alternative splicing patterns *in vitro* has hindered the biochemical characterization of factors. As discussed above, the only factors reported to alter alternative splicing patterns in vertebrates include the SR proteins, U2AF, and some hnRNP proteins (see 3.1). While these studies demonstrate that changes in the levels of various components can alter splice-site selection, it remains to be determined if this mechanism is relevant to the physiological regulation of specific genes.

4. Specific systems

There are numerous characterized genes that utilize alternative RNA splicing for the generation of protein isoform diversity, and many of these genes have been catalogued previously (4, 5). Some examples include genes encoding contractile proteins, immunoglobulins, neuropeptides, and extracellular matrix proteins. In *Drosophila*, genetic approaches have facilitated the identification and subsequent biochemical characterization of developmentally regulated proteins that specifically alter splicing patterns. In contrast to studies in *Drosophila*, much less is known about the cellular factors that mediate alternative splicing in vertebrate systems, although a great deal has been learned about the *cis*-acting elements that are involved in alternative splice-site selection. The following examples from *Drosophila* and vertebrate systems have provided insights into mechanisms involved in splice-site regulation.

4.1 Alternative splicing in *Drosophila*

In the *Drosophila* genes of the sex determination pathway, P-element transposase, and suppressor-of-white-apricot, alternative splicing is subject to regulation by factors that either inhibit or activate the use of alternative 5′ or 3′ splice sites. Based on this kind of regulation, functional or nonfunctional gene products are made.

4.1.1 Sex determination pathway

Sex determination in *Drosophila melanogaster* involves a cascade of regulated alternative splicing events in three genes, namely, *Sex-lethal* (*Sxl*), *transformer* (*tra*), and *doublesex* (*dsx*) (Fig. 2). In these three genes, the female-specific splice of each of the pre-mRNAs appears to be the regulated splicing event, whereas the non-sex-

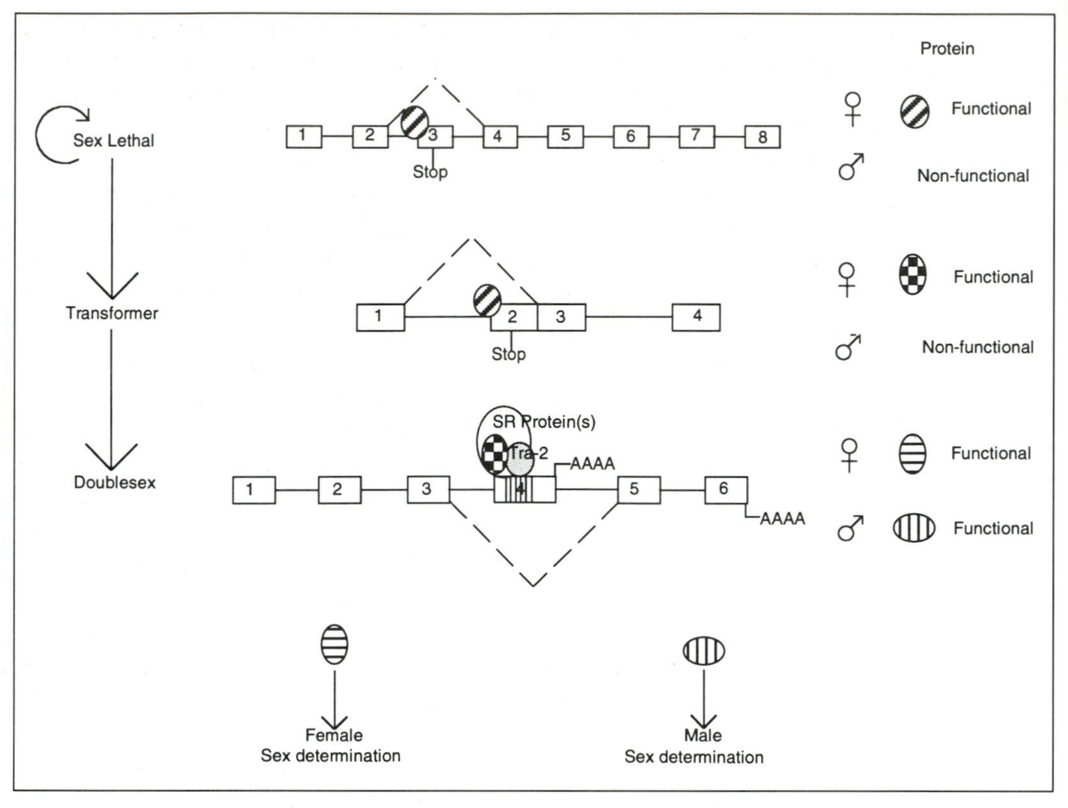

Fig. 2 Alternative splicing involved in the sex determination cascade in *Drosophila*. The pathway on the left depicts the relations among the three regulatory genes, *Sex-lethal*, *transformer*, and *doublesex*; the self-regulation and the influences they exert on the downstream gene are denoted by arrows. The pre-mRNAs corresponding to these genes are shown in the middle panel, with boxes representing exons and lines representing introns. Only the regulated splicing events are shown in all three cases by dash lines. The female pathway is depicted above the boxes and the male/default pathway below the boxes. On the right panel corresponding to the pre-mRNA are the functional and non-functional proteins produced in each case.

specific splice appears to require only the general splicing machinery. Thus, in the absence of the female-specific splicing regulators, the splicing patterns occurring in male flies represent the 'default' patterns. This is partly due to the fact that the splicing signals in the alternative 5′ or 3′ splice sites used in males match the consensus splicing signals better than the regulated (female-specific) splice sites (98–100). As a result, the general splicing machinery recognizes these splice sites more favourably.

The *Sxl* gene is the first one functioning in the sex determination cascade. Its expression is determined in early development in response to the primary determination signal, the X chromosome-to-autosome ratio (reviewed in 101). The *Sxl* gene is turned on in females by the initiation of an auto-regulatory feedback loop, in which the female Sxl protein promotes its own synthesis by directing the female-specific splicing of *Sxl* pre-mRNA (102, 103). The Sxl protein produced in females, which contains two copies of the RNA-recognition motif but lacks any RS domain, blocks the use of the male-specific 3′ splice site of exon 3, leading to the exclusion of this exon (99, 104). In males, inclusion of exon 3 results in the use of a stop codon within this exon, which leads to the production of a non-functional protein (99). At present it is unclear if Sxl alone is sufficient for regulating the female-specific splicing, as other factors have also been implicated in this regulated splicing (105, 106). The female *Sxl* gene product not only regulates alternative splicing of its own pre-mRNA but also directly regulates the downstream gene in the cascade, the *tra* gene (107).

The *tra* pre-mRNA is alternatively spliced, such that the inclusion of exon 2 in males results in a nonfunctional truncated protein, due to the presence of a stop codon in exon 2 (98). The female Sxl protein inhibits inclusion of exon 2 in the *tra* pre-mRNA in the same way as it regulates its own pre-mRNA splicing, by blocking the 3′ splice site of the non-sex-specific exon 2 (108, 109). This allows the use of a competing downstream 3′ splice site and the production of a protein required for female-specific differentiation. Binding of the Sxl protein to the stronger polypyrimidine tract of the 3′ splice site of exon 2 blocks the binding of the essential splicing factor U2AF to this site. As a result, U2AF binds to the downstream female-specific 3′ splice site, for which it has lower affinity (21). It has been shown that when the RS domain of $U2AF^{65}$ is fused to the Sxl protein, the resulting protein no longer blocks the 3′ splice site of exon 2 but instead activates it (21).

In *Sxl* and *tra* pre-mRNAs, selection of the female-specific splice sites occurs in the absence of the competing splice sites (108, 109). In the case of *dsx* pre-mRNA, however, the use of the female-specific splice site is not simply due to a *cis*-competition, but requires positive control by the *tra* and *tra-2* gene products (Tra and Tra-2) (78, 110–114). The doublesex pre-mRNA is alternatively spliced such that exon 4 is included only in females. Since male flies do not produce functional Tra, exon 4 is not included and a male-specific *dsx* gene product is generated. The molecular basis for this regulated splicing event has been characterized in detail. Tra contains an arginine–serine domain (RS domain), while Tra-2 contains two RS domains and an RNA-recognition motif (RRM) (for review, see 115; 98, 116, 117). One of the RS domains of Tra-2 is essential and is required for protein–protein interactions. Tra-2 has been found to interact with itself, Tra, and SF2/ASF *in vitro* and in the yeast two-hybrid system (118). The RRM of Tra-2 is required for regulating alternative splicing of *dsx* pre-mRNA *in vivo*, and is necessary *in vitro* for binding to *tra-2* and *dsx* pre-mRNAs (118). The 3′ splice site of exon 5 of *dsx* pre-mRNA is a relatively strong splice site and represents the default pathway, whereas the 3′ splice site of exon 4 is relatively weak, because of the presence of a number of purine residues in its polypyrimidine tract. Mutating the purines to pyrimidines can activate the splice site and cause exon 4 to be included (100). Exon 4 of the *dsx* gene contains six copies of a 13-nucleotide sequence (100), which is required for female-specific splicing and female-specific polyadenylation (112). Tra-2 is believed to bind directly to these exon sequences together with Tra, leading to the recruitment of general splicing factors (SR proteins and U2AF) to form a splicing enhancer complex which may then commit the pre-mRNA to the female-specific splicing pathway (114). Competitor RNA containing the 13-nucleotide repeat can titrate away Tra and Tra-2 when added to splicing reactions, resulting in the inhibition of female-specific splicing. However, when *dsx* pre-mRNA is pre-incubated with Tra, Tra-2, and SR proteins, it becomes resistant to competition from repeat-containing RNAs, showing that the repeat element participates in the formation of a commitment complex for female-specific splicing (78). The SR proteins have been shown to differ in their ability to cooperate with Tra and Tra-2 in the formation of this commitment complex (78). The repeat element can function constitutively when placed closer to the 3′ splice site in the absence of Tra and Tra-2, suggesting that these proteins function as bridging/recruitment factors (119).

4.1.2. P-element transposase

P-elements are a family of transposable elements found in *Drosophila melanogaster*, whose transposition is limited to the germline cells. P-element pre-mRNA three introns, all of which have to be excised to express the 87 kDa transposase protein in germline tissues (120). On the other hand, in somatic tissues, intron 3 (IVS3) is retained, leading to the expression of a 66 kDa protein that represses transposition (121, 122). The *cis*-acting regulatory elements required for germline regulation have been localized to sequences within exon 3 and IVS3 with the upstream exonic *cis*-element consisting of two pseudo 5′ splice sites (123, 124). *In vitro* experiments using somatic cell extracts show that this region is involved in binding a large multiprotein complex. The consequence of this is the blockage of U1 snRNP binding to the accurate 5′ splice site and, instead, stable interaction with one of the inactive pseudo splice sites is seen (125). Mutations in the *cis*-elements disrupt binding of the complex (125) and activate IVS3 splicing *in vivo* (126) and *in vitro* (127). Members of the multiprotein inhibitory complex include proteins of 97, 65, and 50 kDa. The 50 kDa protein has been revealed to be an hnRNP protein, hrp48 (128). The 97 kDa protein has also been characterized and is referred to as the P-element somatic inhibitor (PSI) (128). Several lines of evidence indicate that PSI is a necessary component of the inhibitory complex. Antibodies against PSI activate splicing of IVS3 in somatic extracts and can be reversed upon addition of recombinant PSI to the reactions (128). Experiments using whole mount embryo-staining show that while hrp48 is ubiquitously expressed, PSI is present in somatic nuclei and is not detectable in the germ line cells (129). There remains the possibility that other factors are also involved in somatic-specific inhibition of IVS3 splicing. Also, the presence of a germline-specific protein that promotes efficient IVS3 splicing cannot yet be excluded.

4.1.3 Suppressor-of-white-apricot

The *suppressor-of-white-apricot* (*su(w)*[a], now referred to as *SWAP*) gene product acts by reverting the effect of a copia retrotransposon insertion allele into the second intron of the white locus (130–132). The *SWAP* gene is auto-regulated by controlling the splicing of its own pre-mRNA (133). The pre-mRNA has seven introns, all of which have to be removed for the production of a functional protein (134). SWAP protein acts by repressing the removal of at least the first intron from the pre-mRNA, causing the accumulation of blocked RNAs (133). Thus, the amount of functional RNA present is controlled at the level of splicing. The SWAP protein also controls alternative splicing of pre-mRNA of the *white-apricot* (*wA*) mutant *white* allele. The presence of functional SWAP protein causes aberrant splicing of the *wA* pre-mRNA and so does not allow the expression of high levels of *wA* mRNA. A 120 amino acid RS domain is found at the C-terminus of the SWAP protein (134). This domain has been shown to be necessary for SWAP protein function, and is interchangeable with the RS domain from the Tra protein (135). It is also required for targeting the protein to the nuclear speckled compartment (135), which has been shown to be enriched in several constitutive splicing components (136, 137, and references therein; Chapter 2).

4.2 Alternative splicing in vertebrates

In vertebrates, alternative RNA splicing is commonly used for generating different tissue- or developmental stage-specific protein isoforms from the same gene. Different mechanisms of alternative RNA splicing have been observed in vertebrates. They include differential selection of 5′ or 3′ splice sites, which may result in alternative exon-skipping or alternative use of mutually exclusive exons. Selection between a poly(A) site and an upstream 3′ splice site, as well as differential intron retention have also been demonstrated as mechanisms of alternative RNA splicing. These mechanisms are illustrated in several gene systems (Fig. 3), which have been extensively studied with respect to how their expression is regulated by alternative RNA splicing.

4.2.1 Tropomyosins

Tropomyosins (TMs) are a diverse group of actin filament-binding proteins found in all eukaryotic cells, with distinct isoforms found in muscle (skeletal, cardiac, and smooth) and non-muscle cells (reviewed in 138). Alternative RNA processing for the generation of TM isoforms has been found in invertebrates (*Caenorhabditis elegans* and *Drosophila*) and vertebrates, indicating that alternative splicing for the generation of TM isoforms occurred relatively early in metazoan evolution (138). Among the four known vertebrate TM genes, the α and β genes are the most intensively studied. Studies of the α and β-TM genes have focused on the use of two mutually exclusive exons. Two fundamental problems arise in these studies of mutually exclusive exons:

(1) what *cis*-acting elements and cellular factors contribute to the choice of each alternative exon;

(2) what prevents the two mutually exclusive exons from being spliced together?

The β-TM gene has been used as a model system for developmental control of alternative splice-site selection during myogenesis. Exon 6 is used in non-muscle cell types (for example, myoblasts, fibroblasts, and HeLa cells) as well as in smooth muscle cells, whereas exon 7 is used exclusively in differentiated myocytes or myotubes of skeletal muscle (Fig. 3; 49, 139). A number of *cis*-acting elements and cellular factors have been implicated in regulation of β-TM pre-mRNA splicing in rat and chicken. The use of exons 6 and 7 (also referred to as exons 6a and 6b, respectively, in the chicken gene) in non-muscle cells is not due to a simple *cis*-competition, as mutations in exon 6 do not result in the use of exon 7 in non-muscle cells (39, 50). On the other hand, studies of the chicken gene demonstrated that exon 6a was recognized in myotubes when the splice sites of the skeletal muscle-specific exon, exon 6b, were destroyed, indicating that a cell-type specific factor is not required for the use of exon 6, and that the use of this exon is not blocked in skeletal muscle cells (50). These studies have given rise to the hypothesis that use of the skeletal muscle-specific exon is blocked in non-muscle cells. The *cis*-acting elements involved in blocking the use of the skeletal muscle exon 7 in non-muscle cells are confined to sequences within exon 7 and the upstream intron (37–39, 47). Interestingly, in both rat and chicken genes, use of the 3′ splice site of the skeletal muscle exon is found to involve the use of a branchpoint located an unusually long distance (133–153 nucleotides) from the 3′ splice site AG dinucleotide (34–35). Two distinct *cis*-acting elements are present in the region between the 3′ splice site of the skeletal muscle exon and the distant branchpoints (37). The first element comprises a polyprimidine tract immediately downstream of the branchpoint, which is important in defining the branchpoint. The second element is located between the polypyrimidine tract and the downstream 3′ splice site. The latter is termed intron regulatory element because deletions or clustered point mutations in this region resulted in the use of the skeletal muscle-specific exon in non-muscle cells (37–39). The cellular factors and the mechanism responsible for exon 7 blockage are poorly understood. At least one factor, polypyrimidine tract binding protein (PTB), is found to bind to the *cis*-acting elements upstream of the skeletal muscle exon 7 (97), but thus far no functional assays have demonstrated a direct role for these interactions in splice-site regulation. More recent studies of the skeletal muscle exon 7 have found that the major site of blockage appears to involve the 3′ splice site of exon 7 (140). Similar observations have also been made for a human α-TM gene (74). Although a blockage mechanism has been implicated for preventing the use of the skeletal muscle-specific exon in non-muscle cells, it remains to be determined if inclusion of the exon in muscle cells results from absence of the blocking activity or expression of a skeletal muscle-specific activator.

The α-TM gene has provided a useful system to study alternative RNA splicing in smooth muscle versus all other cell types. Exon 2 is used exclusively in smooth muscle cells, whereas exon 3 is the preferred choice in all other cell types. The preference for exon 3 appears to be due to the strong pyrimidine tract associated with the 3′ splice site of exon 3 (17). There is a correlation between the preferential use of the 3′ splice site of exon 3 and a 200-fold higher affinity of the associated pyrimidine tract for the splicing factor U2AF, as compared with the pyrimidine tract associated with the use of exon 2 (81). Exon 2 is used in cells when the competing exon 3 is absent (17). However, in the absence of exon 2, exon 3 is not included in smooth muscle cells, suggesting that a blocking mechanism prevents the use of exon 3 in a smooth muscle cell environment (17, 141). Two elements that show sequence conservation between avian and mammalian genes have been identified in the introns flanking exon 3, and are essential for this negative regulation of exon 3 (141).

4.2.2 Troponin

The rat skeletal muscle and chicken cardiac troponin T genes have provided another useful model for studies of alternative splice-site selection (4, 142). Troponin T is part of the troponin complex (including troponins T, I, and C), which plays a critical role in regulating the calcium-sensitive interactions of actin and myosin in skeletal and cardiac muscles (reviewed in 143). Studies of the chicken cardiac troponin T gene have focused on a cassette type splice in which one exon, exon 5, is included in fetal heart and skeletal muscles, but excluded in adult heart muscle (Fig. 3). Transient transfection studies have demonstrated that sequences between exon 5 and the flanking common exons contain sufficient *cis*-acting information required for alternative splice-site selection (144, 145). Studies of the alternatively spliced exon have shown that sequences within the exon contain an important determinant for splice-site use, and that these sequences play a role in removal of the upstream intron but not the downstream intron (25). Further characterization of these *cis*-acting elements has revealed that the internal region of exon 5 contains a purine-rich motif, termed exon splicing element (ESE), which facilitates splicing of both alternative and constitutive exons from heterologous genes (75). Interestingly, the ESE is not required for the preferential inclusion of the alternatively spliced exon 5 in primary skeletal muscle cultures. Thus, the ESE may serve as a general splicing element, and the sequences in the flanking introns appear to play a role in the developmental regulation of the alternative exon (75). A recent study has shown that the splicing enhancer activity can be localized to a nine-nucleotide motif, GAGGAAGAA, in exon 5, and the *in vitro* binding of a subset of SR proteins (SRp30a, SRp40, SRp55, and SRp75) to this specific sequence correlates with the splicing enhancer activity (77).

4.2.3 Immunoglobulin μ heavy chain

The immunoglobulin (Ig) μ heavy-chain gene, encoding the heavy chain of the protein IgM, is among the first cellular genes whose expression was found to be regulated by alternative RNA processing. During the development of humoral immunity, B lymphocytes, which express membrane-bound Ig, differentiate to Ig-secreting plasma cells, resulting in the switch of antibody production from the cell surface form to the secreted form. This switch depends on alternative RNA processing at the 3′ end of the primary transcript of the Ig μ gene (146). The 3′ structure of the Ig μ heavy chain gene is shown in Fig. 3. To produce the membrane-bound Ig μ (μ_m) mRNA, the pre-mRNA is spliced between the fourth constant region domain (Cμ4) and M1 exon, and is cleaved and polyadenylated at the μ_m poly(A) site.

Hence μ_m mRNA includes exons M1 and M2 at the 3′ end, which encode a membrane anchor sequence at the carboxy-terminus of IgM. If cleavage and polyadenylation occur at the μ_s poly(A) site, thereby eliminating the Cμ4-M1 splice, μ_s mRNA coding for the secreted form of IgM is produced. The alternative RNA processing patterns and the regulation of μ_s/μ_m mRNA levels have been intensively studied. These studies have been recently reviewed (147), and are thus only briefly summarized here.

Several mechanisms or models have been considered and tested for the regulation of the μ_s/μ_m mRNA ratio between B cells and plasma cells. These mechanisms include differential transcriptional termination, differential RNA stability, competition between the μ_m poly(A) site and the μ_s poly(A) site, as well as competition between the Cμ4-M1 splice and the μ_s poly(A) site. Although changes in the transcriptional termination region from downstream to upstream of the μ_m-specific exons have been implicated in the regulation of the μ_s/μ_m mRNA levels (148, 149), there is descrepancy among the analyses using different cell lines (150–155). The cumulative data favour the model that the μ_s poly(A) site and the Cμ4-M1 splice compete in Ig μ pre-mRNA processing. Mutations or deletions that weaken the μ_s poly(A) site result in increased production of μ_m mRNA in B and plasma cells (156, 157), thereby demonstrating the importance of the μ_s poly(A) site. On the other hand, when the μ_m poly(A) site is substituted with other, weaker poly(A) sites, there is no effect on μ_s mRNA, suggesting that the μ_m poly(A) site is not in competition with the μ_s poly(A) site (12). The Cμ4 5′ splice site is evolutionarily conserved and has a suboptimal efficiency because it contains three mismatched nucleotides when compared to the consensus sequence. Mutation of the Cμ4 5′ splice site to a consensus one improves the Cμ4-M1 splice efficiency. When such a mutated μ gene is expressed in B or plasma cells, only μ_m mRNA is made (12). These results show that the Cμ4-M1 splice is in competition with the μ_s polyadenylation, and that there needs to be a balance between these two events, both of which have suboptimal efficiencies. This balanced competition allows the gene to respond to the changing cellular environment during B cell differentiation. To date, no specific *cis*-acting element has been implicated in regulation of the shift in μ_s/μ_m expression between B and plasma cells.

In terms of the *trans*-acting factors, the data are consistent with a model in which changes in the relative concentration or activity of a general splicing factor(s) mediate the switch in alternative RNA processing of the Ig μ gene during B cell differentiation. Peterson *et al.* (158) recently reported that the size of the Cμ4 exon affects the competition between the Cμ4-M1 splice and μ_s polyadenylation. When the Cμ4 exon is made shorter, more spliced μ_m mRNA is produced, whereas an expanded Cμ4 exon results in higher level of μ_s mRNA (158). They thus suggest that the μ_s poly(A) site–Cμ4 splice competition model can be considered from another angle. That is, it can be a competition between defining the Cμ4 exon as an internal exon in μ_m mRNA or as a terminal exon in μ_s mRNA. They also raise the possibility that there could be a factor in B cells that decreases the communication between the 3′ splice site and the poly(A) site of a terminal exon (in this case, the 3′

splice site of the Cμ4 exon and the μ_s poly(A) site), so that the Cμ4 exon is not recognized as a terminal exon. Alternatively, B cells could have a factor that enhances the communication between the 3′ and 5′ splice sites of an exon, so that the Cμ4 exon is recognized as an internal exon. In this regard, another group has reported that an activity detected in B cell extracts specifically destabilizes the μ_s polyadenylation complex, although these authors have interpreted the results as relevant to the poly(A) site choice model (159).

A purine-rich sequence with a splicing enhancer activity has been identified in exon M2 of mouse IgM pre-mRNA, which stimulates splicing at the 3′ splice site immediately upstream, and is termed exon-recognition sequence (23). Its function is similar to that of the purine-rich ESE subsequently found in other genes, such as the troponin T and fibronectin genes, and it is considered to be a general splicing element (160). Although the M1–M2 splice is required for efficient Cμ4-M1 splicing *in vitro* (10), the exon recognition sequence in exon M2 has not been shown to be involved in the regulation of the μ_s/μ_m mRNA ratio.

4.2.4 Fibronectin

Fibronectins are a group of extracellular matrix glycoproteins. Two alternatively spliced cassette type exons, EIIIA and EIIIB (also known as ED1 and ED2, or EDA and EDB), are regulated in a cell type-specific manner. For example, both exons are excluded in adult liver, whereas both are included in all fibronectin mRNAs expressed during early embryonic development (for a review, see 161). Studies of the alternative splicing of these two exons have provided some important insights into their regulation. These studies have been recently reviewed by Ffrench-Constant (162), and thus only some of the findings are summarized here. For EIIIA, the regulation appears to involve mainly exon sequences, whereas, for EIIIB, sequences in the downstream intron play a critical role in the regulation. Early studies demonstrated that mutations in EIIIA resulted in exon skipping (163). Further analysis of the sequences within EIIIA identified a purine-rich motif that functions as a splicing enhancer for the 3′ splice site of the exon (26, 27), as well as a second element that functions as a negative modulator for exon recognition (26). The splicing enhancer element appears to serve as a target for SR proteins (27), while the factor(s) that interacts with the negative element has not been identified. The regulation of the other exon, EIIIB, of the alternative exon cassette appears to involve somewhat different *cis*-acting elements. Unlike EIIIA, EIIIB is subject to regulation via sequences in the downstream flanking intron, termed the intronic control region. Sequences in this region are required for cell-type-specific EIIIB inclusion (164). EIIIB is a weak exon, as a consequence of suboptimal 5′ and 3′ splice sites and exon sequences. Replacement with stronger splice sites, or substitution of exon sequences, led to increased EIIIB inclusion *in vivo*. Furthermore, there seems to be a balanced competition between the splice sites of EIIIB and those of the flanking exons. The suboptimal nature of the splicing signals of EIIIB and its flanking exons is important in the maintenance of this balance and the proper regulation of EIIIB inclusion (164).

4.2.5 Calcitonin/CGRP

The pre-mRNA of the calcitonin/CGRP (calcitonin gene-related peptide) gene is alternatively processed to produce calcitonin and CGRP mRNAs in a tissue-specific manner (165, 166). Calcitonin is produced in thyroid C-cells (167), and CGRP in the brain and the peripheral nervous system (168). Calcitonin mRNA contains exons 1–4, whereas CGRP mRNA contains exons 1–3, 5, and 6 (Fig. 3). Both *in vitro* splicing assays and transfections of cultured cell lines have been used to characterize the mechanisms controlling the cell type-specific processing of calcitonin/CGRP pre-mRNA. Although the *in vitro* systems tested have had difficulty in reproducing the cell type-specific patterns (169–171), the *in vitro* studies have suggested a mechanism based on cell type-specific selection of the poly(A) site (169). However, *in vivo* studies involving transfection of model cell lines with minigene constructs have demonstrated that poly(A) site selection of calcitonin/CGRP pre-mRNA is not regulated (172, 173), but instead the use of the 3′ splice site of the calcitonin-specific exon 4 is the critical point of regulation (173). The branchpoint nucleotide preceding exon 4 is a uridine instead of an adenosine (174). Mutation of this uridine to the branchpoint consensus adenosine results in an increase of exon 4 recognition (175). Extensive mutations made in rat calcitonin/CGRP transcription units and analysed in permanently transfected cells did not reveal any unique *cis*-acting element that might be responsible for the regulation of tissue-specific splicing (176). Thus, the authors concluded that tissue-specific splicing of rat calcitonin/CGRP pre-mRNA is a result of differential recognition of suboptimal constitutive intron and exon sequences near the 3′ splice site of the exon 4 by the splicing machinery (176). However, in the human calcitonin/CGRP gene, sequence elements residing in the calcitonin-specific exon 4 have been shown to be required for the recognition of this exon (177, 178). One of these *cis*-acting elements identified in exon 4 contains a purine-rich sequence element similar to the ESE found in other genes (see Section 2.1). Another *cis*-acting element required for recognition of the calcitonin-specific exon 4 is in intron 4, approximately 150 nucleotides downstream of the exon 4 poly(A) signal (179). Interestingly, this intronic enhancer contains a conserved 5′ splice site sequence which is essential for the enhancer function (179). The mechanism by which this intronic enhancer functions to promote exon inclusion remains to be investigated.

4.2.6 Preprotachykinin

The preprotachykinin I gene contains seven exons and encodes three distinct proteins, which function as neurotransmitters or neuromodulators, namely substance P, neurokinin A, and related tachykinin peptides. These proteins result from alternative splicing of exon 4 (E4) or exon 6 (E6) of the preprotachykinin pre-mRNA (180, 181). The regulation of E4 has been extensively studied. In HeLa cells, minigenes containing exons E3–E5 are spliced such that E4 is excluded *in vivo* and *in vitro* (11). Interestingly, E4 is efficiently spliced to E3 if it is first spliced to E5. Thus, there is a preferential order of intron removal. In addition, unlike the example of β-

TM pre-mRNA discussed in Section 4.2.1, the critical element in E5 that activates the 3′ splice site of E4 is the 5′ splice site of E5, rather than sequences within the exon. Alternatively, if the 5′ splice site of E4, which deviates from the consensus sequence at two positions, is mutated to a perfect match with the consensus U1 snRNA binding site, then E4 is spliced to E3 even without first joining to E5 (11). Subsequent studies demonstrated that the binding of U1 snRNP to the downstream 5′ splice site of E5 is responsible for activating the upstream 3′ splice site of E4 (30, 31). These results are in agreement with the exon definition model, which proposes that exons are selected as a result of coordinate recognition of the 5′ and 3′ splice sites of the same exon (29). The mechanism by which U1 snRNP bound to a downstream 5′ splice site enhances the use of the upstream 3′ splice site appears to involve the recruitment of U2AF. Thus, the 3′ splice site of exon E4, which shows a weak U2AF-binding ability, may be used via a mechanism in which U1 snRNP bound to the downstream 5′ splice site facilitates, across the exon, the binding of U2AF (20). Alternatively, a factor could bind to sequences within E4 itself and activate the use of the upstream 3′ splice site. It remains to be determined whether the use of E4 of preprotachykinin pre-mRNA requires the sequential pathway of splicing (that is, E5–E4 splice occurs before E4–E3 splice) as elucidated in HeLa cells (11), since no studies have yet addressed the alternative splicing mechanism of this pre-mRNA in the proper cellular context.

4.2.7 c-*src*

The c-*src* proto-oncogene, which encodes a tyrosine protein kinase, contains an 18-nucleotide exon, N1, that is included in the src mRNA in neuronal cells, but excluded in all other cell types (Fig. 3; 182, 183). The gene contains another neurone-specific exon, N2, occasionally inserted between exons N1 and 4 (184), which has not been studied at the level of splice-site selection. Studies of regulation of the N1 exon using transient transfection in HeLa and LA-N-5 neuroblastoma cells have demonstrated inclusion of exon N1 only in the neuroblastoma cells (185). This neurone-specific pattern of exon N1 inclusion has been reconstructed *in vitro* using nuclear extracts from HeLa cells and WERI-1 retinoblastoma cells (185). The sequences within exon N1 and its flanking introns are sufficient for neurone-specific splicing when placed between two heterologous exons derived from adenovirus (186). Increasing the length of exon N1 to 109 nucleotides resulted in efficient use of this exon in HeLa cells, which led to the hypothesis that exon N1 exclusion in non-neuronal cells may be due to steric interference between spliceosome complexes assembling on both sides of the short exon (186). Thus, mutations that improve the 3′ splice site of exon N1 lead to its inclusion in non-neuronal cells, presumably by overcoming the steric interference through strengthening the assembly of factors on the 3′ splice site (186). Subsequent analyses identified a positive *cis*-acting element in the intron downstream of exon N1. This element, located 38–142 nucleotides downstream from the 5′ splice site of exon N1 serves to activate the N1 exon in a neurone-specific manner. *In vitro* studies suggest that it plays a role in promoting the excision of the intron downstream of exon N1 (185). Mutational

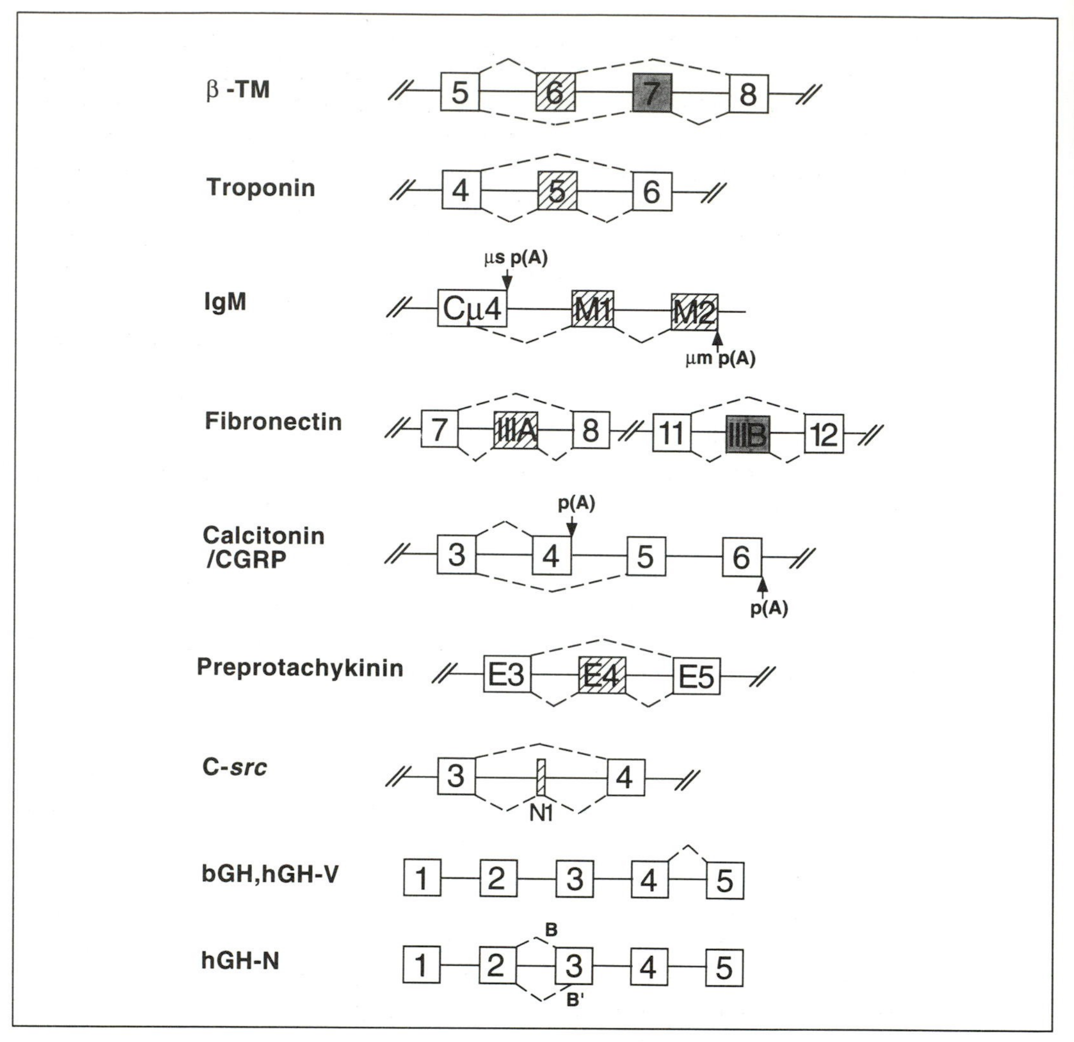

Fig. 3 Schematic representation of alternative splicing pathways of vertebrate pre-mRNAs. Solid lines are introns, open boxes common exons, and hatched and shaded boxes alternative exons. The exons and introns are not drawn to scale. Polyadenylation sites are denoted by p(A). The alternative splicing pathways are respectively shown by dash lines above and below the introns. The β-tropomyosin (β-TM) pre-mRNA diagram can also represent α-TM pre-mRNA, in which the alternative exons are exons 2 and 3, flanked by common exons 1 and 4. The hGH-N pre-mRNA is shown with its two alternative 3′ splice sites, known as B and B′ sites. CGRP, calcitonin gene-related peptide; bGH, bovine growth hormone; hGH-V, human growth hormone variant; hGH-N, human normal growth hormone.

analysis has determined that this positive element region appears to contain multiple positive elements of redundant functions. Interestingly, comparison of this region between the chicken and mouse *c-src* genes revealed several regions of homology, suggesting that this element might also function in the avian gene (185). The 5′ portion of this region (nucleotides 38–70 downstream of exon N1), termed downstream control sequence, binds to a complex of proteins, including a neurone-specific 75 kDa protein and several constitutive proteins, one of which is hnRNP F (187). The binding of this complex to the *src* pre-mRNA in the neuronal cell nuclear extract is required to activate N1 exon splicing *in vitro* (187). The identities of the neurone-specific protein and of most of the other proteins that bind to the downstream control sequence are not known, nor is it known how they facilitate the inclusion of exon N1. The inability of exon N1 to be spliced in non-neuronal cells appears to depend on specific intron sequences flanking the exon. Multiple copies of conserved CUCUCU elements within these intron regions are required for the repression of exon N1 splicing *in vitro* (188). Whether these sequence elements bind to any splicing repressor and how they function in splicing repression remain to be studied.

4.2.8 Growth hormone

The bovine growth hormone (bGH) and human growth hormone variant (hGH-V) genes contain five exons and four introns (Fig. 3). The last intervening sequence (intron D) is subject to alternative splicing. Retention of this intron in bGH mRNA results in the inclusion of an additional 274-nucleotide sequence in the mRNA with an open reading frame that has the termination codon located 50 nucleotides into exon 5 (189). In human placenta, a fraction of the hGH-V mRNA also retains intron D (190). This mode of alternative splicing, which is based on the inefficient splicing of an intron, might be more primitive than other patterns of alternative splicing. Such type of alternative splicing is also used by retroviruses (see Section 4.3.1), and is the only type of alternative splicing observed in yeast (see Chapter 7). Studies of the bGH gene using transient transfection in COS cells have identified a 115-nucleotide sequence beginning 73 nucleotides downstream of the 3′ splice site of exon 5, which is required for use of the 3′ splice site. Within this element, a 10-nucleotide inverted repeat with the sequence CTTCCGGAAG was identified, which can function as a purine-rich exonic splicing enhancer (28, 76, 189). *In vitro* splicing of the bGH pre-mRNA containing exons 4 and 5 is inhibited by the addition of a competitor RNA containing this 115-nucleotide exon sequence, but not by other nonspecific RNA (28). UV cross-linking identified a 35 kDa protein that specifically binds to the splicing enhancer sequences and has subsequently been identified as the general splicing factor SF2/ASF (191). Thus, the binding of SF2/ASF to the purine-rich exonic splicing enhancer within exon 5 stimulates the use of the upstream 3′ splice site. Interestingly, hnRNP A1 has been found to counteract the action of SF2/ASF, suggesting that the relative levels of SF2/ASF and hnRNP A1 may be responsible for the level of retention of intron 4 *in vivo* (191).

Human normal growth hormone (hGH-N) pre-mRNA contains two active 3′ splice sites in exon 3. The major site (B site) is at the 5′ end of the exon, whereas the minor site (B′ site) is 45 nucleotides inside the exon (192, 193). The use of these two 3′ splice sites, at a ratio of 9:1, results in the production of a major growth hormone isoform of 22 kDa and a minor isoform of 20 kDa with different insulin-like properties (reviewed in 194). A mutational analysis of the splicing pattern of hGH-N pre-mRNA has revealed that two bases in the vicinity of the B site facilitate the use of the downstream competing B′ site. Further site-specific mutations demonstrate that these two bases function by stabilizing a specific stem–loop structure, which selectively encompasses the upstream B site along with its branchpoint sequences, thereby inhibiting the selection of this 3′ splice site, and allowing the use of the downstream B′ site (195).

4.3 Viral alternative splicing

In order to generate multiple gene products from compact viral genomes, most viral transcription units can produce multiple mRNAs via alternative RNA processing by the cellular machinery of the host cells. Given the wide variety of viral genomes and the ample research results on viral RNA splicing, as well as the focus of this chapter on cellular pre-mRNA alternative splicing, we will only summarize some of the studies regarding viral RNA splicing, and the reader will be directed to pertinent reviews for detailed discussions. Among the viruses in which alternative RNA splicing has been extensively studied are adenoviruses, simian virus 40 (SV40), and retroviruses, such as human immunodeficiency virus type 1 (HIV-1) and avian sarcoma virus. The post-transcriptional control of adenovirus gene expression, including alternative RNA splicing and poly(A) site selection, has been extensively reviewed by Imperiale *et al.* (196) and, therefore, will not be discussed here.

4.3.1 Retroviruses

Retroviruses are single-stranded RNA viruses that replicate through a DNA intermediate, which integrates into the host genome (reviewed in 197). Unlike the majority of cellular pre-mRNAs, which are spliced efficiently and completely before exiting the nucleus, a fraction of retroviral primary transcripts are unspliced and transported to the cytoplasm, where they serve as genomic RNA for progeny virions, and as the mRNA for structural proteins and enzymes. The remainder of the primary transcripts are alternatively spliced to generate mRNAs for other proteins (for reviews, see 198, 199). Therefore, in the life cycle of retroviruses, it is essential to maintain a balance between spliced and unspliced mRNA in the cytoplasm. In order to maintain this balance, the splice sites of retroviruses are typically very inefficient substrates for splicing (16, 200–202), and different mechanisms seem to be employed to regulate alternative splicing. In the case of Rous sarcoma virus (RSV), an avian retrovirus, multiple *cis*-acting elements and suboptimal 3′ splice sites act additively to control its RNA splicing (203). The RSV genome encodes four genes, namely, *gag*, *pol*, *env*, and *src* genes. It contains a major 5′ splice site and two alternative 3′ splice sites, which are used for the generation of *env* and *src* mRNAs, respectively. One negative *cis*-acting RNA sequence, termed the negative regulator of splicing (NRS) and located in the *gag* gene, regulates splicing at the *env* and the *src* 3′ splice sites (204, 205). It also inhibits splicing when placed within the intron of a cellular gene (206). This inhibitory function of the NRS, when tested within heterologous introns *in vitro* and *in vivo*, correlates with its binding to U11 and U12 snRNPs along with the spliceosomal snRNPs in large RNP complexes that do not mature into spliceosomes (207). Another negative *cis*-acting element is referred to as *src* suppressor of splicing (SSS), and is located more than 70 nucleotides upstream of the *src* 3′ splice site (202, 203). A recent study suggests that the SSS region may represent a binding site for a negative splicing factor present in chicken embryo fibroblasts, but not in non-permissive human cells (208). The *env* exon contains a positive element that acts in *cis* to enhance splicing at the upstream *env* 3′ splice site (16). This exon splicing enhancer has a purine-rich sequence similar to that found in some cellular pre-mRNAs, which have been shown to selectively bind to some SR proteins (see Sections 2.1 and 3.1.1).

HIV-1 is a more complex retrovirus (reviewed in 209). Its approximately 9 kb genome contains 20 exons, which are alternatively spliced to form more than 40 mRNAs (210–213). Using the 9.2 kb primary transcript as mRNA, the Gag and Pol proteins are generated. The Env, Vif, Vpr, and Vpu proteins are encoded by the singly spliced mRNAs, whereas the regulatory proteins Tat, Rev, and Nef are produced from the multiply spliced mRNAs (reviewed in 199). Maintenance of the proper balance of these mRNAs requires the complex interplay of different mechanisms. Most splice sites in the HIV-1 genome are very inefficient. Using transient transfection assays and a splice site swapping strategy, O'Reilly *et al.* (214) showed that, except for two efficient 5′ splice sites, all the other tested 5′ and 3′ splice sites of HIV-1 RNA are suboptimal and have variable efficiencies, which may contribute to the regulation of HIV-1 RNA splicing. The HIV-1 pre-mRNA does not appear to have inhibitory sequences analogous to the RSV NRS in the *gag* or *pol* genes (214). A negative *cis*-acting element termed exon splicing silencer (ESS) in the first *tat* coding exon (*tat* exon 2) acts to inhibit splicing at the flanking upstream 3′ splice site (215). Another *cis*-acting element has been detected within the *tat-rev* exon 3. This bipartite element consists of an ESS, whose sequence and function are similar to the ESS in *tat* exon 2, as well as a purine-rich ESE sequence, which enhances splicing at the upstream *tat-rev* 3′ splice site (216, 217). The significance of having a negative element juxtaposed to a positive element for the regulation of HIV-1 RNA splicing remains to be understood. Besides suboptimal splice sites and *cis*-acting elements, an additional level of regulation of HIV-1 RNA is exerted by a viral regulatory protein, Rev, which functions to allow transport of unspliced or partially spliced mRNA to the cytoplasm. The functions of Rev and the mechanism by which Rev regulates HIV-1 RNA have been reviewed recently (218).

4.3.2 Simian virus 40

SV40 is a simple virus with a closed circular double-stranded DNA genome of 5243 base pairs (reviewed in 219). The early region of SV40 has been used as a model for studying the control of alternative 5′ splice site choice, due to its relatively simple organization. The SV40 early pre-mRNA is differentially spliced, using a shared 3′ splice site, a proximal 5′ splice site used for generating mRNA for the small t antigen, and a distal 5′ splice for the large T antigen. Early studies showed that because of the short size (66 nucleotides) of the small t intron, large T splicing is more efficient than small t splicing, resulting in a three- to five-fold higher level of large T mRNA than small t mRNA in HeLa cells and several other mammalian cell lines (220). However, in 293 cells, a human embryonic kidney cell line transformed with the early region of adenovirus 5 (221), the ratio of small t to large T mRNAs is 10 to 15-fold greater than in many other mammalian cells (15). Using an *in vitro* complementation assay, an alternative splicing factor, ASF, was purified from 293 cell nuclear extracts, and shown to enhance splicing at the proximal small t 5′ splice site (33). ASF is identical to a member of the SR protein family, the essential splicing factor SF2, which was independently characterized and purified from HeLa cell nuclear extracts (58; see Section 3.1.1). More recently, two other SR proteins, SRp40 and SRp55, have been shown to promote splicing at the distal large T 5′ splice site *in vitro* (70). A recent *in vivo* study unexpectedly revealed an inhibitory effect of SF2/ASF transient overexpression on splicing of SV40 pre-mRNA at both 5′ splice sites, with the small t 5′ splice site being more sensitive than the large T site (222). The reason for this discrepancy between the observed *in vitro* and *in vivo* effects of SF2/ASF on SV40 RNA splicing is not yet clear.

5. Evolutionary considerations

While alternative splicing is well established as a widespread mechanism for the regulation of gene expression, many questions remain to be answered regarding the evolution of alternative RNA processing. For one, some gene families comprise a large number of family members without employing alternative RNA processing for the generation of protein isoform diversity; these include intermediate filament proteins, actins, and G proteins. In contrast, other gene families exhibit complex patterns of alternative RNA splicing for the generation of diversity; these include tropomyosins, troponins, and N-CAMs (neuronal cell adhesion molecules). What factors during evolution led to an increase in complexity via alternative RNA splicing, as opposed to expansion of the number of members in a gene family by gene duplication? The explanations for the existence of multigene families include a number of hypotheses encompassing a need for functional redundancy, a requirement for distinct *cis*-elements to regulate gene expression at the level of transcription, and a need for isoform-specific functions in different cell-types. While examples can be found in nature to support each hypothesis, for those genes that employ alternative RNA splicing, the requirement for isoform-specific functions appears to be a major factor for the evolution of this mechanism. The different isoforms that are expressed via alternative RNA splicing appear to be required for the physiological needs of various cells and tissues, although the function of each isoform for a given gene is just beginning to emerge. The expression of a diverse group of isoforms in a highly tissue-specific manner strongly suggests that each isoform may be required to carry out specific cellular functions.

Although alternative RNA splicing is most often associated with generation of protein diversity, many examples found in *Drosophila* indicate that alternative RNA splicing also functions in the on/off control of gene expression. It is not known why these genes are regulated at the post-transcriptional levels rather than simply at the transcriptional level, which would seem to be more 'economical' to the organism. It remains to be determined whether vertebrates also use alternative RNA splicing for the on/off control of gene expression. Another question regarding the evolution of RNA splicing (reviewed in 223), particularly alternative RNA splicing, is how the *cis*-acting elements and cellular factors that are involved in the regulation of alternative splicing 'co-evolved'. In order to regulate a particular splice choice occurring in different tissues and cell types, some cellular factors must also be regulated in a cell type-specific manner. In addition, although many organisms have the same organization of orthologous genes, it is unclear if regulatory mechanisms are conserved across evolutionary boundaries. For example, recent studies of the β-tropomyosin genes from amphibians, birds and mammals demonstrated that proper regulation of the alternative splicing patterns is not maintained when the genes are transferred and expressed in different species (224). These studies suggest that mechanisms of regulation might not be the same among different organisms, even though the genes are organized in a similar fashion. Furthermore, these studies indicate that caution must be exercised with experimental systems involving substrates or factors from different species, although heterologous systems may be useful to identify the factors involved in regulated splicing. Understanding the evolution of RNA processing and alternative RNA splicing will no doubt provide greater insights into the complex questions of molecular evolution.

6. Summary and perspectives

Despite great advances concerning our understanding of splice-site selection, many problems remain. The total number of genes that exhibit alternative RNA splicing is not yet known, and it will be some time before we have a complete picture regarding how many proteins in a cell owe their origins to this process. As regulatory pathways for specific transcripts are elucidated, it will be interesting to determine whether certain regulatory mechanisms of a particular cell type are shared between various pre-mRNAs, and whether certain factors mediate global control of a variety of splicing pathways within one type of cell. In some cases, the same alternative splicing patterns are exibited in different cell types, although it is not known if the same *cis*-acting elements and cellular factors are involved. For example, in the case of the rat β-tropomyosin gene, the same two alternative exons are used in fibroblasts and smooth muscle cells, while the other two alternative exons are used in skeletal and fetal cardiac muscles. Whether the same splice choices are mediated by the same factors in different cell types and at different developmental stages remains to be determined. Since alternative RNA splicing and general splicing are fundamental processes, it is conceivable that some diseases or abnormalities may be associated with changes or defects in the cellular splicing machinery. As the cellular factors involved in RNA splicing are identified and suitable nucleic acid and antibody probes are developed, it will be possible to determine if any splicing component is involved in a genetic disease. This information will be useful in terms of diagnostic and preventive intervention, and in the development of new therapeutic agents. Furthermore, the cell type-specific nature of these processes may offer a target for pharmacological intervention.

Acknowledgements

D.M.H. is the recipient of grants GM43049 and CA58607 from the NIH and is an Established Investigator of the American Heart Association. Y.-C.W. was supported by a postdoctoral fellowship from the American Heart Association.

References

1. Green, M. R. (1991) Biochemical mechanisms of constitutive and regulated pre-mRNA splicing. *Annu. Rev. Cell Biol.*, **7**, 559.
2. Rio, D. C. (1992) RNA processing. *Curr. Opin. Cell Biol.*, **4**, 444.
3. Ahn, A. H. and Kunkel, L. M. (1993) The structural and functional diversity of dystrophin. *Nat. Genet.*, **3**, 283.
4. Breitbart, R. E., Andreadis, A., and Nadal-Ginard, B. (1987) Alternative splicing: A ubiquitous mechanism for the generation of multiple protein isoforms from single genes. *Annu. Rev. Biochem.*, **56**, 467.
5. Smith, C. W. J., Patton, J. G., and Nadal-Ginard, B. (1989) Alternative splicing in the control of gene expression. *Annu. Rev. Genet.*, **23**, 527.
6. Maniatis, T. (1991) Mechanisms of alternative pre-mRNA splicing. *Science*, **251**, 33.
7. McKeown, M. (1992) Alternative RNA splicing. *Annu. Rev. Cell Biol.*, **8**, 133.
8. Norton, P. A. (1994) Alternative pre-mRNA splicing: factors involved in splice site selection. *J. Cell Sci.*, **107**, 1.
9. Helfman, D. M., Ricci, W. M., and Finn, L. A. (1988) Alternative splicing of tropomyosin pre-mRNAs *in vitro* and *in vivo*. *Genes Dev.*, **2**, 1627.
10. Watakabe, A., Sakamoto, H., and Shimura, Y. (1991) Repositioning of an alternative exon sequence of mouse IgM pre-mRNA activates splicing of the preceding intron. *Gene Expression*, **1**, 175.
11. Nasim, F. H., Spears, P. A., Hoffmann, H. M., Kuo, H.-C., and Grabowski, P. J. (1990) A sequential splicing mechanism promotes selection of an optional exon by repositioning a downstream 5′ splice site in preprotachykinin pre-mRNA. *Genes Dev.*, **4**, 1172.
12. Peterson, M. L. and Perry, R. P. (1989) The regulated production of μ_m and μ_s mRNA is dependent on the relative efficiencies of μ_s poly(A) site usage and the Cμ4-to-M1 splice. *Mol. Cell. Biol.*, **9**, 726.
13. Reed, R. and Maniatis, T. (1988) The role of the mammalian branchpoint sequence in pre-mRNA splicing. *Genes Dev.*, **2**, 1268.
14. Zhuang, Y., Goldstein, A. M., and Weiner, A. M. (1989) UACUAAC is the preferred branch site for mammalian mRNA splicing. *Proc. Natl Acad. Sci. USA*, **86**, 2752.
15. Fu, X.-Y., Ge, H., and Manley, J. L. (1988) The role of the polypyrimidine stretch at the SV40 early pre-mRNA 3′ splice site in alternative splicing. *EMBO J.*, **7**, 809.
16. Fu, X.-D., Katz, R. A., Skalka, A. M., and Maniatis, T. (1991) The role of branchpoint and 3′ exon sequences in the control of balanced splicing of avian retrovirus RNA. *Genes Dev.*, **5**, 211.

17. Mullen, M. P., Smith, C. W. J., Patton, J. G., and Nadal-Ginard, B. (1991) α-tropomyosin mutually exclusive exon selection: competition between branchpoint/polypyrimidine tracts determines default exon choice. *Genes Dev.*, **5**, 642.
18. Ruskin, B., Zamore, P. D., and Green, M. R. (1988) A factor, U2AF, is required for U2 snRNP binding and splicing complex assembly. *Cell*, **52**, 207.
19. Singh, R., Valcárcel, J., and Green, M. R. (1995) Distinct binding specificities and functions of higher eukaryotic polypyrimidine tract-binding proteins. *Science*, **268**, 1173.
20. Hoffman, B. E. and Grabowski, P. J. (1992) U1 snRNP targets an essential splicing factor, U2AF65, to the 3′ splice site by a network of interactions spanning the exon. *Genes Dev.*, **6**, 2554.
21. Valcárcel, J., Singh, R., Zamore, P. D., and Green, M. R. (1993) The protein sex-lethal antagonizes the splicing factor U2AF to regulate the alternative splicing of *transformer* pre-mRNA. *Nature*, **362**, 171.
22. Lin, C.-H. and Patton, J. G. (1995) Regulation of alternative 3′ splice site selection by constitutive splicing factors. *RNA*, **1**, 234.
23. Watakabe, A., Tanaka, K., and Shimura, Y. (1993) The role of exon sequences in splice site selection. *Genes Dev.*, **7**, 407.
24. Tsukahara, T., Casciato, C., and Helfman, D. M. (1994) Alternative splicing of β-tropomyosin pre-mRNA: multiple *cis*-elements can contribute to the use of the 5′- and 3′-splice sites of the non-muscle/smooth muscle exon 6. *Nucleic Acids Res.*, **22**, 2318.
25. Cooper, T. A. (1992) *In vitro* splicing of the cardiac troponin T precursors. *J. Biol. Chem.*, **267**, 5330.
26. Caputi, M., Casari, G., Guenzi, S., Tagliabue, R., Sidoli, A., Melo, C. A., *et al.* (1994) A novel bipartite splicing enhancer modulates the differential processing of the human fibronectin EDA exon. *Nucleic Acids Res.*, **22**, 1018.
27. Lavigueur, A., La Branche, H., Kornblihtt, A. R., and Chabot, B. (1993) A splicing enhancer in the human fibronectin alternate ED1 exon interacts with SR proteins and stimulates U2 snRNP binding. *Genes Dev.*, **7**, 2405.
28. Sun, Q., Hampson, R. K., and Rottman, F. M. (1993) *In vitro* analysis of bovine growth hormone pre-mRNA alternative splicing: Involvement of exon sequences and *trans*-acting factor(s). *J. Biol. Chem.*, **268**, 15659.
29. Robberson, B. L., Cote, G. J., and Berget, S. M. (1990) Exon definition may facilitate splice site selection in RNAs with multiple exons. *Mol. Cell. Biol.*, **10**, 84.
30. Grabowski, P. J., Nasim, F. H., Kuo, H.-C., and Burch, R. (1991) Combinatorial splicing of exon pairs by two-site binding of U1 small nuclear ribonucleoprotein particle. *Mol. Cell. Biol.*, **11**, 5919.
31. Kuo, H.-C., Nasim, F. H., and Grabowski, P. J. (1991) Control of alternative splicing by the differential binding of U1 small nuclear ribonucleoprotein particle. *Science*, **251**, 1045.
32. Niwa, M. and Berget, S. M. (1991) Mutation of the AAUAAA polyadenylation signal depresses *in vitro* splicing of proximal but not distal introns. *Genes Dev.*, **5**, 2086.
33. Gattoni, R., Schmitt, P., and Stévenin, J. (1988) *In vitro* splicing of adenovirus E1A transcripts: characterization of novel reactions and of multiple branch points abnormally far from the 3′ splice site. *Nucleic Acids Res.*, **16**, 2389.
34. Helfman, D. M. and Ricci, W. M. (1989) Branch point selection in alternative splicing of tropomyosin pre-mRNAs. *Nucleic Acids Res.*, **17**, 5633.
35. Goux-Pelletan, M., Libri, D., d'Aubenton-Carafa, Y., Fiszman, M., Brody, E., and Marie, J. (1990) *In vitro* splicing of mutually exclusive exons from the chicken β-tropomyosin

gene: role of the branch point location and very long pyrimidine stretch. *EMBO J.*, **9**, 241.
36. Smith, C. W. J. and Nadal-Ginard, B. (1989) Mutually exclusive splicing of α-tropomyosin exons enforced by an unusual lariat branch point location: Implications for constitutive splicing. *Cell*, **56**, 749.
37. Helfman, D. M., Roscigno, R. F., Mulligan, G. J., Finn, L. A., and Weber, K. S. (1990) Identification of two distinct intron elements involved in alternative splicing of β-tropomyosin pre-mRNA. *Genes Dev.*, **4**, 98.
38. Libri, D., Goux-Pelletan, M., Brody, E., and Fiszman, M. Y. (1990) Exon as well as intron sequences are *cis*-regulating elements for the mutually exclusive alternative splicing of the β-tropomyosin gene. *Mol. Cell. Biol.*, **10**, 5036.
39. Guo, W., Mulligan, G. J., Wormsley, S., and Helfman, D. M. (1991) Alternative splicing of β-tropomyosin pre-mRNA: cis-acting elements and cellular factors that block the use of a skeletal muscle exon in nonmuscle cells. *Genes Dev.*, **5**, 2096.
40. Norton, P. A. and Hynes, R. O. (1990) *In vitro* splicing of fibronectin pre-mRNAs. *Nucleic Acids Res.*, **18**, 4089.
41. Spikes, D. and Bingham, P. M. (1992) Analysis of spliceosome assembly and the structure of a regulated intron in *Drosophila in vitro* splicing extracts. *Nucleic Acids Res.*, **20**, 5719.
42. Khoury, G., Gruss, P., Dhar, R., and Lai, C. J. (1979) Processing and expression of early SV40 mRNA: a role for RNA conformation in splicing. *Cell*, **18**, 85.
43. Munroe, S. H. (1984) Secondary structure of splice sites in adenovirus mRNA precursors. *Nucleic Acids Res.*, **12**, 8437.
44. Solnick, D. (1985) Alternative splicing caused by RNA secondary structure. *Cell*, **43**, 667.
45. Eperon, L. P., Graham, I. R., Griffiths, A. D., and Eperon, I. C. (1988) Effects of RNA secondary structure on alternative splicing of pre-mRNA: Is folding limited to a region behind the transcribing RNA polymerase? *Cell*, **54**, 393.
46. Edlind, T. D., Cooley, T. E., and Ihler, G. M. (1987) A conserved base pairing involving an alternative splice site of SV40 and polyoma late RNA. *Nucleic Acids Res.*, **15**, 8566.
47. Libri, D., Piseri, A., and Fiszman, M. Y. (1991) Tissue specific splicing *in vivo* of the β-tropomyosin gene: dependence on an RNA secondary structure. *Science*, **252**, 1842.
48. Clouet-d'Orval, B., d'Aubenton-Carafa, Y., Sirand-Pugnet, P., Gallego, M., Brody, E., and Marie, J. (1991) RNA secondary structure repression of a muscle-specific exon in HeLa cell nuclear extracts. *Science*, **252**, 1823.
49. Libri, D., Lemonnier, M., Meinnel, T., and Fiszman, M. Y. (1989) A single gene codes for the β subunits of smooth and skeletal muscle tropomyosin in the chicken. *J. Biol. Chem.*, **264**, 2935.
50. Libri, D., Balvay, L., and Fiszman, M. Y. (1992) *In vivo* splicing of the β-tropomyosin pre-mRNA: a role for branch point and donor site competition. *Mol. Cell. Biol.*, **12**, 3204.
51. Séraphin, B., Simon, M., Boulet, A., and Faye, G. (1989) Mitochondrial splicing requires a protein from a novel helicase family. *Nature*, **337**, 84.
52. Thomas, J. O., Glowacka, S. K., and Szer, W. (1983) Structure of complexes between a major protein of heterogeneous nuclear ribonucleoprotein particles and polyribonucleotides. *J. Mol. Biol.*, **171**, 439.
53. Valentini, O., Biamonti, G., Pandolfo, M., Morandi, C., and Riva, S. (1985) Mammalian single-stranded DNA binding proteins and heterogeneous nuclear RNA proteins have common antigenic determinants. *Nucleic Acids Res.*, **13**, 337.

54. Dreyfuss, G. (1986) Structure and function of nuclear and cytoplasmic ribonucleoprotein particles. *Annu. Rev. Cell Biol.*, **2**, 459.
55. Mermoud, J. E., Calvio, C., and Lamond, A. I. (1994) Uncovering the role of Ser/Thr protein phosphorylation in nuclear pre-mRNA splicing. *Adv. Prot. Phosphatases*, **8**, 99.
56. Cardinali, B., Cohen, P. T., and Lamond, A. I. (1994) Protein phosphatase 1 can modulate alternative 5′ splice site selection in a HeLa splicing extract. *FEBS Lett.*, **352**, 276.
57. Cáceres, J. F. and Krainer, A. R. (1993) Functional analysis of pre-mRNA splicing factor SF2/ASF structural domains. *EMBO J.*, **12**, 4715.
58. Krainer, A. R., Conway, G. C., and Kozak, D. (1990) Purification and characterization of pre-mRNA splicing factor SF2 from HeLa cells. *Genes Dev.*, **4**, 1158.
59. Ge, H. and Manley, J. L. (1990) A protein factor, ASF, controls cell-specific alternative splicing of SV40 early pre-mRNA *in vitro*. *Cell*, **62**, 25.
60. Champlin, D. T., Frasch, M., Saumweber, H., and Lis, J. T. (1991) Characterization of a *Drosophila* protein associated with boundaries of transcriptionally active chromatin. *Genes Dev.*, **5**, 1611.
61. Fu, X.-D. and Maniatis, T. (1992) Isolation of a complementary DNA that encodes the mammalian splicing factor SC35. *Science*, **256**, 535.
62. Zahler, A. M., Lane, W. S., Stolk, J. A., and Roth, M. B. (1992) SR proteins: a conserved family of pre-mRNA splicing factors. *Genes Dev.*, **6**, 837.
63. Zahler, A. M., Neugebauer, K. M., Stolk, J. A., and Roth, M. B. (1993) Human SR proteins and isolation of cDNA encoding SRp75. *Mol. Cell. Biol.*, **13**, 4023.
64. Kim, S.-H., Smith, J., Claude, A., and Lin, R.-J. (1992) The purified yeast pre-mRNA splicing factor PRP2 is an RNA-dependent NTPase. *EMBO J.*, **11**, 2319.
65. Screaton, G. R., Cáceres, J. F., Mayeda, A., Bell, M. V., Plebanski, M., Jackson, D. G., *et al.* (1995) Identification and characterization of three members of the human SR family of pre-mRNA splicing factors. *EMBO J.*, **14**, 4336.
66. Krainer, A. R., Conway, G. C., and Kozak, D. (1990) The essential pre-mRNA splicing factor SF2 influences 5′ splice site selection by activating proximal sites. *Cell*, **62**, 35.
67. Krainer, A. R., Mayeda, A., Kozak, D., and Binns, G. (1991) Functional expression of cloned human splicing factor SF2: homology to RNA-binding proteins, U1 70K, and *Drosophila* splicing regulators. *Cell*, **66**, 383.
68. Ge, H., Zuo, P., and Manley, J. L. (1991) Primary structure of the human splicing factor ASF reveals similarities with *Drosophila* regulators. *Cell*, **66**, 373.
69. Fu, X.-D. and Maniatis, T. (1992b) The 35-kDa mammalian splicing factor SC35 mediates specific interactions between U1 and U2 small nuclear ribonucleoprotein particles at the 3′ splice site. *Proc. Natl Acad. Sci. USA*, **89**, 1725.
70. Zahler, A. M., Neugebauer, K. M., Lane, W. S., and Roth, M. B. (1993) Distinct functions of SR proteins in alternative pre-mRNA splicing. *Science*, **260**, 219.
71. Mayeda, A. and Krainer, A. R. (1992) Regulation of alternative pre-mRNA splicing by hnRNP A1 and splicing factor SF2. *Cell*, **68**, 365.
72. Mayeda, A., Helfman, D. M., and Krainer, A. R. (1993) Modulation of exon skipping and inclusion by heterogeneous nuclear ribonucleoprotein A1 and pre-mRNA splicing factor SF2/ASF. *Mol. Cell. Biol.*, **13**, 2993.
73. Cáceres, J. F., Stamm, S., Helfman, D. M., and Krainer, A. R. (1994) Regulation of alternative splicing *in vivo* by overexpression of antagonistic splicing factors. *Science*, **265**, 1706.
74. Graham, I. R., Hamshere, M., and Eperon, I. C. (1992) Alternative splicing of a human α-tropomyosin muscle-specific exon: identification of determining sequences. *Mol. Cell. Biol.*, **12**, 3872.

75. Xu, R., Teng, J., and Cooper, T. A. (1993) The cardiac troponin T alternative exon contains a novel purine-rich positive splicing element. *Mol. Cell. Biol.*, **13**, 3660.
76. Dirksen, W. P., Hampson, R. K., Sun, Q., and Rottman, F. M. (1994) A purine-rich exon sequence enhances alternative splicing of bovine growth hormone pre-mRNA. *J. Biol. Chem.*, **269**, 6431.
77. Ramchatesingh, J., Zahler, A. M., Neugebauer, K. M., Roth, M. B., and Cooper, T. A. (1995) A subset of SR proteins activates splicing of the cardiac troponin T alternative exon by direct interactions with an exonic enhancer. *Mol. Cell. Biol.*, **15**, 4898.
78. Tian, M. and Maniatis, T. (1993) A splicing enhancer complex controls alternative splicing of *doublesex* pre-mRNA. *Cell*, **74**, 105.
79. Wu, J. Y. and Maniatis, T. (1993) Specific interactions between proteins implicated in splice site selection and regulated alternative splicing. *Cell*, **75**, 1061.
80. Zamore, P. D. and Green, M. R. (1991) Biochemical characterization of U2 snRNP auxiliary factor: an essential pre-mRNA splicing factor with a novel intranuclear distribution. *EMBO J.*, **10**, 207.
81. Zamore, P. D., Patton, J. G., and Green, M. R. (1992) Cloning and domain structure of the mammalian splicing factor U2AF. *Nature*, **355**, 609.
82. Zhang, M., Zamore, P. D., Carmo-Fonseca, M., Lamond, A. I., and Green, M. R. (1992) Cloning and intracellular localization of the U2 small nuclear ribonucleoprotein auxillary factor small subunit. *Proc. Natl Acad. Sci. USA*, **89**, 8769.
83. Dreyfuss, G., Matunis, M. J., Piñol-Roma, S., and Burd, C. G. (1993) hnRNP proteins and the biogenesis of mRNA. *Annu. Rev. Biochem.*, **62**, 289.
84. Swanson, M. S. and Dreyfuss, G. (1988) RNA binding specificity of hnRNP proteins: a subset bind to the 3′ end of introns. *EMBO J.*, **7**, 3519.
85. Buvoli, M., Cobianchi, F., Biamonti, G., and Riva, S. (1990) Recombinant hnRNP protein A1 and its N-terminal domain show preferential affinity for oligodeoxynucleotides homologous to intron/exon acceptor sites. *Nucleic Acids Res.*, **18**, 6595.
86. Bennett, M., Piñol-Roma, S., Staknis, D., Dreyfuss, G., and Reed, R. (1992) Differential binding of heterogeneous nuclear ribonucleoproteins to mRNA precursors prior to spliceosome assembly *in vitro*. *Mol. Cell. Biol.*, **12**, 3165.
87. Yang, X., Bani, M. R., Lu, S. J., Rowan, S. J., Ben-David, Y., and Chabot, B. (1994) The A1 and $A1^B$ proteins of heterogeneous nuclear ribonucleoparticles modulate 5′ splice site selection *in vivo*. *Proc. Natl Acad. Sci. USA*, **91**, 6924.
88. Shen, J., Zu, K., Cass, C. L., Beyer, A. L., and Hirsh, J. (1995) Exon skipping by overexpression of a *Drosophila* heterogeneous nuclear ribonucleoprotein *in vivo*. *Proc. Natl Acad. Sci. USA*, **92**, 1822.
89. Mayeda, A., Munroe, S. H., Cáceres, J. F., and Krainer, A. R. (1994) Function of conserved domains of hnRNP A1 and other hnRNP A/B proteins. *EMBO J.*, **13**, 5483.
90. Ghetti, A., Piñol-Roma, S., Michael, W. M., Morandi, C., and Dreyfuss, G. (1992) hnRNP I, the polypyrimidine tract-binding protein: distinct nuclear localization and association with hnRNAs. *Nucleic Acids Res.*, **20**, 3671.
91. Garcia-Blanco, M. A., Jamison, S. F., and Sharp, P. A. (1989) Identification and purification of a 62,000-dalton protein that binds specifically to the polypyrimidine tract of introns. *Genes Dev.*, **3**, 1874.
92. Bothwell, A. L., Ballard, D. W., Philbrick, W. M., Lindwall, G., Maher, S. E., Bridgett, M. M., *et al.* (1991) Murine polypyrimidine tract binding protein. Purification, cloning, and mapping of the RNA binding domain. *J. Biol. Chem.*, **266**, 24567.
93. Gil, A., Sharp, P. A., Jamison, S. F., and Garcia-Blanco, M. A. (1991) Characterization of cDNAs encoding the polypyrimidine tract-binding protein. *Genes Dev.*, **5**, 1224.

94. Patton, J. G., Mayer, S. A., Tempst, P., and Nadal-Ginard, B. (1991) Characterization and molecular cloning of polypyrimidine tract-binding protein: a component of a complex necessary for pre-mRNA splicing. *Genes Dev.*, **5**, 1237.
95. Wang, J. and Pederson, T. (1990) A 62,000 molecular weight spliceosome protein crosslinks to the intron polypyrimidine tract. *Nucleic Acids Res.*, **18**, 5995.
96. Patton, J. G., Porro, E., Galceran, J., Tempst, P., and Nadal-Ginard, B. (1993) Cloning and characterization of PSF, a novel pre-mRNA splicing factor. *Genes Dev.*, **7**, 393.
97. Mulligan, G. J., Guo, W., Wormsley, S., and Helfman, D. M. (1992) Polypyrimidine tract binding protein interacts with sequences involved in alternative splicing of β-tropomyosin pre-mRNA. *J. Biol. Chem.*, **267**, 25480.
98. Boggs, R. T., Gregor, P., Idriss, S., Belote, J. M., and McKeown, M. (1987) Regulation of sexual differentiation in *D. melanogaster* via alternative splicing of RNA from the *transformer* gene. *Cell*, **50**, 739.
99. Bell, L. R., Maine, E. M., Schedl, P., and Cline, T. W. (1988) Sex-lethal, a *Drosophila* sex determination switch gene, exhibits sex-specific RNA splicing and sequence similarity to RNA binding proteins. *Cell*, **55**, 1037.
100. Burtis, K. C. and Baker, B. S. (1989) *Drosophila doublesex* gene controls somatic sexual differentiation by producing alternatively spliced mRNAs encoding related sex-specific polypeptides. *Cell*, **56**, 997.
101. Baker, B. S. (1989) Sex in flies: the splice of life. *Nature*, **340**, 521.
102. Bell, L. R., Horabin, J. I., Schedl, P., and Cline, T. W. (1991) Positive autoregulation of sex-lethal by alternative splicing maintains the female determined state in *Drosophila*. *Cell*, **65**, 229.
103. Keyes, L. N., Cline, T. W., and Schedl, P. (1992) The primary sex determination signal of *Drosophila* acts at the level of transcription. *Cell*, **68**, 933.
104. Horabin, J. I. and Schedl, P. (1993) Regulated splicing of the *Drosophila* sex-lethal male exon involves a blockage mechanism. *Mol. Cell. Biol.*, **13**, 1408
105. Granadino, B., Campuzano, S., and Sanchez, L. (1990) The *Drosophila melanogaster* fl(2)d gene is needed for the female-specific splicing of Sex-lethal RNA. *EMBO J.*, **9**, 2597.
106. Sakamoto, H., Inoue, K., Higuchi, I., Ono, Y., and Shimura, Y. (1992) Control of *Drosophila* Sex-lethal pre-mRNA splicing by its own female-specific product. *Nucleic Acids Res.*, **20**, 5533.
107. McKeown, M., Belote, J. M., and Boggs, R. T. (1988) Ectopic expression of the female transformer gene product leads to female differentiation of chromosomally male *Drosophila*. *Cell*, **48**, 489.
108. Sosnowski, B. A., Belote, J. M., and McKeown, M. (1989) Sex-specific alternative splicing of RNA from the transformer gene results from sequence-dependent splice site blockage. *Cell*, **58**, 449.
109. Inoue, K., Hoshijima, K., Sakamoto, H., and Shimura, Y. (1990) Binding of the *Drosophila* sex-lethal gene product to the alternative splice site of transformer primary transcript. *Nature*, **344**, 461.
110. Nagoshi, R. N., Mckeown, M., Burtis, K. C., Belote, J. M., and Baker, B. S. (1988) The control of alternative splicing at genes regulating sexual differentiation in *D. melanogaster*. *Cell*, **53**, 229.
111. Hoshijima, K., Inoue, K., Higuchi, I., Sakamoto, H., and Shimura, Y. (1991) Control of doublesex alternative splicing by transformer and transformer-2 in *Drosophila*. *Science*, **252**, 833.
112. Hedley, M. L. and Maniatis, T. (1991) Sex-specific splicing and polyadenylation of *dsx* pre-mRNA requires a sequence that binds specifically to tra-2 protein *in vitro*. *Cell*, **65**, 579.

113. Ryner, L. C. and Baker, B. S. (1991) Regulation of doublesex pre-mRNA processing occurs by 3′-splice site activation. *Genes Dev.*, **5**, 2071.
114. Tian, M. and Maniatis, T. (1992) Positive control of pre-mRNA splicing *in vitro*. *Science*, **256**, 237.
115. Bandziulis, R. J., Swanson, M. S., and Dreyfuss, G. (1989) RNA-binding proteins as developmental regulators. *Genes Dev.*, **3**, 431.
116. Amrein, H., Gorman, M., and Nothiger, R. (1988) The sex determining gene tra-2 of *Drosophila* encodes a putative RNA binding protein. *Cell*, **55**, 1025.
117. Goralski, T. J., Edstrom, J. E., and Baker, B. S. (1989) The sex determination locus transformer-2 of *Drosophila* encodes a polypeptide with similarity to RNA binding proteins. *Cell*, **56**, 1011.
118. Amrein, H., Hedley, M. L., and Maniatis, T. (1994) The role of specific protein–RNA and protein–protein interactions in positive and negative control of pre-mRNA splicing by transformer 2. *Cell*, **76**, 735.
119. Tian, M. and Maniatis, T. (1994) A splicing enhancer exhibits both constitutive and regulated activities. *Genes Dev.*, **8**, 1703.
120. Laski, F. A., Rio, D. C., and Rubin, G. M. (1986) Tissue specificity of *Drosophila* P-element transposition is regulated at the level of mRNA splicing. *Cell*, **44**, 7.
121. Robertson, H. M. and Engels, W. R. (1989) Modified P elements that mimic the P cytotype in *Drosophila melanogaster*. *Genetics*, **123**, 815.
122. Misra, S. and Rio, D. C. (1990) Cytotype control of *Drosophila* P-element transposition: The 66 kd protein is a repressor of transposase activity. *Cell*, **62**, 269.
123. Laski, F. A. and Rubin, G. M. (1989) Analysis of the *cis*-acting requirements for germ-line specific splicing of the P-element ORF2-ORF3 intron. *Genes Dev.*, **3**, 720.
124. Siebel, C. W. and Rio, D. C. (1990) Regulated splicing of the *Drosophila* P transposable element third intron *in vitro*: somatic repression. *Science*, **248**, 1200.
125. Siebel, C. W., Fresco, L. D., and Rio, D. C. (1992) The mechanism of somatic inhibition of *Drosophila* P-element pre-mRNA splicing: multiprotein complexes at an exon pseudo-5′ splice site control U1 snRNP binding. *Genes Dev.*, **6**, 1386.
126. Chain, A. C., Zollmann, S., Tseng, J. C., and Laski, F. A. (1991) Identification of a *cis*-acting sequence required for germ line-specific splicing of the P-element ORF2-ORF3 intron. *Mol. Cell. Biol.*, **11**, 1538.
127. Tseng, J. C., Zollmann, S., Chain, A. C., and Laski, F. A. (1991) Splicing of the *Drosophila* P element ORF2-ORF3 intron is inhibited in a human cell extract. *Mechanisms Dev.*, **35**, 65.
128. Siebel, C. W., Kanaar, R., and Rio, D. C. (1994) Regulation of tissue-specific P-element pre-mRNA splicing requires the RNA-binding protein PSI. *Genes Dev.*, **8**, 1713.
129. Siebel, C. W., Admon, A., and Rio, D. C. (1995) Soma-specific expression and cloning of PSI, a negative regulator of P-element pre-mRNA splicing. *Genes Dev.*, **9**, 269.
130. Bingham, P. M. and Judd, B. H. (1981) A copy of the copia transposable element is very tightly linked to the w(a) allele at the white locus of *D. Melanogaster*. *Cell*, **25**, 705.
131. Pirrotta, V. and Brockl, C. (1984) Transcription of the *Drosophila* white locus and some of its mutants. *EMBO J.*, **3**, 563.
132. Levis, R., O'Hare, K., and Rubin, G. M. (1984) Effects of transposable element insertion on RNA encoded by the white gene of *Drosphila*. *Cell*, **38**, 471.
133. Zachar, Z., Chou, T. B., and Bingham, P. M. (1987) Evidence that a regulatory gene autoregulates splicing of its transcript. *EMBO J.*, **6**, 4105.
134. Chou, T. B., Zachar, Z., and Bingham, P. M. (1987) Developmental expression of a regulatory gene is programmed at the level of splicing. *EMBO J.*, **6**, 4095.

135. Li, H. and Bingham, P. M. (1991) Arginine/serine-rich domains of the su(wa) and tra RNA processing regulators target proteins to a subnuclear compartment implicated in splicing. *Cell*, **67**, 335.
136. Fu, X.-D. and Maniatis, T. (1990) Factor required for mammalian spliceosome assembly is localized to discrete regions in the nucleus. *Nature*, **343**, 437.
137. Spector, D. L. (1990) Higher order nuclear organization: three-dimensional distribution of small nuclear ribonucleoprotein particles. *Proc. Natl Acad. Sci. USA*, **87**, 147.
138. Lees-Miller, J. P. and Helfman, D. M. (1991) The molecular basis for tropomyosin isoform diversity. *BioEssays*, **13**, 429.
139. Helfman, D. M., Cheley, S., Kuismanen, E., Finn, L. A., and Yamawaki-Kataoka, Y. (1986) Nonmuscle and muscle tropomyosin isoforms are expressed from a single gene by alternative RNA splicing and polyadenylation. *Mol. Cell. Biol.*, **6**, 3582.
140. Guo, W. and Helfman, D. M. (1993) *Cis*-elements involved in alternative splicing in the rat β-tropomyosin gene: the 3′-splice site of the skeletal muscle exon 7 is the major site of blockage in nonmuscle cells. *Nucleic Acids Res.*, **21**, 4762.
141. Gooding, C., Roberts, G. C., Moreau, G., Nadal-Ginard, B., and Smith, C. W. J. (1994) Smooth muscle-specific switching of α-tropomyosin mutually exclusive exon selection by specific inhibition of the strong default exon. *EMBO J.*, **13**, 3861.
142. Cooper, T. A. and Ordahl, C. P. (1985) A single cardiac troponin T gene generates embryonic and adult isoforms via developmentally regulated alternate splicing. *J. Biol. Chem.*, **260**, 11140.
143. Adelstein, R. S. and Eisenberg, E. (1980) Regulation and kinetics of the actin–myosin–ATP interaction. *Annu. Rev. Biochem.*, **49**, 921.
144. Cooper, T. A., Cardone, M. H., and Ordahl, C. P. (1988) *Cis* requirements for alternative splicing of the cardiac troponin T pre-mRNA. *Nucleic Acids Res.*, **16**, 8443.
145. Cooper, T. A. and Ordahl, C. P. (1989) Nucleotide substitutions within the cardiac troponin T alternative exon disrupt pre-mRNA alternative splicing. *Nucleic Acids Res.*, **17**, 7905.
146. Guise, J. W., Galli, G., Nevins, J. R., and Tucker, P. W. (1989) Developmental regulation of secreted and membrane forms of immunoglobulin mu chain. In *Immunoglobulin genes*. Honjo, T. *et al.* (ed.). Academic Press, New York, p. 275.
147. Peterson, M. L. (1994) RNA processing and expression of immunoglobulin genes. In *Handbook of B and T lymphocytes*, Snow, E. C. (ed.). Academic Press, San Diego, p. 321.
148. Yuan, D. and Dang, T. (1989) Regulation of μ_m vs μ_s mRNA expression in an inducible B cell line. *Mol. Immunol.*, **26**, 1059.
149. Weiss, E. A., Michael, A., and Yuan, D. (1989) Role of transcriptional termination in the regulation of μ mRNA expression in B lymphocytes. *J. Immunol.*, **143**, 1046.
150. Kelley, D. E. and Perry, R. P. (1986) Transcriptional and posttranscriptional control of immunoglobulin mRNA production during B lymphocyte development. *Nucleic Acids Res.*, **14**, 5431.
151. Ruether, J. E., Maderious, A., Lavery, D., Logan, J., Fu, S. M., and Chen-Kiang, S. (1986) Cell-type-specific synthesis of murine immunoglobulin μ RNA from an adenovirus vector. *Mol. Cell. Biol.*, **6**, 123.
152. Galli, G., Guise, J. W., McDevitt, M. A., Tucker, P. W., and Nevins, J. R. (1987) Relative position and strengths of poly(A) sites as well as transcription termination are critical to membrane versus secreted μ-chain expression during B-cell development. *Genes Dev.*, **1**, 471.
153. Guise, J. W., Lim, P. L., Yuan, D., and Tucker, P. W. (1988b) Alternative expression of

secreted and membrane forms of immunoglobulin μ-chain is regulated by transcriptional termination in stable plasmacytoma transfectants. *J. Immunol.*, **140**, 3988.

154. Law, R., Kuwabara, M. D., Briskin, M., Fasel, N., Hermanson, G., Sigman, D. *et al.* (1987) Protein binding site at the immunoglobulin $\mu_{membrane}$ polyadenylation signal: possible role in transcription termination. *Proc. Natl Acad. Sci. USA*, **84**, 9160.
155. Tisch, R., Kondo, N., and Hozumi, N. (1990) Parameters that govern the regulation of immunoglobulin δ heavy-chain gene expression. *Mol. Cell. Biol.*, **10**, 5340.
156. Danner, D. and Leder, P. (1985) Role of an RNA cleavage/poly(A) addition site in the production of membrane-bound and secreted IgM mRNA. *Proc. Natl Acad. Sci. USA*, **82**, 8658.
157. Nishikura, K. and Vuocolo, G. A. (1984) Synthesis of two mRNAs by utilization of alternative polyadenylation sites: expression of SV40-mouse immunoglobulin μ chain gene recombinants in Cos monkey cells. *EMBO J.*, **3**, 689.
158. Peterson, M. L., Bryman, M. B., Peiter, M., and Cowan, C. (1994) Exon size affects competition between splicing and cleavage-polyadenylation in the immunoglobulin gene. *Mol. Cell. Biol.*, **14**, 77.
159. Yan, D.-H., Weiss, E. A., and Nevins, J. R. (1995) Identification of an activity in B-cell extracts that selectively impairs the formation of an immunoglobulin μ_s poly(A) site processing complex. *Mol. Cell. Biol.*, **15**, 1901.
160. Tanaka, K., Watakabe, A., and Shimura, Y. (1994) Polypurine sequences within a downstream exon function as a splicing enhancer. *Mol. Cell. Biol.*, **14**, 1347.
161. Hynes, R. O. (1990) *Fibronectins.* Springer-Verlag, New York.
162. Ffrench-Constant, C. (1995) Alternative splicing of fibronectin-many different proteins but few different functions. *Exp. Cell Res.*, **221**, 261.
163. Mardon, H. J., Sebastio, G., and Baralle, F. E. (1987) A role for exon sequences in alternative splicing of the human fibronectin gene. *Nucleic Acids Res.*, **15**, 7725.
164. Huh, G. S. and Hynes, R. O. (1993) Elements regulating an alternatively spliced exon of the rat fibronectin gene. *Mol. Cell. Biol.*, **13**, 5301.
165. Amara, S. G., Jonas, V., Rosenfeld, M. G., Ong, E. S., and Evans, R. M. (1982) Alternative RNA processing in calcitonin gene expression generates mRNAs encoding different polypeptide products. *Nature*, **298**, 240.
166. Rosenfeld, M. G., Mermod, J., Amara, S. G., Swanson, L. W., Sawchenco, P. E., Rivier, J., *et al.* (1983) Production of a novel neuropeptide encoded by the calcitonin gene via tissue-specific RNA processing. *Nature*, **304**, 129.
167. Zaidi, M., Moonga, B. S., Bevis, P. J., Alam, A. S., Legon, S., Wimalawansa, S., *et al.* (1991) Expression and function of the calcitonin gene products. *Vit. Horm.*, **46**, 87.
168. Hökfelt, T., Arvidsson, U., Ceccatelli, S., Cortes, R., Cullheim, S., Dagerlind, A., *et al.* (1992) Calcitonin gene-related peptide in the brain, spinal cord, and some peripheral systems. *Ann. N. Y. Acad. Sci.*, **657**, 119.
169. Bovenberg, R. A., Adema, G. J., Jansz, H. S., and Bass, P. D. (1988) Model for tissue-specific calcitonin/CGRP-I RNA processing from *in vitro* experiments. *Nucleic Acids Res.*, **16**, 7867.
170. Bovenberg, R. A., Moen, T. C., Jansz, H. S., and Bass, P. D. (1989) *In vitro* splicing analysis of mini-gene constructs of the alternatively processed human calcitonin/CGRP-I pre-mRNA. *Biochem. Biophys. Acta*, **1008**, 223.
171. Cote, G. J., Nguyen, I. N., Lips, C. J. M., Berget, S. M., and Gagel, R. F. (1991) Validation of an *in vitro* RNA processing system for CT/CGRP precursor mRNA. *Nucleic Acids Res.*, **19**, 3601.

172. Leff, S. E., Evans, R. M., and Rosenfeld, M. G. (1987) Splice commitment dictates neuron-specific alternative RNA processing in calcitonin/CGRP gene expression. *Cell*, **48**, 517.
173. Emeson, R. B., Hedjran, F., Yeakley, J. M., Guise, J. W., and Rosenfeld, M. J. (1989) Alternative production of calcitonin and CGRP mRNA is regulated at the calcitonin-specific splice acceptor. *Nature*, **341**, 76.
174. Adema, G. J., Bovenberg, R. A. L., Jansz, H. S., and Baas, P. D. (1988) Unusual branch point selection involved in splicing of the alternatively processed calcitonin/CGRP-1 pre-mRNA. *Nucleic Acids Res.*, **16**, 9513.
175. Adema, G. J., van Hulst, K. L., and Baas, P. D. (1990) Uridine branch acceptor is a *cis*-acting element involved in regulation of the alternative processing of calcitonin/CGRP-1 pre-mRNA. *Nucleic Acids Res.*, **18**, 5365.
176. Yeakley, J. M., Hedjran, F., Morfin, J.-P., Merillat, N., Rosenfeld, M. G., and Emeson, R. B. (1993) Control of calcitonin/calcitonin gene-related peptide pre-mRNA processing by constitutive intron and exon elements. *Mol. Cell. Biol.*, **13**, 5999.
177. Cote, G. J., Stolow, D. T., Peleg, S., Berget, S. M., and Gagel, R. F. (1992) Identification of exon sequences and an exon binding protein involved in alternative RNA splicing of calcitonin/CGRP. *Nucleic Acids Res.*, **20**, 2361.
178. van Oers, C. C. M., Adema, G. J., Zandberg, H., Moen, T. C., and Baas, P. D. (1994) Two different sequence elements within exon 4 are necessary for calcitonin-specific splicing of the human calcitonin/calcitonin gene-related peptide 1 pre-mRNA. *Mol. Cell. Biol.*, **14**, 951.
179. Lou, H., Yang, Y., Cote, G. J., Berget, S. M., and Gagel, R. F. (1995) An intron enhancer containing a 5′ splice site sequence in the human calcitonin/calcitonin gene-related peptide gene. *Mol. Cell. Biol.*, **15**, 7135.
180. Nawa, H., Kotani, H., and Nakanishi, S. (1984) Tissue-specific generation of two preprotachykinin mRNAs from one gene by alternative RNA splicing. *Nature*, **312**, 729.
181. Krause, J. E., Chirgwin, J. M., Carter, M. S., Xu, Z. S., and Hershey, A. D. (1987) Three rat preprotachykinin mRNAs encode the neuropeptides substance P and neurokinin A. *Proc. Natl Acad. Sci. USA*, **84**, 881.
182. Levy, J. B., Dorai, T., Wang, L. H., and Brugge, J. S. (1987) The structurally distinct form of pp60$^{c\text{-}src}$ detected in neuronal cells is encoded by a unique c-*src* mRNA. *Mol. Cell. Biol.*, **7**, 4142.
183. Martinez, R., Mathey-Prevot, B., Bernards, A., and Baltimore, D. (1987) Neuronal pp60$^{c\text{-}src}$ contains a six-amino acid insertion relative to its non-neuronal counterpart. *Science*, **237**, 411.
184. Pyper, J. M. and Bolen, J. B. (1990) Identification of a novel neuronal c-*src* exon expressed in human brain. *Mol. Cell. Biol.*, **10**, 2035.
185. Black, D. L. (1992) Activation of c-*src* neuron-specific splicing by an unusual RNA element *in vivo* and *in vitro*. *Cell*, **69**, 795.
186. Black, D. L. (1991) Does steric interference between splice sites block the splicing of a short c-*src* neuron-specific exon in non-neuronal cells? *Genes Dev.*, **5**, 389.
187. Min, H., Chan, R. C., and Black, D. L. (1995) The generally expressed hnRNP F is involved in a neural-specific pre-mRNA splicing event. *Genes Dev.*, **9**, 2659.
188. Chan, R. C. and Black, D. L. (1995) Conserved intron elements repress splicing of a neuron-specific c-*src* exon *in vitro*. *Mol. Cell. Biol.*, **15**, 6377.
189. Hampson, R. K., La Follette, L., and Rottman, F. M. (1989) Alternative processing of bovine growth hormone mRNA is influenced by downstream exon sequences. *Mol. Cell. Biol.*, **9**, 1604.

190. Liebhaber. S. A., Urbanek, M., Ray, J., Tuan, R. S., and Cooke, N. E. (1989) Characterization and histologic localization of human growth hormone-variant gene expression in the placenta. *J. Clin. Invest.*, **83**, 1985.
191. Sun, Q., Mayeda, A., Hampson, R. K., Krainer, A. R., and Rottman, F. M. (1993) General splicing factor SF2/ASF promotes alternative splicing by binding to an exonic splicing enhancer. *Genes Dev.*, **7**, 2598.
192. DeNoto, F. M., Moore, D. C., and Goodman, H. M. (1981) Human growth hormone DNA sequence and mRNA structure: possible alternative splicing. *Nucleic Acids Res.*, **9**, 3719.
193. Cooke, N. E., Ray, J., Watson, M. A., Estes, P. A., Kuo, B. A., and Liebhaber, S. A. (1988) Human growth hormone gene and the highly homologous growth hormone variant gene display different splicing patterns. *J. Clin. Invest.*, **82**, 270.
194. Lewis, U. J., Singh, R. N. P., Tutwiler, G. F., Sigel, M. B., VanderLaan, E. F., and VanderLaan, W. P. (1980) Human growth hormone: a complex of proteins. *Rec. Prog. Hormone Res.*, **36**, 477.
195. Estes, P. A., Cooke, N. E., and Liebhaber, S. A. (1992) A native RNA secondary structure controls alternative splice-site selection and generates two human growth hormone isoforms. *J. Biol. Chem.*, **267**, 14902.
196. Imperiale, M. J., Akusjärvi, G., and Leppard, K. N. (1995) Post-transcriptional control of adenovirus gene expression. *Curr. Top. Microbiol. Immunol.*, **199**, 139.
197. Coffin, J. M. (1990) Retroviridae and their replication. In *Field's virology*, Fields, B. N. and Knipe, D. M. (ed.). Raven Press, New York, p. 1437.
198. Stoltzfus, C. M. (1988) Synthesis and processing of avian sarcoma retrovirus RNA. *Adv. Virus Res.*, **35**, 1.
199. Rosenblatt, J. D., Miles, S., Gasson, J. C., and Prager, D. (1995) *Curr. Top. Microbiol. Immunol.*, **193**, 25.
200. Chang, D. D. and Sharp, P. A. (1989) Regulation by HIV-1 *rev* depends upon recognition of splice sites. *Cell*, **59**, 789.
201. Katz, R. A. and Skalka, A. M. (1990) Control of retroviral RNA splicing through maintenance of suboptimal processing signals. *Mol. Cell. Biol.*, **10**, 696.
202. Berberich, S. L. and Stoltzfus, C. M. (1991) Mutations in the regions of the Rous sarcoma virus 3′ splice sites: implications for regulation of alternative splicing. *J. Virol.*, **65**, 2640.
203. McNally, M. T. and Beemon, K. (1992) Intronic sequence and 3′ splice sites control Rouse sarcoma virus RNA splicing. *J. Virol.*, **66**, 6.
204. Arrigo, S. and Beemon, K. (1988) Regulation of Rous sarcoma virus RNA splicing and stability. *Mol. Cell. Biol.*, **8**, 4858.
205. Stoltzfus, C. M. and Fogartty, S. J. (1989) Multiple regions in the Rous sarcoma virus *src* gene intron act in *cis* to affect accumulation of unspliced RNA. *J. Virol.*, **63**, 1669
206. McNally, M. T., Gontarek, R. R., and Beemon, K. (1991) Characterization of Rous sarcoma virus intronic sequences that negatively regulate splicing. *Virology*, **185**, 99.
207. Gontarek, R. R., McNally, M. T., and Beemon, K. (1993) Mutation of an RSV intronic element abolishes both U11/U12 snRNP binding and negative regulation of splicing. *Genes Dev.*, **7**, 1926.
208. Amendt, B. A., Simpson, S. B., and Stoltzfus, C. M. (1995) Inhibition of RNA splicing at the Rous sarcoma virus *src* 3′ splice site is mediated by an interaction between a negative *cis* element and a chicken embryo fibroblast nuclear factor. *J. Virol.*, **69**, 5068.
209. Cullen, B. R. (1991) Human immunodeficiency virus as a prototypic complex retrovirus. *J. Virol.*, **65**, 1053.

210. Muesing, M. A., Smith, D. H., Cabradilla, C. D., Benton, C. V., Lasky, L. A., and Capon, D. J. (1985) Nucleic acid structure and expression of the human AIDS/lymphadenopathy retrovirus. *Nature*, **313**, 450.
211. Purcell, D. F. and Martin, M. (1993) Alternative splicing of human immunodeficiency virus type 1 mRNA modulates virus protein expression, replication, and infectivity. *J. Virol.*, **67**, 6365.
212. Robert-Guroff, M., Popovic, M., Gartner, S., Markham, P., Gallo, R. C., and Reitz, M. S. (1990) Structure and expression of *tat-*, *rev-*, and *nef*-specific transcripts of human immunodeficiency virus type 1 in infected lymphocytes and macrophages. *J. Virol.*, **64**, 3391.
213. Schwartz, S., Felber, K., Benco, D. M., Fenyo, E.-M., and Pavlaski, G. N. (1990) Cloning and functional analysis of multiply spliced mRNA species of human immunodeficiency virus type 1. *J. Virol.*, **64**, 2519.
214. O'Reilly, M. M., McNally, M. T., and Beemon, K. (1995) Two strong 5′ splice sites and competing, suboptimal 3′ splice sites involved in alternative splicing of human immunodeficiency virus type 1 RNA. *Virology*, **213**, 373.
215. Amendt, B. A., Hesslein, D., Chang, L.-J., and Stoltzfus, C. M. (1994) Presence of negative and positive *cis*-acting RNA splicing elements within and flanking the first *tat* coding exon of human immunodeficiency virus type 1. *Mol. Cell. Biol.*, **14**, 3960.
216. Amendt, B. A., Si, Z.-H., and Stoltzfus, C. M. (1995) Presence of exon splicing silencers within human immunodeficiency virus type 1 *tat* exon 2 and *tat-rev* exon 3: evidence for inhibition mediated by cellular factors. *Mol. Cell. Biol.*, **15**, 4606.
217. Staffa, A. and Cochrane, A. (1995) Identification of positive and negative splicing regulatory elements within the terminal *tat-rev* exon of human immunodeficiency virus type 1. *Mol. Cell. Biol.*, **15**, 4597.
218. Hope, T. and Pomerantz, R. J. (1995) The human immunodeficiency virus type 1 Rev protein: a pivotal protein in the viral life cycle. *Curr. Top. Microbiol. Immunol.*, **193**, 91.
219. Tooze, J. (1981) *DNA tumor viruses*, 2nd edn. Cold Spring Harbor Laboratory Press.
220. Manley, J. L., Noble, J. C. S., Chaudhuri, M., Fu, X.-Y., Michaeli, T., and Prives, C. (1986) The pathway of SV40 early mRNA splicing. *Cancer Cells*, **4**, 259.
221. Graham, F. L., Smiley, J., Russell, W. C., and Nairn, R. (1977) Characteristics of a human cell line transformed by DNA from human adenovirus type 5. *J. Gen. Virol.*, **36**, 59.
222. Wang, J. and Manley, J. L. (1995) Overexpression of the SR proteins ASF/SF2 and SC35 influences alternative splicing *in vivo* in diverse ways. *RNA*, **1**, 335.
223. Gesteland, R. F. and Atkins, J. F. (eds). (1993) *The RNA world*. Cold Spring Harbor Laboratory Press.
224. Balvay, L., Pret, A.-M., Libri, D., Helfman, D. M., and Fiszman, M. Y. (1994) Splicing of the alternative exons of the chicken, rat, and *Xenopus* β-tropomyosin transcripts requires class-specific elements. *J. Biol. Chem.*, **269**, 19675.

9 | Capping, methylation, and 3′-end formation of pre-mRNA

ELMAR WAHLE and WALTER KELLER

1. Introduction

This chapter deals with three covalent modifications of mRNA precursors that do not change the RNA's coding sequence: capping of the 5′ end; cleavage at the 3′ end followed, in nearly all cases, by the addition of a poly(A) tail; and methylation of internal bases. Both terminal modifications—cap and poly(A) tail—are thought to be involved in the same set of reactions, splicing, RNA transport, translation initiation, and RNA stabilization or regulation of mRNA decay. Interestingly, there are indications of a direct or indirect interaction between 5′ end and 3′ end (1, 2). The function of internal methylations is unknown.

2. Modification of the 5′ end: capping

In all eukaryotic cells studied, the 5′ ends of all pre-mRNAs are converted to so-called cap structures (reviewed in 3, 4). The cap is required for splicing (5), the export of RNA from the nucleus (6), translation initiation (7), and protection of mRNA against 5′-3′-exonucleases (2).

The 5′-triphosphate end of RNA polymerase II transcripts, pppN, is first converted to a diphosphate, ppN, by a RNA 5′-triphosphatase. A mRNA guanylyltransferase then adds guanosine monophosphate, derived from GTP, generating an unusual 5′-5′-triphosphate linkage, G(5′)ppp(5′)N. 5′-triphosphatase and guanylyltransferase are usually tightly linked, either in two associated polypeptides or in two domains of the same polypeptide. In the latter case, separate domains responsible for the two activities can often be prepared in an active form by partial proteolysis (3). Guanylyltransferase is encoded by an essential gene in yeast (8). The reaction proceeds through a covalent enzyme–GMP intermediate in all enzymes studied (3). The guanylate residue is linked by a phosphoamide bond to the ε-amino group of a lysine within the sequence Lys-Thr-Asp-Gly, which is conserved not only in guanylyl transferases but also in DNA and RNA ligases (9–12). In the latter enzymes, the equivalent lysine residue serves in the formation of the covalent enzyme–adenylyl intermediate. As pointed out before (9), the similarities of the two classes of enzymes also extend to the fate of the covalently bound purine nucleotide: transfer to the 5′-phosphate of a polynucleotide generates a phosphoanhydride bond—GpppN in the case of capping enzymes, AppN in the case of ligases.

Transfer of a methyl group from *S*-adenosylmethionine to the N^7 of the guanine base results in m^7 GpppN, also called cap 0, which is found at the 5′ ends of yeast and plant mRNAs. Whereas cellular (guanine-7)-methyltransferases are easily separated from the guanylyltransferases, the two activities from vaccinia virus are tightly associated in a heterodimeric complex. 5′-triphosphatase and guanylyltransferase acitivities are found in the isolated 95 kDa subunit, but methyltransferase activity requires an association of the 31 kDa subunit and a C-terminal domain of the large subunit (13, 14, and references therein). The active site is located in the latter (15, 16).

The cap 0 can be further modified by up to three additional methyl transfers from *S*-adenosylmethionine. Methylation of the 2′-OH group of the ribose of the first encoded nucleotide generates cap 1, additional methylation of the 2′-OH group of the second nucleotide generates cap 2. The structure of cap 2 is shown in Fig. 1. Both cap 1 and 2 are found in mammalian cells. Methylations are catalysed by two

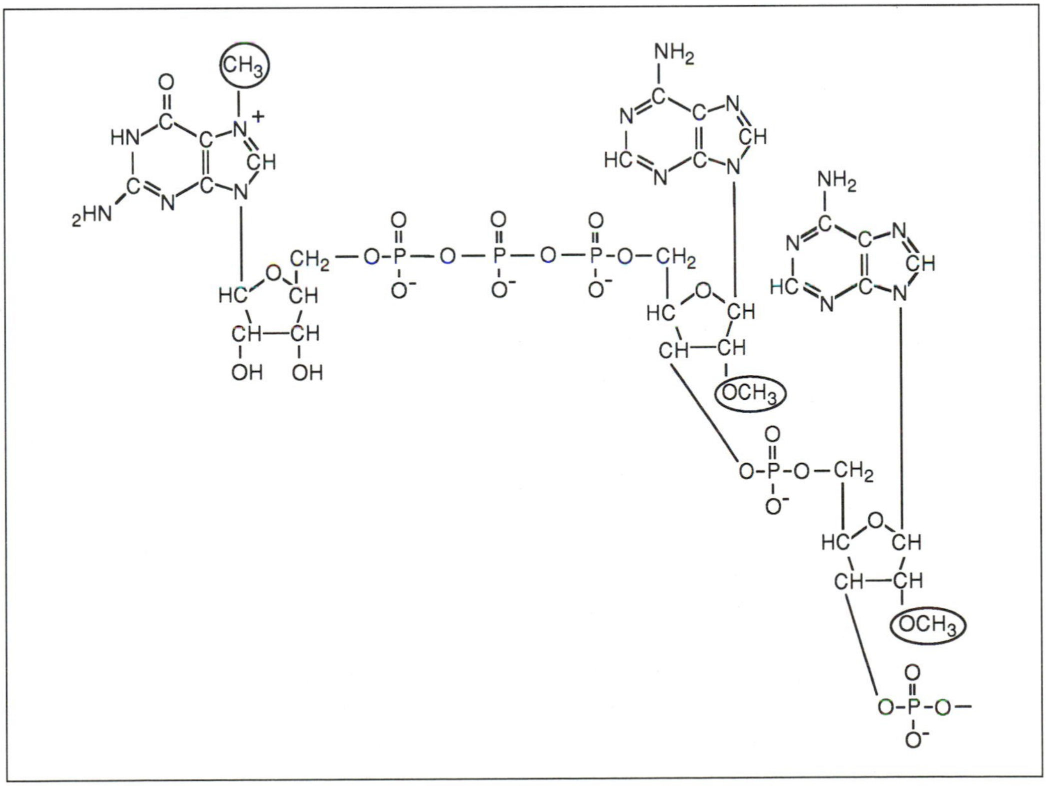

Fig. 1 Structure of cap 2. The three methyl groups are marked by circles.

distinct 2′-*O*-methyltransferases. One such enzyme encoded by vaccinia virus also serves as a stimulatory subunit in the viral poly(A) polymerase (17; see below). If the first encoded base in a mammalian mRNA is an adenine, its N^6 position can also be methylated.

Capping is limited to RNA polymerase II transcripts and occurs very soon after transcription initiation. In the case of vaccinia virus, the basis for this tight coupling is a direct association of the capping enzyme, which also serves as a transcription termination factor, with the viral RNA polymerase (18).

3. Internal methylation

In addition to the cap, internal nucleotides in mRNA are also methylated (reviewed in 4). The most frequent internal methylation is at the N^6 position of adenosine, but m^5C is also found. A consensus sequence for adenine methylation is RGACU in single-stranded RNA (19). However, a given site is not methylated in every RNA molecule, and many potential sites are never methylated. The average number of methyl groups is between one and three per mRNA molecule. Methylated sites occur in translated and untranslated regions and in introns as well as in exons. Sequence-specific RNA methylation has been reproduced *in vitro* (20). *S*-adenosylmethionine is the methyl donor. The methyltransferase has been partially purified from HeLa cell nuclear extract. The enzyme is surprisingly complex, three different fractions being required for the activity (21).

4. Cleavage and polyadenylation of the 3′ end

4.1 Overview

In mammalian and most if not all other eukaryotic cells, the 3′ ends of mRNAs are generated by post-transcriptional processing of longer precursors (Fig. 2). The pre-mRNA is first cleaved endonucleolytically downstream of the coding sequence. Almost all mRNAs then receive a poly(A) tail, which in mammalian cells is 200–250 nucleotides long. Once the RNA has been transported to the cytoplasm, the tail is gradually shortened.

The poly(A) tail plays an essential role in translation initiation (1, 22; reviewed in 23–25). It also functions in the control of mRNA stability (reviewed in 24, 26, 27) and possibly in the transport of mRNA from the nucleus to the cytoplasm (28, 29). Although the poly(A) tail probably does not play a direct role in splicing, polyadenylation and splicing seem to be linked *in vivo* (see Section 4.2.3). Thus, the poly(A) tail and the monomethyl-G cap are involved in the same set of reactions. These two terminal modifications are found on all mRNAs (for the only exception, see Section 5), and only on mRNAs. They may thus serve to identify a RNA molecule as a messenger and guide it through all the steps of its processing, transport, function, and degradation.

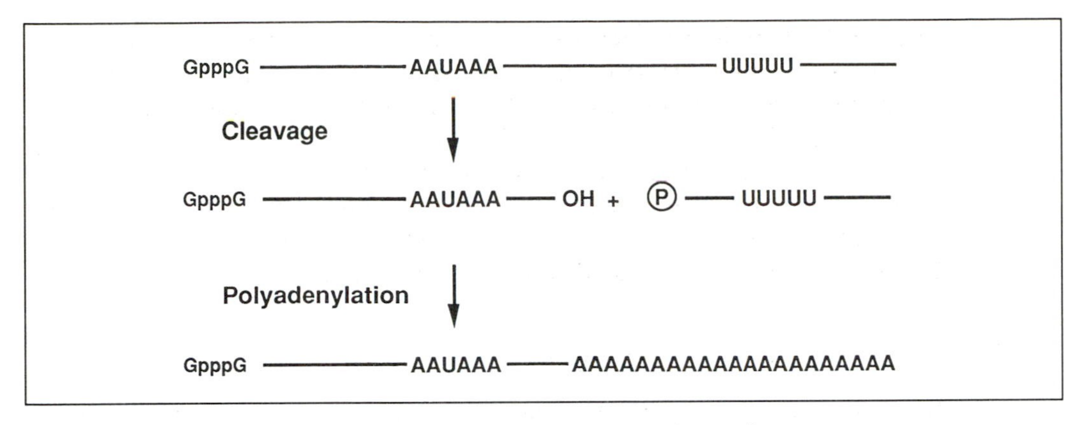

Fig. 2 Outline of 3′ processing in vertebrate cells.

4.2 Polyadenylation in metazoan cells

4.2.1 Polyadenylation signals

The site of cleavage and poly(A) addition in the pre-mRNA is determined by three types of sequences. The first of these is the hexanucleotide AAUAAA, which is located 10–30 nucleotides upstream of the poly(A) site (reviewed in 30). Human haemoglobin deficiencies caused by point mutations in the hexanucleotide sequences of globin genes (31, 32) clearly show that this sequence is essential for polyadenylation, and that polyadenylation is essential for gene expression. The AAUAAA sequence is very highly conserved, perfect copies being found in about 90% of all genes. Variants, however, do exist, and some genes do not contain any recognizable hexanucleotide element at all (reviewed in 30). It is conceivable that the expression of such genes is limited by a low polyadenylation efficiency. However, other sequences may compensate for a poor hexanucleotide element (33). The possibility that a transcription termination site increases the use of a suboptimal upstream polyadenylation site has been discussed (34–36; see also Section 4.2.3).

A less conserved sequence element or group of elements is found within 50 nucleotides downstream of the poly(A) site. One version of these is a short run of U residues, whereas another is GU-rich (reviewed in 30). It is not clear whether these are two variants of the same signal or two different elements (see also 4.2.2.3). A given poly(A) site may contain more than one downstream element (37, 38), and others appear not to contain either of the two elements (39, 40). Although point mutations in these sequences can affect processing efficiency, the effects are not very strong (41), and a modification–interference study of 3′-end processing did not find single-base modifications downstream of the cleavage site that were able to abolish the reaction (42). Some downstream elements could only be inactivated by large deletions (see for example, 43). Since the downstream elements are much more variable than the AAUAAA sequence, they may partially determine the efficiency of a polyadenylation site (see 4.2.2.3).

The combination of the hexanucleotide sequence and the downstream element is necessary and sufficient to induce 3′-end formation (44). However, a third class of signals exists upstream of the AAUAAA sequence at least in some genes. Signals of this type have been found in the L1, L3, and L4 sites of adenovirus and the late polyadenylation site of SV40 (39, 45–47). The L1 upstream sequences are distinguished by showing an effect only in a competitive situation between two poly(A) sites in the same transcription unit; in contrast to other upstream elements, they do not affect processing efficiency in transcription units with a single poly(A) site (45). U-rich motifs such as AUUUGURA (48) or UUCUUUUU (47) are important components of the upstream sequences. Thus, these are related in function and in structure to the cytoplasmic polyadenylation elements (CPEs), which direct this process during oocyte maturation (see Section 4.2.4). Upstream elements are most prominent among retroviruses (49). Since the replication of retroviruses requires the presence of terminally redundant sequences, polyadenylation signals are often found at both ends of the genome. Upon transcription of the integrated viral DNA, the polyadenylation signal close to the promoter has to be ignored, whereas the signal at the 3′ end has to be used. In many retroviruses, the unique sequences upstream of the promoter-distal poly(A) site favour the use of this site (33, 50–52). The potential promoter-proximal site lacks such upstream elements. A second mechanism that contributes to the exclusive use of the downstream site is the inhibition of polyadenylation at the upstream site by the proximity of the promoter (53, 54). The same mechanism has also been described in a plant virus, cauliflower mosaic virus (55). Occlusion by the promoter appears to be specific for retroviral poly(A) sites (53, 54, 56). The human immunodeficiency virus (HIV) uses both mechanisms, and conflicting results have been reported as to which is more important (57, 58). It has been argued that it is not promoter proximity itself that suppresses the use of the upstream poly(A) site, but the absence of splice sites between promoter and poly(A) site. This may prevent processing of the majority of RNAs and destabilize the minority of RNAs that is processed (58; for a discussion of the connection between splicing and polyadenylation, see Section 4.2.3). However, this model provides no simple explanation for the fact that non-retroviral poly(A) sites are usually not sensitive to promoter proximity (see above). Poly(A) site strength does not seem to be correlated with sensitivity to a nearby promoter (56). Occlusion of the upstream poly(A) site of HIV by the promoter depends on the activation of the promoter by the Tat protein (59).

The preferred base at the poly(A) site itself is an A, and a C residue preceding it is also moderately conserved. Mutations in the A change the exact position of the endonucleolytic scission, with a preference for adenines in the new sites, but do not affect the efficiency of processing (60). A second factor influencing the location of the cleavage site is the position of the downstream element: a shift in the position of an essential downstream run of Us leads to a shift of the preferred cleavage site, apparently to another CA dinucleotide (61, 62).

4.2.2 Polyadenylation *in vitro*

4.2.2.1 Polyadenylation in nuclear extract

An *in vitro* system for 3′-end processing of pre-mRNA permitted the first analysis of the reaction mechanism (63, 64). Extract from HeLa cell nuclei processes not only RNA synthesized by endogenous RNA polymerase II but also exogenous transcripts added to the reaction. Processing depends on the sequence elements discussed above, including the upstream element, if present (65, 66). Analysis of this system confirmed that 3′-end processing is initiated by endonucleolytic cleavage of the pre-mRNA at the poly(A) site, as had been proposed on the basis of *in vivo* experiments (63, 64). The endonucleolytic cut is normally followed by rapid polyadenylation of the upstream fragment and degradation of the downstream fragment; the cleaved intermediate never accumulates to detectable levels.

Addition of EDTA to the extract permits cleavage but inhibits both degradation of the downstream fragment and polyadenylation, thus allowing a direct analysis of the two initial cleavage products that is not complicated by the addition of the poly(A) tails (64). The structures of the two RNA fragments are most easily explained by a precise endonucleolytic scission at the poly(A) site, leaving a 3′-hydroxyl and a 5′-phosphate end, followed by exonucleolytic degradation of the downstream fragment from its 5′ end (67, 68) (Fig. 2). No cleavage is observed in the absence of ATP, but analogues with non-hydrolysable α-β- or β-γ-bonds can substitute (64, 67). ATP is essential for the formation of 3′-processing complexes (see below); beyond that, its role in the cleavage reaction has not been investigated. ATP analogues lacking the 3′-hydroxyl group essential for chain elongation or containing a non-hydrolysable α-β bond support cleavage but inhibit polyadenylation (64). Thus, they provide an alternative to EDTA when a direct detection of the cleaved reaction intermediate is desired.

AAUAAA-dependent polyadenylation may be examined independently of cleavage through the use of 'pre-cleaved' substrate RNA that already ends at the poly(A) site (69). The independent assays of cleavage and polyadenylation have been important for the purification and analysis of the factors that carry out 3′-end processing, as described below.

4.2.2.2 3′-end processing factors

By chromatographic fractionation and reconstitution of the *in vitro* processing system, six factors have been identified that are involved in 3′-end cleavage and polyadenylation. Four of them have been purified to homogeneity so far. We will first give a summary of these factors and then discuss their activities in more detail below.

The essential polyadenylation signal AAUAAA is bound by the cleavage and polyadenylation specificity factor (CPSF). CPSF is required for 3′-cleavage as well as for polyadenylation. The factor has been purified from calf thymus and HeLa cells as a complex of four subunits of 160, 100, 73, and 30 kDa (70). These four polypeptides co-purify in an approximately equimolar ratio through all steps examined and are co-precipitated by monoclonal and polyclonal antibodies (70; A. Jenny *et al.*, unpublished). The native molecular weight of the complex is consistent with a heterotetrameric structure (70). A more recent purification of CPSF confirmed the presence of the three largest subunits but did not detect the 30 kDa polypeptide (71). Both preparations were active in AAUAAA-dependent polyadenylation as well as in pre-mRNA cleavage assays (70, 71; our unpublished data). A cDNA encoding the 100 kDa subunit of CPSF shows no obvious relationship to known sequences (A. Jenny *et al.*, submitted).

Cleavage stimulation factor (CStF) (72) or cleavage factor 1 (CF1) (73, 74) is required only for cleavage of the pre-mRNA but not for polyadenylation. CStF/CF1 binds the downstream element (see Section 4.2.2.3). It is composed of polypeptides of 77, 64, and 45 kDa, which co-sediment through glycerol gradients. Monoclonal antibodies against two subunits inhibit the activity and precipitate the heterotrimer (72). A cDNA sequence for the 64 kDa subunit (75) reveals the presence of the so-called RNP- or RRM-domain common to a number of RNA-binding proteins (76). A cDNA for the 50 kDa subunit has also been cloned (77).

Two additional factors are also required specifically for the endonucleolytic scission but dispensable for polyadenylation (73, 78, 79). These have not been further characterized. A protein, DSEF 1, that binds to a GU-rich downstream element has been partially purified (80). A function in 3′-end processing remains to be demonstrated.

The poly(A) tail is synthesized by poly(A) polymerase. The enzyme is also required for the cleavage reaction, although the degree of poly(A) polymerase dependence varies for different poly(A) sites (81, 82). Two homogeneous preparations of poly(A) polymerase from calf thymus have been reported (83, 84). cDNA cloning (85, 86) revealed that the 60 kDa protein obtained in these purifications was a proteolysis product lacking a C-terminal domain of 20 kDa. Both the full length protein purified from various expression systems and the 60 kDa proteolysis product are functional in both cleavage and polyadenylation assays, but quantitative differences exist (84, 86; our unpublished data). C-terminal deletions leaving less than 57 kDa are inactive in both assays (85, 87; our unpublished data). The N-terminus of poly(A) polymerase contains sequences resembling the RNP- or RRM-domain. Mutations of these sequences reduce or abolish activity, but it has not been demonstrated directly that they affect primer binding (87; our unpublished data). A similarity to a 'polymerase module' found in many template-dependent polymerases (85) is probably coincidental as mutagenesis of many of the important amino acids does not reduce enzymatic activity (87; our unpublished data). Poly(A) polymerase contains two nuclear localization signals in its C-terminal serine- and threonine-rich domain (87). This domain is also phosphorylated (87, 88). Phosphorylation appears to affect the enzyme's activity in polyadenylation assays (our unpublished data). Poly(A) polymerase exclusively uses ATP as the precursor. Pyrophosphate is released as the second product (83, 84).

Poly(A)-binding protein II (PAB II) stimulates poly(A) polymerase during the extension of poly(A) tails (89). This protein is clearly distinct (89, 90) from the better characterized 70 kDa poly(A)-binding protein, PAB I (24). PAB II binds specifically to single-stranded poly(rA) or poly(rG), probably by means of an RNP-domain (89, 90).

Investigations with antibodies against CStF, CPSF, poly(A) polymerase, and PAB II have all shown the anticipated nuclear localization of 3′-processing factors (72, 87, 91; A. Jenny *et al.*, submitted for publication).

Despite earlier indications to the contrary, no snRNP appears to be essential for 3′-processing (70, 79).

4.2.2.3 Recognition of the RNA substrate

Since polyadenylation often precedes splicing (92), the correct poly(A) site has to be recognized with high precision within a very large precursor RNA. A single gene can have more than one poly(A) site, and, in some cases, the choice of the site determines which protein product is made (see Section 4.2.4). Also, poly(A) sites can differ in their intrinsic strength (46, 47, 93), probably owing to a variable affinity for processing factors. Since polyadenylation is essential for gene expression, the strength of a poly(A) site may determine the amount of gene product made. Thus, recognition of the RNA substrate is likely to be of regulatory importance.

In nuclear extract, RNA molecules containing the signals for cleavage and polyadenylation are rapidly assembled into large heparin-resistant protein–RNA complexes, which have been analysed by glycerol gradient centrifugation or gel electrophoresis under non-denaturing conditions. Complex formation depends on the presence of both the AAUAAA signal and the downstream element, and protects both signals from nuclease digestion. ATP, but not its hydrolysis, is required. Processing has been demonstrated to occur within the complexes, and electrophoretic and other properties of the complexes change as the reaction proceeds (reviewed in 30).

Since not all the factors involved in 3′-end processing have been purified, the assembly of the processing complexes has not been fully characterized. However, the studies that have been carried out revealed that the stability of the complex depends on cooperative interactions between several components.

Purified CPSF binds the polyadenylation signal AAUAAA in gel retardation assays (70, 94, 95). Binding requires all six nucleotides in AAUAAA but none outside (95), although sequences outside AAUAAA seem to be able to modulate binding (see Section 4.2.4). CPSF contacts not only bases but also riboses (94, 96). A 160 kDa polypeptide, identified as the largest CPSF subunit by co-purification and immunoprecipitation, can be UV cross-linked to AAUAAA-containing RNA (95); A. Jenny *et al.*, submitted). By the same criteria, the 30 kDa polypeptide present in CPSF is also judged to bind the substrate RNA. However, this polypeptide appears not to be an essential subunit (71).

The CPSF–RNA complexes are unstable as shown by their immediate decay upon addition of competitor RNA (97). The complexes are stabilized by the cleavage factor CStF/CF1 (73, 98). In agreement with the presence of an RNP domain predicted from its cDNA sequence, the 64 kDa subunit of CStF can be UV cross-linked to RNA. RNA-binding and cross-linking is stimulated by CPSF (71, 72, 74,

75, 99, 100). The cross-link has been mapped to a run of Us acting as a downstream element (62). This is consistent with the observation that the stabilizing influence of CStF on the CPSF–RNA complex depends on the presence of the downstream sequences (73, 98). Gel retardation of RNA by CStF in the absence of CPSF has also been reported (71), but the specificity of the interaction has not been examined. The isolated 64 kDa subunit of CStF binds RNA without apparent specificity for downstream polyadenylation signals (62, 75). Thus, the specificity of binding probably depends on the two other subunits and additional factors like CPSF. Although a run of Us is sufficient to bind CStF in the context of the entire processing complex (62), it is unknown whether other downstream sequences function in the same manner.

In a study of downstream element mutations, the stability of the CPSF–CStF–RNA complex formed *in vitro* was correlated with the efficiency of the poly(A) site *in vivo*. It was proposed that the affinity of the downstream element for CStF determines the stability of a 3′-end processing complex and, since the downstream element is the variable part of a polyadenylation signal, the strength of a poly(A) site *in vivo* (98). Results relating the stability of the 3′-processing complex to the efficiency of processing have also been obtained by others. However, in these cases upstream sequences and sequences immediately surrounding AAUAAA were responsible rather than downstream elements (47, 66, 93).

Poly(A) polymerase by itself, in the presence of Mg^{2+}, is nearly inactive and requires either CPSF or PAB II for efficient poly(A) synthesis. The activity appears to be limited by weak binding to RNA as concluded from the absence of detectable complexes between the enzyme and RNA (95) and by functional data (see below). The enzyme itself also has little or no specificity with regard to the primer RNA. In particular, it does not recognize the AAUAAA sequence (78, 81, 84). However, like CStF, poly(A) polymerase clearly binds to the CPSF–RNA complex, stabilizing it and further retarding its migration in native gels (71, 97). From this, one may infer that CPSF activates poly(A) polymerase by increasing its affinity for the 3′ end to be elongated. Stabilization of the CPSF–RNA interaction by poly(A) polymerase might also account for the requirement for poly(A) polymerase in the cleavage reaction. A direct interaction between CPSF and the polymerase is likely but remains to be demonstrated.

PAB II binds the growing poly(A) tail; this binding is required for the stimulation of poly(A) polymerase (89). The minimum site required for high affinity binding of PAB II is A_{12}. The K_D for this site is 2×10^{-9} M and the half-life of the complex is on the order of seconds (90). A stabilization of the poly(A) polymerase–RNA interaction by PAB II can be inferred from functional data (see below) but has not been shown directly. PAB II appears to have little effect on the stability of the CPSF–RNA complex. However, it does stabilize the CPSF–poly(A) polymerase–RNA complex. Thus, CPSF and PAB II together activate poly(A) polymerase by increasing its affinity for the RNA through the formation of a quaternary complex with an oligoadenylated RNA (97).

In summary, while all four 3′-end processing factors characterized so far bind the substrate RNA directly, none of them does so very strongly. Instead, the processing complexes are stabilized by multiple interactions of the proteins with each other and the RNA. These cooperative interactions serve to hold poly(A) polymerase at the 3′-end of the RNA and probably also direct the unknown endonuclease to its site of action.

Since the remaining 3′-end processing factors have not been purified, their contributions to substrate recognition and the stability of the processing complex are unknown.

The basis for the stimulatory effect of sequences upstream of AAUAAA (see Section 4.2.1) is still controversial. It has been proposed that a component of the U1 snRNP, the 32 kDa U1A protein, recognizes the upstream element in the SV40 late site (101). However, a role of this protein has so far not been confirmed by its addition to a reconstituted polyadenylation reaction. A study of upstream sequences that play a role during cytoplasmic polyadenylation in *Xenopus* oocytes (see Section 4.2.4) suggested that they are bound by CPSF itself (102). However, with the purity of current CPSF preparations, a role of a contaminating RNA-binding protein cannot be excluded. Finally, one has to keep in mind that different upstream sequences do not necessarily have to work by the same mechanism.

4.2.2.4 Reaction mechanism

Since 3′ cleavage has not been reconstituted with purified components, not much is known about the mechanism of the reaction beyond what has been learned from crude nuclear extract (see 4.2.2.1). The CPSF–CStF complex described above appears to straddle the cleavage site by interaction with AAUAAA and the downstream element (Fig. 3). It has been proposed that this complex determines the cleavage site with CStF acting as a 'molecular ruler' (62). While a displacement of the CStF binding site does indeed displace the preferred cleavage site, there are also preferences for certain nucleotides at the cleavage site itself (see Section 4.2.1). Also, it is not yet clear how the idea of a clearly defined CStF binding site at a fixed distance from the cleavage site can be reconciled with the diffuse and redundant nature of downstream elements. The endonuclease responsible for 3′ cleavage has not been identified.

The mechanism of polyadenylation, as far as it is known, can essentially be deduced from the mechanism of substrate recognition outlined above. Although poly(A) polymerase is able to catalyse polyadenylation in the absence of any other factor, the reaction is nonspecific with respect to the RNA sequence and very inefficient due to the high K_M for the RNA primer in the presence of Mg^{2+}. Nonspecific polyadenylation is strongly stimulated by Mn^{2+}ions, which decrease the K_M for RNA. V_{max} is similar with both ions (84). The low affinity of poly(A) polymerase for RNA is also reflected in an entirely distributive mode of action, that is, the enzyme dissociates from its primer after the addition of every single nucleotide. This is true even in the presence of Mn^{2+} (84). Gel retardation assays can detect complexes between poly(A) polymerase and substrate RNA only at high enzyme concentrations (G. Martin and W. K., unpublished). The limitation of polymerase activity by a

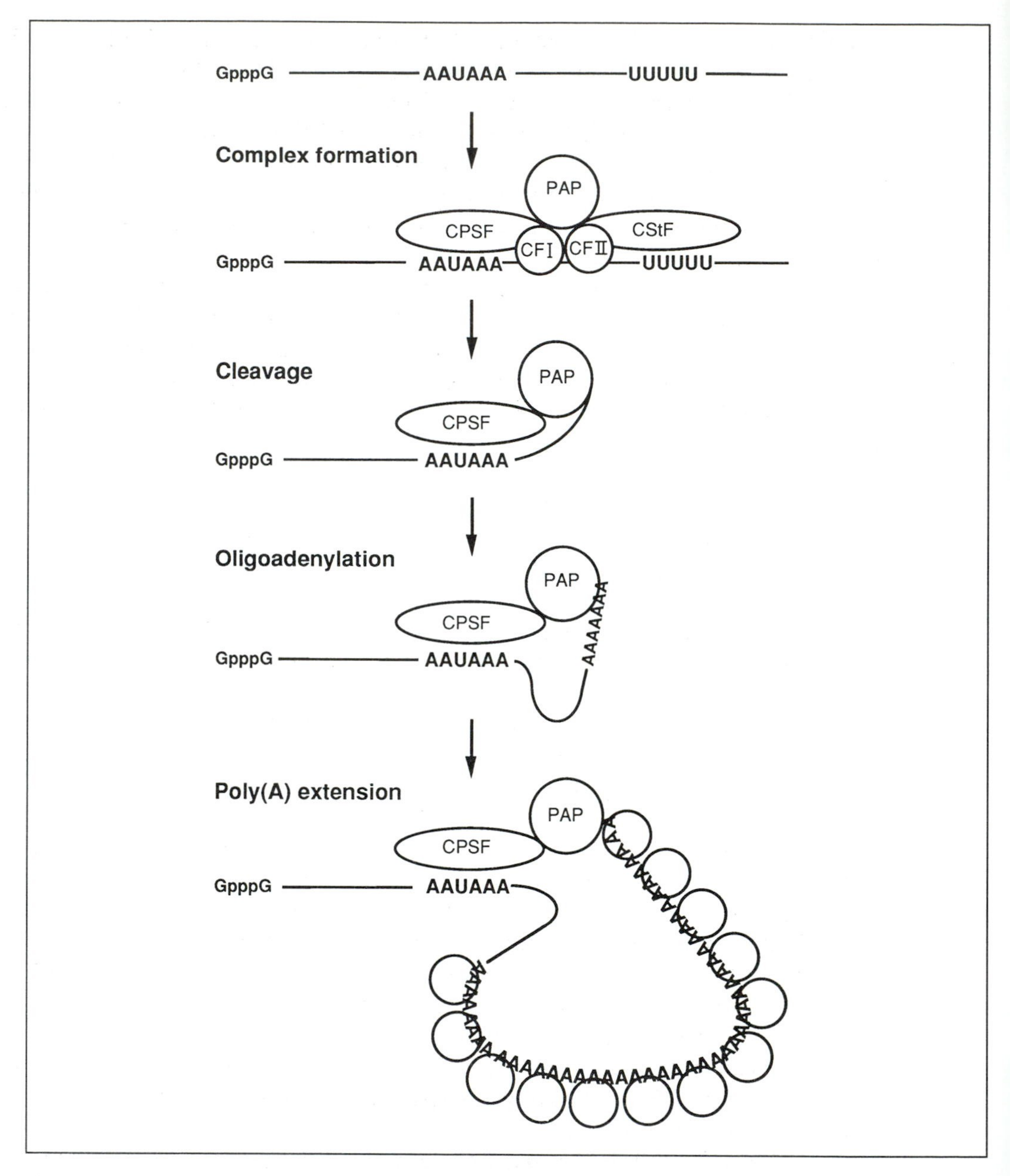

Fig. 3 Mechanism of 3′-end formation in vertebrates. CF I, CF II, and PAP stand for cleavage factor I, cleavage factor II, and poly(A) polymerase, respectively. While the interaction of CPSF and CStF with AAUAAA and downstream sequences has been established (see text), the arrangement of CF I, CF II, and poly(A) polymerase is arbitrary. The release of factors after the cleavage event indicated in the cartoon has also not been demonstrated. Direct interactions between proteins are hypothetical.

low affinity for RNA is consistent with the proposal that both CPSF and PAB II act by stabilizing the interaction between the enzyme and its substrate.

Poly(A) polymerase is activated and endowed with specificity for AAUAAA-containing RNAs through the interaction with CPSF, as outlined above. The two proteins are sufficient for the initial extension of a precleaved RNA. CPSF effects a small increase in the processivity of poly(A) polymerase to up to 10 nucleotides polymerized per binding event, consistent with the stabilization of the polymerase–RNA interaction (97).

Poly(A) polymerase can also be activated by PAB II and thereby acquires a specificity for oligoadenylated RNA (89). An RNA containing at least 10 to 11 A residues can be extended independently of CPSF and its binding site, AAUAAA. A slightly increased processivity of poly(A) polymerase (at most five nucleotides) in the presence of PAB II suggests that this factor also stabilizes the binding of the enzyme to RNA (97).

On a substrate RNA containing both the AAUAAA sequence and an oligo(A) tail, PAB II and CPSF can stimulate poly(A) polymerase simultaneously—oligo(A) elongation in the presence of both factors is faster than with either factor alone. This is in agreement with the formation of the quaternary complex described above. As a result, polyadenylation proceeds with biphasic kinetics: addition of the first 10 adenylate residues depends only on CPSF and is slow, whereas the further extension depends on both CPSF and PAB II and is fast (89). In contrast to the barely processive action of poly(A) polymerase combined with either CPSF or PAB II alone, the quaternary complex, containing all three proteins and the RNA, is competent for the essentially completely processive synthesis of a full length poly(A) tail (97). As poly(A) polymerase is required for the initial endonucleolytic cleavage (see 4.2.2.2), it may be incorporated into the 3′-processing complex at the very beginning of the reaction, establishing an interaction with CPSF. Whereas cleavage factors are probably released from the complex once they have served their purpose, and PAB II is not included until the stage of poly(A) extension, it is likely that CPSF and poly(A) polymerase remain associated with an RNA molecule throughout the entire 3′-processing reaction, from initial cleavage until synthesis of the poly(A) tail has been completed. In this context, it also seems likely that the slow initial phase of polyadenylation exists only when the reaction is reconstituted with pre-cleaved RNA. In the coupled reaction, proteins involved in the cleavage step may assist CPSF in holding poly(A) polymerase in the complex until PAB II has bound the growing tail.

Once the poly(A) tail has reached its normal length of approximately 200–250 nucleotides, further elongation in the reconstituted system becomes very slow (89, 97). Both with respect to biphasic polyadenylation and length control, the reconstituted system reproduces properties of the polyadenylation reaction first observed in crude nuclear extracts (103). The essence of length control is a switch from the processive elongation of short poly(A) tails to a distributive, and therefore slow, mechanism. This switch depends on a true measurement of poly(A) tail length, possibly through the number of PAB II molecules bound to it (E. W., unpublished).

The nature of the changes taking place in the polyadenylation complex once the final length has been reached is unknown.

4.2.3 Relationship of 3′ cleavage/polyadenylation to transcription and splicing

In the *in vitro* system discussed so far, 3′-end processing occurs on short pre-made RNA substrates, independent of transcription and splicing. Likewise, splicing can be carried out *in vitro* without transcription or polyadenylation (see Chapters 4 and 6). This proves that there is no obligatory coupling between these processes. However, *in vivo* 3′-end processing seems to be coupled to both transcription and splicing.

A link between transcription and polyadenylation seems plausible, since polyadenylation occurs before RNA polymerase has reached the end of the transcription unit (92). Although the issue is somewhat controversial, several studies concluded that transcripts initiated from RNA polymerase I or III promoters, or from the snRNA type of RNA polymerase II promoters, cannot undergo 3′-end cleavage and polyadenylation, even if they contain the proper sequence elements (104–107). Kinetic data are also in agreement with a connection between transcription and polyadenylation. In a typical reaction in nuclear extract, processing of 50–80% of the substrate will take on the order of 30–60 minutes although the extract does not appear to be saturated with RNA. Yet it has been estimated that *in vivo* an RNA is polyadenylated within one minute after its synthesis (108). These data suggest that, although polyadenylation does not absolutely depend on transcription, it may be kinetically coupled to it. Evidence for a kinetic coupling has been obtained in HeLa cell nuclear extract: transcripts made by endogenous RNA polymerase II were processed with hardly any time lag, whereas exogenous RNA or transcripts made in the extract from appropriate templates by added T7 RNA polymerase were processed very slowly or not at all (109). Similar observations were also made in a yeast *in vitro* system (110).

A second link between 3′-processing and transcription concerns transcription termination. *In vivo* studies have shown that termination can be prevented by the inactivation of polyadenylation signals and induced by the introduction of efficient poly(A) sites (111; earlier work reviewed in 30). A transcriptional pause site is also involved in termination and enhances the use of the poly(A) site (112, 113).

There is also a connection between splicing and polyadenylation. The presence of introns in a transcription unit greatly stimulates 3′ cleavage and polyadenylation (114, 115). Histone mRNAs, the only mRNAs that do not have a poly(A) tail, do not contain introns. Introduction of an intron into such an RNA induces the use of a cryptic polyadenylation site (116). Several studies found a link between the use of a 3′ splice site and a downstream poly(A) site (117–123). These data may be rationalized in terms of the 'exon definition' model (124), which suggests that splice sites are defined by an interaction of splicing factors not across the intron to be removed, but across the exon to be retained in the message. According to this model, the splicing apparatus scans from the 3′ splice site at the 5′ boundary of an exon to the next 5′ splice site downstream. Since 3′-terminal exons do not have a downstream 5′ splice site, the polyadenylation site is believed to substitute. An interaction between 3′-processing complexes and some component of the splicing machinery is thus predicted by the model. The U1 snRNP has been suggested as a component responsible for coupling of splicing and polyadenylation (125).

4.2.4 Regulation of polyadenylation

Polyadenylation can be regulated in several ways:

(1) by competition between two or more cleavage/polyadenylation sites;

(2) by competition between cleavage/polyadenylation and an incompatible splicing pattern;

(3) by the extent to which a single processing site is used;

(4) by the length of the poly(A) tail added to a particular mRNA.

A good example of the first kind of regulation is provided by adenovirus. Early during infection, transcripts from the major late promoter contain three polyadenylation sites—L1, L2, and L3—of which L1 is used predominantly. Late in infection, transcripts contain the additional sites L4 and L5, and all sites are used with some preference for L3. It has been suggested that the change in poly(A) site usage is caused by a decrease in the activity of CStF/CF1 (see Section 4.2.2.2). Since L1 is a weak site (47, 93), its use early in infection would be expected to be favoured by a high concentration of CStF/CF1 and the site's proximal position in the transcript. A decrease in CStF/CF1 activity late in infection should favour the use of stronger polyadenylation sites (126). This model, however, does not explain the *cis*-effect that has been observed for poly(A) site choice: when cells in the late stage of adenovirus infection are superinfected with a second virus, the latter shows the poly(A) site choice typical for early infection (127).

Competition between a poly(A) site and an incompatible splice site determines the choice between membrane-bound and secreted forms of immunoglobulins. A balance of efficiencies of the competing sites is essential and sufficient for regulation; specific sequences do not seem to be involved. The shift to the secreted form occurring during B-cell development appears to be caused by a more efficient 3′-processing reaction in these cells. However, the biochemical basis for this change is not known (128–130).

Regulation of the frequency with which a single poly(A) site is used occurs in the gene for the A protein of the U1 snRNP. The U1-A protein mRNA contains a complex binding site for the protein in its 3′ untranslated region. Binding of U1-A does not inhibit cleavage of the mRNA precursor at the poly(A) site but inhibits polyadenylation of the cleaved RNA by an interaction with poly(A) polymerase (131). This autoregulatory mechanism presumably leads to degradation of the cleaved intermediate.

Changes in poly(A) tail length appear to be the most important regulatory events at the mRNA's 3′ end. As mentioned above, the poly(A) tail is involved in an as yet undefined manner in the initiation of translation. There is an ever growing list of cases in which this property of poly(A) tails is exploited for translational control of gene expression. Most examples come from oocyte maturation and early development of various animal species (reviewed in 132,133).

Translation in oocytes and early embryos is directed by mRNAs deposited during oogenesis. Some mRNAs are translated only in immature oocytes. Upon oocyte maturation, translation of these RNAs ceases because their poly(A) tails are degraded. In oocytes, deadenylation is a default pathway—it does not require any particular RNA sequence (134,135). In contrast, deadenylation of a particular RNA in fertilized eggs was found to depend on specific sequences (136). A different class of mRNAs are stored in immature oocytes in a translationally inactive state. These RNAs carry short oligo(A) tails generated by shortening of originally full-length poly(A) tails. This shortening reaction, like the one in fertilized eggs, depends on sequences in the RNA and thus differs from the default deadenylation mentioned above (137). During oocyte maturation, the oligo(A) tails are extended and, at the same time, translation of the messages begins. Extension of the oligo(A) tail is necessary and sufficient to induce translation of a previously dormant message (138–141). Oligo(A) extension is a cytoplasmic event (142). It is directed by two RNA sequences: the cytoplasmic polyadenylation element (CPE) which is related to the sequence UUUUUAU in both *Xenopus* and mouse oocytes, and the regular polyadenylation signal AAUAAA (139, 142, 143). The reaction is catalysed by a poly(A) polymerase, which has so far been found to be indistiguishable from the nuclear enzyme discussed above, and an RNA-binding activity that is specific for both AAUAAA and the CPE (144). Surprisingly, CPE-dependent polyadenylation can be reproduced *in vitro* with *E. coli*-expressed mammalian poly(A) polymerase and CPSF purified from calf thymus. CPSF itself appears to prefer RNA substrates that carry a CPE in addition to AAUAAA (102). UV cross-linking has identified a protein of 58 kDa that binds specifically to a CPE (145). It remains to be seen how binding of this protein and binding of CPSF can be reconciled. The cyclin-activated p34-cdc2 kinase may be involved in triggering cytoplasmic oligo(A) extension (145, 146).

4.3 Polyadenylation in other systems

4.3.1 Vaccinia virus

Vaccinia virus encodes a heterodimeric poly(A) polymerase (147,148). The 55 kDa subunit has catalytic activity. Surprisingly, its sequence is not obviously related to the other known poly(A) polymerase sequences. The 39 kDa subunit is a poly(A)-binding protein involved in the elongation phase of poly(A) synthesis much like the cellular protein PAB II (148). The same polypeptide also serves as a 2′-*O*-methyltransferase in cap synthesis (17, 149; see Section 2). The isolated 55 kDa subunit elongates non-poly(A) sequences in a single processive event by about 30 nucleotides. Further extension is distributive and slow (150). This is due to the fact that the 55 kDa subunit requires uridylate residues for binding. Upon translocation of the protein along the growing poly(A) tail, which lacks uridylates, affinity for the RNA drops (151). The 39 kDa subunit stimulates a second phase of polyadenylation (152, 153) by allowing the polymerase to bind to the U-less poly(A) tail.

4.3.2 Yeast

Cell-free extracts from *Saccharomyces cerevisiae* process pre-mRNAs extending beyond their mature 3′ ends by precise endonucleolytic cleavage followed by polyadenylation (154–156). Cleavage and polyadenylation depend on sequence elements that are indispensable for 3′-end formation *in vivo*. Thus, the basic mechanism of 3′-end formation is similar to the one in higher eukaryotes. Transcripts extending beyond the polyadenylation site have also been detected *in vivo* both in *S. cerevisiae* (110) and *Schizosaccharomyces pombe* (157). Sequences specifying RNA 3′ ends are interchangeable between these two yeasts (158). *In vitro*, cleavage requires a nucleoside triphosphate with a hydrolysable β-γ bond. Polyadenylation is specific for the upstream cleavage product. It is suppressed if CTP is substituted for ATP (154). In contrast to mammalian extracts, the cleaved intermediate lacking poly(A) is prominent even under normal reaction conditions.

Sequences directing 3′-end formation in yeast are poorly conserved and poorly understood. The involvement of a bipartite signal upstream of the poly(A) site, UAG ... UAUGUA, initially proposed on the basis of sequence comparisons, has now been firmly established by point mutations and analysis of revertants (159, 160; earlier work reviewed in 30, 49). The bipartite motif and related sequences like UAUAUA or UUUUUAUA are thought to function as upstream elements enhancing the use of poly(A) sites further downstream (161, 162). The poly(A) site is determined by a second motif at varying distances downstream from the first class of sequences and approximately 20 nucleotides upstream from the poly(A) site. Sequences functional in this position include UUAAGUUC (161) and A_8 (163). Finally, nucleotides surrounding the site itself are important as well (161, 164). Signals downstream of the processing site have been suggested in *S. cerevisiae* but not defined (165). Such a signal has been described in more detail in *S. pombe* (157). Polyadenylation sites in yeast are heterogeneous with respect to the orientation-dependence of their function; the majority seems to be orientation-independent (164, 166, 167). This might be due to the degeneracy of the signals and the function of invertable sequences, such as UAUAUA. Secondary structures might also be involved (165, 167). Poly(A) addition sites for a given gene can be scattered over a fairly large region (164).

Poly(A) polymerase from *S. cerevisiae* has been purified to homogeneity (168). Its properties are similar to those of the mammalian enzyme: it is specific for ATP but non-specific for the RNA, acts distributively, and is more active in the presence of Mn^{2+}. The 64 kDa enzyme is encoded by an essential gene, PAP1 (169). A temperature-sensitive allele leads to a loss of poly(A) tails at the restrictive temperature *in vivo* (170). The N-terminal 400 amino acids of yeast poly(A) polymerase, including the potential RNP-domain, are 47% identical to those of the mammalian enzyme, whereas the C-termini of the proteins have no obvious similarity. Fractionation of yeast extract has led to the identification of four factors required for 3′-end cleavage and polyadenylation (171). Two of these factors, CF1 and CF2, carry out the cleavage reaction. Poly(A) polymerase is not required for this activity. Three fractions are necessary for the polyadenylation of the upstream cleavage product. One contains poly(A) polymerase and can be replaced by a homogeneous enzyme purified from an overexpressing *E. coli* strain. A second fraction is CF1, which is also required for cleavage and, therefore, resembles mammalian CPSF. A third fraction, PF1, functions only in polyadenylation. These factors have not been purified to homogeneity. However, the products of two previously described genes, RNA14 and RNA15 (172), have been identified as subunits of CF1 (L. Minvielle *et al.*, submitted). The RNA15 gene product is an RNA-binding protein. A gene encoding a 35 kDa subunit of PF1 has also been isolated (P. Preker *et al.*, in preparation). The protein interacts directly with poly(A) polymerase. A prp20 mutation affects synthesis or maintenance of poly(A) tails through unknown mechanisms (173).

4.3.3 Plants

3′-end formation of plant mRNAs has recently been reviewed (174). Plant polyadenylation signals have at least a superficial similarity with those of yeast. They are composed of three sequence elements: a far-upstream element, which can be more than 100 nucleotides upstream of the polyadenylation site; a near-upstream element around 20 to 30 nucleotides upstream; and the polyadenylation site itself. Downstream sequences are not strictly required. Also, as in yeast, the sequences are poorly conserved and redundant: the far-upstream element can be inactivated only by large deletions. Repeats of the sequence UUUGUA are important in this element (174, 175). The near-upstream element is sensitive to linker scanning but relatively tolerant to point mutations (174, 175). An AAUAAA sequence can be part of the near-upstream element, but its conservation and sensitivity to mutation is far from that found in mammals. The degeneracy of plant polyadenylation signals has been emphasized in a study showing that the insertion of any of five different AU-rich fragments between 113 bp and 441 bp long, one of them from bacteriophage λ, caused 3′-end formation in a test mRNA (176). These authors argued that mere AU-richness might suffice as a polyadenylation signal. Finally, as in yeast, poly(A) addition often occurs at multiple sites spread over a distance of up to 150 nucleotides. The induction of 3′-end formation in yeast by a DNA fragment from cauliflower mosaic virus appeared to confirm the similarity of polyadenylation signals in yeast and plants. However, a mutational analysis showed that the sequences responsible were different from those used in plants (159). In general, polyadenylation signals are not interchangeable between yeast and plants (174). Due to the lack of an *in vitro* system, the mechanism of 3′-end formation in plants has not been established, and with the exception of poly(A) polymerase, the factors responsible are unknown.

4.3.4 Bacteria

Poly(A) polymerase and polyadenylated RNA have been found in bacteria (177 and references therein). The sequence of the polymerase (178) has no apparent similarity to other known poly(A) polymerases. Poly(A) polymerase is not essential in *Escherichia coli*, but one interesting biological function has been described: The gene encoding poly(A) polymerase, *pcnB*, was originally identified by mutations reducing the copy number of ColE1-type plasmids. Replication of these plasmids is controlled by the intracellular concentration of a small anti-sense RNA, RNA I. RNA I is normally polyadenylated. The absence of the poly(A) tail in *pcnB* mutants leads to a stabilization of RNA I, an increase of its intracellular concentration, and a more efficient repression of plasmid replication (179). This is in surprising contrast to eukaryotes, where poly(A) is generally thought to have a stabilizing influence on RNA (27).

5. Processing of histone pre-mRNA 3′ ends

Messenger RNAs encoding the major replication-dependent histones are the only known exceptions to the polyadenylation of mRNA. 3′-end processing of histone pre-mRNAs in metazoans consists of a single endonucleolytic cut immediately downstream of a stem–loop structure (reviewed in 180). A conserved purine-rich element downstream of the cleavage site is essential under all conditions tested, whereas the stem–loop structure has been found to be merely stimulatory in some studies (181–183) and essential in another (184). Both its secondary structure and sequence are required. In experiments in which the downstream element was moved away from the stem–loop, the processing site also moved, with cleavage always occurring at a fixed distance 5′ of the downstream element (185).

Histone 3′-end processing can be carried out *in vitro* in nuclear extract from mammalian cells. Detection of both the upstream and the downstream cleavage product demonstrated the endonucleolytic nature of the reaction. Processing proceeds in the presence of high concentrations of EDTA and appears to be ATP-independent (186). Three factors are thought to be involved. The downstream element of the histone pre-mRNA is complementary to a sequence near the 5′ end of the U7 snRNA. U7 is one of the minor snRNAs, approximately 65 nucleotides in length, containing a trimethylated cap and packaged in an Sm-precipitable snRNP (187–190; see Chapter 5). U7 is required for histone 3′-end processing (188, 191). Mutations in the downstream element of the precursor RNA that abolish processing can be rescued by compensatory mutations in U7. Thus, U7 is involved in 3′-end processing through base pairing with the precursor RNA (192, 193). The U7 snRNP contains a full set of the core snRNP proteins (see Chapter 5) and two U7-specific polypeptides of 14 kDa and 50 kDa (194). The isolated snRNP by itself is not competent for histone 3′-end processing.

When nuclear extracts are depleted of the U7 snRNP by treatment with appropriate antibodies, micrococcal nuclease, or oligonucleotide-targeted degradation by RNase H (see Chapter 5), the histone 3′-end processing activity is destroyed. The activity is also sensitive to a mild heat treatment of the extract. Heat-inactivated extracts can complement snRNP-depleted extracts, suggesting that two different components are affected. The heat-labile factor co-purifies with the U7-snRNP through a number of chromatographic steps but can be partially separated by gel filtration. It may thus be a loosely associated snRNP protein (195).

A third factor is thought to interact with the hairpin structure just upstream of the cleavage site (181, 189, 196). Although the hairpin-binding factor is precipitable by Sm-antisera, it is distinguished from U7 by its resistance to micrococcal nuclease and can be separated from the snRNP by chromatography. It is also believed to be distinct from the heat-labile factor because the latter does not react with Sm-antisera (195). The hairpin-binding factor has so far only been identified in binding studies and indirect assays. Direct functional assays to demonstrate an involvement of this factor in the processing reaction will require its separation from both the heat-labile factor and the U7 snRNP.

The major histone genes are regulated in a cell-cycle dependent manner (reviewed in 197). 3′-end processing is one of the regulated events, in addition to transcription initiation and RNA degradation. The activity of the heat-labile factor appears to be responsible for the cell-cycle regulation (198), possibly through a modulation of the U7 snRNP structure (199).

In yeast and in plants, histone mRNAs are polyadenylated.

References

1. Gallie, D. R. (1991) The cap and poly(A) tail function synergistically to regulate mRNA translational efficiency. *Genes Dev.*, **5**, 2108.
2. Muhlrad, D., Decker, C. J., and Parker, R. (1994) Deadenylation of the unstable mRNA encoded by the yeast MFA2 gene leads to decapping followed by 5′–3′ digestion of the transcript. *Genes Dev.*, **8**, 855.
3. Mizumoto, K. and Kaziro, Y. (1987) Messenger RNA capping enzymes from eukaryotic cells. *Prog. Nucl. Acid Res. Mol. Biol.*, **34**, 1.
4. Narayan, P. and Rottman, F. M. (1992) Methylation of mRNA. *Adv. Enzymol.*, **65**, 255.
5. Konarska, M. M., Padgett, R. A., and Sharp, P. A. (1984) Recognition of cap structure in splicing *in vitro* of mRNA precursors. *Cell*, **38**, 731.
6. Izaurralde, E., Stepinski, J., Darzynkiewicz, E., and Mattaj, I. W. (1992) A cap binding protein that may mediate nuclear export of RNA polymerase II-transcribed RNAs. *J. Cell Biol.*, **118**, 1287.
7. Rhoads, R. E. (1991) Initiation: mRNA and 60S subunit binding. In *Translation in eukaryotes*, Trachsel, H. (ed.). CRC Press, Boca Raton, p. 109.
8. Shibagaki, Y., Itoh, N., Yamada, H., Nagata, S., and Mizumoto, K. (1992) mRNA capping enzyme: Isolation and characterization of the gene encoding mRNA guanylyltransferase subunit from *Saccharomyces cerevisiae*. *J. Biol. Chem.*, **267**, 9521.
9. Cong, P. and Shuman, S. (1993) Covalent catalysis in nucleotidyl transfer: A KTDG motif essential for enzyme-GMP complex formation by mRNA capping enzyme is conserved at the active sites of RNA and DNA ligases. *J. Biol. Chem.*, **268**, 7256.
10. Niles, E. G. and Christen, L. (1993) Identification of the vaccinia virus mRNA guanylyltransferase active site lysine. *J. Biol. Chem.*, **268**, 24986.
11. Schwer, B. and Shuman, S. (1994) Mutational analysis of yeast mRNA capping enzyme. *Proc. Natl Acad. Sci. USA*, **91**, 4328.
12. Fresco, L. D. and Buratowski, S. (1994) Active site of the mRNA-capping enzyme

guanylyltransferase from *Saccharomyces cerevisiae*: similarity to the nucleotidyl attachment motif of DNA and RNA ligases. *Proc. Natl Acad. Sci. USA*, **91**, 6624.
13. Cong, P. and Shuman, S. (1992) Methyltransferase and subunit association domains of vaccinia virus mRNA capping enzyme. *J. Biol. Chem.*, **267**, 16424.
14. Higman, M. A., Bourgeois, N., and Niles, E. G. (1992) The vaccinia virus mRNA (guanine-N^7)-methyltransferase requires both subunits of the mRNA capping enzyme for activity. *J. Biol. Chem.*, **267**, 16430.
15. Higman, M. A., Christen, L. A., and Niles, E. G. (1994) The mRNA (guanine-7-)methyltransferase domain of the vaccinia virus mRNA capping enzyme: Expression in *Escherichia coli* and structural and kinetic comparison to the intact capping enzyme. *J. Biol. Chem.*, **269**, 14974.
16. Higman, M. A. and Niles, E. G. (1994) Location of the *S*-adenosyl-L-methionine binding region of the vaccinia virus mRNA (guanine-7)methyltransferase. *J. Biol. Chem.*, **269**, 14982.
17. Schnierle, B. S., Gershon, P. D., and Moss, B. (1992) Cap-specific mRNA (nucleoside-$O^{2'}$-)-methyltransferase and poly(A) polymerase stimulatory activities of vaccinia virus are mediated by a single protein. *Proc. Natl Acad. Sci. USA*, **89**, 2897.
18. Hagler, J. and Shuman, S. (1992) A freeze-frame view of eukaryotic transcription during elongation and capping of nascent mRNA. *Science*, **255**, 983.
19. Narayan, P., Ludwiczak, R. L., Goodwin, E. C., and Rottman, F. M. (1994) Context effects on N^6-adenosine methylation sites in prolactin mRNA. *Nucleic Acids Res.*, **22**, 419.
20. Narayan, P. and Rottman, F. M. (1988) An *in vitro* system for accurate methylation of internal adenosine residues in messenger RNA. *Science*, **242**, 1159.
21. Bokar, J. A., Rath-Shambaugh, M. E., Ludwiczak, R., Narayan, P., and Rottman, F. (1994) Characterization and partial purification of mRNA N^6-adenosine methyltransferase from HeLa cell nuclei. Internal mRNA methylation requires a multisubunit complex. *J. Biol. Chem.*, **269**, 17697.
22. Sachs, A. B. and Davis, R. W. (1989) The poly(A) binding protein is required for poly(A) shortening and 60S ribosomal subunit-dependent translation initiation. *Cell*, **58**, 857.
23. Munroe, D. and Jacobson, A. (1990) Tales of poly(A): a review. *Gene*, **91**, 151.
24. Sachs, A. (1990) The role of poly(A) in the translation and stability of mRNA. *Curr. Opinion Cell Biol.*, **2**, 1092.
25. Jackson, R. J. and Standart, N. (1990) Do the poly(A) tail and 3′ untranslated region control mRNA translation? *Cell*, **62**, 15.
26. Sachs, A. B. (1993) Messenger RNA degradation in eukaryotes. *Cell*, **74**, 413.
27. Decker, C. J. and Parker, R. (1994) Mechanisms of mRNA degradation in eukaryotes. *Trends Biochem. Sci.*, **19**, 336.
28. Wickens, M. and Stephenson, P. (1984) Role of the conserved AAUAAA sequence: four AAUAAA point mutants prevent messenger RNA 3′ end formation. *Science*, **226**, 1045.
29. Eckner, R., Ellmeier, W., and Birnstiel, M. L. (1991) Mature mRNA 3′ end formation stimulates RNA export from the nucleus. *EMBO J.*, **10**, 3513.
30. Wahle, E. and Keller, W. (1992) The biochemistry of 3′-end cleavage and polyadenylation of messenger RNA precursors. *Annu. Rev. Biochem.*, **61**, 419.
31. Higgs, D. R., Goodbourn, S. E. Y., Lamb, J., Clegg, J. B., Weatherall, D. J., and Proudfoot, N. J. (1983) α-Thalassaemia caused by a polyadenylation signal mutation. *Nature*, **306**, 398.

32. Orkin, S. H., Cheng, T.-C., Antonarakis, S. E., and Kazazian, H. H. (1985) Thalassemia due to a mutation in the cleavage-polyadenylation signal of the human β-globin gene. *EMBO J.*, **4**, 453.
33. Russnak, R. and Ganem, D. (1990) Sequences 5′ to the polyadenylation signal mediate differential poly(A) site use in hepatitis B viruses. *Genes Dev.*, **4**, 764.
34. Dorsett, D. (1990) Potentiation of a polyadenylylation site by a downstream protein–DNA interaction. *Proc. Natl Acad. Sci. USA*, **87**, 4373.
35. Enriquez-Harris, P., Levitt, N., Briggs, D., and Proudfoot, N. J. (1991). A pause site for RNA polymerase II is associated with termination of transcription. *EMBO J.*, **10**, 1833.
36. Rabbitts, K. G. and Morgan, G. T. (1992) Alternative 3′ processing of *Xenopus* α-tubulin mRNAs: efficient use of a CAUAAA polyadenylation signal. *Nucleic Acids Res.*, **20**, 2947.
37. Gil, A. and Proudfoot, N. J. (1987) Position-dependent sequence elements downstream of AAUAAA are required for efficient rabbit β-globin mRNA 3′ end formation. *Cell*, **49**, 399.
38. Chen, J. S. and Nordstrom, J. L. (1992) Bipartite structure of the downstream element of the mouse beta globin (major) poly(A) signal. *Nucleic Acids Res.*, **20**, 2565.
39. Sittler, A., Gallinaro, H., and Jacob, M. (1994) Upstream and downstream *cis*-acting elements for cleavage at the L4 poladenylation site of adenovirus-2. *Nucleic Acids Res.*, **22**, 222.
40. Goodwin, E. C. and Rottman, F. M. (1992) The 3′-flanking sequence of the bovine growth hormone gene contains novel elements required for efficient and accurate polyadenylation. *J. Biol. Chem.*, **267**, 16330.
41. McDevitt, M. A., Hart, R. P., Wong, W. W., and Nevins, J. R. (1986) Sequences capable of restoring poly(A) site function define two distinct downstream elements. *EMBO J.*, **5**, 2907.
42. Conway, L. and Wickens, M. (1987) Analysis of mRNA 3′ end formation by modification interference: the only modifications which prevent processing lie in AAUAAA and the poly(A) site. *EMBO J.*, **6**, 4177.
43. Zarkower, D. and Wickens, M. (1988) A functionally redundant downstream sequence in SV40 late pre-mRNA is required for mRNA 3′-end formation and for assembly of a precleavage complex in vitro. *J. Biol. Chem:*, **263**, 5780.
44. Levitt, N., Briggs, D., Gil, A., and Proudfoot, N. J. (1989) Definition of an efficient synthetic poly(A) site. *Genes Dev.*, **3**, 1019.
45. DeZazzo, J. D. and Imperiale, M. J. (1989) Sequences upstream of AAUAAA influence poly(A) site selection in a complex transcription unit. *Mol. Cell. Biol.*, **9**, 4951.
46. Carswell, S. and Alwine, J. C. (1989) Efficiency of utilization of the simian virus 40 late polyadenylation site: effects of upstream sequences. *Mol. Cell. Biol.*, **9**, 4248.
47. Prescott, J. and Falck-Pedersen, E. (1994) Sequence elements upstream of the 3′ cleavage site confer substrate strength to the adenovirus L1 and L3 polyadenylation sites. *Mol. Cell. Biol.*, **14**, 4682.
48. Schek, N. B., Cooke, C., and Alwine, J. C. (1992) Definition of the upstream efficiency element of the simian virus 40 late polyadenylation signal by using *in vitro* analyses. *Mol. Cell. Biol.*, **12**, 5386.
49. Proudfoot, N. (1991) Poly(A) signals. *Cell*, **64**, 671.
50. DeZazzo, J. D., Kilpatrick, J. E., and Imperiale, M. J. (1991) Involvement of long terminal repeat U3 sequences overlapping the transcription control region in human immunodeficiency virus type 1 mRNA 3′ end formation. *Mol. Cell. Biol.*, **11**, 1624.

51. Kurkulos, M., Weinberg, J. M., Pepling, M. E., and Mount, S. M. (1991) Polyadenylylation in *copia* requires unusually distant upstream sequences. *Proc. Natl Acad. Sci. USA*, **88**, 3038.
52. Valsamakis, A., Zeichner, S., Carswell, S., and Alwine, J. C. (1991) The human immunodeficiency virus type 1 polyadenylation signal: a 3′ long terminal repeat element upstream of the AAUAAA necessary for efficient polyadenylation. *Proc. Natl Acad. Sci. USA*, **88**, 2108.
53. Iwasaki, K. and Temin, H. M. (1990) The efficiency of RNA 3′-end formation is determined by the distance between the cap site and the poly(A) site in spleen necrosis virus. *Genes Dev.*, **4**, 2299.
54. Weichs an der Glon, C., Monks, J., and Proudfoot, N. J. (1991) Occlusion of the HIV poly(A) site. *Genes Dev.*, **5**, 244.
55. Sanfacon, H. and Hohn, T. (1990) Proximity to the promoter inhibits recognition of cauliflower mosaic virus polyadenylation signal. *Nature*, **346**, 81.
56. Iwasaki, K. and Temin, H. M. (1992) Multiple sequence elements are involved in RNA 3′ end formation in spleen necrosis virus. *Gene Expression*, **2**, 7.
57. Cherrington, J. and Ganem, D. (1992) Regulation of polyadenylation in human immunodeficiency virus (HIV): contributions of promoter proximity and upstream sequences. *EMBO J.*, **11**, 1513.
58. DeZazzo, J. D., Scott, J. M., and Imperiale, M. J. (1992) Relative roles of signals upstream of AAUAAA and promoter proximity in regulation of human immunodeficiency virus type 1 mRNA 3′ end formation. *Mol. Cell. Biol.*, **12**, 5555.
59. Weichs an der Glon, C., Ashe, M., Eggermont, J., and Proudfoot, N. J. (1993) Tat-dependent occlusion of the HIV poly(A) site. *EMBO J.*, **12**, 2119.
60. Sheets, M. D., Ogg, S. C., and Wickens, M. P. (1990) Point mutations in AAUAAA and the poly(A) addition site: effects on the accuracy and efficiency of cleavage and polyadenylation *in vitro*. *Nucleic Acids Res.*, **18**, 5799.
61. Chou, Z.-F., Chen, F., and Wilusz, J. (1994) Sequence and position requirements for uridylate-rich downstream elements of polyadenylation signals. *Nucleic Acids Res.*, **22**, 2525.
62. MacDonald, C. C., Wilusz, J., and Shenk, T. (1994) The 64-kilodalton subunit of the CStF polyadenylation factor binds to pre-mRNAs downstream of the cleavage site and influences cleavage site location. *Mol. Cell. Biol.*, **14**, 6647.
63. Moore, C. L. and Sharp, P. A. (1984) Site-specific polyadenylation in a cell-free reaction. *Cell*, **36**, 581.
64. Moore, C. L. and Sharp, P. A. (1985) Accurate cleavage and polyadenylation of exogenous RNA substrate. *Cell*, **41**, 845.
65. Valsamakis, A., Schek, N., and Alwine, J. C. (1992) Elements upstream of the AAUAAA within the human immunodeficiency virus polyadenylation signal are required for efficient polyadenylation in vitro. *Mol. Cell. Biol.*, **12**, 3699.
66. Gilmartin, G. M., Fleming, E. S., and Oetjen, J. (1992) Activation of HIV-1 pre-mRNA 3′-end processing *in vitro* requires both an upstream element and TAR. *EMBO J.*, **11**, 4419.
67. Moore, C. L., Skolnik-David, H., and Sharp, P. A. (1986) Analysis of RNA cleavage at the adenovirus-2 L3 polyadenylation site. *EMBO J.*, **5**, 1929.
68. Sheets, M. D., Stephenson, P., and Wickens, M. P. (1987) Products of in vitro cleavage and polyadenylation of simian virus 40 late pre-mRNAs. *Mol. Cell. Biol.*, **7**, 1518.
69. Zarkower, D., Stephenson, P., Sheets, M., and Wickens, M. (1986) The AAUAAA

sequence is required both for cleavage and for polyadenylation of simian virus 40 pre-mRNA *in vitro*. *Mol. Cell. Biol.*, **6**, 2317.
70. Bienroth, S., Wahle, E., Suter-Crazzolara, C., and Keller, W. (1991) Purification of the cleavage and polyadenylation factor involved in the 3′-processing of messenger RNA precursors. *J. Biol. Chem.*, **266**, 19768.
71. Murthy, K. G. K. and Manley, J. L. (1992) Characterization of the multisubunit cleavage-polyadenylation specificity factor from calf thymus. *J. Biol. Chem.*, **267**, 14804.
72. Takagaki, Y., Manley, J. L., MacDonald, C. C., Wilusz, and Shenk, T. (1990) A multisubunit factor, CStF, is required for polyadenylation of mammalian pre-mRNAs. *Genes Dev.*, **4**, 2112.
73. Gilmartin, G. M. and Nevins, J. R. (1989) An ordered pathway of assembly of components required for polyadenylation site recognition and processing. *Genes Dev.*, **3**, 2180.
74. Gilmartin, G. M. and Nevins, J. R. (1991) Molecular analyses of two poly(A) site-processing factors that determine the recognition and efficiency of cleavage of the pre-mRNA. *Mol. Cell. Biol.*, **11**, 2432.
75. Takagaki, Y., MacDonald, C. C., Shenk, T., and Manley, J. L. (1992) The human 64-kDa polyadenylylation factor contains a ribonucleoprotein-type RNA binding domain and unusual auxiliary motifs. *Proc. Natl Acad. Sci. USA*, **89**, 1403.
76. Burd, C. G. and Dreyfuss, G. (1994) Conserved structures and diversity of functions of RNA-binding proteins. *Science*, **265**, 615.
77. Takagaki, Y. and Manley, J. L. (1992) A human polyadenylation factor is a G protein β-subunit homologue. *J. Biol. Chem.*, **267**, 23471.
78. Christofori, G. and Keller, W. (1988) 3′-cleavage and polyadenylation of mRNA precursors in vitro requires a poly(A) polymerase, a cleavage factor and a snRNP. *Cell*, **54**, 875.
79. Takagaki, Y., Ryner, L., and Manley, J. L. (1989) Four factors are required for 3′-end cleavage of pre-mRNAs. *Genes Dev.*, **3**, 1711.
80. Qian, Z. and Wilusz, J. (1991) An RNA-binding protein specifically interacts with a functionally important domain of the downstream element of the simian virus 40 late polyadenylation signal. *Mol. Cell. Biol.*, **11**, 5312.
81. Takagaki, Y., Ryner, L., and Manley, J. L. (1988) Separation and characterization of a poly(A) polymerase and a cleavage/specificity factor required for pre-mRNA polyadenylation. *Cell*, **52**, 731.
82. Christofori, G. and Keller, W. (1989) Poly(A) polymerase purified from HeLa cell nuclear extract is required for both cleavage and polyadenylation of pre-mRNA *in vitro*. *Mol. Cell. Biol.*, **9**, 193.
83. Tsiapalis, C. M., Dorson, J. W., and Bollum, F. J. (1975) Purification of terminal riboadenylate transferase from calf thymus gland. *J. Biol. Chem.*, **250**, 4486.
84. Wahle, E. (1991) Purification and characterization of a mammalian polyadenylate polymerase involved in the 3′ end processing of messenger RNA precursors. *J. Biol. Chem.*, **266**, 3131.
85. Raabe, T., Bollum, F. J., and Manley, J. L. (1991) Primary structure and expression of bovine poly(A) polymerase. *Nature*, **353**, 229.
86. Wahle E., Martin, G., Schiltz, E., and Keller, W. (1991) Isolation and expression of cDNA clones encoding mammalian poly(A) polymerase. *EMBO J.*, **10**, 4251.
87. Raabe, T., Murthy, K. G. K., and Manley, J. L. (1994) Poly(A) polymerase contains multiple functional domains. *Mol. Cell. Biol.*, **14**, 2946.

88. Thuresson, A.-C., Aström, J., Aström, A., Grönvik, K.-O., and Virtanen, A. (1994) Multiple forms of poly(A) polymerases in human cells. *Proc. Natl Acad. Sci. USA*, **91**, 979.
89. Wahle, E. (1991) A novel poly(A)-binding protein acts as a specificity factor in the second phase of messenger RNA polyadenylation. *Cell*, **66**, 759.
90. Wahle, E., Lustig, A., Jenö, P., and Maurer, P. (1993) Mammalian poly(A)-binding protein II: physical properties and binding to polynucleotides. *J. Biol. Chem.*, **268**, 2937.
91. Krause, S., Fakan, S., Weis, K., and Wahle, E. (1994) Immunodetection of poly(A) binding protein II in the cell nucleus. *Exp. Cell Res.*, **214**, 75.
92. Nevins, J. R. (1983) The pathway of eukaryotic mRNA formation. *Annu. Rev. Biochem.*, **52**, 441.
93. Prescott, J. C. and Falck-Pedersen, E. (1992) Varied poly(A) site efficiency in the adenovirus major late transcription unit. *J. Biol. Chem.*, **267**, 8175.
94. Bardwell, V. J., Wickens, M., Bienroth, S., Keller, W., Sproat, B. S., and Lamond, A. I. (1991) Site-directed ribose methylation identifies 2′-OH groups in polyadenylation substrates critical for AAUAAA recognition and poly(A) addition. *Cell*, **65**, 125.
95. Keller, W., Bienroth, S., Lang, K., and Christofori, G. (1991) Cleavage and polyadenylation factor CPF specifically interacts with the pre-mRNA 3′ processing signal AAUAAA. *EMBO J.*, **10**, 4241.
96. Wigley, P. L., Sheets, M. D., Zarkower, D. A., Whitmer, M. E., and Wickens, M. (1990) Polyadenylation of mRNA: minimal substrates and a requirement for the 2′ hydroxyl of the U in AAUAAA. *Mol. Cell. Biol.*, **10**, 1705.
97. Bienroth, S., Keller, W., and Wahle, E. (1993) Assembly of a processive polyadenylation complex. *EMBO J.*, **12**, 585.
98. Weiss, E. A., Gilmartin, G., and Nevins, J. R. (1991) Poly(A) site efficiency reflects the stability of complex formation involving the downstream element. *EMBO J.*, **10**, 215.
99. Wilusz, J. and Shenk, T. (1988) A 64 kd nuclear protein binds to RNA segments that include the AAUAAA polyadenylation motif. *Cell*, **52**, 221.
100. Wilusz, J., Shenk, T., Takagaki, Y., and Manley, J. (1990) A multicomponent complex is required for the AAUAAA-dependent cross-linking of a 64-kilodalton protein to polyadenylation substrates. *Mol. Cell. Biol.*, **10**, 1244.
101. Lutz, C. S. and Alwine, J. C. (1994) Direct interaction of the U1 snRNP-A protein with the upstream efficiency element of the SV40 late polyadenylation signal. *Genes Dev.*, **8**, 576.
102. Bilger, A., Fox, C. A., Wahle, E., and Wickens, M. (1994) Nuclear polyadenylation factors recognize cytoplasmic polyadenylation elements. *Genes Dev.*, **8**, 1106.
103. Sheets, M. D. and Wickens, M. (1989) Two phases in the addition of a poly(A) tail. *Genes Dev.*, **3**, 1401.
104. Smale, S. T. and Tjian, R. (1985) Transcription of herpes simplex virus tk sequences under the control of wild-type and mutant RNA polymerase I promoters. *Mol. Cell. Biol.*, **5**, 352.
105. Neumann de Vegvar, H. E., Lund, E., and Dahlberg, J. E. (1986) 3′ end formation of U1 snRNA precursors is coupled to transcription from snRNA promoters. *Cell*, **47**, 259.
106. Sisodia, S. S., Sollner-Webb, B., and Cleveland, D. W. (1987) Specificity of RNA maturation pathways: RNAs transcribed by RNA polymerase III are not substrates for splicing or polyadenylation. *Mol. Cell. Biol.*, **7**, 3602.
107. Dahlberg, J. E. and Schenborn, E. T. (1988). The human U1 snRNA promoter and enhancer do not direct synthesis of messenger RNA. *Nucleic Acids Res.*, **16**, 5827.

108. Salditt-Georgieff, M., Harpold, M., Sawicki, S., Nevins, J., and Darnell, J. E. (1980) Addition of poly(A) to nuclear RNA occurs soon after RNA synthesis. *J. Cell Biol.*, **86**, 844.
109. Mifflin, R. C. and Kellems, R. E. (1991) Coupled transcription–polyadenylation in a cell-free system. *J. Biol. Chem.*, **266**, 19593.
110. Hyman, L. E. and Moore, C. L. (1993) Termination and pausing of RNA polymerase II downstream of yeast polyadenylation sites. *Mol. Cell. Biol.*, **13**, 5159.
111. Edwalds-Gilbert, G., Prescott, J., and Falck-Pedersen, E. (1993) 3′ RNA processing efficiency plays a primary role in generating termination-competent RNA polymerase II elongation complexes. *Mol. Cell. Biol.*, **13**, 3472.
112. Enriquez-Harris, P., Levitt, N., Briggs, D., and Proudfoot, N. J. (1991) A pause site for RNA polymerase II is associated with termination of transcription. *EMBO J.*, **10**, 1833.
113. Eggermont, J. and Proudfoot, N. J. (1993) Poly(A) signals and transcriptional pause sites combine to prevent interference between RNA polymerase II promoters. *EMBO J.*, **12**, 2539.
114. Huang, M. T. F. and Gorman, C. M. (1990) Intervening sequences increase efficiency of RNA 3′ processing and accumulation of cytoplasmic RNA. *Nucleic Acids Res.*, **18**, 937.
115. Chiou, H. D., Dabrowski, C., and Alwine, J. C. (1991) Simian virus 40 late mRNA leader sequences involved in augmenting mRNA accumulation via multiple mechanisms, including increased polyadenylation efficiency. *J. Virol.*, **65**, 6677.
116. Pandey, N. B., Chodchoy, N., Liu, T.-J., and Marzluff, W. F. (1990) Introns in histone genes alter the distribution of 3′-ends. *Nucleic Acids Res.*, **18**, 3161.
117. Luo, Y. and Carmichael, G. G. (1991) Splice site choice in a complex transcription unit containing multiple inefficient polyadenylation signals. *Mol. Cell. Biol.*, **11**, 5291.
118. Niwa, M., Rose, S. D., and Berget, S. M. (1990) In vitro polyadenylation is stimulated by the presence of an upstream intron. *Genes Dev.*, **4**, 1552.
119. Niwa, M. and Berget, S. (1991) Mutation of the AAUAAA polyadenylation signal depresses *in vitro* splicing of proximal but not distal introns. *Genes Dev.*, **5**, 2086.
120. Niwa, M., MacDonald, C. C., and Berget, S. M. (1992) Are vertebrate exons scanned during splice-site selection? *Nature*, **360**, 277.
121. Liu, X. and Mertz, J. E. (1993) Polyadenylation site selection cannot occur *in vivo* after excision of the 3′-terminal intron. *Nucleic Acids Res.*, **21**, 5256.
122. Nesic, D., Cheng, J., and Maquat, L. E. (1993) Sequences within the last intron function in RNA 3′-end formation in cultured cells. *Mol. Cell. Biol.*, **13**, 3359.
123. Nesic, D. and Maquat, L. E. (1994) Upstream introns influence the efficiency of final intron removal and RNA 3′-end formation. *Genes Dev.*, **8**, 363.
124. Robberson, B. L., Cote, G., and Berget, S. M. (1990) Exon definition may participate in splice site recognition in multi-exon RNAs. *Mol. Cell. Biol.*, **10**, 84.
125. Montzka Wassarman, K. and Steitz, J. (1993) Association with terminal exons in pre-mRNAs: a new role for the U1 snRNP? *Genes. Dev.*, **7**, 647.
126. Mann, K. P., Weiss, E. A., and Nevins, J. R. (1993) Alternative poly(A) site utilization during adenovirus infection coincides with a decrease in the activity of a poly(A) site processing factor. *Mol. Cell. Biol.*, **13**, 2411.
127. Falck-Pedersen, E. and Logan, J. (1989) Regulation of poly(A) site selection in adenovirus. *J. Virol.*, **63**, 532.
128. Peterson, M. L. and Perry, R. P. (1989) The regulated production of μ_m and μ_s mRNA is dependent on the relative efficiencies of μ_s poly(A) site usage and the Cμ4-to-M1 splice. *Mol. Cell. Biol.*, **9**, 726.

129. Peterson, M. L., Gimmi, E. R., and Perry, R. P. (1991) The developmentally regulated shift from membrane to secreted μ mRNA production is accompanied by an increase in cleavage–polyadenylation efficiency but no measurable change in splicing efficiency. *Mol. Cell. Biol.*, **11**, 2324.
130. Peterson, M. L. (1992) Balanced efficiencies of splicing and cleavage–polyadenylation are required for μ_s and μ_m mRNA regulation. *Gene Expression*, **2**, 319.
131. Gunderson, S. I., Beyer, K., Martin, G., Keller, W., Boelens, W. C., and Mattaj, I. W. (1994) The human U1A snRNP protein regulates polyadenylation via a direct interaction with poly(A) polymerase. *Cell*, **76**, 531.
132. Wickens, M. (1990) In the beginning is the end: regulation of poly(A) addition and removal during early development. *Trends Biochem. Sci.*, **15**, 320.
133. Wickens, M. (1992) Forward, backward, how much, when: mechanisms of poly(A) addition and removal and their role in early development. *Sem. Dev. Biol.*, **3**, 399.
134. Varnum, S. M. and Wormington, W. M. (1990) Deadenylation of maternal mRNAs during *Xenopus* oocyte maturation does not require specific *cis*-sequences: a default mechanism for translational control. *Genes Dev.*, **4**, 2278.
135. Fox, C. A. and Wickens, M. (1990) Poly(A) removal during oocyte maturation: a default reaction selectively prevented by specific sequences in the 3′ UTR of certain maternal mRNAs. *Genes Dev.* **4**, 2287.
136. Bouvet, P., Omilli, F., Arlot-Bonnemains, Y., Legagneux, V., Roghi, C., Bassez, T., *et al.* (1994) The deadenylation conferred by the 3′ untranslated region of a developmentally controlled mRNA in *Xenopus* embryos is switched to polyadenylation by deletion of a short sequence element. *Mol. Cell. Biol.*, **14**, 1893.
137. Huarte, J., Stutz, A., O'Connell, M. L., Gubler, P., Belin, D., Darrow, A. L., *et al.* (1992) Transient translational silencing by reversible mRNA deadenylation. *Cell*, **69**, 1021.
138. Vassalli, J.-D., Huarte, J., Belin, D., Gubler, P., Vassalli, A., O'Connell, M. L., *et al.* (1989) Regulated polyadenylation controls mRNA translation during meiotic maturation of mouse oocytes. *Genes Dev.*, **3**, 2163.
139. McGrew, L. L., Dworkin-Rastl, E., Dworkin, M. B., and Richter, J. D. (1989) Poly(A) elongation during *Xenopus* oocyte maturation is required for translational recruitment and is mediated by a short sequence element. *Genes Dev.*, **3**, 803.
140. Paris, J. and Richter, J. D. (1990) Maturation-specific polyadenylation and translational control: diversity of cytoplasmic polyadenylation elements, influence of poly(A) tail size, and formation of stable polyadenylation complexes. *Mol. Cell. Biol.*, **10**, 5634.
141. Sheets, M. D., Fox, C. A., Hunt, T., Vande Woude, G., and Wickens, M. (1994) The 3′-untranslated regions of c-*mos* and cyclin mRNAs stimulate translation by regulating cytoplasmic polyadenylation. *Genes Dev.*, **8**, 926.
142. Fox, C. A., Sheets, M. D., and Wickens, M. P. (1989) Poly(A) addition during maturation of frog oocytes: distinct nuclear and cytoplasmic activities and regulation by the sequence UUUUUAU. *Genes Dev.*, **3**, 2151.
143. Salles, F. J., Darrow, A. L., O'Connell, M. L., and Strickland, S. (1992) Isolation of novel murine maternal mRNAs regulated by cytoplasmic polyadenylation. *Genes Dev.*, **6**, 1202.
144. Fox, C., Sheets, M., Wahle, E., and Wickens, M. (1992) Polyadenylation of maternal mRNA during oocyte maturation: poly(A) addition in vitro requires a regulated RNA binding activity and a poly(A) polymerase. *EMBO J.*, **11**, 5021.
145. Paris, J., Swenson, K., Piwnica-Worms, H., and Richter, J. D. (1991) Maturation-specific

polyadenylation: in vitro activation by p34[cdc2] and phosphorylation of a 58-kD CPE-binding protein. *Genes Dev.*, **5**, 1697.

146. McGrew, L. L. and Richter, J. D. (1990) Translational control by cytoplasmic polyadenylation during *Xenopus* oocyte maturation: characterization of *cis* and *trans* elements and regulation by cyclin/MPF. *EMBO J.*, **9**, 3743.
147. Moss, B., Rosenblum, E. N., and Gershowitz, A. (1975) Characterization of a polyriboadenylate polymerase from vaccinia virions. *J. Biol. Chem.*, **250**, 4722.
148. Gershon, P., Ahn, B.-Y., Garfield, M., and Moss, B. (1991) Poly(A) polymerase and a dissociable polyadenylation stimulatory factor encoded by vaccinia virus. *Cell*, **66**, 1269.
149. Schnierle, B. S., Gershon, P. D., and Moss, B. (1994) Mutational analysis of a multifunctional protein, with mRNA 5′ cap-specific (nucleoside-2′-*O*-)-methyltransferase and 3′ adenylyltransferase stimulatory activities, encoded by vaccinia virus. *J. Biol. Chem.*, **269**, 20700.
150. Gershon, P. D. and Moss, B. (1992) Transition from rapid processive to slow nonprocessive polyadenylation by vaccinia virus poly(A) polymerase catalytic subunit is regulated by the net length of the poly(A) tail. *Genes Dev.*, **6**, 1575.
151. Gershon, P. D. and Moss, B. (1993) Uridylate-containing RNA sequences determine specificity for binding and polyadenylation by the catalytic subunit of vaccinia virus poly(A) polymerase. *EMBO J.*, **12**, 4705.
152. Shuman, S. and Moss, B. (1988) Vaccinia virus poly(A) polymerase: Specificity for nucleotides and nucleotide analogs. *J. Biol. Chem.*, **263**, 8405.
153. Gershon, P. D. and Moss, B. (1993) Stimulation of poly(A) tail elongation by the VP39 subunit of the vaccinia virus-encoded poly(A) polymerase. *J. Biol. Chem.*, **268**, 2203.
154. Butler, J. S. and Platt, T. (1988) RNA processing generates the mature 3′ end of yeast CYC1 messenger RNA in vitro. *Science*, **242**, 1270.
155. Butler, J. S., Sadhale, P., and Platt, T. (1990) RNA processing *in vitro* produces mature 3′ ends of a variety of *Saccharomyces cerevisiae* mRNAs. *Mol. Cell. Biol.*, **10**, 2599.
156. Sadhale, P. P., Sapolsky, R., Davis, R. W., Butler, J. S., and Platt, T. (1991) Polymerase chain reaction mapping of yeast GAL7 mRNA polyadenylation sites demonstrates that 3′ end processing *in vitro* faithfully reproduces the 3′ ends observed *in vivo*. *Nucleic Acids Res.*, **19**, 3683.
157. Humphrey, T., Birse, C. E., and Proudfoot, N. J. (1994) RNA 3′ end signals of the *S. pombe* ura4 gene comprise a site determining and efficiency element. *EMBO J.*, **13**, 2441.
158. Humphrey, T., Sadhale, P., Platt, T., and Proudfoot, N. (1991) Homologous mRNA 3′ end formation in fission and budding yeast. *EMBO J.*, **10**, 3503.
159. Irniger, S., Sanfacon, H., Egli, C. M., and Braus, G. H. (1992) Different sequence elements are required for function of the cauliflower mosaic virus polyadenylation site in *Saccharomyces cerevisiae* compared with in plants. *Mol. Cell. Biol.*, **12**, 2322.
160. Irniger, S. and Braus, G. H. (1994) Saturation mutagenesis of a polyadenylylation signal reveals a hexanucleotide element essential for mRNA 3′ end formation in *Saccharomyces cerevisiae*. *Proc. Natl Acad. Sci. USA*, **91**, 257.
161. Russo, P., Li, W.-Z., Guo, Z., and Sherman, F. (1993) Signals that produce 3′ termini in CYC1 mRNA of the yeast *Saccharomyces cerevisiae*. *Mol. Cell. Biol.*, **13**, 7836.
162. Hou, W., Russnak, R., and Platt, T. (1994) Poly(A) site selection in the yeast Ty retroelement requires an upstream region and sequence-specific titratable factor(s) *in vitro*. *EMBO J.*, **13**, 446.

163. Heidmann, S., Schindewolf, C., Stumpf, G., and Domdey, H. (1994) Flexibility and interchangeability of polyadenylation signals in *Saccharomyces cerevisiae*. *Mol. Cell. Biol.*, **14**, 4633.
164. Heidmann, S., Obermaier, B., Vogel, K., and Domdey, H. (1992) Identification of pre-mRNA polyadenylation sites in *Saccharomyces cerevisiae*. *Mol. Cell. Biol.*, **12**, 4215.
165. Hyman, L. E., Seiler, S. H., Whoriskey, J., and Moore, C. L. (1991) Point mutations upstream of the yeast ADH2 poly(A) site significantly reduce the efficiency of 3′-end formation. *Mol. Cell. Biol.*, **11**, 2004.
166. Irniger, S., Egli, C. M., and Braus, G. H. (1991) Different classes of polyadenylation sites in the yeast *Saccharomyces cerevisiae*. *Mol. Cell. Biol.*, **11**, 3060.
167. Sadhale, P. P. and Platt, T. (1992) Unusual aspects of *in vitro* RNA processing in the 3′ regions of the GAL1, GAL7, and GAL10 genes in *Saccharomyces cerevisiae*. *Mol. Cell. Biol.*, **12**, 4262.
168. Lingner, J., Radtke, I., Wahle, E., and Keller, W. (1991) Purification and characterization of poly(A) polymerase from *Saccharomyces cerevisiae*. *J. Biol. Chem.*, **266**, 8741.
169. Lingner, J., Kellermann, J., and Keller, W. (1991) Cloning and expression of the essential gene for poly(A) polymerase from *S. cerevisiae*. *Nature*, **354**, 496.
170. Patel, D. and Butler, J. S. (1992) Conditional defect in mRNA 3′ end processing caused by a mutation in the gene for poly(A) polymerase. *Mol. Cell. Biol.*, **12**, 3297.
171. Chen, J. and Moore, C. (1992) Separation of factors required for cleavage and polyadenylation of yeast pre-mRNA. *Mol. Cell. Biol.*, **12**, 3470.
172. Minvielle-Sebastia, L., Winsor, B., Bonneaud, N., and Lacroute, F. (1991) Mutations in the yeast RNA14 and RNA15 genes result in an abnormal mRNA decay rate; sequence analysis reveals an RNA-binding domain in the RNA15 protein. *Mol. Cell. Biol.*, **11**, 3075.
173. Forrester, W., Stutz, F., Rosbash, M., and Wickens, M. (1992) Defects in mRNA 3′-end formation, transcription initiation, and mRNA transport associated with the yeast mutation *prp20*: possible coupling of mRNA processing and chromatin structure. *Genes Dev.*, **6**, 1914.
174. Hunt, A. G. (1994) Messenger RNA 3′ end formation in plants. *Annu. Rev. Plant Physiol. Plant Mol. Biol.*, **45**, 47.
175. Rothnie, H. M., Reid, J., and Hohn, T. (1994) The contribution of AAUAAA and the upstream element UUUGUA to the efficiency of mRNA 3′-end formation in plants. *EMBO J.*, **13**, 2200.
176. Luehrsen, K. R. and Walbot, V. (1994) Intron creation and polyadenylation in maize are directed by AU-rich RNA. *Genes Dev.*, **8**, 1117.
177. Cao, G.-J. and Sarkar, N. (1992) Poly(A) RNA in *Escherichia coli*: nucleotide sequence at the junction of the 1pp transcript and the polyadenylate moiety. *Proc. Natl Acad. Sci. USA*, **89**, 7546.
178. Cao, G.-J. and Sarkar, N. (1992) Identification of the gene for an *Escherichia coli* poly(A) polymerase. *Proc. Natl Acad. Sci. USA*, **89**, 10380.
179. Xu, F., Lin-Chao, S., and Cohen, S. N. (1993) The *Escherichia coli pcnB* gene promotes adenylylation of antisense RNAI of ColE1-type plasmids *in vivo* and degradation of RNAI decay intermediates. *Proc. Natl Acad. Sci. USA*, **90**, 6756.
180. Birnstiel, M. L. and Schaufele, F. J. (1988) Structure and function of minor snRNPs. In *Structure and function of major and minor small nuclear ribonucleoprotein particles*, Birnstiel, M. L. (ed.). Springer, Berlin, p. 155.
181. Vasserot, A. P., Schaufele, F. J., and Birnstiel, M. L. (1989) Conserved terminal hairpin sequences of histone mRNA precursors are not involved in duplex formation with the

U7 RNA but act as a target site for a distinct processing factor. *Proc. Natl Acad. Sci. USA*, **86**, 4345.

182. Mowry, K. L., Oh, R., and Steitz, J. A. (1989) Each of the conserved sequence elements flanking the cleavage site of mammalian histone pre-mRNAs has a distinct role in the 3′-end processing reaction. *Mol. Cell. Biol.*, **9**, 3105.
183. Streit, A., Wittop Koning, T., Soldati, D., Melin, L., and Schümperli, D. (1993) Variable effects of the conserved RNA hairpin element upon 3′ end processing of histone pre-mRNA in vitro. *Nucleic Acids Res.*, **21**, 1569.
184. Pandey, N. B., Williams, A. S., Sun, J-H., Brown, V. D., Bond, U., and Marzluff, W. F. (1994) Point mutations in the stem-loop at the 3′ end of mouse histone mRNA reduce expression by reducing the efficiency of 3′ end formation. *Mol. Cell. Biol.*, **14**, 1709.
185. Scharl, E. C. and Steitz, J. A. (1994) The site of 3′ end formation of histone messenger RNA is a fixed distance from the downstream element recognized by the U7 snRNP. *EMBO J.*, **13**, 2432.
186. Gick, O., Krämer, A., Keller, W., and Birnstiel, M. L. (1986) Generation of histone mRNA 3′ ends by endonucleolytic cleavage of the pre-mRNA in a snRNP-dependent in vitro reaction. *EMBO J.*, **5**, 1319.
187. Strub, K., Galli, G., Busslinger, M., and Birnstiel, M. L. (1984) The cDNA sequences of the sea urchin U7 small nuclear RNA suggest specific contacts between histone mRNA precursor and U7 RNA during RNA processing. *EMBO J.*, **3**, 2801.
188. Strub, K. and Birnstiel, M. L. (1986) Genetic complementation in the *Xenopus* oocyte: co-expression of sea urchin histone and U7 RNAs restores processing of H3 pre-mRNA in the oocyte. *EMBO J.*, **5**, 1675.
189. Mowry, K. L. and Steitz, J. A. (1987) Identification of the human U7 snRNP as one of several factors involved in the 3′ end maturation of histone premessenger RNA's. *Science*, **238**, 1682.
190. Cotten, M., Gick, O., Vasserot, A., Schaffner, G., and Birnstiel, M. L. (1988) Specific contacts between mammalian U7 snRNA and histone precursor RNA are indispensable for the in vitro 3′ RNA processing reaction. *EMBO J.*, **7**, 801.
191. Galli, G., Hofstetter, H., Stunnenberg, H. G., and Birnstiel, M. L. (1983) Biochemical complementation with RNA in the *Xenopus* oocyte: a small RNA is required for the generation of 3′ histone mRNA termini. *EMBO J.*, **3**, 823.
192. Schaufele, F., Gilmartin, G. M., Bannwarth, W., and Birnstiel, M. L. (1986) Compensatory mutations suggest that base-pairing with a small nuclear RNA is required to form the 3′ end of H3 messenger RNA. *Nature*, **323**, 777.
193. Bond, U. M., Yario, T. A., and Steitz, J. A. (1991) Multiple processing-defective mutations in a mammalian histone pre-mRNA are suppressed by compensatory changes in U7 RNA both *in vivo* and *in vitro*. *Genes Dev.*, **5**, 1709.
194. Smith, H. O., Tabiti, K., Schaffner, G., Soldati, D., Albrecht, U., and Birnstiel, M. L. (1991) Two-step affinity purification of U7 small nuclear ribonucleoprotein particles using complementary biotinylated 2′-*O*-methyl oligoribonucleotides. *Proc. Natl Acad. Sci. USA*, **88**, 9784.
195. Gick, O., Krämer, A., Vasserot, A., and Birnstiel, M. L. (1987) Heat-labile regulatory factor is required for 3′ processing of histone precursor mRNAs. *Proc. Natl Acad. Sci. USA*, **84**, 8937.
196. Mowry, K. L. and Steitz, J. A. (1987) Both conserved signals on mammalian histone pre-mRNAs associate with small nuclear ribonucleoproteins during 3′ end formation *in vitro*. *Mol. Cell. Biol.*, **7**, 1663.

197. Schümperli, D. (1988) Multilevel regulation of replication-dependent histone genes. *Trends Genet.*, **4**, 187.
198. Lüscher, B. and Schümperli, D. (1987) RNA 3′ processing regulates histone mRNA levels in a mammalian cell cycle mutant: A processing factor becomes limiting in G1-arrested cells. *EMBO J.*, **6**, 1721.
199. Hoffmann, I. and Birnstiel, M. L. (1990) Cell cycle-dependent regulation of histone pre-cursor mRNA processing by modulation of U7 snRNA accessibility. *Nature*, **346**, 665.

10 | *Trans*-splicing

TIMOTHY W. NILSEN

1. Introduction

Intermolecular (*trans*-) splicing is by definition an RNA-processing reaction which precisely joins exons from separately transcribed RNAs. As discussed elsewhere in this volume, there are four types of *cis*-splicing reactions:

(1) excision of group I introns (in many cases by autocatalysis) (see Chapter 1);

(2) excision of group II introns (in some cases by autocatalysis) (see Chapter 1);

(3) snRNP-mediated splicing of nuclear pre-mRNAs (see Chapters 4–7); and

(4) enzymatic removal of introns from pre-tRNAs (reviewed in 1, 2).

Various experimental approaches (both *in vivo* and *in vitro*) have demonstrated that all four types of splicing can also occur in *trans* (that is, with exons separated on different RNAs). However, in nature, only two types of *trans*-splicing have been well documented. The first type is directly analogous to *cis*-splicing of group II introns and is represented by 'split' group II introns in plant and fungal organelles. The second type of *trans*-splicing found in nature is termed spliced-leader (SL) addition *trans*-splicing. SL-addition *trans*-splicing has been demonstrated in a variety of lower eukaryotes and has been most thoroughly studied in the trypanosomatid protozoans and in nematodes.

The existence of *trans*-splicing poses many challenging mechanistic as well as functional questions and adds an unanticipated layer of complexity to the understanding of eukaryotic mRNA maturation. The intent of this chapter is to provide a summary of the current state of knowledge concerning *trans*-splicing and to highlight important remaining questions, of which there are many. Several reviews covering various aspects of *trans*-splicing have appeared in recent years and the reader is referred to these for additional information and alternative perspectives (1–12).

2. *Trans*-splicing of organellar group II introns

As reviewed in Chapter 1, the organellar genomes of plants and fungi contain numerous group I and group II introns. These introns are typed based upon their characteristic secondary structure features. Group II introns, in particular, conform to a structure composed of six stem–loop domains radiating from a central core (13). The first evidence that such introns could exist in pieces came from analysis of transcripts encoding the tobacco chloroplast ribosomal protein S12 (14). Here, the mature mRNA contains three exons. Exons 2 and 3 and flanking sequence are co-transcribed; these exons are separated by a conventional group II intron. Remarkably, exon 1 and its 3′ flanking sequence are transcribed from the other strand of the chloroplast genome as a distinct RNA. Analysis of the transcribed sequences downstream of exon 1 and upstream of exon 2 indicated that the two separate RNAs could, in combination, form the diagnostic group II secondary structure. Briefly, the two RNAs could associate via a base-pairing interaction predicted to form the stem of domain 3. Following annealing, exon 1 could be linked to exon 2 via the group II splicing pathway (14).

Following the initial description of group II *trans*-splicing in the tobacco chloroplast, numerous examples of similar split introns have emerged (see 12, 15–19 and references therein). With one exception, these introns are 'broken' at a single point which separates the base-pairing partners of a stem–loop domain. *Trans*-splicing of these introns can be explained by annealing of appropriate complementary sequences. It is at present unknown if this annealing requires the participation of protein factors. Recently, a more complex example of group II *trans*-splicing has been described. The *psaA* gene of *Chlamydomonas reinhardtii* has a very unusual structure where three exons are encoded in dispersed regions of the chloroplast genome. The mature *psaA* mRNA is assembled by *trans*-splicing of three separate transcripts (17). Genetic evidence indicated that a specific locus distinct from those encoding the three transcripts was necessary for *trans*-splicing of exons 1 and 2. Remarkably, transformation-rescue experiments have shown that the locus encodes a small (about 0.7 kb) RNA (17). When combined with the RNAs encoding exons 1 and 2, the small RNA restores a typical group II intron secondary structure. As with the examples discussed above, association of the small RNA with exon-containing RNAs is thought to occur via base-pairing interactions. These observations are exciting because they suggest a possible scenario for the evolution of the U snRNAs which are required for nuclear pre-mRNA splicing (11, 12, 20; see Chapters 1 and 5). Furthermore, analysis of this type of complex split group II intron may provide important insights into the mechanism of group II splicing itself.

It should be noted here that in all the examples of naturally occurring *trans*-spliced group II introns studied to date, association of participating RNAs is presumed to occur through interaction of complementary sequences. *In vitro* experiments examining a self-splicing group II intron indicate that separated RNAs can associate via other mechanisms (presumably tertiary interactions) (21). It will be of considerable interest if analogous split group II introns are discovered in nature.

Finally, recent analysis has revealed a fascinating relationship between *trans*-splicing of group II introns and RNA editing. As discussed in Chapter 11, RNA editing is a post-transcriptional RNA processing event which changes the informational content of a primary RNA transcript. Usually RNA editing is thought of in terms of remodeling the coding sequence of mRNAs to yield translatable molecules. It is widespread in plant mitochondria and is clearly necessary for production of translatable mRNAs (reviewed in 22). Remarkably, RNA editing is also apparently necessary for the productive *trans*-splicing of some plant mitochondrial group II introns. Here, the primary transcripts of split exons are edited both in exons and in introns. Editing of intronic sequence is required to create the appropriate complementarities needed for association of the separately transcribed RNAs (19, 22, 23). These are (at present) the only examples of editing of intron sequence and they raise intriguing questions relevant to the origin and function of RNA editing in general.

In summary, the discovery and analysis of *trans*-spliced introns in organellar genomes has provided strong support for secondary structure models of group II introns. Additionally, studies of these split introns have suggested intriguing (if speculative) models for the evolution of U snRNAs (11, 12).

3. Spliced-leader addition *trans*-splicing: overview

3.1 Discovery and characterization

SL addition *trans*-splicing was initially discovered in trypanosomes as a result of experiments designed to characterize the genes and mRNAs encoding variant surface glycoproteins (VSGs) (reviewed in 3). Briefly, it was found that cDNA and genomic clones encoding a VSG were identical except for a 39 nucleotide sequence found at the 5′ end of the cDNA clone. Nuclease mapping experiments placed the site of divergence between the clones to a likely splice-acceptor site, suggesting that the 39 nucleotide sequence was attached to the mRNA via splicing. However, hybridization analysis revealed that the 39 nucleotide sequence was not encoded near the VSG gene, but instead was reiterated about 100 times at another genomic locus. Further experiments revealed that this sequence was transcribed as the 5′ terminus of a small (about 140 nucleotides RNA—the SL RNA. In the SL RNA the 39 nucleotide sequence was immediately upstream of what appeared to be a conventional 5′ splice site. Collectively, these observations suggested the possibility that the 39 nucleotide sequence could be affixed to the VSG mRNA via *trans*-splicing, although several other possible mechanisms of attachment could not be excluded (see 3 for review).

Definitive evidence for *trans*-splicing as the mechanism for SL addition came from the demonstration *in vivo* of intermediates predicted from a *trans*-splicing reaction (see Fig. 1) (24, 25). As reviewed in Chapter 1, *cis*-splicing proceeds through two successive *trans*-esterification reactions. In the first step, cleavage of the 5′ splice site is accompanied by formation of a 2′–5′ phosphodiester bond between the branchpoint adenosine and the 5′-terminal guanosine of the intron, producing the lariat intermediate. By analogy, if *trans*-splicing were to proceed through the same two-step reaction, it was reasoned that the first step should produce a Y-form intermediate instead of a lariat (see Fig. 1). The characterization of such intermediates in trypanosome RNA not only established *trans*-splicing as a mechanism of pre-mRNA maturation but also showed that it proceeded through a reaction pathway analogous to that of *cis*-splicing.

Soon after the demonstration of *trans*-splicing in trypanosomes it was shown that

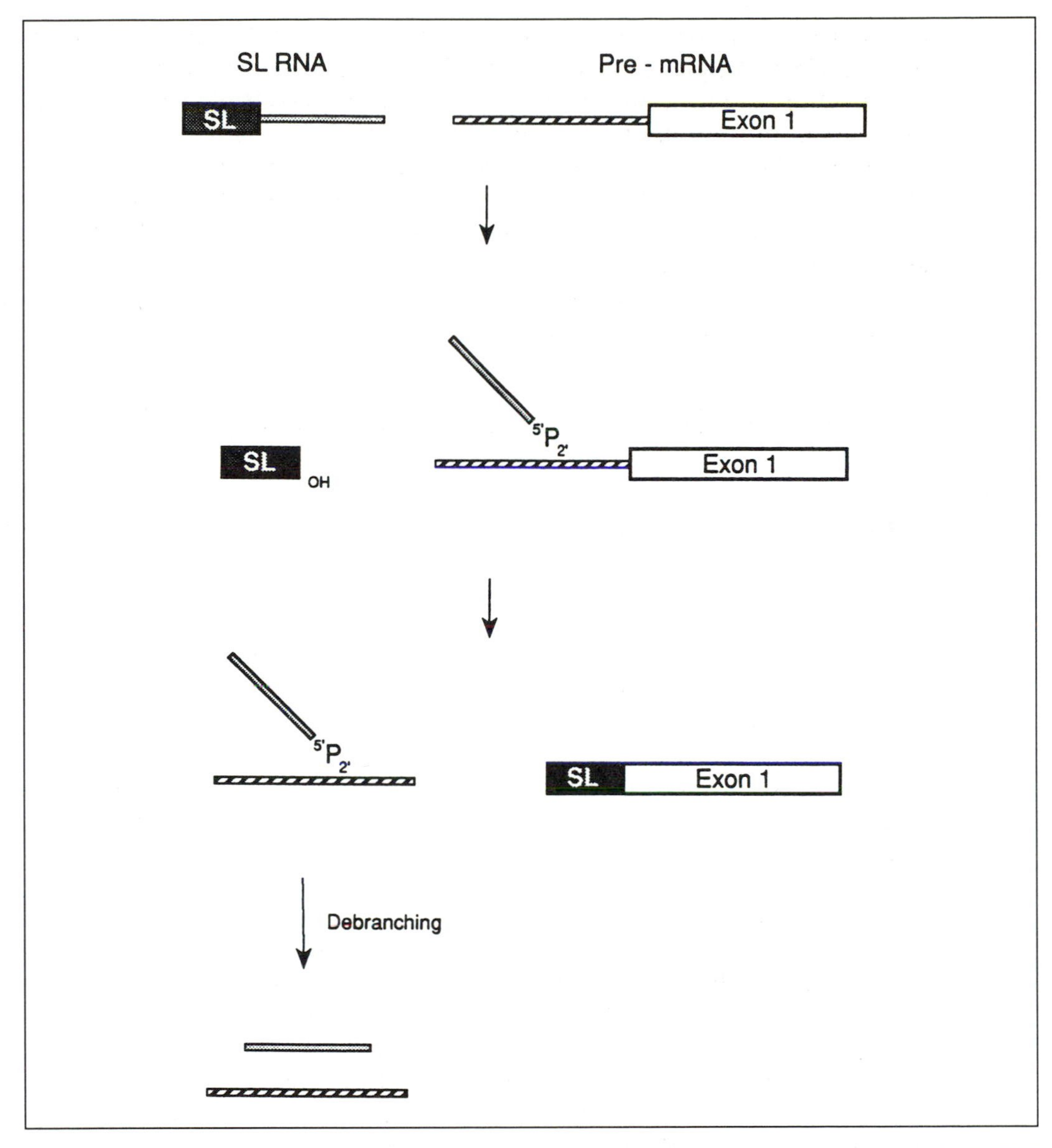

Fig. 1 SL-addition *trans*-splicing. In the first step of *trans*-splicing, cleavage at the 5′ splice site in the SL RNA yields the free SL with a 3′-hydroxyl terminus. Concomitant with this cleavage is the formation of a new 2′–5′-phosphodiester bond between the 5′ end of the 'intron' portion of the SL RNA and an adenosine residue upstream of the *trans* 3′ splice site. In the second step, the 3′-hydroxyl on the free SL attacks the 3′ splice junction phosphate creating the *trans*-spliced product and releasing the Y intron product. Subsequently, the 2′–5′ linkage in the Y intron product is hydrolysed by debranching enzyme, producing two linear intron molecules derived from the SL RNA and acceptor pre-mRNA.

all (within the limits of detection) mRNAs in these organisms possess the 39 nucleotide sequence derived from the SL RNA (26). Furthermore, *trans*-splicing appears to be the only mechanism of splicing used in trypanosomes since no convential (*cis*) introns have been found (reviewed in 4).

For some time after its discovery, it was thought that *trans*-splicing was an oddity of trypanosome gene expression, perhaps reflecting the early evolutionary branching of these protozoa. Thus, it came as a surprise when an essentially identical phenomenon was described in the free-living nematode *Caenorhabditis elegans*. Here, analysis of the genes and mRNAs encoding actins revealed that some nematode mRNAs contained a common 5′-terminal 22 nucleotide sequence acquired by *trans*-splicing (27). The nematode 22 nucleotide SL is derived from a small RNA similar to the trypanosome SL RNA. Following its discovery in *C. elegans*, it soon became clear that *trans*-splicing is ubiquitous in the nematode phylum (28–32). There are two major biological differences between *trans*-splicing in nematodes and trypanosomes. First, many, but not all, nematode pre-mRNAs are *trans*-spliced (8, 27, 28); and second, nematode pre-mRNAs (including those that are *trans*-spliced) contain conventional introns processed by *cis*-splicing.

Recently, two additional examples have been added to the list of organisms that carry out SL-addition *trans*-splicing: the flat worm *Schistosoma mansoni*, which is in the Trematode phylum, (33) and the photosynthetic protist, *Euglena gracilis* (34). *Trans*-splicing in these organisms has not been extensively studied. However, they appear more similar to nematodes than to trypanosomes in that not all mRNAs are *trans*-spliced and that *trans*-spliced pre-mRNAs contain cis-introns (33, 34). To date, *trans*-splicing appears to be restricted to certain lower eukaryotes. However, the prevalence of SL *trans*-splicing among primitive eukaryotes (if in fact, it is restricted to these organisms) is an important question which remains to be answered.

3.2 SL RNAs are specialized U snRNAs

As discussed in Chapter 5, *cis*-spliceosomal U snRNAs do not function as naked RNAs but rather as ribonucleoproteins. With the exception of U6 snRNA, the nucleoplasmic U snRNAs possess two features in common, a trimethylated guanosine cap structure and an Sm binding site which in general conforms to the consensus sequence $RAU_{n\geq4}$ GR. The Sm binding site serves as a recognition sequence for the association of a group of core proteins common to Sm snRNPs.

Soon after the discovery of *trans*-splicing in nematodes, it became apparent that their SL RNAs had the two diagnostic features of U snRNAs. Both immunoprecipitation with antibodies specific for trimethylguanosine (m_3G) caps and direct chemical analysis established that the nematode SL RNA possessed an m_3G cap (35–39). Furthermore, all nematode SL RNAs sequenced to date contain the sequence AAUUUUGG, which conforms to the consensus Sm binding sequence (8, 40). Several experiments established that this sequence promoted the association of Sm proteins. SL RNPs were immunoprecipitated by human Sm antisera from extracts of either *C. elegans* or *Ascaris* (36, 41). Furthermore, naked SL RNAs (prepared by *in vitro* transcription with bacteriophage RNA polymerase) efficiently assemble into Sm snRNPs when incubated in extracts prepared from higher eukaryotic cells or *Ascaris* embryos (35, 36).

The SL RNAs of schistosomes and *Euglena* also appear to resemble U snRNAs (Fig. 2). Immunoprecipitation of the schistosome SL RNA suggests that it contains an m_3G cap and the RNA has a potential Sm binding sequence (AGUUUUCUUUGG) although it has yet to be shown whether this sequence binds proteins with Sm antigenic determinants (33). The cap structure and protein composition of the *Euglena* SL RNPs have not been determined; but these RNAs also contain a consensus Sm binding site (AAUUUUGG) (34).

Until recently, the relationship of trypanosome SL RNAs to U snRNAs was unclear. It was shown some time ago that trypanosome SL RNAs existed as RNPs and that they contained a single-stranded region reminiscent of an Sm binding site (see Fig. 2). However, trypanosomes do not contain proteins that cross-react with Sm antisera, so straightforward immunoprecipitation analysis was not possible. Furthermore, although trypanosome U snRNAs possess the m_3G cap characteristic of eukaryotic U snRNAs, trypanosomatid SL RNAs have a unique hypermodified cap structure in which the capping nucleotide is 7-methylguanosine and the first few bases contain extensive post-transcriptional modifications (42, 43). The exact nature of these modifications has been resolved recently using mass spectroscopy (44).

Despite the difference in cap structure, it is now clear that trypanosome SL RNPs are indeed related to U snRNPs. Using an oligonucleotide affinity approach, Bindereif and colleagues have purified the trypanosome SL RNP as well as U snRNPs and have shown that these RNPs contain common (as well as specific) protein components (45) that are related to the protein constituents of higher eukaryotic snRNPs (46).

In summary, the combined observations in organisms which carry out SL-addition *trans*-splicing indicate that SL RNAs are (as originally suggested by Sharp) (9) a specialized form of U snRNA in which an exon (the SL) is affixed to an snRNA-like sequence. Unlike the U snRNAs that are necessary co-factors for *cis*-splicing, the SL RNA is consumed during the *trans*-splicing reaction.

4. Mechanism of SL-addition *trans*-splicing

4.1 Systems for analysis

Of the organisms that carry out SL addition *trans*-splicing, only trypanosomes and nematodes have been extensively analysed. Thus, the bulk of the discussion of mechanism is devoted to these systems. As discussed in Chapters 2–8, much of our current knowledge of the mechanism of pre-mRNA *cis*-splicing has been obtained from biochemical analysis of cell-free systems derived from higher eukaryotic cells and the combined biochemical and genetic approaches available in yeast.

To date, in trypanosomes, it has not been possible to analyse *trans*-splicing *in*

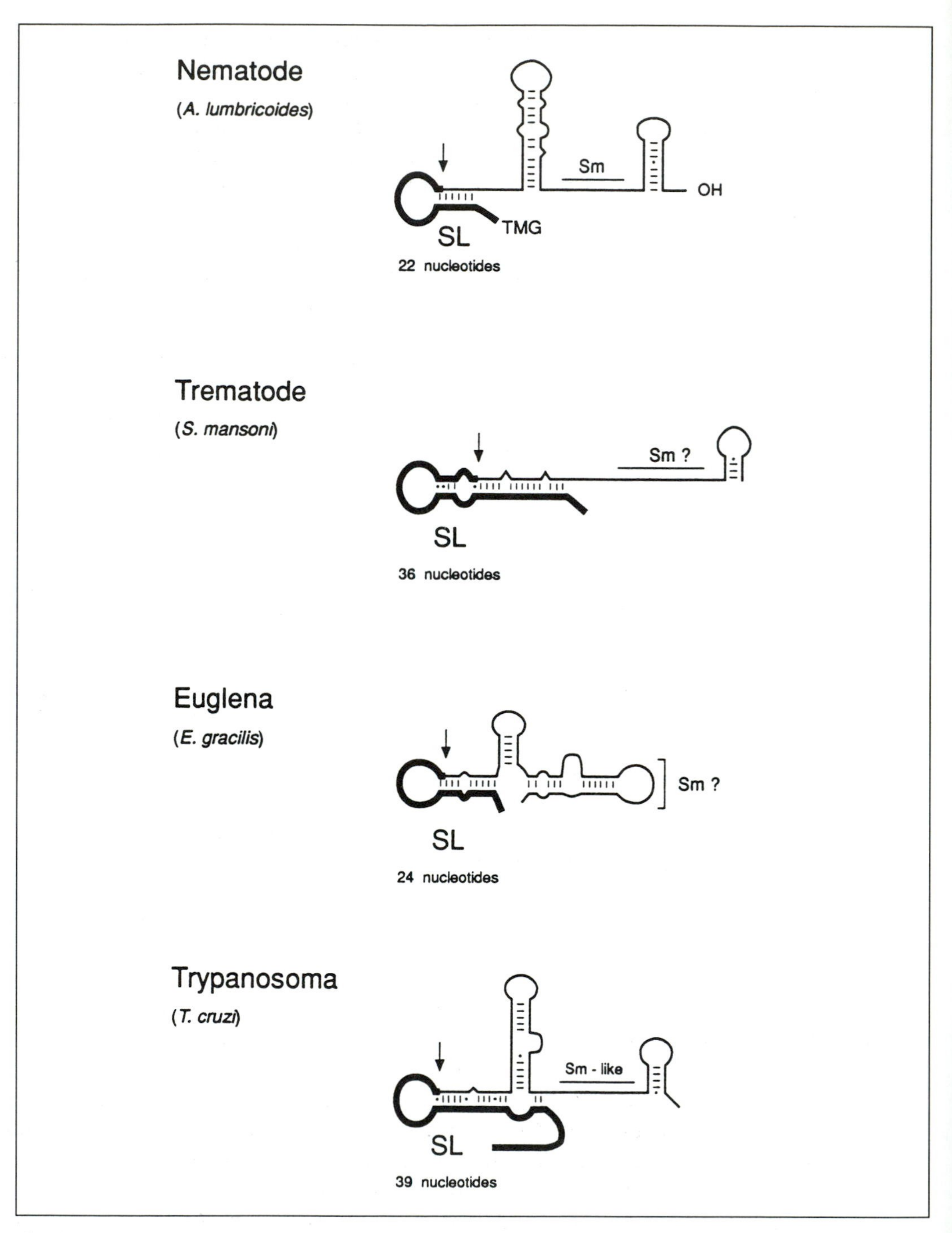

Fig. 2 Predicted secondary structures of *trans*-spliced leader RNAs. With the exception of the trematode SL RNA, each structure contains three stem–loops (see text). In stem–loop I sequences within SLs (thick line) are predicted to base pair with the 5′ splice site. The single-stranded region between stem–loops II and III contains the Sm binding site in the nematode SL RNA and potential Sm binding or Sm-like binding sites in the other SL RNAs. The length of each SL in nucleotides is indicated. TMG denotes the trimethylguanosine cap.

vitro, because suitable cell-free systems have not been developed. A surrogate (semi-*in vitro*) cell-free system consisting of permeabilized cells (47) has (as discussed below) provided important insights into snRNP co-factors required for trypanosome *trans*-splicing. The remainder of our understanding of trypanosome *trans*-splicing has come from *in vivo* approaches, including transfection and crosslinking (48–55). Additional information has accrued from the purification and characterization of components known to be important in the process (47, 56–58). However, these types of studies are, in general, technically difficult to carry out and in some cases hard to interpret, since functional *in vitro* assays are not available.

In nematodes, a cell-free system from synchronous *Ascaris* embryos has been developed which efficiently catalyses *trans*-splicing (59). Availability of this cell-free system, in which both substrates of *trans*-splicing (the SL RNA and acceptor pre-mRNA) can be added exogenously, has permitted the use of standard mutational approaches to assess which sequences within the donor and acceptor are required for function (36, 59). To date, the acceptor has not been analysed extensively. A thorough dissection of the SL RNA has been carried out and the results of these experiments are summarized in the next section.

4.2 Sequences required for nematode SL RNA function in *trans*-splicing

As described above, the SL RNAs of nematodes share two features in common with U snRNAs: a trimethylguanosine cap structure and a functional Sm binding site. To determine if these properties were necessary for SL RNA function, point mutations were introduced into the SL RNA's Sm binding site. Such mutations prevented both binding of Sm proteins and cap trimethylation (36). (It had previously been shown in other systems that cap trimethylation of the U snRNAs depended upon a functional Sm binding site; 60.) These mutations also completely prevented participation of the mutant SL RNAs in *trans*-splicing. To separate the potential roles of snRNP assembly and cap trimethylation, *trans*-splicing with wild-type SL RNAs was assayed in the presence of the methylation inhibitor *S*-adenosyl homocysteine. In these experiments, SL RNAs with unmethylated caps functioned perfectly well in *trans*-splicing (36). Thus, in order to participate in *trans*-splicing, the SL RNA must assemble into an Sm snRNP particle; however the hypermethylated cap structure is not relevant for *trans*-splicing *in vitro*.

This observation stands in apparent contrast to results obtained in the trypanosome system. Here, two groups have shown that SL RNAs with undermethylated cap structures do not function in *trans*-splicing (61, 62). As noted above, the trypanosome SL RNA cap contains a unique array of base and sugar methylations. It will be of considerable interest if this novel cap is required for *trans*-splicing, either directly, or by serving as the recognition site for a factor. A less interesting, but formally possible explanation for the lack of function of undermethylated SL RNAs in *trans*-splicing is that they are not properly localized to the nucleus. (In other eukaryotes, the m_3G cap serves as part of the nuclear localization signal of snRNAs; 63.) Resolution of these possibilities awaits a cell-free splicing system from trypanosomes or elucidation of the maturation pathway of the trypanosome SL RNA. To date, it has not been possible to determine if assembly of the SL RNA into an RNP is a nuclear or cytoplasmic process.

4.3 Specific exon sequences are not required for nematode SL RNA function

As discussed above, SL RNAs can be considered to be chimeric molecules composed of an exon domain (in nematodes, the 22 nucleotide SL) and an snRNA-like domain (9, 35, 39). It seemed likely that the exon domain would be important for SL RNA function, since it has been perfectly conserved in all nematodes. Furthermore, the SL domain is predicted to participate in an intramolecular base-pairing interaction which spans the 5′ splice site (see Fig. 2). This secondary structure feature is found in SL RNAs from all organisms studied to date (31, 33–35) and it was proposed that this intramolecular base pairing in the SL RNA could be functionally important by mimicking the intermolecular base pairing between U1 snRNA and *cis* 5′ splice sites (35) (see below). Thus, it was surprising that mutational disruption of this base pairing did not affect the function of the SL RNA in *trans*-splicing *in vitro* (64). Even more surprisingly, SL RNAs containing nearly complete truncations of the exon domain (to two nucleotides) functioned (with reduced efficiency) in *trans*-splicing (64). These results excluded the notion that the exon sequence contributed significantly to SL RNA function *in vitro* and suggested the possibility that the snRNA-like domain could deliver heterologous exons to an appropriate acceptor via *trans*-splicing. Indeed, when SL RNAs were constructed with a variety of artificial exons ranging in size from 29–246 nucleotides, each exon was used as a donor in *trans*-splicing (64). The observations described above indicated that determinants of SL RNA function resided within the snRNA-like domain of the molecule.

4.4. Sequences in the snRNA-like domain of the nematode SL RNA required for function

Chemical modification–interference analysis has been used to define purine residues within the snRNA-like domain critical for SL RNA activity. This experiment identified nine individual purine residues that were clustered in two regions of the molecule, a short stretch on the 3′ side of stem II and the single-stranded Sm binding site region between stems II and III (40). The importance of the sequence elements identified by modification interference was confirmed by mutation and by their ability, in combination, to confer SL RNA function to a fragment of an *Ascaris* U1 snRNA (40).

The functional significance of the single-stranded region containing the Sm binding site and adjacent nucleotides was examined further. Somewhat surprisingly, chemical modification of any one of the Sm binding site purines did not inhibit assembly of the SL RNA into an Sm snRNP. This observation, coupled with the observation that a functional Sm binding site derived from the *Ascaris* U1 snRNA (AAUUUUUGC) could not substitute for the analogous sequence in the SL RNA (AAUUUUGG), suggested that the Sm binding region of the SL RNA had a role in addition to directing assembly into an Sm snRNP. To address the possibility that the significance of this sequence might (at least in part) reside in its ability to interact with another RNA, cross-linking experiments using aminomethyltrioxsalen (AMT) were performed (40). These experiments identified cross-linked species when labelled synthetic SL RNA was incubated in extract. Cross-linked molecules of identical mobility were also observed with endogenous SL RNA (40). Synthetic SL RNAs altered in critical nucleotides 3′ of the Sm binding site efficiently assembled into Sm RNPs but failed to function in *trans*-splicing. These same mutant SL RNAs failed to cross-link, suggesting that the interaction identified by cross-linking could be functionally significant.

To determine if the cross-link resulted from an interaction between the SL RNA and a known U snRNA, cross-linked species were digested with RNase H using a panel of oligodeoxynucleotides complementary to U1, U2, U4, U5, and U6 snRNAs. The cross-linked moieties were sensitive to RNase H in the presence of two separate oligodeoxynucleotides complementary to U6 snRNA and insensitive to digestion when any of the other oligodeoxynucleotides were used (40). Inspection of the SL RNA and U6 snRNA sequences revealed a striking complementarity of 18 consecutive base pairs with one bulged nucleotide. In the SL RNA, the region of complementarity included the Sm binding site and extended 10 bases 3′ of the Sm binding site (Fig. 3). These results suggested that an SL RNA–U6 snRNA base-pairing interaction was important for nematode *trans*-splicing. The potential significance of this interaction will be discussed below.

4.5. U snRNAs and *trans*-splicing

4.5.1 Lessons from *cis*-splicing

Extensive studies in *cis*-splicing systems have revealed a complex array of snRNP–pre-mRNA and snRNP–snRNP interactions critical for assembly of the spliceosome and subsequent catalysis. These are described in detail in Chapters 4–7, but are reviewed briefly here for comparison to *trans*-splicing systems.

The first step in *cis*-splicing is the recognition of the 5′ splice site by U1 snRNP, which involves a base-pairing interaction between the 5′ end of U1 and the 5′ splice site. Subsequent to U1 binding, U2 snRNP binds to the pre-mRNA at the branch site. Again this recognition involves base pairing between U2 snRNA and the pre-mRNA. Following U1 and U2 binding, U4, U5, and U6 join the pre-splicing complex as a triple snRNP with U4 and U6 extensively base paired to each other. In addition to its base pairing with U4 snRNA, U6 snRNA makes base pairing contacts with U2 snRNA (65). This involves bases near the 5′ end of U2 snRNA and a region near the 3′ end of U6 snRNA (65). Compensatory mutagenesis has established that

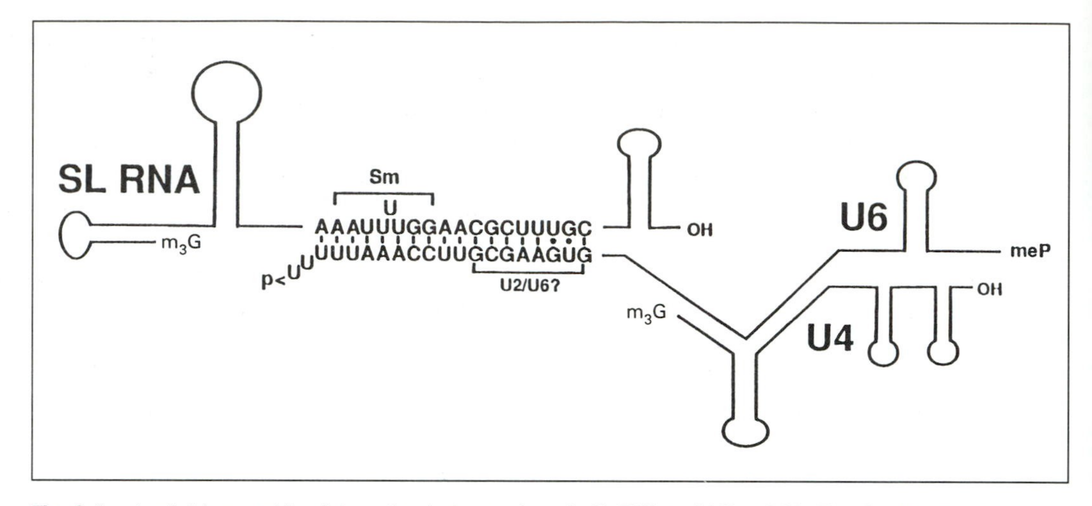

Fig. 3 A potential base-pairing interaction between *Ascaris* SL RNA and U6 snRNA. The Sm binding site in the SL RNA (AAUUUUGG) is bracketed, as is the region of U6 snRNA which has been shown to interact with U2 snRNA in mammalian cells (see text). Cap structures are shown to indicate polarity. $m_3{}^G$, trimethylguanosine (in SL RNA and U4 snRNA); meP, γ-monomethyl phosphate (in U6 snRNA). The AMT cross-link described in the text connects the 5′-most uridine of the SL RNA's Sm binding site and the uridine residue (U98) five residues upstream from the 3′ end of U6 snRNA.

this U2/U6 interaction is essential for mammalian *cis*-splicing (66, 67) but does not appear to be required for yeast *cis*-splicing (68). This interaction may serve to stabilize, or correctly position the triple snRNP U4/U6·U5 in the spliceosome.

Recent analysis from several groups has provided new and exciting insights into additional snRNA–snRNA interactions within the fully formed complex (69–73; reviewed in 74). These observations have provided the basis for a model of *cis*-splicing which is described briefly below. Before the first step of splicing, exons to be spliced are juxtaposed through base-pairing interactions with both U1 snRNA and U5 snRNA. Genetic evidence in fission yeast suggests that U1 contacts conserved intron sequences at both the 5′ and 3′ splice sites (75). Perhaps U5 snRNP can simultaneously make base-pairing contacts with exon sequences immediately upstream of the 5′ splice site and downstream of the 3′ splice site (72, 76, 77). As suggested by Steitz (78), the combination of these interactions could produce a four-armed structure reminiscent of the Holliday intermediate in DNA recombination. Concurrent with, or following, recognition and juxtaposition of the 5′ and 3′ splice sites there is a dynamic reorganization in the spliceosome, whereupon the U4–U6 interaction is disrupted and U6 forms a new base-pairing interaction with U2 snRNA (69). This pairing, which is designated internal U6–U2, is distinct from the U2–U6 interaction described above.

At this point, U4 can be removed from the spliceosome with no consequence to the subsequent cleavage ligation reactions (79). Thus, U4 is not an active participant in the catalysis of splicing. Although the details are at present somewhat obscure, the U1–pre-mRNA interaction is disrupted and replaced by a U6–pre-mRNA interaction where U6 makes base-pairing contacts with intron sequences just downstream of the 5′ exon (70, 71, 73, 80, 81). U5 remains associated with the pre-mRNA through contacts with exon sequences adjacent to the 5′ splice site (82). An important consequence of the dynamic rearrangement of RNA–RNA interactions in the pre-catalytic spliceosome is that the two participants of the first step (the branchpoint and the 5′ splice site) are brought into physical proximity with each other (see 74 for recent review).

Following reorganization of the spliceosome, the two sequential transesterification reactions are initiated. It is widely assumed that U6 and U2 in combination are involved in catalysis of these reactions by mechanisms analogous to those which catalyse either group I (83) or group II self-catalysed splicing reactions (69; see Chapter 1).

4.5.2 Studies in *trans*-splicing systems

Given the fundamental similarity between *cis*- and *trans*-splicing, it is quite likely that they are catalysed by the same mechanism. Soon after the discovery of *trans*-splicing in trypanosomes, it was demonstrated that these organisms possess homologues of U2, U4, and U6 snRNAs (84, 85). Using a permeabilized cell system (47) and RNase H targeted degradation, Tschudi and Ullu clearly demonstrated that these snRNAs are required for trypanosome *trans*-splicing (86). Similar analysis in the *Ascaris* cell-free system showed that U2, U4, and U6 are also required for nematode *trans*-splicing (87). Thus, it is clear that *cis*- and *trans*-splicing require at least some common components. Furthermore, sequence analysis of both nematode and trypanosome U snRNAs has revealed that their respective U2, U4, and U6 snRNAs can potentially form the same array of base-pairing interactions characterized in *cis*-splicing systems (see 88 for discussion and references). While it remains to be shown that these interactions are necessary for *trans*-splicing, it seems relatively safe to assume that they will be. Here, it should be noted that the base-pairing interaction between U6 snRNA and the nematode SL RNA described above (see Section 4.4) overlaps the region of U6 which in mammalian cells (and trypanosomes, see below) base pairs to the 5′ end of U2 snRNA. Several considerations suggest that the U6–SL interaction is limited to the region of the SL RNA that includes the Sm binding site and three adjacent nucleotides (40) (see Fig. 3). If this is true, U6 could, in the *trans*-spliceosome, simultaneously interact with the SL RNA and the 5′ end of U2 snRNA (see below).

While it is clear that U2, U4, and U6 snRNPs are essential for *trans*-splicing, it is not known whether U1 and U5 snRNPs (known to be essential for *cis*-splicing, see above) are required for this reaction. Curiously, trypanosomes (which carry out only *trans*-splicing) appear to a lack homologue of U1 snRNA (4). At face value, this observation would suggest that U1 is not required for *trans*-splicing and its function, 5′ splice-site identification, would have to be carried out by other factor(s).

Several years ago, it was suggested that the SL RNA might fulfill a role in *trans*-splicing equivalent to that of U1 snRNP in *cis*-splicing (35). Support for this hypothesis came from studies of the processing of chimeric *cis*-splicing substrates containing

SL RNA splice donor sites linked in *cis* to an adenoviral splice acceptor site and 3′ exon (89). Such transcripts, containing either nematode or trypanosomatid SL RNA sequences, were efficiently spliced in Hela cell extracts where the 5′ end of U1 snRNA was ablated by targeted RNase H digestion (89). However, subsequent analysis revealed that the SL RNA sequences did not completely obviate the need for U1 snRNP, since an additional function of U1 snRNP (independent of its base pairing to the 5′ splice site) was necessary for the splicing of the chimeric transcripts (90). It would seem that the *Ascaris* extracts could provide a straightforward system in which to address the role (or lack thereof) of U1 snRNP in *trans*-splicing. However, to date this has not proven to be the case. Since nematodes carry out both *cis*- and *trans*-splicing they contain a full complement of snRNAs including U1 and U5 (91). In *Ascaris* extracts, oligonucleotide-directed RNase H digestion of U1 efficiently inhibited *cis*-splicing but did not affect *trans*-splicing (87). Furthermore, mutations in the 5′ splice site predicted to disrupt a U1–5′ splice site base-pairing interaction did not affect SL RNA function (64). These experiments suggest that U1 may not be involved in nematode *trans*-splicing. However, they cannot be considered definitive since, as in mammalian cells (see above and 92), U1 snRNP may perform roles in splicing independent of its ability to base pair with 5′ splice sites. It is extremely difficult to rule out the possibility that U1 snRNP is performing these (or as yet undiscovered) roles in *trans*-splicing. Therefore, in the absence of a totally reconstituted system, whether or not U1 has a role in nematode *trans*-splicing will remain an open question.

The other U snRNA implicated in splice-site recognition and juxtaposition is U5 snRNA (72, 76, 77, 80). Recently, in trypanosomes, psoralen cross-linking *in vivo* has revealed the presence of a novel RNA—the splice leader associated (SLA) RNA (55). This RNA appears to interact with the SL RNA near the 5′ splice site. The interacting region of the SLA RNA is within a nine nucleotide region of perfect homology to the invariant loop sequence of U5 snRNAs (55). These observations suggest that the SLA RNA may be the trypanosome homologue of U5 snRNA. However, a functional assay for the SLA RNA has yet to be developed.

U5 snRNA is present in nematodes, and it has been suggested that it participates in *trans*-splicing (78). Unfortunately it is difficult to test this notion directly, because U5 is notoriously resistant to targeted degradation and is nearly impossible to deplete from extracts by other means.

The U5–5′ splice-site interaction was revealed through genetic analysis in yeast (76, 77) and more recently by site-specific cross-linking in mammalian extracts (72). Similar cross-linking experiments in *Ascaris* extracts have failed to reveal any U5–SL splice-donor site interaction, but at present the significance of this negative finding is unclear. Thus, as with U1, the role or lack thereof of U5 in nematode *trans*-splicing remains unclear. It seems likely that these questions might best be answered through the direct analysis of the *trans*-splicing equivalent of the *cis*-spliceosome. Though there is little doubt that such a complex exists, the *trans*-spliceosome, to date, has eluded capture. Recent progress in the creation of substrate RNAs with site-specific substitutions which arrest splicing after the first step (93) should yield 'trapped' *trans*-spliceosomes. If U5 is important for *trans*-splicing it should be found in such complexes.

4.6 A working model for SL-addition *trans*-splicing

As detailed above, extensive analysis in a variety of systems has led to an emerging picture of snRNA–snRNA interactions in *cis*-splicing. In nematode *trans*-splicing, the evidence suggests that an additional interaction (between the SL RNA and U6 snRNA) may be functionally important (40). Collectively these observations suggest a plausible (if somewhat speculative) model for *trans*-spliceosome assembly (Fig. 4), in which the SL RNA is base paired to U6 snRNA (interaction a) through the SL RNA's Sm binding region and the 3′ end of U6. At the same time, the 3′ end of U6 is base paired to the 5′ end of U2 snRNA (interaction b) (65). U2 snRNA is in turn base paired with the branch site of the pre-mRNA (interaction c). Following release of U4 snRNA (Fig. 5), the additional internal base-pairing interaction between U6 and U2 (interaction d) becomes possible (69). As discussed above, the role, if any, of U1 and U5 snRNPs in *trans*-splicing is unknown. The following discussion makes the assumption that these snRNPs do not participate in *trans*-splicing.

In this model, the 5′ splice site on the SL RNA is fixed relative to U6 snRNA because the two molecules are held together by base pairing. In this intimate association, it seems possible that U6 snRNA could interact by base pairing with the 5′ splice site (interaction e) without the prior intervention of U1 snRNA. Through the simultaneous interactions of U2 with U6 and U2 with the branchpoint, the 3′ splice site could be brought into proximity with the catalytic centre of the spliceosome. If this were the case, how could the 3′ splice site be recognized? In group II self-catalysed splicing, a single long-range base pair has been implicated in 3′ splice-site recognition (94). Similarly, many group I introns lack recognizable 3′ exon guide sequences (13). Given these examples, it is not inconceivable that 3′ splice-site recognition in *trans*-splicing may be mediated by yet to be discovered secondary (or higher order) structural interactions. Site-specific cross-linking analysis at the 3′ splice site may provide some insight into this question.

Although this model is speculative, it provides a possible scenario for a snRNP-mediated splicing reaction which would not require U1 or U5 snRNPs. More significantly, it also provides a plausible answer to one of the most important questions in *trans*-splicing: how can the SL RNA and pre-mRNA efficiently associate in the absence of sequence complementarity to each other? In this model, U2 and U6 snRNAs serve as connecting bridges between the two substrates of *trans*-splicing through concurrent base-pairing interactions.

The working model of *trans*-splicing draws heavily upon extensive studies in *cis*-splicing systems and a focused analysis of sequences required for function of the nematode SL RNA. Legitimate questions arise in considering which aspects of the model have been experimentally verified and whether the model is nematode-specific or possibly applicable to SL-addition *trans*-splicing in general. The potential importance of the base-pairing interactions depicted in Fig. 4 in *cis*-splicing were

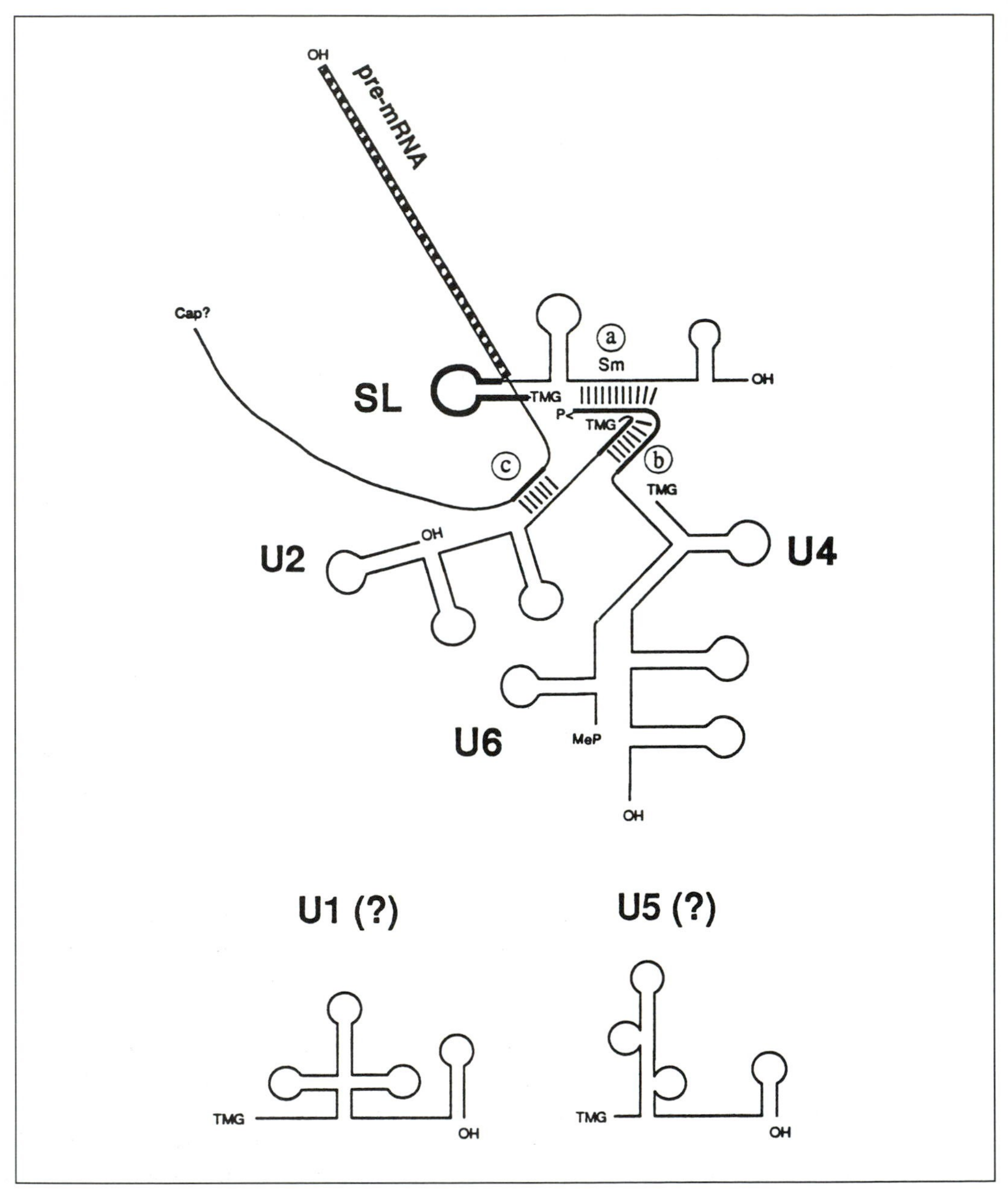

Fig. 4 Model of *trans*-spliceosome organization prior to release of U4 snRNA. As discussed in the text, concurrent base-pairing interactions between the SL RNA and U6 snRNA (interaction a), between U6 snRNA and U2 snRNA (5′ U2–3′ U6, interaction b), and between U2 snRNA and the branchpoint sequence (bold line) on the pre-mRNA (interaction c) could link the pre-mRNA and SL RNA in appropriate geometry for *trans*-splicing. The roles, if any, of U1 snRNA and U5 snRNA are unknown. TMG, trimethylguanosine cap; MeP, γ-monomethylphosphate cap.

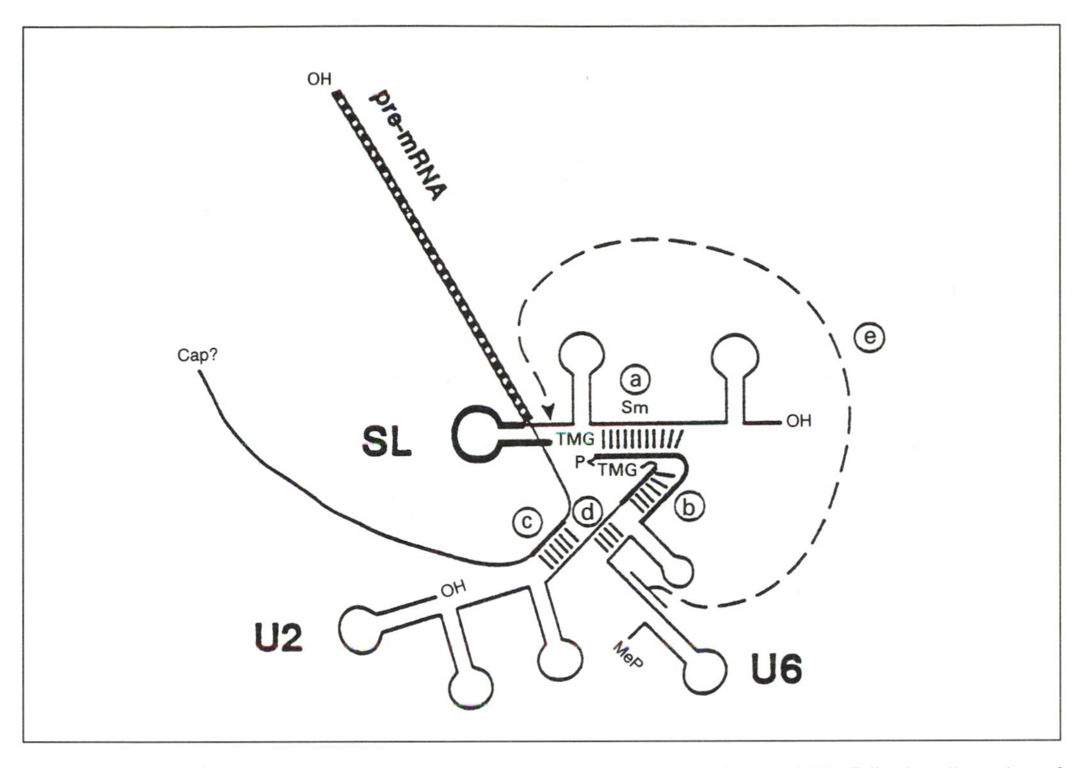

Fig. 5 Possible organization of the *trans*-spliceosome following release of U4 snRNA. Following disruption of U4–U6 base pairing, two additional base-pairing interactions become possible (see text). An internal region of U6 becomes accessible for base pairing with U2 snRNA (interaction d). This interaction involves sequences in U2 that are between the 5′ U2–U6 (b) and U2–branchpoint (c) sites. Concurrently, another internal region of U6 could interact with sequences downstream of the 5′ splice site in the SL RNA (interaction e).

first suggested by phylogenetic conservation. Subsequently, mutagenesis and compensatory mutagenesis stringently proved the relevance of interactions b, c, d, and e (Figs 4 and 5) as well as the U1 snRNA–5′ splice-site pairing interaction in *cis*-splicing systems.

To date, comparable approaches have not been used in any *trans*-splicing system. Thus, the importance of each base-pairing interaction shown in Figs 4 and 5 for *trans*-splicing remains to be established. Nevertheless, several lines of evidence support the idea that interactions important in *cis*-splicing will also be critical for *trans*-splicing. First, the *trans*-splicing reaction yields intermediates and products directly analogous to those seen in *cis*-splicing (see Fig. 1) indicating that these two processes are catalytically similar, if not identical. Second, despite intensive searching, there is no evidence for potential *trans*-splicing-specific versions of U2, U4, or U6 snRNAs in organisms which carry out both *cis*- and *trans*-splicing. Finally, in trypanosomes, which carry out only *trans*-splicing, the potential to form interactions b and d has been conserved and experimental support for interaction b has been obtained via *in vivo* cross-linking experiments (52).

From these considerations, it seems likely that interactions b, d, and possibly e will be essential for *trans*-splicing. However, what of interactions a and c? While not yet proven, it seems probable that the base-pairing interaction between the nematode SL RNA and U6 snRNA (interaction a) will be functionally significant (40). Can these molecules interact by base-pairing in other organisms that carry out leader addition *trans*-splicing? Unfortunately, the sequence of U6 snRNA has not been determined in schistosomes or *Euglena*. In trypanosomes, only one U6 snRNA sequence (that of *T. brucei*) is known. An inspection of potential SL RNA–U6 snRNA pairings in this organism reveals that the two RNAs could pair across the 'Sm analogous' region of the SL RNA. As in nematodes, this region of potential pairing is immediately adjacent to, and partially overlaps, the 3′ region of U6 snRNA which interacts with the 5′ end of U2 snRNA. However, the order of these interactions is inverted with respect to that found in nematodes: the potential SL RNA–U6 snRNA interaction in trypanosomes is 5′ (in U6) to the region of U6 which interacts with U2 snRNA. It does not seem that transposition of the order of these base-pairing interactions would significantly alter the model shown in Fig. 5. Nevertheless, it will clearly be important to determine if the trypanosome SL RNA and U6 snRNA interact by base-pairing. It should be noted here that some time ago it was suggested that the trypanosome SL RNA might be recruited to the *trans*-spliceosome via a base-pairing interaction between regions of its 39 nucleotide SL and U2 snRNA (85). Furthermore, sedimentation analysis suggested that the SL RNA and U2 snRNA might exist together as a particle (53). However, the proposed pairing is not strongly conserved phylogenetically and has not received further experimental support.

The final base-pairing interaction in the model (interaction c, Fig. 5) involves U2 snRNA and the branchpoint sequence of the pre-mRNA. There is no doubt that such a pairing occurs and is required for *cis*-splicing; the existence of this base-pairing has been rigorously established in both yeast and mammalian systems (95–97). Since SL-addition *trans*-splicing requires U2 snRNA and proceeds through a branching reaction directly analogous to lariat formation, it would seem safe to assume that U2 snRNA would, in *trans*-splicing, fulfill the same roles that it does in *cis*-splicing, including pairing to the pre-mRNA branch site. However, this is far from proven and, in fact, the available evidence suggests (surprisingly) that the process of branchpoint recognition may not be the same in *cis*- and *trans*-splicing. In *cis*-splicing, branchpoint recognition involves pairing of the sequence GUAGUA (at about positions 35–40 in U2 snRNA) to the branchpoint. The GUAGUA motif has been stringently conserved in all U2 snRNAs sequenced to date, with the exception of the trypanosomatids. In these organisms (which carry out only *trans*-splicing) the GUAGUA sequence is replaced with a UA-rich sequence (UAUUAA in *T. Brucei* (98, 99). Furthermore, the only *trans*-splice branch sites mapped in trypanosomes do not have any significant complementarity to this variant branch site recognition sequence (100). In nematodes, only one *trans*-splice branch site has been determined and here again there is no significant similarity to the consensus branchpoint sequence, even though nematode U2 snRNAs contain the canonical GUAGUA sequence (59, 91).

Together with the apparent lack of sequence complementarity between U2 snRNA and *trans*-splice branchpoints, there remains an additional significant question concerning branch site recognition in *trans*-splicing: how is U2 positioned near the *trans*-splice acceptor site? In both yeast and mammalian cells, the binding of U1 snRNP to the 5′ splice site commits a pre-mRNA to splicing (101, 102) and a U1 snRNP function (independent of its pairing to the 5′ splice site) is necessary for stable association of U2 snRNP to the branchpoint (92). If U1 snRNP is not involved in *trans*-splicing (probable in nematodes and more certain in trypanosomes) what promotes U2 snRNP association with the branchpoint sequence? Currently, there are no experimental data relevant to this question, but it is tempting to speculate that the SL RNP might fulfill this function in *trans*-splicing.

4.7 Splice-site selection in SL-addition *trans*-splicing

In addition to the problem of *trans*-splice acceptor site identification discussed above, the existence of *trans*-splicing in nematodes, trematodes, and *Euglena* presents another interesting problem regarding splice-site selection. In these organisms, whose pre-mRNAs are processed by both *cis*- and *trans*-splicing, how are internal 3′ splice sites (which function only in *cis*) excluded as sites of SL addition?

At least two hypotheses (either in combination or alone) could account for the exclusion of internal 3′ splice sites in *trans*-splicing. First, *trans* 3′ splice sites could possess positive sequence elements which would promote SL addition or, conversely, *cis* 3′ splice sites could contain negative sequence elements which would prevent the formation of a *trans*-splicing complex. It should be noted that sites of SL addition have been determined for several pre-mRNAs in a variety of nematodes and they always coincide with what appear to be conventional 3′ splice sites. Systematic comparison of these *trans* 3′ splice sites has not revealed any common sequence motifs which could distinguish them from *cis* 3′ splice-site sequences.

A third possible explanation for exclusion of internal 3′ splice sites could be that the presence of a 5′ splice site upstream of a 3′ splice site in some way precludes *trans*-splicing. Experiments conducted *in vivo* in *C. elegans* indicate that this latter explanation is correct. Blumenthal and colleagues introduced a portion of a normal *cis* intron (lacking a 5′ splice site but including a 3′ splice site) into the 5′ untranslated sequence of a gene whose message was normally not *trans*-spliced (103). Upon analysis of mRNAs produced from this gene, it was found that the bulk of the mRNAs contained the SL sequence. These experiments indicated that a normal *cis* 3′ splice site could function as a *trans* 5′ splice site and demonstrated convincingly that it was possible to convert a non-*trans*-spliced mRNA into a *trans*-spliced mRNA *in vivo*. These same workers then did the reciprocal experiment, by inserting a 5′ splice site upstream of a *trans* 3′ splice site. In these constructs, the normally *trans*-spliced mRNAs were found to be exclusively *cis*-spliced (103). Together these experiments argue strongly that an unpaired 3′ splice site in the primary transcript is necessary and sufficient to direct *trans*-splicing. It will be of considerable interest to determine the biochemical basis of suppression of *trans*-splicing by an upstream 5′ splice site.

5. Summary and perspectives

The foregoing discussions have focused extensively on mechanistic questions regarding SL-addition *trans*-splicing. For discussion of the possible biological roles of the *trans*-splicing pathway, the reader is referred to recent reviews (4, 7, 8). While much progress has been made in understanding *trans*-splicing, it should be clear that many important questions remain unanswered. In particular, it is necessary to determine whether U1 and U5 snRNPs play any role in *trans*-splicing in organisms that carry out both *cis*- and *trans*-splicing. Furthermore, the mechanism of splice-site and branchpoint recognition is currently obscure. It will also be important to clarify the mechanistic similarities and/or differences between SL-addition *trans*-splicing in trypanosomes and other organisms which process their mRNAs by both *cis*- and *trans*-splicing. In this regard, a trypanosomatid cell-free system capable of accurate *trans*-splicing would be extremely valuable. Finally, it will be of interest to determine how widespread SL addition *trans*-splicing is in nature.

Thus far, snRNP-mediated *trans*-splicing has only been observed in certain lower eukaryotes. However, early experiments clearly demonstrated that exons encoded on independent transcripts (one containing a 5′ splice site and one containing a 3′ splice site) could be accurately joined in mammalian cells or extracts (104, 105). The efficiency of these *trans*-splicing reactions was enhanced when the 5′ and 3′ half-transcripts contained regions of complementarity (104, 105). Although it remains to be proven, this type of *trans*-splicing may be involved in the maturation of immunoglobulin and c-*myb* mRNAs (see for example (106, 107). More recently, Bruzik and Maniatis (108) have shown that mammalian cells are capable of catalysing SL-addition *trans*-splicing when presented with appropriate heterologous SL-RNAs and acceptor pre-mRNAs. These studies suggest the intriguing possibility that SL-addition *trans*-splicing may not be restricted to lower eukaryotes.

Time will tell if any type of *trans*-splicing naturally occurs in the nuclei of higher eukaryotic cells. However, given the complexity of pre-mRNA maturation in 'primitive' organisms, it will not be surprising if it does.

6. Note added in proof

While this book was in production, significant progress in the study of *trans*-splicing mechanisms has been made, particularly in the area of SL-addition *trans*-splicing. Space constraints preclude either a detailed discussion of these results or a thorough integration with the main text. Nevertheless, a few highlights with selected references are summarized below.

With regard to the mechanism of SL-addition *trans*-splicing, the nematode *trans*-spliceosome has been isolated (109). Analysis of this complex unambiguously revealed the presence of U5 snRNA and the absence of U1 snRNA. Furthermore, functional analysis of U5 snRNA *in vitro* (109) and *in vivo* in trypanosomes (110) leaves no doubt that U5 snRNA is necessary for *trans*-splicing. In addition, it has become clear that SL RNP recruitment to the *trans*-spliceosome cannot be explained solely by RNA/RNA interactions. Purification of a nematode SL RNP (111) and *in vivo* mutational analysis of a trypanosomatid SL RNA (112) indicate that the requirements for SL RNP function are complex and depend upon its specific protein constituents. Elucidation of the role(s) of these proteins is clearly quite important, because they may hold the key to understanding the differences between *cis* and *trans*-splicing (see 111 for discussion). In addition to mechanistic studies, the biological role of *trans*-splicing, particularly in nematodes, has become clearer (see 113, for recent review).

Finally, it now seems apparent that the *trans*-splicing reactions observed in mammalian cells involving SL RNAs (108) are not indicative of the presence of natural SL addition in higher organisms. There is no evidence for endogenous SL RNAs in vertebrates, and the *trans*-splicing reactions do not have the same requirements as those observed in organisms that naturally carry out SL addition (see 111 for discussion). While *trans*-splicing can occur in mammalian cells and cell-free extracts, these reactions are mediated by protein–protein interactions characteristic of *cis*-splicing (114, 115).

In summary, while significant progress has been made, many of the important questions outlined in the main text remain unanswered.

References

1. Blumenthal, T. and Thomas, J. (1988) *Cis* and *trans*-mRNA splicing in *C. elegans*. *Trends Genet.*, **4**, 305.
2. Phizicky, E. M. and Greer, C. L. (1993) Pre-tRNA splicing: variation on a theme or exception to the rule? *Trends Biochem. Sci.*, **18**, 31.
3. Borst, P. (1986) Discontinuous transcription and antigenic variation in trypanosomes. *Annu. Rev. Biochem.*, **55**, 701.
4. Agabian, N. (1990) *Trans*-splicing of nuclear pre-mRNAs. *Cell*, **61**, 1157.
5. Laird, P. (1989) *Trans*-splicing in trypanosomes—archaism or adaptation? *Trends Genet.*, **5**, 204.
6. Nilsen, T. W. (1989) *Trans*-splicing in nematodes. *Exp. Parasitol.*, **69**, 413.
7. Nilsen, T. W. (1992) *Trans*-splicing in protozoa and helminths. *Infectious Agents Dis.*, **1**, 212.
8. Nilsen, T. W. (1993) Nematode *trans*-splicing. *Annu. Rev. Microbiol.*, **47**, 413.
9. Sharp, P. A. (1987) *Trans*-splicing: variation on a familiar theme? *Cell*, **50**, 147.
10. Huang, X.-Y. and Hirsh, D. (1992) RNA *Trans*-splicing. In *Genetic engineering*, Setlow, J. (ed.). Plenum Press, New York, p. 211.
11. Sharp, P. A. (1991) Five easy pieces. *Science*, **254**, 663.
12. Perlman, P. and Butow, R. (1991) Introns in pieces. *Curr. Biol.*, **1**, 331.
13. Michel, F., Umesono, K., and Ozeki, H. (1989) Comparative and functional anatomy of group II catalytic introns—a review. *Gene*, **82**, 5.
14. Koller, B., Fromm, H., Galun, E., and Edelman, M. (1987) Evidence for *in vivo trans*-splicing of pre-mRNAs in tobacco chloroplasts. *Cell*, **48**, 111.
15. Chapdelaine, Y. and Bonen, L. (1991) The wheat mitochondrial gene for subunit I of the NADH dehydrogenase complex: A *trans*-splicing model for this gene in pieces. *Cell*, **65**, 465.
16. Jacquier, A. and Jacquesson-Breuleux, N. (1991) Splice site selection and the role of the lariat in a group II intron. *J. Mol. Biol.*, **219**, 415.

17. Goldschmidt-Clermont, M., Choquet, Y., Girard-Bascov, J., Michel, F., Shrimer-Rahire, M., and Rochaix, J. (1991) A small chloroplast RNA may be required for *trans*-splicing in *Chlamydomonas reinhardtii*. *Cell*, **65**, 135.
18. Kück, U., Choquet, Y., Schneider, M., Dron, M., and Bennoun, P. (1987) Structural and transcription analysis of two homologous genes for the P700 chlorophyll *a* apoproteins in *Chlamydomonas reinhardtii*: evidence for *in vivo trans*-splicing. *EMBO J.*, **6**, 2185.
19. Wissinger, B., Schuster, W., and Brennicke, A. (1991) *Trans*-splicing in *Oenothera* mitochondria: *nadl* mRNAs are edited in exon and *trans*-splicing group II intron sequences. *Cell*, **65**, 473.
20. Parry, H. D., Scherly, D., and Mattaj, I. W. (1989) 'Snurpogenesis': the transcription and assembly of U snRNP components. *Trends Biochem. Sci.*, **14**, 15.
21. Jarrell, K., Dietrich, R. C., and Perlman, P. S. (1988) Group II intron domain 5 facilitates a *trans*-splicing reaction. *Mol. Cell Biol.*, **8**, 2361.
22. Wissinger, B., Brennicke, A., and Schuster, W. (1992) Regenerating good sense. *Trends Genet.*, **8**, 322.
23. Lamattina, L., Weil, J., and Grienenberger, J. (1989) RNA editing at a splicing site of NADH dehydrogenase subunit IV gene transcript in wheat mitochondria. *FEBS Lett.*, **258**, 79.
24. Sutton, R. and Boothroyd, J. C. (1986) Evidence for *trans*-splicing in trypanosomes. *Cell*, **47**, 527.
25. Murphy, W. J., Watkins, K. P., and Agabian, N. (1986) Identification of a novel Y branch structure as an intermediate in trypanosome mRNA processing: evidence for *trans*-splicing. *Cell*, **47**, 517.
26. Walder, J. A., Eder, P. S., Engman, D. M., Brentano, S. T., Walder, R. Y., Knutzon, D. S., *et al.* (1986) The 35-nucleotide spliced leader sequence is common to all trypanosome messenger RNA's. *Science*, **233**, 569.
27. Krause, M. and Hirsh, D. (1987) A *trans*-spliced leader sequence on actin mRNA in *C. elegans*. *Cell*, **49**, 753.
28. Bektesh, S. L., van Doren, K. V., and Hirsh, D. (1988) Presence of the *Caenorhabditis elegans* spliced leader on different mRNAs and in different genera of nematodes. *Genes Dev.*, **2**, 1277.
29. Zeng, W., Alarcon, C. M., and Donelson, J. E. (1990) Many transcribed regions of the *Onchocerca volvulus* genome contain the spliced leader sequence of *Caenorhabditis elegans*. *Mol. Cell. Biol.*, **10**, 2765.
30. Joshua, G. W. P., Chuang, R. Y., Cheng, S. C., Lin, S. F., Tuan, R. S., and Wang, C. C. (1991) The spliced leader gene of *Angiostrongylus cantonesis*. *Mol. Biochem. Parasitol.*, **460**, 209.
31. Nilsen, T. W., Shambaugh, J., Denker, J. A., Chubb, G., Faser, C., Putnam, L., *et al.* (1989) Characterization and expression of a spliced leader RNA in the parasitic nematode, *Ascaris lumbricoides* var *suum*. *Mol. Cell. Biol.*, **9**, 3543.
32. Takacs, A. M., Denker, J. A., Perrine, K. G., Maroney, P. A., and Nilsen, T. W. (1988) A 22-nucleotide spliced leader in the human parasitic nematode *Brugia malayi* is identical to the *trans*-spliced leader exon in *Caenorhabditis elegans*. *Proc. Natl Acad. Sci. USA*, **85**, 7932.
33. Rajkovic, A., Davis, R. E., Simonsen, J. N., and Rottman, F. M. (1990) A spliced leader is present on a subset of mRNAs from the human parasite *Schistosoma mansoni*. *Proc. Natl Acad. Sci. USA*, **87**, 8879.
34. Tessier, L.-H., Keller, M., Chan, R., Fournier, R., Weil, J. H., and Imbault, P. (1991) Short leader sequences may be transferred from small RNAs to pre-mature mRNAs by *trans*-splicing in Euglena. *EMBO J.*, **10**, 2621.

35. Bruzik, J. P., van Doren, K., Hirsh, D., and Steitz, J. A. (1988) *Trans*-splicing involves a novel form of small ribonucleoprotein particles. *Nature*, **335**, 559.
36. Maroney, P. A., Hannon, G. J., Denker, J. A., and Nilsen, T. W. (1990) The nematode spliced leader RNA participates in *trans*-splicing as an Sm snRNP. *EMBO J.*, **9**, 3667.
37. Maroney, P. A., Hannon, G. J., and Nilsen, T. W. (1990) Transcription and cap trimethylation of a nematode spliced leader RNA in a cell-free system. *Proc. Natl Acad. Sci. USA*, **87**, 709.
38. Thomas, J. D., Conrad, R. C., and Blumenthal, T. (1988) The *C. elegans trans*-spliced leader RNA is bound to Sm and has a trimethylguanosine cap. *Cell*, **54**, 533.
39. van Doren, K. and Hirsh, D. (1988) *Trans*-spliced leader RNAs exist as small nuclear ribonucleoprotein particles in *Caenorhabditis elegans*. *Nature*, **335**, 556.
40. Hannon, G. J., Maroney, P. A., Yu, Y.-T., Hannon, G. E., and Nilsen, T. W. (1992) Interaction of U6 snRNA with a sequence required for function of the nematode SL RNA in *trans*-splicing interacts with U6 snRNA. *Science*, **258**, 1775.
41. van Doren, K. and Hirsh, D. (1990) mRNAs that mature through *trans*-splicing in *Caenorhabditis elegans* have a trimethyl guanosine cap at their 5′ terminus. *Mol. Cell. Biol.*, **10**, 1769.
42. Perry, K. L., Watkins, K. P., and Agabian, N. (1987) Trypanosome mRNAs have unusual 'cap 4' structures acquired by addition of a spliced leader. *Proc. Natl Acad. Sci. USA*, **84**, 8190.
43. Freistadt, M. S., Cross, G. A. M., Branch, A. D., and Robertson, H. D. (1987) Direct analysis of the mini-exon donor RNA of *Trypanosoma brucei*: detection of a novel cap structure also present in messenger RNA. *Nucleic Acids Res.*, **11**, 9861.
44. Bangs, J. D., Crain, P. F., Hashizume, T., McCloskey, J. A., and Boothroyd, J. C. (1992) Mass spectrometry of mRNA cap 4 from trypanosomatids reveals two novel nucleosides. *J. Biol. Chem.*, **267**, 9805.
45. Palfi, Z., Günzl, A., Cross, M., and Bindereif, A. (1991) Affinity purification of *Tryanosoma brucei* snRNPs reveals common and specific protein components. *Proc. Natl Acad. Sci. USA*, **88**, 9097.
46. Cross, M., Wieland, B., Palfi Z., Günzl, A., Röthlisberger, U., Lahm, H.-W., *et al.* (1993) The *trans*-spliceosomal U2 snRNP protein 40k of *Trypanosoma brucei*: Cloning and analysis of functional domains reveals homology to a mammalian snRNP protein. *EMBO J.*, **12**, 1239.
47. Ullu, E. and Tschudi, C. (1990) Permeable trypanosome cells as a model system for transcription and trans-splicing. *Nucleic Acids Res.*, **18**, 3319.
48. Curroto de Lafaille, M. A., Laban, A., and Wirth, D. F. (1992) Gene expression in *Leishmania*: analysis of essential 5′ DNA sequences. *Proc. Natl Acad. Sci. USA*, **89**, 2703.
49. Huang, J. and van der Ploeg, L. H. T. (1991) Maturation of polycistronic pre-mRNA in *Trypanosoma brucei*: analysis of *trans*-splicing and poly(A) addition at nascent RNA transcripts from the hsp70 locus. *Mol. Cell. Biol.*, **11**, 3180.
50. Muhich, M. L. and Boothroyd, J. C. (1988) Polycystronic transcripts in trypanosomes and their accumulation during heat shock: evidence for a precursor role in mRNA synthesis. *Mol. Cell. Biol.*, **8**, 3837.
51. Ullu, E., Matthews, K. R., and Tschudi, C. (1993) Temporal order of RNA-processing reactions in trypanosomes: rapid *trans*-splicing precedes polyadenylation of newly-synthesized tubulin transcripts. *Mol. Cell. Biol.*, **13**, 720.
52. Watkins, K. P. and Agabian, N. (1991) *In vivo* UV cross-linking of U snRNAs that participate in trypanosome *trans*-splicing. *Genes Dev.*, **5**, 1859.

53. de Lafaille, M. A. C., Laban, A., and Wirth, D. F. (1992) Gene expression in *Leishmania*: analysis of essential 5′ DNA sequences. *Proc. Natl Acad. Sci. USA*, **89**, 2703.
54. Huang, J. and van der Ploeg, L. H. T. (1991) Requirement of a polypyrimidine tract for *trans*-splicing in trypanosomes: discriminating the PARP promoter from the immediately adjacent 3′ splice acceptor site. *EMBO J.*, **10**, 3877.
55. Watkins, K. P., Dungan, J. M., and Agabian, N. (1994) Identification of a small RNA that interacts with the 5′ splice site of the *Trypanosoma brucei* spliced leader RNA *in vivo*. *Cell*, **67**, 171.
56. Gröning, K., Palfi, Z., Gupta, S., Cross, M., Wolff, T., and Bindereif, A. (1991) A new U6 snRNP-specific protein conserved between *cis*- and *trans*-splicing systems. *Mol. Cell. Biol.*, **11**, 2026.
57. Cross, M., Günzl, A., Palfi, Z., and Bindereif, A. (1991) Analysis of small nuclear ribonucleoproteins in *Trypanosoma brucei*: structural organization and protein composition of the SL RNP. *Mol. Cell. Biol.*, **11**, 5516.
58. Günzl, A., Cross, M., and Bindereif, A. (1992) Domain structure of U2 and U4/U6 snRNPs from *Trypanosoma brucei*: identification of transspliceosomal specific RNA-protein interactions. *Mol. Cell. Biol.*, **12**, 468.
59. Hannon, G. J., Maroney, P. A., Denker, J. A., and Nilsen, T. W. (1990) *Trans*-splicing of nematode pre-messenger RNA *in vitro*. *Cell*, **61**, 1247.
60 Mattaj, I. W. (1986) Cap trimethylation of U snRNA is cystoplasmic and dependent on U snRNP protein binding. *Cell*, **46**, 905.
61. Ullu, E. and Tschudi, C. (1991) *Trans*-splicing in trypanosomes requires methylation of the 5′ end of the spliced leader RNA. *Proc. Natl Acad. Sci. USA*, **88**, 10074.
62. McNally, K. P. and Agabian, N. (1992) *Trypanosoma brucei* spliced leader RNA methylations are required for *trans*-splicing *in vivo*. *Mol. Cell. Biol.*, **12**, 4844.
63. Mattaj, I. W. (1988) U snRNP assembly and transport. In *Small nuclear ribonucleoprotein particles*, Birnstiel, M. L. (ed.). Springer-Verlag, Berlin, p. 100.
64. Maroney, P. A., Hannon, G. J., Shambaugh, J. D., and Nilsen, T. W. (1991) Intramolecular base pairing between the nematode spliced leader and its 5′ splice site is not essential for *trans*-splicing *in vitro*. *EMBO J.*, **10**, 3869.
65. Hausner, T.-P., Giglio, L. M., and Weiner, A. M. (1990) Evidence for base-pairing between mammalian U2 and U6 small nuclear ribonucleoprotein particles. *Genes Dev.*, **4**, 2146.
66. Datta, B. and Weiner, A. M. (1991) Genetic evidence for base pairing between U2 and U6 snRNAs in mammalian mRNA splicing. *Nature*, **352**, 821.
67. Wu, J. A. and Manley, J. L. (1991) Base-pairing between U2 and U6 snRNAs is necessary for splicing of mammalian pre-mRNA. *Nature*, **352**, 818.
68. McPheeters, D. S., Fabrizio, P., and Abelson, J. (1989) *In vitro* reconstitution of functional yeast U2 snRNPs. *Genes Dev.*, **3**, 2124.
69. Madhani, H. D. and Guthrie, C. (1992) A novel base-pairing interaction between U2 and U6 snRNAs suggests a mechanism for catalytic activation of the spliceosome. *Cell*, **71**, 803.
70. Sawa, H. and Abelson, J. (1992) Evidence for a base-pairing interaction between U6 snRNA and the 5′ splice site during the splicing reaction in yeast. *Proc. Natl Acad. Sci. USA*, **89**, 11269.
71. Wassarman, D. A. and Steitz, J. A. (1992) Interactions of small nuclear RNAs with precursor messenger RNA during *in vitro* splicing. *Science*, **257**, 1918.
72. Wyatt, J. R., Sontheimer, E. J., and Steitz, J. (1993) Site-specific crosslinking of mammalian U5 snRNP to the 5′ splice site prior to the first step of premessenger RNA splicing. *Genes Dev.*, **6**, 2542.

73. Sawa, H. and Shimura, Y. (1992) Association of U6 snRNA with the 5′ splice site region of pre-mRNA in the spliceosome. *Genes Dev.*, **6**, 244.
74. Nilsen, T. W. (1994) RNA–RNA interactions in the spliceosome: Unraveling the ties that bind. *Cell*, **78**, 1.
75. Reich, C., Van Hoy, R., Porter, G. L., and Wise, J. A. (1992) Mutations of the 3′ splice site can be suppressed by compensatory base changes in U1 snRNA in fission yeast. *Cell*, **69**, 1159.
76. Newman, A. M. and Norman, C. (1991) Mutations in yeast U5 snRNA alter the specificity of 5′ splice-site cleavage. *Cell*, **65**, 115.
77. Newman, A. M. and Norman, C. (1992) U5 snRNA interacts with exon sequences at 5′ and 3′ splice sites. *Cell*, **68**, 743.
78. Steitz, J. A. (1992) Splicing takes a Holliday. *Science*, **257**, 888.
79. Yean, S.-L. and Lin, R.-J. (1991) U4 small nuclear RNA dissociates from a yeast spliceosome and does not participate in the subsequent splicing reaction. *Mol. Cell. Biol.*, **11**, 5571.
80. Lesser, C. F. and Guthrie, C. (1993) Mutations in U6 snRNA that alter splice site specificity: implications for the active site. *Science*, **262**, 1982.
81. Kandels-Lewis, S. and Séraphin, B. (1993) Role of U6 snRNA in 5′ splice site selection. *Science*, **262**, 2035.
82. Sontheimer, E. J. and Steitz, J. A. (1993) The U5 and U6 small nuclear RNAs as active site components of the spliceosome. *Science*, **262**, 1989.
83. McPheeters, D. S. and Abelson, J. (1992) Mutational analysis of the yeast U2 snRNA suggests a structural similarity to the catalytic core of Group I introns. *Cell*, **71**, 819.
84. Mottram, J., Perry, K. L., Lizardi, P. M., Lührmann, R., Agabian, N., and Nelson, R. G. (1989) Isolation and sequence of four small nuclear U RNA genes of *Trypanosoma brucei* subsp. *brucei*: identification of the U2, U4 and U6 RNA analogs. *Mol. Cell. Biol.*, **9**, 1212.
85. Tschudi, C., Richards, F. F., and Ullu, E. (1986) The U2 RNA analogue of *Trypanosoma brucei gambiense*: implications for splicing mechanisms in trypanosomes. *Nucleic Acids Res.*, **14**, 8893.
86. Tschudi, C. and Ullu, E. (1990) Destruction of U2, U4, or U6 small nuclear RNAs blocks *trans*-splicing in trypanosome cells. *Cell*, **61**, 459.
87. Hannon, G. J., Maroney, P. A., and Nilsen, T. W. (1991) U small nuclear ribonucleoprotein requirements for nematode *cis*- and *trans*-splicing *in vitro*. *J. Biol. Chem.*, **266**, 22792.
88. Yu, Y.-T., Maroney, P., and Nilsen, T. W. (1993) Functional reconstitution of U6 snRNA in nematode *cis* and *trans*-splicing: U6 can serve as both a branch acceptor and a 5′ exon. *Cell*, **75**, 1049.
89. Bruzik, J. P. and Steitz, J. A. (1990) Spliced leader RNA sequences can substitute for the essential 5′ end of U1 RNA during splicing in a mammalian *in vitro* system. *Cell*, **62**, 889.
90. Seiwert, S. D. and Steitz, J. A. (1993) Uncoupling two functions of the 5′ end of U1 snRNA during *in vitro* splicing. *Mol. Cell. Biol.* **13**, 3134.
91. Thomas, J., Lea, K., Zuker-Aprison, E., and Blumenthal, T. (1990) The spliceosomal snRNAs of *Caenorhabditis elegans*. *Nucleic Acids Res.*, **18**, 2633.
92. Barabino, S. M. L., Blencowe, B. J., Ryder, U., Sproat, B. S., and Lamond, A. (1990) Targeted snRNP depletion reveals an additional role for mammalian U1 snRNP in spliceosome assembly. *Cell*, **63**, 293.
93. Moore, M. J. and Sharp, P. A. (1992) Site-specific modification of pre-mRNA: the 2′ hydroxyl groups at the splice sites. *Science*, **256**, 992.
94. Jacquier, A. (1990) Self-splicing group II and nuclear pre-mRNA introns: how similar are they? *Trends Biochem. Sci.*, **15**, 351.

95. Parker, R., Siliciano, P. G., and Guthrie, C. (1987) Recognition of the UACUAAC box during mRNA splicing in yeast involves base-pairing to the U2-like snRNA. *Cell*, **49**, 229.
96. Wu, J. A. and Manley, J. L. (1989) Mammalian pre-mRNA branch site selection by U2 snRNP involves base-pairing. *Genes Dev.*, **3**, 1553.
97. Zhuang, T. and Weiner, A. M. (1989) A compensatory base change in human U2 snRNA can suppress a branch site mutation. *Genes Dev.*, **3**, 1545.
98. Tschudi, E., Williams, S. P., and Ullu, E. (1990) Conserved sequences in the U2 snRNA-encoding genes of Kinetoplastida do not include the putative branchpoint recognition region. *Gene*, **91**, 71.
99. Hartshorne, T. and Agabian, N. (1990) A new U2 RNA secondary structure provided by phylogenetic analysis of trypanosomatid U2 RNAs. *Genes Dev.*, **4**, 2121.
100. Patzelt, E., Perry, K. L., and Agabian, N. (1989) Mapping of branch sites in *trans*-spliced pre-mRNAs of *Trypanosoma brucei*. *Mol. Cell. Biol.*, **9**, 4191.
101. Jamison, S. F., Crow, A., and Garcia-Blanco, M. A. (1992) The spliceosome assembly pathway in mammalian extracts. *Mol. Cell. Biol.*, **10**, 4279.
102. Rosbash, M. and Séraphin, B. (1991) Who's on first? *Trends Biochem. Sci.*, **16**, 187.
103. Conrad, R., Liou, R. F., and Blumenthal, T. (1993) Conversion of a *trans*-spliced gene in *C. elegans* into a conventional gene by introduction of a splice donor site. *EMBO*, **12**, 1249.
104. Solnick, D. (1985) *Trans*-splicing of mRNA precursors. *Cell*, **42**, 157.
105. Konarska, M. M., Padgett, R. A., and Sharp, P. A. (1985) *Trans*-splicing of mRNA precursors *in vitro*. *Cell*, **42**, 165.
106. Vellard, M., Soret, J., Viegas-Pequignot, E., Galibert, F., Van Cong, N., Dutrillaux, B., *et al.* (1991) C-*myb* proto-oncogene: evidence for intermolecular recombination of coding sequences. *Oncogene*, **6**, 505.
107. Shimizu, A., Nussensweig, M., Mizuta, T., Leder, P., and Honjo, T. (1989) Immunoglobulin double-isotype expression by *trans*-mRNA in a human immunoglobulin transgenic mouse. *Proc. Natl Acad. Sci. USA*, **86**, 8020.
108. Bruzik, J. P. and Maniatis, T. (1992) Spliced leader RNAs from lower eukaryotes are *trans*-spliced in mammalian cells. *Nature*, **360**, 692.
109. Maroney, P. A., Yu, Y. -T., Jankowska, M., and Nilsen, T. W. (1996) Direct analysis of nematode *cis*- and *trans*-spliceosomes: A functional role for U5 snRNA in spliced leader addition *trans*-splicing and the identification of novel Sm snRNPs. *RNA* **2**, 735.
110. Dungan, J. M., Watkins, K. P., and Agabian, N. (1996) Evidence for the presence of a small U5-like RNA in active *trans*-spliceosomes of Trypanosoma brucei. *EMBO J.* **15**, 4016.
111. Denker, J. A., Maroney, P. A., Yu, Y. -T., Kanost, R. A., and Nilsen, T. W. (1996) Multiple requirements for nematode spliced leader RNP function in *trans*-splicing. *RNA* **2**, 746.
112. Lucke, S., Xu, G. L., Palfi, Z., Cross, M., Bellofatto, V., and Bindereif, A. (1996) Spliced leader RNA of trypanosomes: *in vivo* mutational analysis reveals extensive and distinct requirements for *trans* splicing and cap4 formation. *EMBO J.* **15**:4380.
113. Blumenthal, T. and Spieth, J. (1996) Gene structure and organization in *Caenorhabditis elegans*. *Curr. Opin. Genet. Dev.* **6**, 692.
114. Bruzik, J. P. and Maniatis, T. (1995) Enhancer-dependent interaction between 5′ and 3′ splice sites in *trans*. *Proc. Natl Acad. Sci. USA* **92**, 7056.
115. Chiara, M. D. and Reed, R. (1995) A two-step mechanism for 5′ and 3′ splice-site pairing. *Nature* **375**, 510.

11 | mRNA editing

LARRY SIMPSON and OTAVIO H. THIEMANN

1. Introduction

RNA editing is a term that was coined by Benne *et al.* (1) to describe the apparently post-transcriptional insertion of four Us at three sites in the coding region of the cytochrome oxidase subunit II (COII) mRNA in two kinetoplastid species, *Crithidia fasciculata* and *Trypanosoma brucei*. This modification overcame a −1 frameshift in the COII gene that had been shown to be evolutionarily conserved in these two species and also in *Leishmania tarentolae* (2). As additional examples of RNA sequence modifications were reported in a variety of organisms (2–10), RNA editing' came to be defined in a broad sense as any process that results in the production of an RNA molecule which differs in nucleotide sequence in coding regions from the DNA template, with the exception of classical *cis*-splicing (11). However, even this definition was rendered obsolete by the discovery of specific sequence changes in rRNAs and tRNAs in several organisms, in one case involving the insertion of cytidine residues (12, 13) and in some cases the substitution of uridine for cytidine (14) or guanosine for adenosine residues (15). In this review we will only discuss the trypanosome type of RNA editing, which has been termed insertion–deletion editing by Bass (16), to distinguish it from the substitution editing found in other cells.

The kinetoplastid protozoa, together with the euglenoids, represent one of the most ancient eukaryotic lineages in terms of rRNA phylogenetic reconstructions (17), and possibly represent one of the surviving initial lineages which possessed a mitochondrion. The kinetoplastids consist of two major branches, the Bodonina (bodonids and cryptobiids) and the Trypanosomatina (trypanosomatids), both of which contain a single mitochondrion with a large mass of DNA. The region of the mitochondrion where the DNA is localized is termed the kinetoplast as a consequence of its association with the flagellum, and its DNA is called kinetoplast DNA (kDNA).

Little was known until recently about the kDNA of the bodonids. The kDNA of *Bodo caudatus*, a free-living bodonid flagellate, was shown in 1986 (18) to consist of heterogeneous sized large molecules of unknown function. More recently, the kDNA of a parasitic cryptobiid flagellate, *Trypanoplasma borreli*, was isolated and partially characterized (19, 20). The kDNA of *T. borreli* consists of a 40–80 kb circular component that contains several structural genes and several larger circular components consisting of tandemly organized 1 kb sequences encoding small RNAs which may be involved in the mechanism of RNA editing (see Section 3).

The kDNA of the trypanosomatids has been studied intensively using species from several genera (21). It consists of thousands of catenated small circular molecules, the minicircles, and a smaller number of catenated larger circular molecules, the maxicircles. The minicircle size varies in different species from approximately 0.5 kb to 2.5 kb and the maxicircle varies from 22 kb to 36 kb. The maxicircle molecule is the homologue of the mitochondrial DNA found in animal cells and the minicircle molecule encodes the guide RNAs (gRNAs) involved in mediating editing of the transcripts of certain maxicircle structural genes (22). These genes whose transcripts are edited are termed 'cryptogenes' (11).

2. Mitochondrial cryptogenes in kinetoplastids

The maxicircle genome is organized into an informational region that contains the two rRNA genes (23–26), at least 11 structural genes, and an apparently non-informational region that contains tandem repeats of varying complexities (21, 27–35) (Fig. 1). The latter region differs in size and sequence in different species and accordingly is termed the 'divergent' or 'variable' region. Several maxicircle genes yield transcripts which are apparently never edited: cytochrome oxidase subunit I (COI); NADH dehydrogenase subunits 1 (ND1), 4 (ND4), and 5 (ND5); and maxicircle unidentified reading frames 1 (MURF1) and 5 (MURF5). MURF5 (previously called ORF10) (31) in *L. tarentolae* is an apparently unedited 100 amino acid open reading frame situated between G2 and ND7, in the same polarity as G2. It exhibits a codon bias characteristic of a coding region and has significant similarity to the translated equivalent open reading frame in *T. brucei* (previously called ORF8) (31), but has no homology with any database sequence. The HR2 nucleotide sequence, which represents a sequence fairly well conserved between *L. tarentolae* and *T. brucei*, lies within this region (31).

Cryptogenes produce transcripts which are edited, and the extent of editing varies from gene to gene. Those cryptogenes, transcripts of which are only slightly or moderately edited in *L. tarentolae* and *C. fasciculata*, include cytochrome b (CYb) (4, 5, 36), cytochrome oxidase subunits II (COII) (1, 2) and III (COIII), ND7, MURF2, and MURF4 (3, 9). In *T. brucei*, the ND7, COIII, and MURF4 genes are not recognizable at the genomic sequence level (31), but the transcripts are extensively edited or 'pan-edited' (6–8) (Section 2.1) to produce potentially translatable sequences which have homology with known genes. In addition to these cryptogenes, there are six G-rich intergenic regions (labelled G1–G6 in *L. tarentolae* and CR1–CR6 in *T. brucei*) which in *T. brucei* and in a recently isolated strain of *L. tarentolae* (LEM125) yield transcripts that are pan-edited (11, 37–40). The G6 transcript is edited to produce an mRNA encoding the ribosomal protein, S12; the edited G1, G2, and G5 transcripts encode components of complex I of the electron transport chain—ND8, ND9, and ND3, respectively; and the edited G3 and G4 transcripts encode hydrophobic proteins of unknown function.

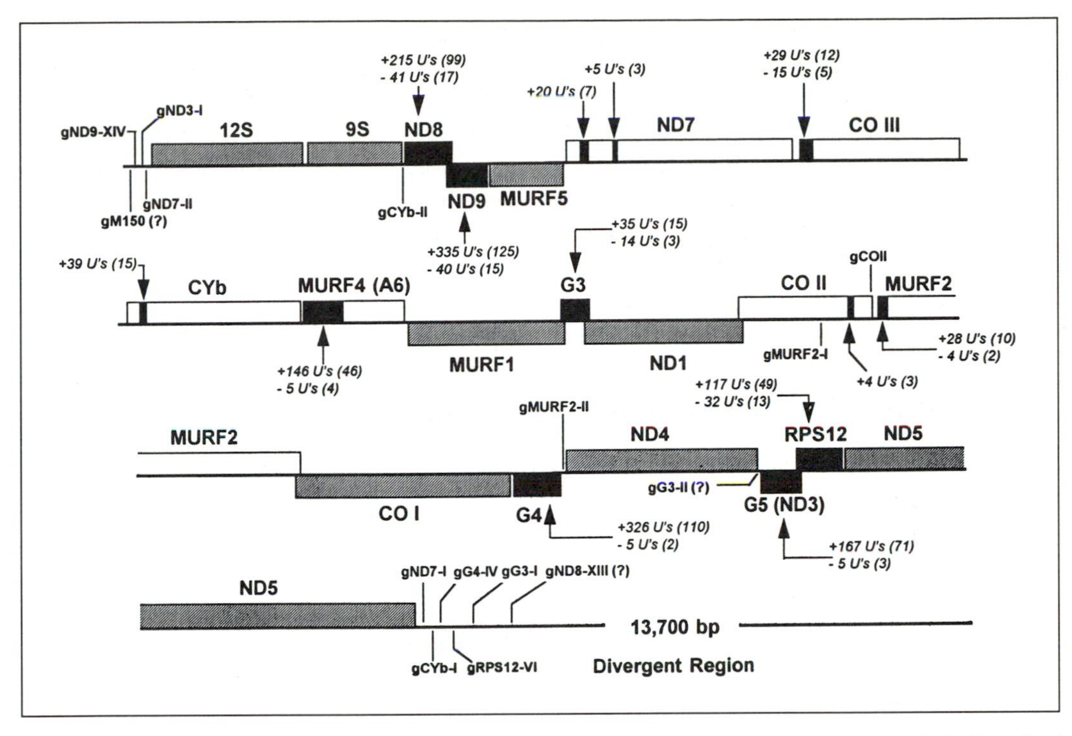

Fig. 1 Localization of RNA editing events in the maxicircle genome of *L. tarentolae*. The maxicircle is linearized at a single *Eco*RI site. Unedited genes are shown as hatched boxes and cryptogenes by open boxes, with the strandedness shown by the box being above or below the line. Pre-edited regions are shown as black areas and the number of Us added and deleted is indicated above each region. gRNA genes are indicated by lines, with a '?' if they are unconfirmed.

2.1 Extent of editing and identification of gene products of edited mRNAs

The extent of editing of cryptogene transcripts ranges from the insertion of four or five Us in three sites in the middle of the gene (internal editing), to the addition of 30–40 Us at 15–20 sites at the 5′ end of the gene (5′ editing), to pan-editing (11) over the entire length of the gene. Initially it seemed that pan-editing was limited to the ND7, COIII, and MURF4 cryptogenes in *T. brucei*. However, the 5′ third of the MURF4 mRNA (6, 41) has been found to be pan-edited in *L. tarentolae*, in addition to all the G1–G6 mRNAs (40).

2.1.1 The MURF4 cryptogene

In *T. brucei*, the 821 nucleotide MURF4 pre-edited mRNA is pan-edited with the addition of 448 Us at 173 sites and the deletion of 28 Us at 12 sites (6, 41). Editing occurs both within the coding region and within the 3′ untranslated region, and terminates one nucleotide upstream of the created AuG methionine translation initiation codon. The 5′ end (32 nucleotides) of the untranslated 5′ flanking sequence and the 3′ terminal portion (37 nucleotides) of the 3′ untranslated sequence remain unedited. It is not known if there are separate editing domains, which are defined as regions of the RNA which are edited independently of other regions, since an analysis of multiple partially edited MURF4 transcripts has not yet been performed.

The editing in the case of the *L. tarentolae* MURF4 transcript is less extensive, but still substantial (6, 41). Pan-editing is limited to the 5′ portion of the mRNA, and involves the addition of 106 Us at 46 sites and the deletion of 5 Us at four sites within a 112 nucleotide region. An AuG methionine initiation codon is created at the same relative position as in *T. brucei*.

The pan-edited RNA from *T. brucei* contains an ORF which shows high similarity with the translated protein from the 5′ pan-edited *L. tarentolae* RNA. The originally reported *T. brucei* MURF4 sequence (6) indicated a substantial divergence of the translated amino acid sequence at the C terminus of the protein. However, a revised *T. brucei* sequence (41) corrected for a frameshift error showed that the similarity in translated amino acid sequences continues to the C terminus, with 65% identity and 85% overall similarity, allowing for conservative replacements. From a multiple sequence alignment with known ATPase 6 proteins, Bhat *et al.* (6, 41) suggested that the MURF4 protein belongs to this family of proteins, although highly diverged. However, the similarity of the kinetoplastid MURF4 amino acid sequences with known ATPase 6 sequences is of marginal statistical significance if one applies a Monte Carlo shuffling analysis (42, 43) (L. Simpson, unpublished). On the other hand, a PROFILESEARCH analysis (44) of the Swiss Protein database gave 12 known ATPase 6 proteins as the best 'hits' with Z values of 5.0–7.7 standard deviation (sd) units from the mean of aligned random sequences (M. Peris, and L. Simpson, unpublished). It is entirely possible that MURF4 represents a highly diverged ATPase 6, but this cannot be confirmed by sequence comparisons alone and rather must be substantiated by sequence analysis of the kinetoplastid ATPase 6 protein.

2.1.2 The G6 (CR6) cryptogene

Another pan-edited cryptogene in *L. tarentolae* is the G6 region. Editing occurs in three separate 'editing domains' (Fig. 2) in both UC strain and LEM125 strain *L. tarentolae*. The mature edited *L. tarentolae* G6 mRNA contains 117 Usinserted at 49 sites and 32 Us deleted at 13 sites (39). The pan-edited mRNA encodes an 85 amino acid polypeptide which appears to represent a highly diverged ribosomal protein S12 (RPS12). In *T. brucei*, 132 Us are inserted and 28 Us are deleted, yielding a predicted 82 amino acid protein with good similarity to the *L. tarentolae* G6 protein sequence (38). The two most conserved regions of the protein sequences correspond to the sites which are known to confer streptomycin resistance and streptomycin dependence. Marginally significant Z values of 6.5–8.9 sd units were obtained for alignments of the G6 sequence with three chloroplast S12 proteins and one eubacterial S12 protein. Statistically insignificant Z values, however, were obtained for the alignments of G6 with mitochondrial S12 proteins from *Paramecium* and *Zea mays*. The conclusion that the G6 protein belongs to the RPS12 family is strengthened by

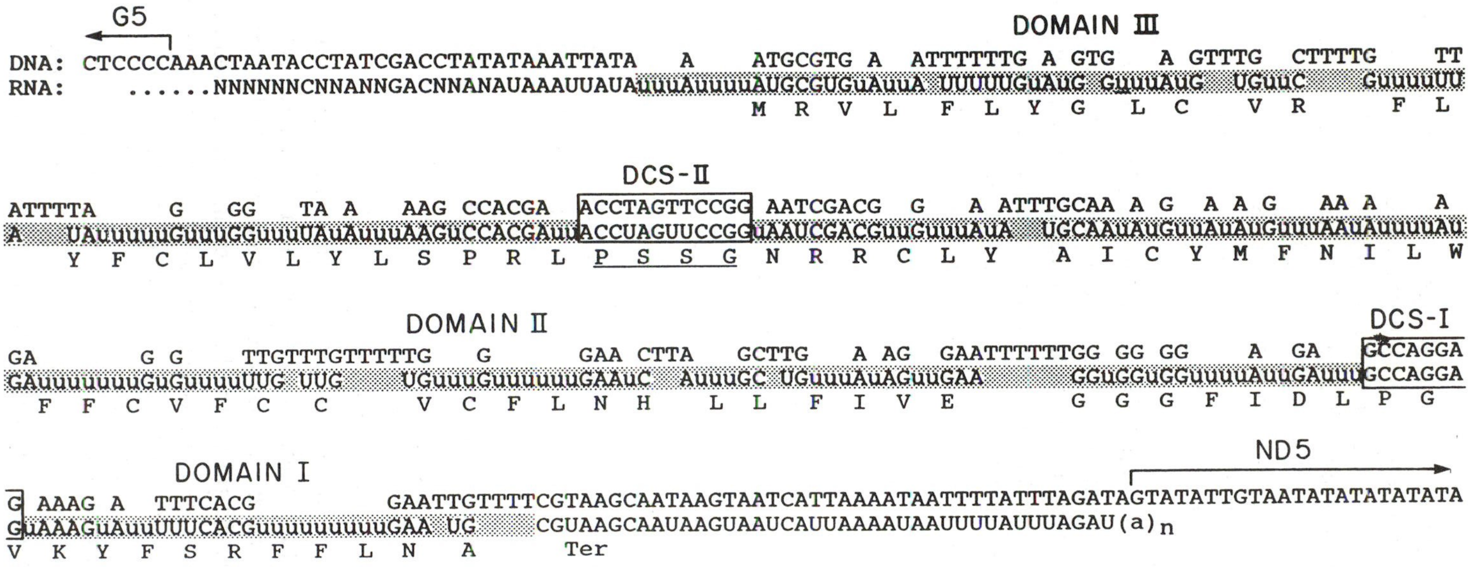

```
  G5                                                                      DOMAIN III
DNA: CTCCCCAAACTAATACCTATCGACCTATATAAATTATA   A    ATGCGTG A  ATTTTTTG A GTG   A GTTTG  CTTTTG     TT
RNA:     ......NNNNNNCNNANNGACNNANAUAAAUUAUAuuuAuuuuAUGCGUGuAuuA UUUUUGuAUG GuuuAUG  UGuuC    GuuuuUU
                                                 M  R  V  L  F  L  Y  G  L  C   V R     F L

                                        DCS-II
ATTTTA     G   GG   TA A   AAG CCACGA ACCTAGTTCCGG AATCGACG  G   A ATTTGCAA A G  A A G   AA A    A
A   UAuuuuuGuuuGGuuuUAuAuuuAAGuCCACGAuuACCUAGUUCCGGuAAUCGACGuuGuuuAuA  UGCAAuAuGuuAuAuGuuuAAuAuuuuAu
   Y  F  C  L  V  L  Y  L  S  P  R  L  P  S  S  G  N  R  R  C  L  Y   A  I  C  Y  M  F  N  I  L  W

                        DOMAIN II                                                            DCS-I
GA        G G    TTGTTTGTTTTTG   G      GAA CTTA    GCTTG    A AG   GAATTTTTTGG GG GG     A  GA  GCCAGGA
GAuuuuuuuGuGuuuuUUG UUG    UGuuuGuuuuuuGAAuC  AuuuGC UGuuuAuAGuuGAA         GGuGGuGGuuuuAuuGAuuuGCCAGGA
  F  F  C  V  F  C   C      V  C  F  L  N  H     L  L  F  I  V  E          G  G  G  F  I  D  L  P  G

                                                                                          ND5
        DOMAIN I
G AAAG A  TTTCACG          GAATTGTTTTCGTAAGCAATAAGTAATCATTAAAATAATTTTATTTAGATAGTATATTGTAATATATATATATA
GuAAAGuAuuUUUCACGuuuuuuuuuGAA UG     CGUAAGCAAUAAGUAAUCAUUAAAAUAAUUUUAUUUAGAU(a)n
V  K  Y  F  S  R  F  F  L  N  A          Ter
```

Fig. 2 The pan-edited RPS12 (G6) cryptogene from *L. tarentolae*. The RPS12 DNA sequence is nucleotides 14 636 to 14 913 in GenBank entry LEIKPMAX. Adjacent portions and polarities of the G5 and ND5 genes are shown by arrows. The sequence of mature edited RPS12 mRNA is shown, with editing domains I, II, and III indicated by stippling. Uridines added by editing are shown as 'u'. The amino acid sequence of the translated RPS12 protein is beneath the edited mRNA sequence. The domain-connection sequences DCS-I and DCS-II are boxed. (Reprinted with permission from 39.)

similarities in hydropathy patterns, but as in the case of MURF4, the amino acid sequence of the RPS12 protein is required to confirm this assignment.

The presence of three editing domains in the RPS12 cryptogene in *L. tarentolae* opens the possibility that independent editing of separate domains may modulate amino acid sequences of proteins in addition to simply creating translatable mRNAs. Another example of a cryptogene with more than one editing domain is the ND7 pan-edited cryptogene of *T. brucei* (7). In *L. tarentolae* and *C. fasciculata*, this gene has an internal frameshift editing site and a 5′ editing site, which also may be considered two separate editing domains, especially since these sites occur at the equivalent of the 5′ termini of the two pan-edited domains in *T. brucei* (2, 3) (Fig. 3).

Alignment of the mature edited RPS12 RNA sequences from *L. tarentolae* and *T. brucei* shows that the two domain-connection sequences (DCSs), which may represent the functionally significant regions of the protein, show absolute sequence conservation at the amino acid level (Fig. 4). However, at the RNA level, the DCSs are edited in *T. brucei,* thereby eliminating the possibility that the existence of separate editing domains in *L. tarentolae* is due to a functional constraint on editing in the DCS regions. This lack of correspondence of editing domains between these two species also makes it less likely that, at least in this case, independent regulation of editing in separate domains is used as a regulatory mechanism. Unedited DCSs were also not present in the pan-edited RPS12 sequence of the cryptobiid, *T. borreli* (19).

2.1.3 The G1 (CR1) cryptogene

The CR1 region of the maxicircle genome of *T. brucei* produces a 361 nucleotide pre-edited transcript that is pan-edited by the addition of 259 Us and the deletion of 46 Us at a total of 127 sites (37). The 5′ terminal 34 nucleotides and the 3′ terminal 27 nucleotides remain unedited, as is the case with all other reported pan-edited mRNAs. Editing continues four sites upstream of the created AuG initiation codon. The open reading frame encodes a 145 amino acid protein that shows significant similarity to a nuclear-encoded subunit of the bovine mitochondrial respiratory complex I, CI-23kD, including a repeated cysteine motif characteristic of a class of non-haem iron sulphur proteins. From this homology, the *T. brucei* protein was identified as NADH dehydrogenase subunit 8 (ND8).

Thiemann *et al.* (40) showed that productive editing of G1 transcripts also occurs in the LEM125 strain of *L. tarentolae.* The mature edited G1 RNA of LEM125 is 520 nucleotides in length, and is pan-edited by the addition of 215 Us in 99 sites and the deletion of 41 Us in 17 sites. An open reading frame (ORF) of 145 amino acid residues which is encoded by the edited transcript is homologous to the ND8 polypeptide encoded by the edited G1 (CR1) RNA in *T. brucei* (37).

2.1.4 The G2 (CR2) cryptogene

A 322 nucleotide transcript of the CR2 cryptogene of *T. brucei* is edited by the addition of 345 Us and the deletion of 20 encoded Us, yielding a mature edited mRNA encoding a 194 amino acid protein. This protein has significant similarity to a sub-

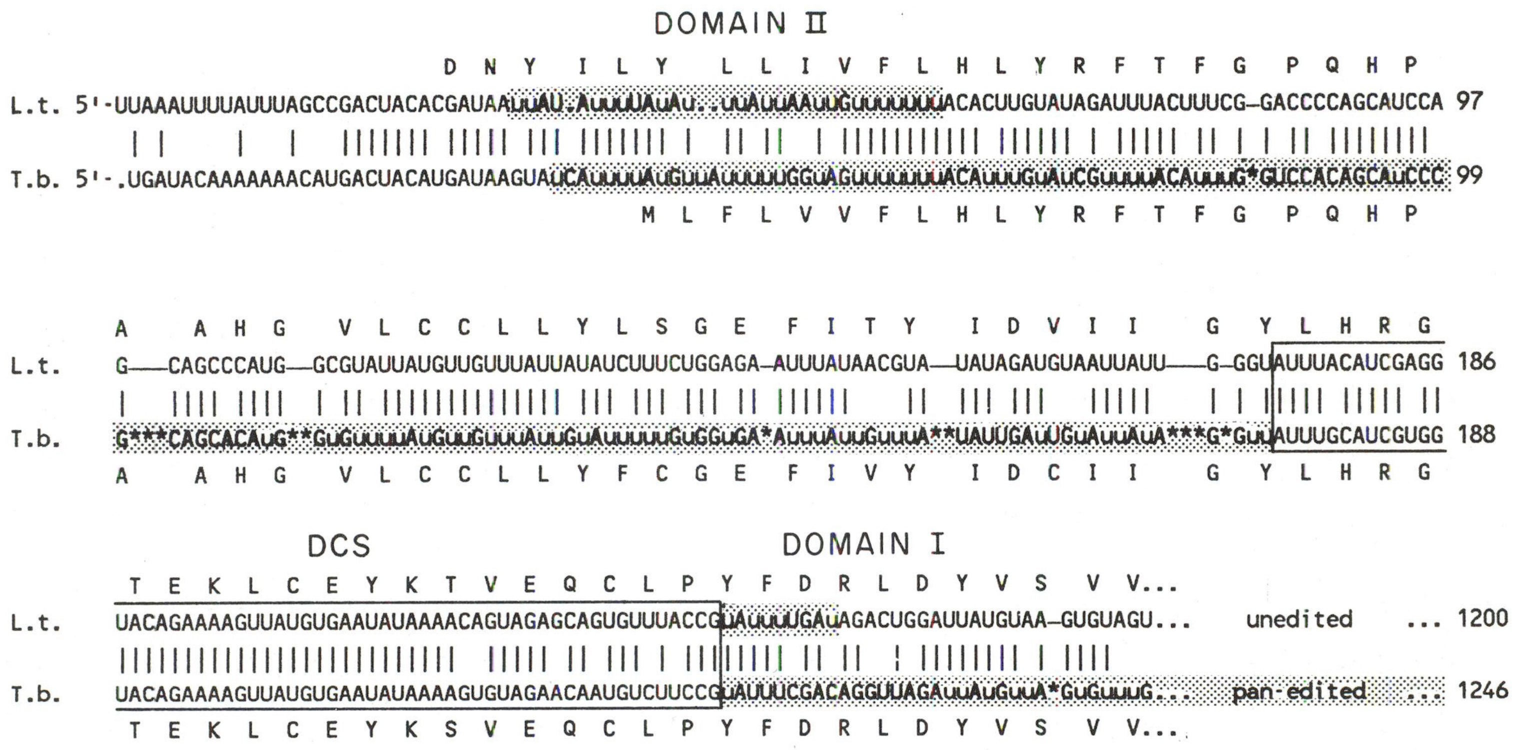

```
                                        DOMAIN II
                          D N Y I L Y  L L I V F L H L Y R F T F G  P Q H P
L.t. 5'-UUAAAUUUUAUUUAGCCGACUACACGAUAAuuAU.AuuuUAuAu..uuAuuAAuuGuuuuuuuACACUUGUAUAGAUUUACUUUCG-GACCCCAGCAUCCA 97
T.b. 5'-.UGAUACAAAAAAACAUGACUACAUGAUAAGUAuCAuuuuAuGuuAuuuuGGuAGuuuuuuuACAuuuGuAuCGuuuuACAuuuG*GUCCACAGCAuCCC 99
                                         M L F L V V F L H L Y R F T F G  P Q H P

      A   A H G  V L C C L L Y L S G E  F I T Y  I D V I I  G Y L H R G
L.t.  G--CAGCCCAUG--GCGUAUUAUGUUGUUUAUUAUAUCUUUCUGGAGA-AUUUAUAACGUA--UAUAGAUGUAAUUAUU---G-GGUAUUUACAUCGAGG 186
T.b.  G***CAGCACAuG**GuGuuuuAuGuuGuuuAuuGuAuuuuuGuGGuGA*AuuuAuuGuuuA**UAUUGAuUGuAuuAuA***G*GuuAUUUGCAUCGUGG 188
      A   A H G  V L C C L L Y F C G E  F I V Y  I D C I I  G Y L H R G

                  DCS                            DOMAIN I
      T E K L C E Y K T V E Q C L P Y F D R L D Y V S  V V...
L.t.  UACAGAAAAGUUAUGUGAAUAUAAAACAGUAGAGCAGUGUUUACCGuAuuuUGAuAGACUGGAUUAUGUAA-GUGUAGU...   unedited   ... 1200
T.b.  UACAGAAAAGUUAUGUGAAUAUAAAAGUGUAGAACAAUGUCUUCCGuAUUUCGACAGGUUAGAuuAuGuuA*GuGuuuG...  pan-edited  ... 1246
      T E K L C E Y K S V E Q C L P Y F D R L D Y V S  V V...
```

Fig. 3 Comparison of editing profiles in mature edited ND7 RNA sequences from *L. tarentolae* (L.t.) and *T. brucei* (T.b.). Matches are shown by vertical lines and dashes indicate the gaps introduced for alignment. The translated amino acid sequences are shown above and below the RNA sequences. Edited regions are stippled, and the DCS sequences are boxed. Only the 5′ portion of domain of the *T. brucei* sequence is shown. (Reprinted with permission from 111.)

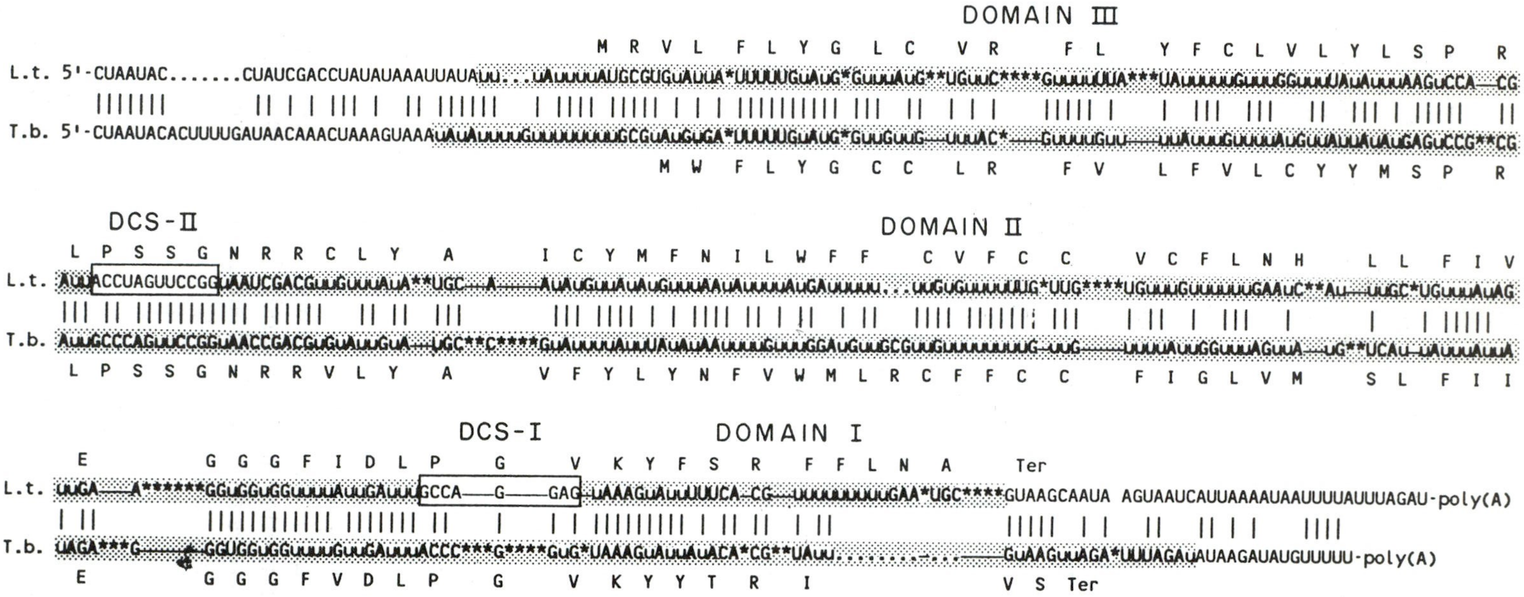

```
                                                   DOMAIN III
                                      M  R  V  L   F  L  Y  G  L  C   V R    F L   Y F C L V L Y L S P  R
L.t. 5'-CUAAUAC.......CUAUCGACCUAUAUAAAUUAUAUU...uAuuuuAUGCGUGuAuuA*UUUUGuAuG*GuuuAuG**UGuuC****GuuuuUUA***UAuuuuuGuuuGGuuuUAuAuuuAAGuCCA—CG
T.b. 5'-CUAAUACACUUUUGAUAACAAACUAAAGUAAAuAuAuuuuGuuuuuuuuGCGuAuGuGA*UUUUGuAuG*GuuGuuG—uuuAC*—GuuuuGuu—uuAuuuGuuuuAuGuuAuuAuAuGAGuCCG**CG
                                      M  W   F  L  Y  G  C  C   L R    F V   L F V L C Y Y M S P  R

  DCS-II                                                  DOMAIN II
  L P S S G N R R C L Y  A    I C Y M F N I L W F F   C V F C C   V C F L N H   L L F I V
L.t. AuuACCUAGUUCCGGuAAUCGACGuuGuuuAuA**UGC—A—AuAuGuuAuAuGuuuAAuAuuuuAuGAuuuuu...uuGuGuuuuuUG*UUG****UGuuuGuuuuuGAAuC**Au—uuGC*UGuuuAuAG
T.b. AuuGCCCAGuuCCGGuAACCGACGuGuAuuGuA—UGC**C****GuAuuuuAuuUAuAuAAuuuuGuuuGGAuGuuGCGuuGuuuuuuuUG—uuG—uuuuAuuGGuuuAGuuA—uG**UCAu—uAuuuAuuA
  L P S S G N R R V L Y  A    V F Y L Y N F V W M L R C F F C C   F I G L V M   S L F I I

            DCS-I              DOMAIN I
  E     G G G F I D L P   G    V K Y F S R   F F L N A   Ter
L.t. uuGA—A******GGuGGuGGuuuuAuuGAuuuGCCA—G—GAG—uAAAGuAuuUUUCA—CG—uuuuuuuuuGAA*UGC****GUAAGCAAUA AGUAAUCAUUAAAAUAAUUUUAUUUAGAU-poly(A)
T.b. UAGA***G—GGUGGuGGuuuuGuuGAuuuACCC***G****GuG*UAAAGuAuuAuACA*CG**UAuu........—GuAAGuuAGA*UUUAGAuAUAAGAUAUGUUUUU-poly(A)
  E     G G G F V D L P   G    V K Y Y T R   I        V S Ter
```

Fig. 4 Alignments of RPS12 sequences from *L. tarentolae* (L.t.) and *T. brucei* (T.b.) Edited regions are stippled, inserted uridines are indicated as 'u', and deleted uridines as '*'. The DCS sequences in *L. tarentolae* are boxed. The amino acid sequences of the translated proteins are shown above and below the edited RNA sequences. (Reprinted with permission from 111.)

unit of complex I from other organisms and was termed NADH dehydrogenase subunit 9 (ND9) (45). The transcript of the G2 cryptogene of *L. tarentolae* LEM125 (40) contains 335 U additions in 125 sites and 40 deletions in 15 sites, producing a mature edited mRNA encoding a 196 amino acid sequence. This sequence shows significant similarity to the ND9 sequence from *T. brucei*.

2.1.5 The G3 (CR3) cryptogene

The G3 genomic sequence is the shortest cryptogene in the maxicircle and, at least in the case of *L. tarentolae* LEM125 (40), shows some unusual features. The transcript is edited by 35 U additions and 14 U deletions. The deletions occur in the last three sites and the intervening genomic sequence is G-rich, suggesting incomplete editing. The edited RNA encodes an ORF of 51 hydrophobic amino acids which has no homology to any database sequence.

2.1.6 The G4 (CR4) cryptogene

A 283 nucleotide transcript of the CR4 cryptogene of *T. brucei* is edited by the insertion of 352 Us and the deletion of 40 Us to yield an edited mRNA encoding two possible ORFs with no detectable homology to any database sequence. In the case of *L. tarentolae* LEM125, the G4 transcript is edited by the addition of 326 Us in 110 sites and the deletion of 5 Us in two sites. A predicted ORF is 169 amino acids long and exhibits a limited similarity with one of the ORFs from the edited *T. brucei* mRNA.

2.1.7 The G5 (CR5) cryptogene

The transcript of the CR5 cryptogene of *T. brucei* is edited in two separate domains separated by an 8 nucleotide unedited DCS (46). The 3′ domain shows several different editing patterns whereas the 5′ domain has a single consensus pattern. A total of 205–217 Us are inserted and 13–16 Us are deleted in both domains. The existence of several different editing patterns in domain I would give rise to multiple carboxy-terminal protein sequences if all were translated, and is reminiscent of a misediting situation. Misediting is the occurence of unexpected editing patterns which differ from the mature edited sequence (47). The edited transcript encodes a potential protein that shows a motif characteristic of iron-sulphur proteins which are present in complex I of the respiratory chain, and exhibits a limited similarity to ND3 sequences from other organisms.

In *L. tarentolae* LEM125, there appears to be a single editing domain in G5, in which 167 Us are inserted at 71 sites and 5 Us are deleted at three sites. The mature edited mRNA encodes an ORF of 115 amino acids which has significant similarity to the *T. brucei* ND3 sequence.

2.2 Addition of Us to 3′ ends of mRNAs

Many maxicircle transcripts occur as two distinct size classes (4, 5, 36, 48–51). This size variation has been shown to be due mainly to variation in the length of the poly(A) tail (41, 52). The transcripts include CYb, COI, and COII from *T. brucei*,

MURF4, CYb, and COIII from *L. tarentolae*, and ND1, MURF2, ND7, and COII from *C. fasciculata*. The two MURF4 transcripts of *L. tarentolae* contain 20 nucleotide and 120 nucleotide poly(A) tails (41, 52). In addition, variable numbers of U residues are found within the poly(A) tails (53). Since the patterns of U insertion differ in different cDNAs and several cDNAs show 3′ terminal Us, it is likely that this U addition is not gRNA-mediated, but is due to a 3′ terminal uridylyl transferase (TUTase) activity which is present in the kinetoplast-mitochondrion (54).

In the initial report of the mitochondrial TUTase activity (54), it was shown that the major endogenous RNA species labeled when isolated organelles are incubated with [α^{32}P]UTP are the 9S and 12S mitochondrial rRNAs (54). This observation was confirmed and extended for *T. brucei* by Adler *et al.* (55) who showed that the 12S RNA has a 3′ tail of 2–17 Us whereas the 9S RNA has a 3′ tail of precisely 11 Us. This suggests the possibility of a more precise mechanism for 3′ U addition in this case. It was speculated that the functional significance of this 3′ U addition to the rRNAs may reside in modifications of the secondary or tertiary structures of the rRNAs, in protection against exonuclease degradation, or may even involve interactions with mRNAs (55).

2.3 Developmental regulation of RNA editing in the life cycle of *T. brucei*

The life cycle of the African trypanosome, *T. brucei*, involves several developmental stages which differ dramatically in terms of mitochondrial physiology, among other features (56). The procyclic trypomastigote forms found in the tsetse fly midgut possess an active mitochondrion that has a functional cytochrome-based respiratory chain, whereas the slender bloodstream trypomastigote forms found in the mammalian bloodstream have an inactive mitochondrion that lacks detectable cytochrome oxidase. Respiration of the latter forms is through a cyanide-insensitive alternate oxidase, and oxidative phosphorylation is absent (57). Several of the mitochondrial cryptogenes—MURF4, MURF2, G4 (CR4), and COIII—are constitutively edited in both life cycle stages, whereas CYb and COII are edited only in the procyclic stage (4, 6–8, 50, 58). A third set of cryptogene transcripts—RPS12, ND8, ND9, and CR3—are fully edited only in the bloodstream stages (37, 38, 59). The ND7 transcript shows a more complex pattern of regulation, in that domain I is fully edited only in the bloodstream stages and domain II is edited in both stages (7). The level of regulation of editing is uncertain. In several cases, the relative gRNA abundances for regulated transcripts do not change between life-cycle stages, suggesting that regulation is not at the level of gRNA abundance (58). However, in the only cases of pan-edited genes in which the entire gRNA cascade is known (the RPS12 and MURF4 genes of *L. tarentolae*) a single late-acting gRNA is in low abundance relative to the other gRNAs, opening the possibility that pan-editing might be regulated by the relative abundance of a single gRNA in the cascade (39).

Developmental regulation of the extent of 3′ polyadenylation of maxicircle transcripts has also been demonstrated in *T. brucei*. A larger fraction of CYb, COI, and COII transcripts have longer poly(A) tails and are more abundant in procyclic than in bloodstream forms (52). On the other hand, a greater number of CR1 transcripts have longer poly(A) tails and are more abundant in bloodstream than in procyclic forms. Edited transcripts have longer poly(A) tails than unedited transcripts. The ND4 and MURF1 transcripts have a similar size distribution of poly(A) tails in both stages. Bhat *et al.* (52) speculated that regulation of polyadenylation may influence mitochondrial gene expression in this species.

Developmental regulation of editing has not been studied in the case of *L. tarentolae* due to the inability to cultivate the lizard stages of the life cycle, and to a lack of knowledge of the biology of the complete life cycle in the lizard host. In the case of monogenetic species such as *C. fasciculata*, which lack a vertebrate host cycle, it would be expected that editing is unregulated. However, there could well be physiological changes of the parasite within the insect host, which involve regulation of editing at some level. This remains to be investigated.

3. Guide RNAs

Guide RNAs (gRNAs) are small mitochondrial RNAs which can form perfect hybrids with mature edited mRNAs if G-U (and, in some cases, A-C) base pairs are allowed (22) (Figs 5 and 6). The region of base complementarily extends 3′ of the pre-edited region for a variable length, and the formation of the 3′ anchor RNA–RNA duplex was hypothesized to represent the initial interaction between a specific gRNA and a specific pre-edited mRNA. The gRNAs also possess non-encoded 3′ oligo(U) tails, 5–30 nucleotides in length, and can be recognized on acrylamide–urea gels by a characteristic multiple banding pattern migrating ahead of tRNAs (60). They also have 5′-di- or tri-phosphates and can be capped *in vitro* by vaccinia virus guanylyl transferase.

Minicircle-encoded gRNAs were actually first visualized on northern blots as contaminants of *L. tarentolae* mitochondrial tRNA preparations, but the significance of this observation was not appreciated at the time (61).

3.1 Maxicircle-encoded gRNAs

The existence of seven maxicircle-encoded gRNAs was initially predicted by a computer analysis of the known 21 kb *L. tarentolae* maxicircle sequence for short sequences that could give rise to transcripts that were complementary to mature edited mRNA sequences if G-U base-pairing was allowed (Table 1). This was confirmed by northern blot hybridization using synthetic oligomer probes and by direct 5′ end sequencing of hybrid-selected isolated gRNAs (22). The locations of the gRNA genes in the maxicircle genome have no relationship to the location of the homologous cryptogenes. Isolated gMURF2-II was shown to possess a di- or triphosphate at the 5′ end and therefore may represent a primary transcript (60). All

```
                                  gND8-II (mc)
      3'..C----UagCaaCaagaUaCgaU---AAACUaagaAUAAACACAAA-5'
          |     :|:|||||||:||||:||    |||||||:|||||||||||||          gND8-I(mc)
                                                 3'..gUagaUaCaagaUagaUCgagUCaCaUaG--AGCUCUUCAUAUA-5'
                                                     :||:|:||||:|:|:|||:|:||||||||| |||||||||||||
DNA 5'...   GTTTTG   G   G    A G  ATTTTTTGA     A   G G    G   AG    AG G A CTTTCGAGAAGTATATTTGATTAATAATTTAATAAATTAATTTT  3'
RNA 5'...uuG****GuuGuuGuuuuAuGuuA***UUUGAuuuuuAuuuGuGuuuuGuuuAGuuuAGuGuAuC**UCGAGAAGUAUAUUUGAUUAAUAAUUUAAUAAAUUAAUUUU(A)n
          18     17   16  15   14    12  11          10    9  8   7    6      5  4  3  2  1
```

Fig. 5 Two overlapping gRNAs mediating the editing of blocks I and II of ND8 mRNA. The duplex anchor sequences are shaded. G-U base-pairs are indicated by: and A-U and G-C base-pairs by |. Editing sites are numbered 3′ to 5′. Deletions are indicated by *. The 3′-oligo(U) tails of the gRNAs are not shown.

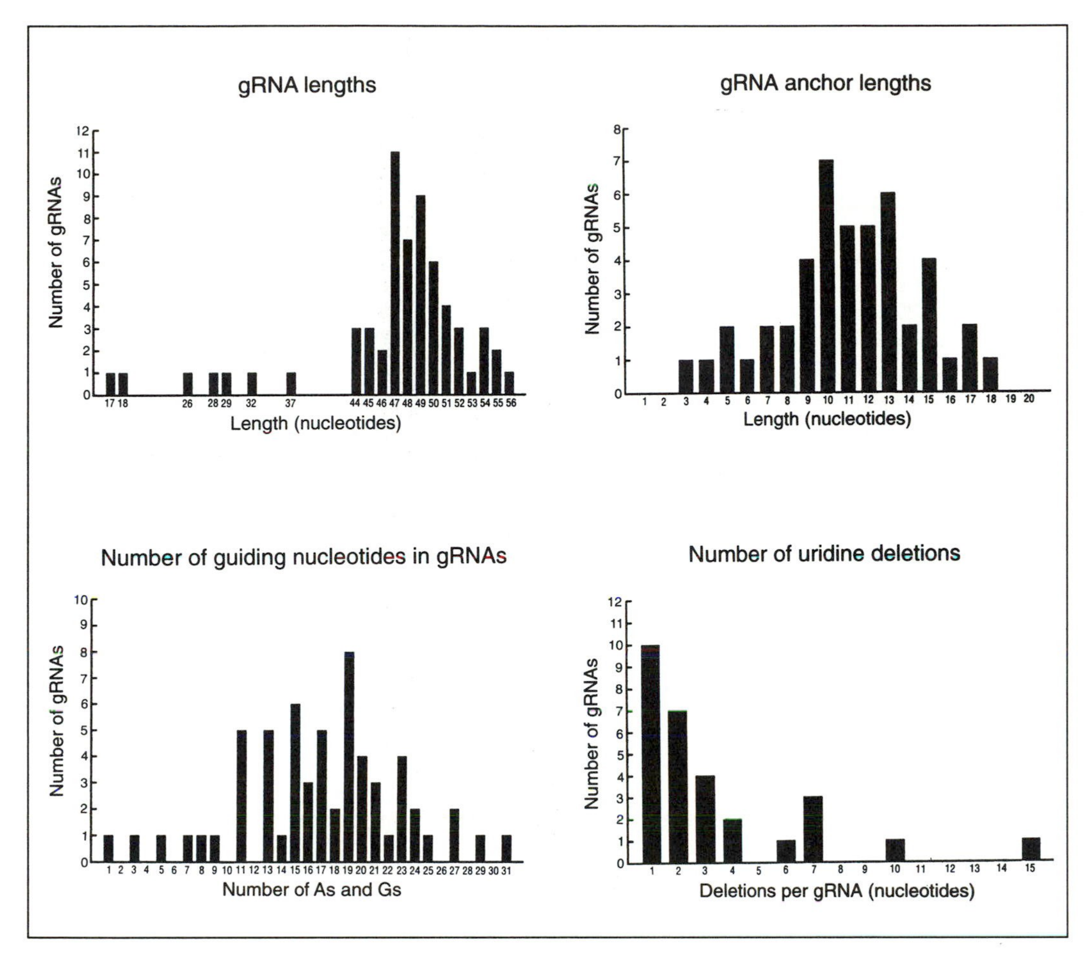

Fig. 6 Properties of *L. tarentolae* gRNAs. A summary of the size distribution and editing capacity of the identified gRNAs from the UC and LEM125 stains of *L. tarentolae*. See Table 2 for a list of the gRNAs included in this figure.

maxicircle-encoded gRNAs in *L. tarentolae* appear to terminate, or to be 3′-processed, at a string of encoded U residues, but a low percentage of 3′-truncated gRNAs were also found in steady-state kRNA. An additional maxicircle-encoded gRNA (M150) of unknown function was detected in the UC strain of *L. tarentolae* due to its presence in a misedited COIII mRNA–gRNA chimeric molecule (47) (see below). Five additional *L. tarentolae* maxicircle-encoded gRNAs—gND9-XIV, gG3-I, gG3-II, gG4-IV, and gND3-I—were recently identified by computer analysis (40). These gRNAs were confirmed by northern analysis and primer-extension sequencing (40).

The gRNA for COII was found to be in *cis* at the 3′ terminus of the mRNA, and a fold-back mechanism for mediation of editing was proposed in this case (22).

Seven gRNA genes located at identical relative positions in the maxicircle

Table 1 Identified gRNAs from *L. tarentolae* UC and LEM125 strains

RPS12 (G6)[a]	
gRPS12-I	mc[b]
gRPS12-II	mc
gRPS12-III	mc
gRPS12-IV	mc
gRPS12-V	mc
gRPS12-VI	Mc (17020–16975)[c]
gRPS12-VII	mc
gRPS12-VIII	mc
MURF2	
gMURF2-I	Mc (9908–9893)
gMURF2-II	Mc (13087–13146)
MURF4 (ATPase 6)	
gMURF4-I	mc
gMURF4-II	mc
gMURF4-III	mc
gMURF4-IV	mc
gMURF4-V	mc
gMURF4-VI	mc
COII	
gCOII	Mc (10120–10148)
COIII	
gCOIII-I	mc
gCOIII-II	mc
CYb	
gCYb-I	Mc (16803–16767)
gCYb-II	Mc (2290–2239)
ND3 (G5)	
gND3-I	Mc (304–350)
gND3-II	mc
gND3-III	mc
gND3-V	mc
gND3-VI	mc
gND3-IX	mc
ND7	
gND7-I	Mc (16724–16752)
gND7-II	Mc (395–346)
ND8 (G1)	
gND8-I	mc
gND8-II	mc
gND8-III	mc
gND8-IV	mc
gND8-VI	mc
gND8-VII	mc
gND8-IX	mc
gND8-X	mc
gND8-XII	mc
gND8-XIII[d]	Mc (17364–17391)
ND9 (G2)	
gND9-II	mc
gND9-III	mc
gND9-V	mc
gND9-VI	mc
gND9-VII	mc
gND9-VIII	mc
gND9-IX	mc
gND9-XII	mc
gND9-XIV	Mc (219–262)
gND9-XV	mc
G3	
gG3-I	Mc (17199–17215)
gG3-II[e]	Mc (14472–14445)
gG3-III	mc
G4	
gG4-I	mc
gG4-II	mc
gG4-III	mc
gG4-IV	Mc (16881–16931)
gG4-V	mc
gG4-VI	mc
gG4-VIII	mc
gG4-IX	mc
gG4-X	mc
gG4-XIV	mc
Unassigned	
gM150[f]	Mc (150–102)

[a] The cryptogenes are indicated in bold.
[b] mc: minicircle-encoded gRNA.
[c] Mc: maxicircle-encoded gRNA. The position of the gene in the *L. tarentolae* maxicircle (LEIKPMAX) sequence is indicated in parenthesis.
[d] Putative maxicircle-encoded gRNA for ND8 (G1) Block XIII.
[e] Putative maxicircle-encoded gRNA for G3 Block II.
[f] Putative gRNA found in a gRNA–mRNA misguided chimera.

genome were identified by sequence analysis alone for *C. fasciculata* (62). The presence of several compensatory base changes in anchor regions that preserve base-pairing with mRNA transcripts from the same species provides strong evidence for the existence of these gRNAs and for the importance of the anchor region for the editing process itself (62). In the *T. brucei* maxicircle sequence, only the gMURF2-I and II and gCOII genes are present, with the gND7-I and II, and the gCYb-I and II genes being missing (62).

A computer algorithm was developed based on the pairwise local similarity algorithm of Smith and Waterman for finding gRNAs given the cryptogene sequence (63). A test of this method with four known cryptogenes from *L. tarentolae* showed that additional information was required for accurate identification of gRNA sequences. The statistical distribution of the longest candidate gRNA sequences showed that the average expected length of gRNAs should be approximately 35–43 nucleotides, a value which is entirely consistent with the observed length distribution values in Fig. 6.

3.2 Minicircle-encoded gRNAs

Each kDNA network contains approximately 5000–10 000 minicircles linked together by catenation. The size of the minicircle is species-specific and varies from 400 bp to 2500 bp. The minicircle molecule is organized, depending on the species, into one, two, or four conserved regions and a corresponding number of variable regions. The conserved region contains the origins of replication for both strands. In most kinetoplastid species there are multiple minicircle sequence classes, which are defined by the variable region sequences, within a single network. The number of different sequence classes in the African trypanosome, *T. brucei*, is over 300 (21).

The genetic role of this enigmatic molecule was finally uncovered when it was shown that the *L. tarentolae* D12 minicircle encodes a gRNA which potentially mediates the editing of sites 1–8 of the COIII mRNA (64). In *L. tarentolae*, each minicircle encodes a single gRNA located approximately 150 bp from the end of the conserved region and the 'bend' in the DNA (65) (Fig. 7). In *T. brucei*, each minicircle encodes at least three gRNAs which are located between 'cassettes' of 18 bp inverted repeats (66) (Fig. 7). However, four genes encoding redundant gRNAs mediating the editing of block 1 of CYb were shown to be localized outside the 18 bp inverted repeat cassettes in different *T. brucei* minicircles (67).

The extent of minicircle complexity in the UC strain of *L. tarentolae* was analysed by screening several large minicircle DNA libraries for different gRNAs by a process of negative colony hybridization selection (68). A total of 17 minicircle sequence classes was shown to exist, each of which was found to encode a gRNA (Tables 1 and 2). The editing role of all but two of these gRNAs was determined by comparison with known edited mRNA sequences. The relative abundance of the different minicircles was determined by dot blot analysis to vary from as few as 30 copies to as many as 3000 copies per network of 10 000 minicircles. The relative abundance of the steady-state gRNAs encoded by these minicircles was measured

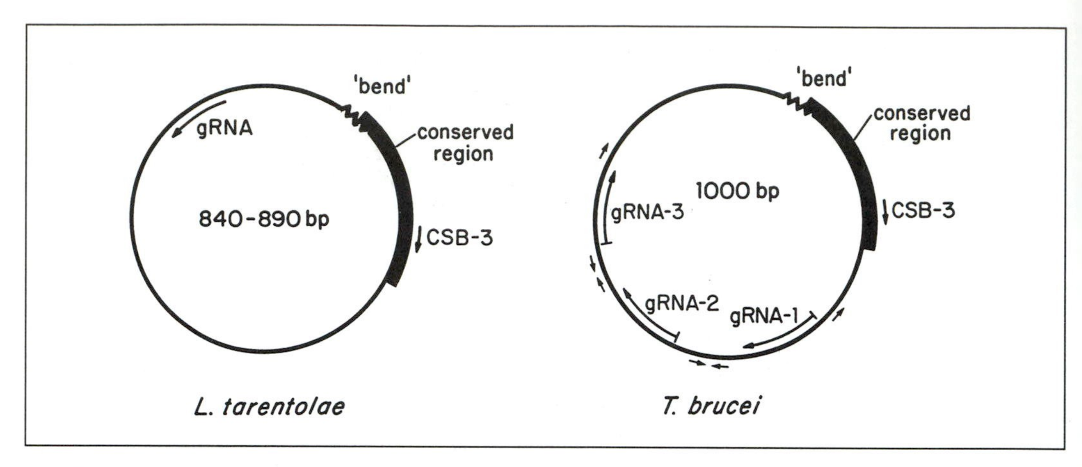

Fig. 7 Genomic organization of kDNA minicircles from two trypanosomatid species, showing locations of the conserved region (black box), the DNA 'bend', and the gRNA genes (arrows show polarity). CSB-3 is a conserved 12-mer involved in replication initiation. The 18-mer inverted repeats flanking most of the gRNA genes in *T. brucei* are indicated by arrows.

by northern blot analysis and shown generally to have no correlation with the minicircle copy number. This indicates that steady-state gRNA abundance is determined by relative promoter strength or turnover rather than by a gene dosage effect.

In *T. brucei*, each 1 kb minicircle encodes three gRNAs usually situated within three cassettes of 18 bp inverted repeats (5′-GAAATAAGATAATAGATA—~110 bp—TATTTATTATTTTATTTT-3′) in the same polarity (66) (Fig. 7). The gRNA transcripts can be capped *in vitro* by vaccinia virus guanylyl transferase, and therefore may represent primary transcripts, although this has not yet been experimentally demonstrated. Transcription appears to initiate at a purine within the sequence, 5′-AYAYA-3′, 32 bp from the upstream inverted repeat. As in the case of *L. tarentolae* gRNAs, the *T. brucei* gRNAs have non-encoded 3′ oligo(U) tails of heterogeneous length. In *T. brucei*, the minicircle sequence heterogeneity is more than 10-fold larger than that in *L. tarentolae*, suggesting that the total gRNA complement comprises more than 700–900 different sequences. This large repertoire of gRNAs can only partially be explained by the existence of three pan-edited cryptogenes—ND7, COIII, and MURF4—which are only 5′- and internal-edited in *L. tarentolae* and *Crithidia*. To date, seven gRNAs have been reported for ND7 in *T. brucei* (69), 10 for ND8 (70), 11 for A6 (MURF4) (70), five for CR1 (G1) (37), and 17 for COIII (70), but in no case has a complete set of overlapping gRNAs yet been obtained. These gRNAs represent approximately 30% of the known edited sequence information in this species. Of the 50 identified gRNAs, 32 show some level of overlap with another gRNA, with the extent of overlap ranging from 2 to 52 nucleotides. In some cases, completely overlapping 'redundant' gRNAs were observed, differing in sequence but encoding the same editing information (70). This gRNA redundancy is probably the major factor contributing to the large number of gRNAs in this species. The reason for this is not known but may be

Table 2 Guide RNA complexity in *L. tarentolae* UC and LEM125 strains

	Number of gRNAs encoded by:		
Cryptogenes	**UC + LEM125 maxicircle DNA**	**UC + LEM125 minicircle DNA**	**Total (expected)**
COII	1	0	1
COIII	0	2	2
ND7	2	0	2
CYb	2	0	2
MURF2	2	0	2
MURF4 (A6)	0	6	6
RPS12 (G6)	1	7	8
	UC + LEM125	**LEM125**	
ND8 (G1)	1[a]	9	9 (14)
ND9 (G2)	1	8	9 (17)
G3	2[b]	1	3 (6)
G4	1	9	10 (15)
ND3 (G5)	1[c]	5	6 (9)
Total:	13	47	60 (83)

[a] gND8-XII, a putative maxicircle-encoded gRNA with several mismatches.
[b] gG3-II, a putative maxicircle-encoded gRNA which was not detectable by northern and primer extension analysis.
[c] gM150, a putative gRNA found in a gRNA–mRNA misguided chimera.
(Adapted from 40.)

related to the developmental regulation of editing that occurs during the complex life cycle of the African trypanosome. In this regard, the defective African trypanosomes, *T. equiperdum, T. evansi*, and *T. equinum*, which lack the insect stage of the life cycle, have homogeneous minicircle populations which encode three gRNAs (71, 72). However, in these species, the maxicircle DNA either has large deletions and rearrangements (73) or is absent, and editing does not occur.

The 3′ terminus of gCYb-II from *L. tarentolae* (60) was mapped by S1 analysis. The majority of molecules had the 3′ ends predicted from the alignment of gRNA with edited mRNA, but a low percentage of truncated molecules were also observed, which were suggested to be due to premature termination. Another method for analysis of gRNA 3′ ends is to obtain sequences of gRNA–mRNA chimeric molecules (see below). By this method, evidence for a high proportion of 3′ truncated gRNAs was obtained in the case of *T. brucei*, and the suggestion was made that truncation may occur during the editing process (69). Similar results were obtained with gRNA–mRNA chimeric molecules from *C. fasciculata*, and the suggestion was made that the apparent 3′ to 5′ progression of editing determined by the alignment of partially edited clones (47) may be an artifact of editing mediated by 3′-truncated gRNAs (74).

However, this is definitely not the situation in the case of the partially edited *L. tarentolae* CYb transcripts analysed by Sturm and Simpson (75). The 3′ ends of all identified *L. tarentolae* gRNAs from both the UC and the LEM125 strains were sequenced directly by RACE-PCR during the construction of a gRNA library (O. Thiemann and L. Simpson, unpublished; 40), and were found to have a slight amount of 3′ end heterogeneity, but truncated molecules such as observed in *C. fasciculata* were not observed.

3.3 Complete sets of overlapping gRNAs for the CYb, MURF2, COIII, MURF4, and RPS12 mRNAs in *L. tarentolae*

Two overlapping maxicircle-encoded gRNAs mediate the 5′-editing of the CYb mRNA in *L. tarentolae* (22), but the duplex anchor for gCYb-II would be only 5 base pairs. In *C. fasciculata* (76), there are also two overlapping gRNAs for the editing of CYb. However, in this case, a stable duplex anchor for gCYb-II is created by the editing mediated by gCYb-I. If there had been a sequence error for the *L. tarentolae* gCYb-I involving a single U residue in a string of 5 Us, then a stable 13-nucleotide *L. tarentolae* gCYb-II anchor sequence would also be created by block I editing (G. Connell, personal communication).

Two maxicircle-encoded gRNAs are involved in the editing of the MURF2 mRNA in *L. tarentolae*. gMURF2-I is unusual in that it only encodes a single U-addition.

Two minicircle-encoded gRNAs mediate the editing of the 5′-edited COIII RNA in *L. tarentolae* (68). The anchor sequence for gCOIII-I is complementary to unedited mRNA sequence, whereas the anchor for gCOIII-II covers editing sites 5–8 of gCOIII-I.

Six overlapping minicircle-encoded gRNAs mediate the editing of the 5′ pan-edited MURF4 mRNA in *L. tarentolae* (68). The 3′-most gRNA (gMURF4-I) forms an anchor duplex with unedited sequence, but all the other gRNAs form anchors with edited sequences. There is a possibility that gMURF4-V forms a six base-pair anchor with unedited sequence, and therefore would create a second editing domain, but this must be verified by analysis of partially edited RNAs.

Eight overlapping gRNAs mediate editing of the RPS12 mRNA in *L. tarentolae* (68) (Fig. 8). Seven of these are minicircle-encoded and one is maxicircle-encoded (gRPS12-VI). The initial gRNA in each of the three domains forms an anchor with unedited sequences, whereas all other gRNAs form anchors with edited sequences.

The precise overlapping of gRNAs in these three pan-edited cryptogenes in this species is strikingly economical and provides a simple explanation for the observed 3′ to 5′ polarity of editing within a domain (Figs 5 and 8).

Analysis of these gRNA–mRNA hybrids showed that the G-U base pairs are limited to regions of the edited mRNA upstream of the anchor sequence (68), and there are few in anchor duplexes. This suggests a role for the relatively weak G-U base pairs in allowing breathing of the edited mRNA–gRNA duplex block, which would permit formation of the adjacent upstream mRNA–gRNA anchor duplex. This would not, however, eliminate the possible need for an RNA helicase to de-

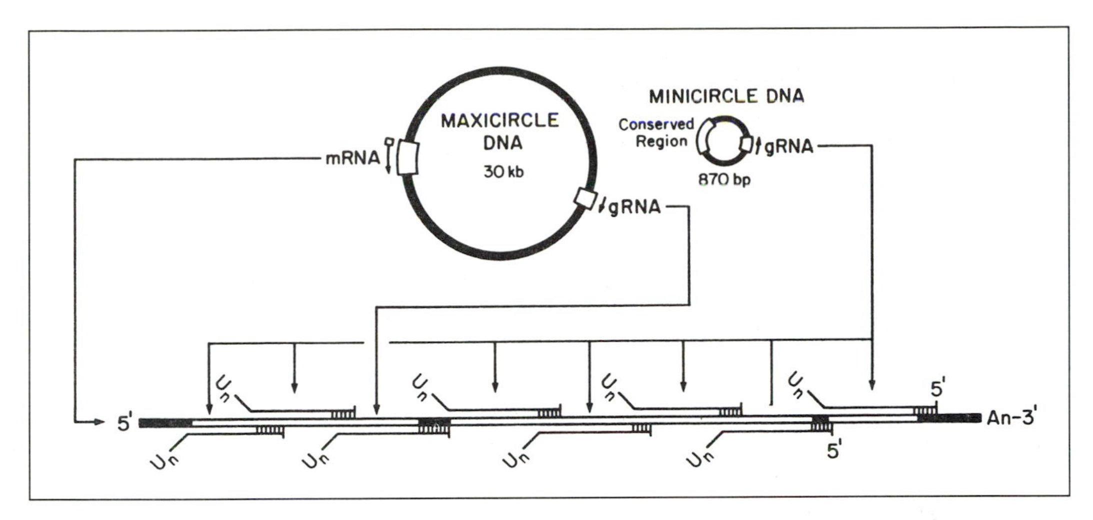

Fig. 8 Diagram of overlapping gRNAs for editing of RPS12 mRNA. The unedited mRNA sequences used for anchors by the initial gRNAs in each domain are indicated by black bars and the gRNA/mRNA anchor duplexes by straight lines. The genomic origins of the pre-edited mRNA and each gRNA are indicated by arrows.

stabilize the terminal gRNA–edited mRNA duplex to allow translation, and such an activity has recently been reported in mitochondrial extracts from *T. brucei* (77).

3.4 Sets of overlapping gRNAs for the G1–G5 mRNAs in the LEM125 strain of *L. tarentolae*

As discussed above, a complete set of productively edited mRNAs for G1–G5 was obtained from the LEM125 strain of *L. tarentolae*. A gRNA library was constructed for this strain (40), and gRNAs mediating the editing of these transcripts were selected by a process of negative selective hybridization, using DNA oligonucleotide probes for gRNAs from the UC strain and sequencing of remaining clones. By this method, 30 new minicircle-encoded gRNAs and five new maxicircle-encoded gRNAs were identified (Tables 1 and 2). Many of the gRNAs were found to overlap and the overlap was restricted to the anchor region, as was described previously for the MURF4 and RPS12 editing cascade in *L. tarentolae* UC strain. It is likely that a more extensive search would yield a complete set of overlapping gRNAs for these transcripts, and we therefore conclude that these transcripts are productively edited in this strain.

One example of a redundant gRNA was detected for the editing of block III of the ND3 transcript (40). This indicates that redundant gRNAs are not restricted to the African trypanosomes. However, it is clear that the abundance of redundant gRNAs is much lower in *L. tarentolae* than in *T. brucei*.

Two previously unassigned gRNAs from the UC strain of *L. tarentolae*—gLt19 and gB4 (68)—proved to be identical homologues of the gRNAs which mediate editing of G4 block III and ND3 block IX in the LEM125 strain (40). It is of some interest that the minicircles encoding these two gRNAs represent the highest copy number minicircles in the UC strain, but are significantly lower copy number in the LEM125 strain.

3.5 Loss of minicircle-encoded gRNAs in an old laboratory strain of *L. tarentolae*

In the case of *L. tarentolae*, productive editing of transcripts of the G1–G5 cryptogenes was only observed in the recently isolated LEM125 strain (40). Attempts in our laboratory to RT-PCR amplify partially and fully edited RNAs from the UC strain, which has been in culture in various laboratories for over 50 years, were uniformly unsuccessful, except in the case of G5, which gave rise to a percentage of clones containing correctly edited block I sequence and incorrectly edited upstream sequences (see Section 6). Examination of the total gRNA complement of the UC strain (Table 2) led to the conclusion that this failure to properly edit these transcripts was most likely to be a result of the absence of specific minicircle sequence classes encoding gRNAs for these specific editing cascades. At least 32 new minicircle-encoded gRNAs for the editing of the G1–G5 transcripts have been shown to be present in the LEM125 strain and to be absent in the UC strain (Tables 1 and 2) (40). We have speculated that specific minicircle sequence classes encoding these gRNAs have been lost during the long cultural history of the UC strain, possibly due to mis-segregation of minicircles at division of the single kDNA network or to the selective amplification of minor sequence classes which occurs in the phenomenon of transkinetoplastidy (78). The loss of these genes is tolerated since the encoded components of complex I of the respiratory chain are apparently not required during this stage of the life cycle (79).

The loss of gRNA genes in the UC strain has created several 'pseudocryptogenes', which are transcribed, but the transcripts are not productively edited. The observed correlation of the loss of specific sets of gRNA genes with defects in editing of these transcripts provides the first *in vivo* evidence for the genetic role of gRNAs in RNA editing.

4. Models for the mechanism of RNA editing and the involvement of gRNAs

The sequence information for addition and deletion of Us is contained in the gRNA molecules, but the precise mechanism by which this information is transferred to the mRNA is still unresolved. The enzyme cascade model (22) (Fig. 9) was based on the existence of several enzyme activities in purified kinetoplast-mitochondria fractions of *L. tarentolae*—a 3′ TUTase, an RNA ligase, and a specific RNA cleavage activity (54, 80). The initial event was proposed to be the hybridization of the 5′ portion of the gRNA to the mRNA anchor sequence just downstream of the pre-edited region. Then a specific cleavage was postulated at the 3′-most mismatched

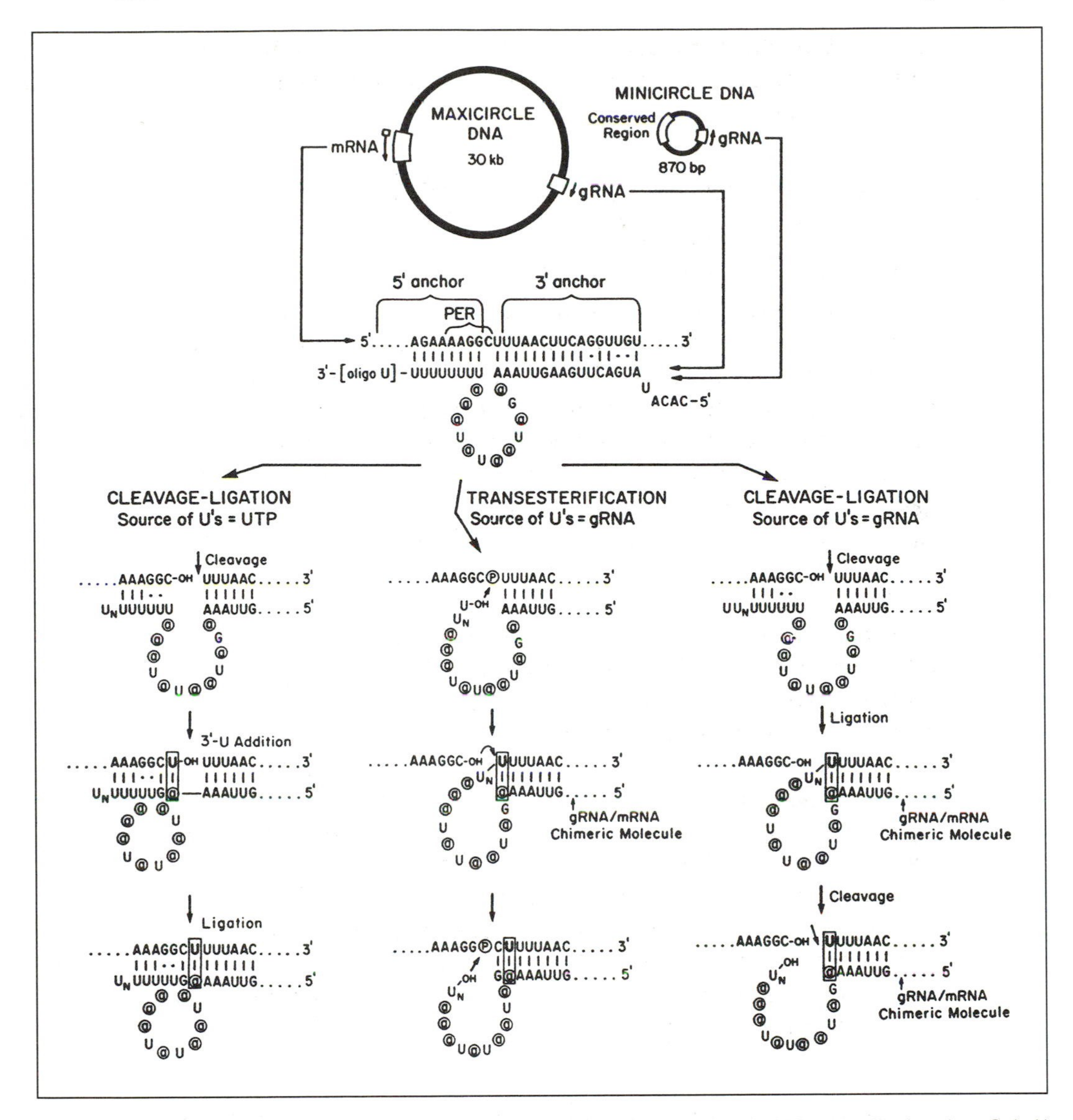

Fig. 9 Models for the mechanism of RNA editing in kinetoplastid mitochondria. PER, pre-edited region. Only U insertions are shown. The guiding a and g nucleotides in the gRNA are shown in lower case and circled. The gRNA shown is gCYb-I.

base on the mRNA, liberating an internal 3′-OH. The addition of a U residue to this 3′-OH was then hypothesized, which would form a base pair with a guiding A or G residue on the gRNA. The final event would be the religation of the 5′ mRNA fragment with the 3′ mRNA fragment. Another cycle of editing at the next upstream mismatched base would then ensue. Deletions were presumed to be due to exonuclease trimming of an exposed non-base-paired 3′ U residue in the mRNA. Evidence for this model is mainly the existence of all three postulated enzymatic activities in the mitochondrion, and the co-localization of several of these activities in ribonucleoprotein complexes in mitochondrial extracts from *T. brucei* (81).

The presence of a heterogeneous 3′ oligo(U) tail on the gRNA presents a puzzle. We initially suggested that this tail could form a duplex structure with the pre-edited region which is mainly composed of G and A residues, and thereby increase the stability of the initial mRNA–gRNA hybrid (60). In the enzyme cascade model the role of the TUTase is both to add Us at the editing site and to add the Us to the 3′ ends of the gRNAs. A more direct role for the oligo(U) tail was proposed in the transesterification model of editing (82, 83) (Fig. 9). In this model, the role of the TUTase is to maintain the oligo(U) tail of the gRNA for use as a donor of U residues to the editing site. The initial event is a cleavage–ligation attack by the 3′-OH of the gRNA at the mismatch, giving rise to a transient gRNA–mRNA chimeric molecule. This initial transesterification is activated by hybridization of the 3′-terminal U residues of the gRNA with the internal guide sequence of the gRNA in a manner similar to that occurring in the ribosomal RNA self-splicing intron of *Tetrahymena* (84–89; Chapter 1). A second transesterification between the 3′-OH of the upstream mRNA fragment and the next mismatch liberates the gRNA and produces the transfer of at least one U residue from the gRNA into the mRNA at an editing site (Fig. 9). Deletions of Us could occur either by the initial transesterification being 5′ of the mismatched U residue with the second transesterification being 3′ of that residue, or by an exonuclease trimming of the exposed 3′ U residue after the initial transesterification at a normal site.

This mechanism is formally very similar to the reversal of RNA splicing of group II introns, which has been observed in *in vitro* models (84; Chapter 1). Evidence for this model mainly rests on the existence of the predicted gRNA–mRNA chimeric molecules, which will be discussed below.

5. Chimeric gRNA–mRNA molecules

A prediction of the transesterification model for editing is the existence of transient gRNA–mRNA chimeric molecules. Such molecules were isolated from *L. tarentolae* kinetoplast RNA by PCR amplification, using a 3′ primer specific for the mRNA downstream of the pre-edited version and 5′ primer specific for the gRNA (82). These chimeric molecules have gRNAs usually attached at normal editing sites by 8–26 U residues, with the downstream editing sites being completely edited. Chimeric molecules for ND7, COIII, and COII were observed. In the COII chimeras, the connecting nucleotides are Us and As, which is consistent with the fact that the COII gRNA is in *cis* at the 3′ terminus of the mRNA.

The majority of gRNAs were found to be attached at the 3′-most editing sites or at sites with a large number of U additions. The 3′-most encoded nucleotides in the gRNA portion are in most cases consistent with the 3′ ends predicted by the largest gRNA–edited mRNA duplex that can be formed. However, several examples of 3′-truncated gRNAs were observed in the case of ND7. Chimeric molecules for the pan-edited MURF4 cryptogene were also observed in *T. brucei* mitochondrial RNA

(69). In this case, the majority of the encoded gRNA sequences are 3′ truncated and the connecting oligo(U) sequence is more heterogeneous and shorter than in *L. tarentolae*.

It should be emphasized that the existence of these molecules is consistent with the transesterification model but does not prove it, since chimeric molecules could also be generated by adventitious ligation of the 3′ end of the gRNA to the 3′ mRNA fragment produced in the enzyme cascade model. The possibility of PCR-generated *in vitro* homologous recombination (90) is made unlikely by the fact that chimeric molecules from *T. brucei* have also been detected in non-amplified cDNA libraries (K. Stuart, personal communication) and as a minor high molecular weight band representing approximately 5% of the gRNA abundance in total mitochondrial RNA by northern blot analysis in *L. tarentolae* (82).

6. Generation of misedited patterns: random editing or misguiding?

It was initially shown for the pan-edited COIII cryptogene that editing proceeded generally 3′ to 5′ (91). Analysis of a large number of partially edited molecules for CYb and COIII in *L. tarentolae* indicated that editing also proceeds 3′ to 5′ within a single editing block (75). Almost all of the CYb partially edited RNAs showed a precise 3′ to 5′ polarity, but 42% of the COIII partially edited RNAs showed unexpected editing patterns at the junction regions. We have presented evidence that unexpected editing patterns are produced by misediting due to specific events of misguiding through the interaction of inappropriate gRNAs, or of appropriate gRNAs in an inappropriate fashion (47). The basic concept is that of a 'guiding frame': misediting represents 3′ to 5′ editing occurring in an incorrect guiding frame. We have proposed four possible misguiding mechanisms and have presented several examples of each type (Fig. 10). Chimeric molecules consisting of misedited mRNAs attached to heterologous gRNAs which had the potential of guiding the editing events were also observed. Misguiding would represent mechanistically correct editing, but misedited mRNAs must always be corrected by in-frame editing mediated by the appropriate gRNA to obtain the mature edited mRNA. It is of course also possible that this mechanism may not account for all observed misedited sequences and that a certain amount of stochastic editing does occur in addition.

The preferential localization of misediting to junction regions can be explained as a natural consequence of the misguiding hypothesis. The scattered clusters of misedited sequences observed at a low frequency in the *T. brucei* CR1 partially edited RNAs could be due to the greater diversity of redundant gRNAs in this species, which may make pan-editing inherently inaccurate (37).

Several illustrative examples of misediting were recently obtained for the ND3 (G5) cryptogene in the UC strain of *L. tarentolae* (40, 92). Analysis of clones of partially edited G5 transcripts from the UC strain showed a subset that possessed a

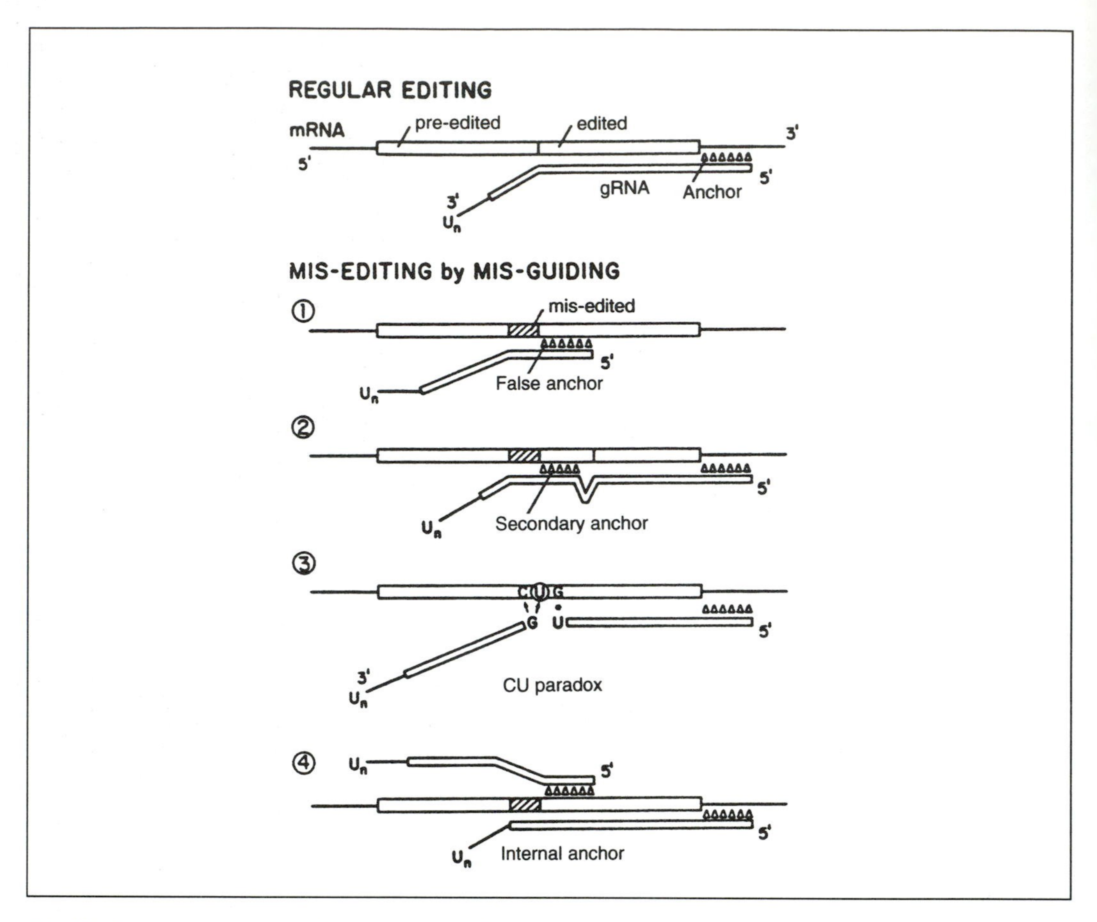

Fig. 10 Schematic representation of normal editing and misediting produced by misguiding. Edited and pre-edited sequences of mRNA are shown as open boxes. Misedited sequences are indicated by cross-hatching and anchor sequences are represented by carets. (Reprinted from 47 with permission.)

correctly edited block I sequence followed by a variety of misedited upstream sequences (Fig. 11). The presence of a maxicircle-encoded gRNA in the UC strain explains the correct editing of block I. The gRNA for block II in the LEM125 strain is minicircle-encoded and is absent in the UC strain. Several examples of misedited sequences extending from a correctly edited block I sequence were observed, in which the misediting appeared to be mediated by non-cognate gRNAs (92). In a few cases, the misediting created false anchor sequences for additional non-cognate gRNAs which further extended the misediting. Another group of clones with polyadenylation sites up to six nucleotides from the anchor for block I editing showed misediting of block I also, suggesting the possibility of some minimal space requirements for the assembly of an editing complex on the 3′ end of a pre-edited mRNA.

An analysis of partially edited RNAs from the CYb and COIII genes of *T. brucei* showed a high abundance of misedited patterns at junction regions (93). These

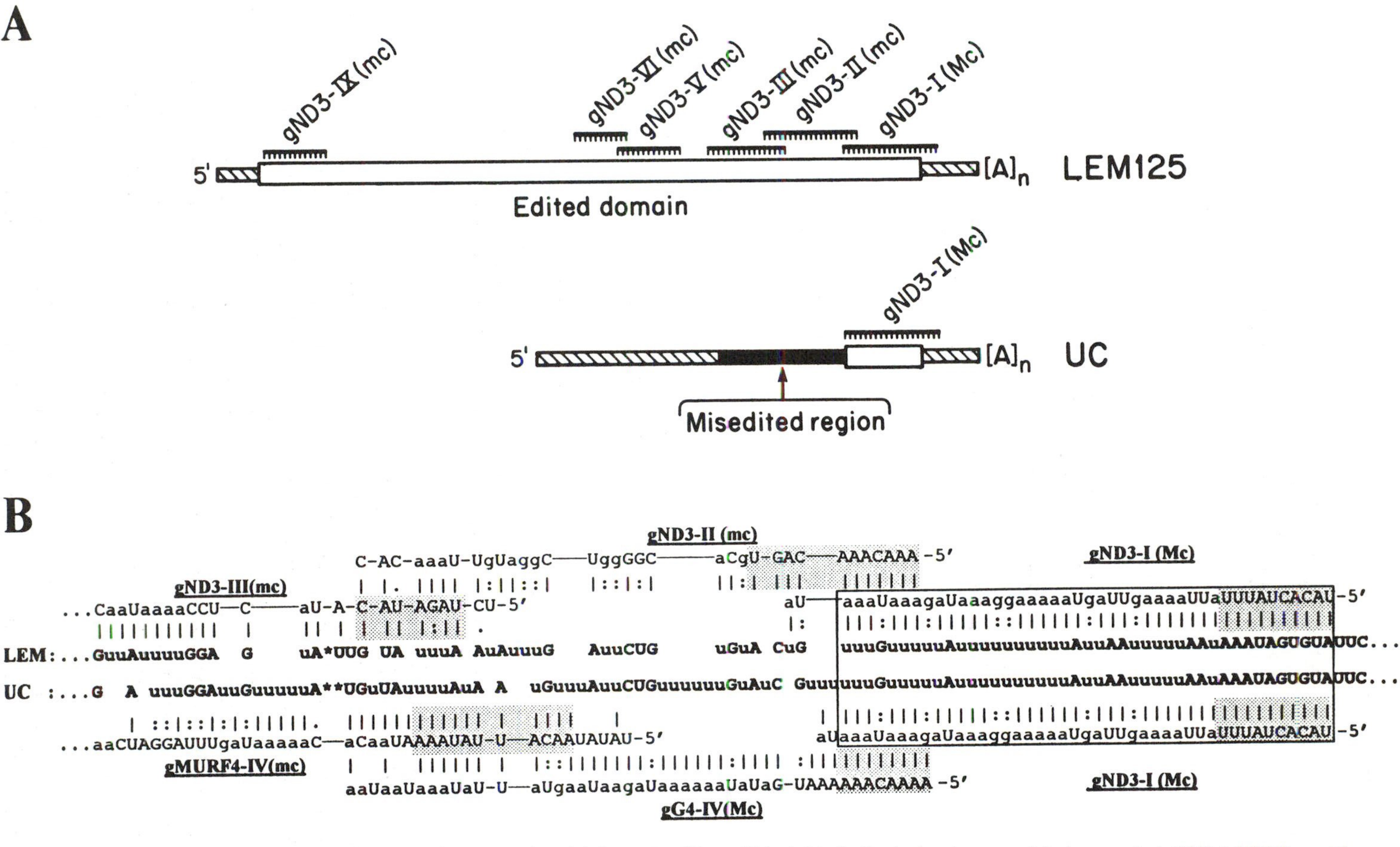

Fig. 11 Misediting of G5 transcripts in UC strain of *L.tarentolae*. (a) Correct editing of block I in both strains by a maxicircle-encoded gRNA (gND3-I), and incorrect editing of the upstream sequences in the UC strain due to an absence of the cognate minicircle-encoded gRNAs. (b) Example of one partially edited G5 RNA from the UC strain which has correct editing of block I and misediting of upstream sequence due to the sequential actions of two non-cognate misguiding gRNAs. The correctly edited block I is boxed, and the gRNAs are shown above and below the edited RNA sequence. The mature edited LEM125 G5 sequence and the cognate gRNAs are shown for comparison. (Reprinted from 40 with permission.)

results were interpreted as evidence that editing does not proceed strictly 3′ to 5′, but is stochastic in nature—in other words, that editing is essentially completely random within a domain, and that formation of the correct duplex with the gRNA 'freezes' editing in that region (93). In this model, the secondary structure of the pre-edited mRNA itself determines the sites of editing by presenting a single-stranded loop region to a single strand-specific endonuclease (80, 94). Subsequent events could involve either the enzyme cascade steps or the transesterification steps for the U additions and deletions. gRNA–mRNA chimeric molecules could also be generated in this model by ligation of the 3′ end of the gRNA to the mRNA at the cleavage site, but these would represent aberrant products unless a second cleavage–ligation occurred, releasing the shortened gRNA (Fig. 9).

The 'dynamic interaction' model (95) is also an attempt to explain the existence of misedited sequences as a normal consequence of editing. The model suggests a series of progressively more stable, but incompletely base-paired, mRNA–gRNA interactions in the junction region. This model is essentially identical to one type of misguiding in which there is a loopout of either the mRNA or the gRNA leading to a loss of guiding frame (47). Editing of sites would occur multiple times and overall editing would not proceed strictly 3′ to 5′. The model proposes a progressive realignment of gRNA–edited mRNA sequences to form the final mature edited mRNA, and is compatible with both the enzymatic and the transesterification mechanisms of editing.

The misguiding hypothesis of Sturm *et al.* (47) can explain the origin of a substantial number of misedited sequences in *L. tarentolae*, and a knowledge of the total gRNA complement in *T. brucei* should allow a test of the validity of this and other hypotheses for *T. brucei*.

7. Enzymatic activities possibly associated with RNA editing

TUTase and RNA ligase activities were first reported in total cell extracts of *T. brucei* (74). Mitochondrial TUTase and RNA ligase activities were demonstrated in *L. tarentolae* (54). The TUTase adds U residues to the 3′ OH of most substrate RNAs, but the mitochondrial gRNAs represent exceptionally good substrates. The activity is inhibited by heparin, and is specific for UTP. A TUTase activity was also observed in *T. brucei* kinetoplast-mitochondrial fractions isolated by Percoll density gradient centrifugation (81, 96). The mitochondrial RNA ligase activity in *L. tarentolae* is ATP-dependent. Both the TUTase and the RNA ligase activities were solubilized, in the case of *L. tarentolae*, with Triton X-100.

A ribonuclease activity in a 100 000 g supernatant of a Triton lysate of a mitochondrial fraction from *L. tarentolae* was shown to be activated by incubation with heparin or by predigestion of the lysate with proteinase K (80). *In vitro*-transcribed CYb mRNA is cleaved at several sites within putative single-stranded regions of the mRNA. The major cleavage occurs two nucleotides upstream from the initial

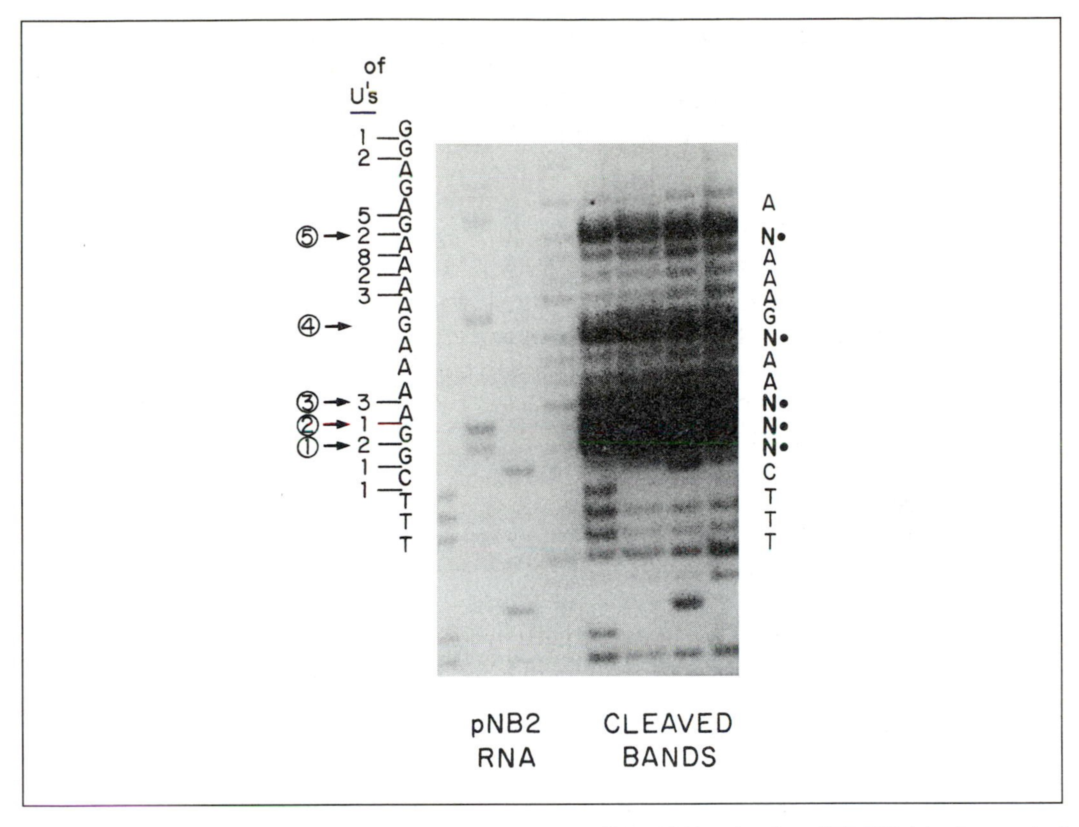

Fig. 12 Localization of sites of cleavage within the synthetic CYb mRNA molecule. pNB2 RNA is a fragment of pre-edited CYb mRNA with some vector sequence at the 5′ end. At the left side is a sequencing ladder of intact pNB2 RNA, obtained using a [^{32}P]-labeled oligonucleotide primer. Sites of editing and number of Us added in the mature RNA sequence are shown by arrows, The pNB2 RNA was incubated in the presence of heparin with mitochondrial extract from *L. tarentolae*, which was pre-digested with proteinase K, and the two major cleavage fragments were gel-isolated. A sequencing ladder of the fragments is shown on the right, with the nucleotides adjacent. The sites with nucleotides in all lanes are interpreted as sites of cleavage and are indicated by arrows on the left and by N. on the right. (Reprinted with permission from 80.)

editing site (Fig. 12). The cleavage activity is inhibited by SDS or extraction with phenol/chloroform and therefore probably represents a protease-resistant protein, rather than an RNA ribozyme. The activity was sized between 10 kDa and 30 kDa by ultrafiltration. The results of micrococcal nuclease digestion were equivocal, in that digestion inhibits the protease-induced activity but not the heparin-induced activity. A similar activity was observed in *T. brucei* mitochondrial extracts, using synthetic CYb, COII, and COII mRNAs (94). Pre-edited CYb RNA is specifically cleaved at the identical site as in *L. tarentolae*, but mature edited RNA is not affected. A specific cleavage just downstream of the PER in the COII mRNA and multiple cleavages within the large PER of the pan-edited COIII mRNA were also observed. Since it was also observed that the single-strand-specific nuclease, mung bean nuclease, cleaved the CYb RNA at identical locations, the suggestion was made that the mitochondrial activity represents a single strand-specific endonuclease. However, in all cases, the cleavage activity had to be activated by inhibition of the TUTase with heparin, digestion with proteinase K, or depletion of UTP by preincubation of the isolated mitochondria.

The activation effect is not a definitive proof of the enzymatic mechanism since it could possibly also be explained in terms of the transesterification model for editing. In group I splicing, hydrolysis at normal sites of transesterification occurs in the absence of the guanosine residue which provides the attacking 3′-OH group (Chapter 1). The observed cleavages of pre-edited mRNAs could represent hydrolysis catalysed by a protease-resistant catalytic core in the absence of activated gRNA (80).

8. *In vitro* formation of gRNA–mRNA chimeric molecules

Several reports have shown the presence of an activity in mitochondrial extracts that promotes the formation of gRNA–mRNA chimeric molecules *in vitro*. Harris and Hajduk (97) used labeled synthetic gRNA or mRNA from *T. brucei* to monitor the formation of chimeric molecules for the CYb cryptogene by gel electrophoresis. Synthetic pre-edited CYb mRNA was used, together with a synthetic gRNA covering the first three editing sites. The sequence of the gRNA was taken from the *L. tarentolae* gCYb-I gRNA. Harris and Hajduk (97) were unable to PCR-amplify the *in vitro* products and therefore attempted to directly sequence 5′-end-labeled chimeric molecules. Only very limited direct sequence information was obtained from the *in vitro* products and the site of attachment of the gRNA remained ambiguous.

Koslowsky *et al.* (98) employed PCR amplification to analyse chimeric molecules formed with synthetic A6 (MURF4) pre-edited mRNA and synthetic gA6–14 gRNA from *T. brucei*. However, the amplified *in vitro* chimeras were unusual in that they lacked a stretch of U residues linking the gRNA to the mRNA and most of the gRNAs were truncated at variable sites at the 3′ end, suggesting that the process was aberrant in some respect. In both studies, specificity was demonstrated by showing that a variety of heterologous RNAs did not form chimeras with added gRNA. In addition, a requirement for a gRNA 3′-terminal hydroxyl group was demonstrated and addition of protease-sensitive mitochondrial extract was required for chimeric molecule formation.

Blum and Simpson (99) have obtained similar results in the case of *L. tarentolae*. Synthetic pre-edited messenger RNA and synthetic gRNA for the ND7 cryptogene from *L. tarentolae* form chimeric molecules upon incubation in the presence of an extract from sonicated mitochondria. These chimeric molecules consist of the gRNAs covalently linked to the mRNAs by short oligo(U) tails at normal editing sites in most cases (Fig. 13). Unlike the *in vivo* chimeras in steady-state kinetoplast RNA, the *in vitro* chimeras showed no editing of downstream editing sites. The

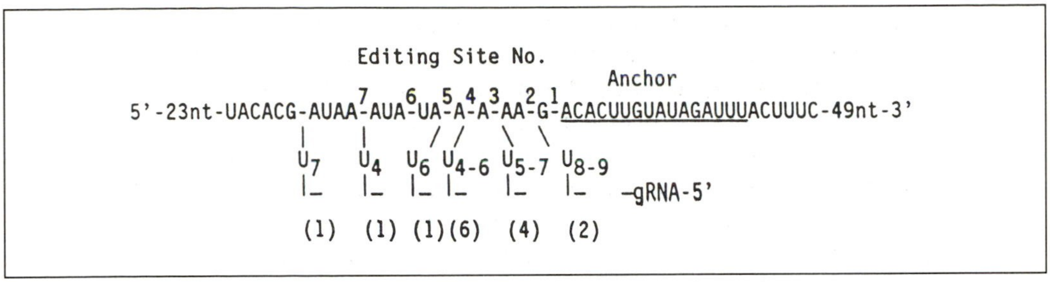

```
                          Editing Site No.
                                                   Anchor
                               7    6  5 4 3  2 1
5'-23nt-UACACG-AUAA-AUA-UA-A-A-AA-G-ACACUUGUAUAGAUUUACUUUC-49nt-3'
                |     |    / /   \ \
                U7    U4 U6 U4-6 U5-7 U8-9
                |_    |_ |_ |_   |_   |_    -gRNA-5'
               (1)   (1) (1)(6)  (4)  (2)
```

Fig. 13 *In vitro* formation of gRNA–mRNA chimeric molecules. Sequences of gRNA–mRNA chimeric molecules obtained by PCR amplification of a gel-isolated band, produced after incubation of synthetic mRNA and synthetic gRNA with mitochondrial extract from *L. tarentolae*. The attachments of the gRNAs are indicated together with the number of Us connecting the two molecules. Editing sites are numbered and the anchor sequences are underlined. (Reprinted with permission from 99.)

formation of chimeras requires ATP and is dependent on the formation of a gRNA–mRNA anchor duplex 3′ of the pre-edited region, as shown by *in vitro* mutagenesis of the mRNA and restoration of activity by compensatory base changes in the gRNA. mRNA sequences 3′ and 5′ of the pre-edited region also affect the efficiency of the chimera-forming activity.

One possible explanation for the synthesis of aberrant chimeric molecules *in vitro* is a failure to rejoin the two separated mRNA fragments after chimera formation. Alternatively, correctly edited products may actually be formed during the incubation, but may undergo rapid hydrolysis (80). This would be consistent with the observed lack of chimeras with gRNAs attached at editing site 1, which is the preferred attachment site for *in vivo* chimeras (82). The remaining stable chimeras would then represent aberrant products accumulating during the incubation. Nevertheless, the observed *in vitro* formation of chimeras may still accurately reflect the initial step of RNA editing, consistent with the transesterification model for RNA editing. Alternatively, chimeric molecules could be created by site-specific cleavage (81, 94) and adventitious ligation of the 3′ oligo(U) tail of the gRNA to the cleaved mRNA 3′ fragment, in line with the enzyme cascade model. The development of an accurate and complete *in vitro* editing system will be required to definitively distinguish between these models, but the availability of an *in vitro* system for the formation of chimeric gRNA–mRNA molecules should already allow a precise dissection of the sequence requirements for chimera formation and also a fractionation of the extract components required for this activity.

9. Ribonucleoprotein complexes containing putative components of the editing machinery

A variety of mitochondrial ribonucleoprotein (RNP) complexes containing putative components of the editing machinery have been reported. Two classes of gRNA-containing RNP complexes in a mitochondrial extract from *T. brucei* can be detected by sedimentation in glycerol gradients: a 19S complex which contains gRNA, TUTase, RNA ligase, and chimera-forming activity, and a 35S complex which has in addition pre-edited RNA, but lacks tightly bound TUTase (81). Pollard *et al.* (81) suggested that the cosedimentation of the RNA ligase and chimeric-forming activity indicates that the mechanism for chimeric formation does not involve trans-esterification but rather cleavage–ligation. Gel retardation analysis has also been used to detect several mitochondrial RNP complexes from *T. brucei* which interact with exogenous synthetic gRNAs, but the function of these complexes remains unknown (100–102).

Peris *et al.* (103) have described two classes of RNP complexes in a mitochondrial extract from *L. tarentolae*. The 'T-complexes' contain gRNAs and mRNAs and are operationally defined as being labeled with [α-^{32}P]UTP by an endogenous TUTase activity. These complexes sediment at 10–13S in glycerol and migrate in native gels as about six bands. T-complexes may represent gRNA-maturation complexes, in regard to maintenance of the 3′ oligo(U) tails of the gRNAs. The 'G-complexes' also contain gRNAs, but sediment at 25S in glycerol, and exhibit an *in vitro* RNA editing-like activity, which is not found in the T-complex region of the gradient. This RNA editing-like activity involves the incorporation of uridine residues into the pre-edited region of a synthetic mRNA (104) (see Section 10).

10. *In vitro* RNA editing-like activities

Seiwert and Stuart (105) have described an *in vitro* activity in a mitochondrial extract from *T. brucei* which accurately removes uridines from the first editing site of the A6 (MURF4) pre-edited mRNA. This deletion activity is dependent on the presence of ATP and exogenous guide RNA for this editing block (gA6[14]). Mutations in both the gRNA and the mRNA editing site indicate strongly that the number of Us deleted *in vitro* is controlled by base-pairing interactions. This represents the first *in vitro* evidence confirming the predicted genetic role of gRNA in specifying the sequence information for editing, at least in the case of U-deletions. However U-additions were not observed in this *in vitro* system, raising the possibility that they have a different mechanism than U-deletions.

Frech *et al.* (104) showed that a mitochondrial extract from *L. tarentolae* could direct the incorporation of U residues derived from [α-^{32}P]UTP within synthetic CYb mRNA. The U-incorporation is imprecise but is confined to the pre-edited region. No direct evidence for the involvement of endogenous gRNAs was obtained, and, interestingly, the addition of exogenous gRNAs (or other nonspecific RNAs) inhibits the incorporation. However, specific inhibition by digestion of the extract with micrococcal nuclease suggests a requirement for some type of endogenous RNA. A low level of incorporation of C residues was found to occur at the same sites as U residues. This activity sediments in glycerol at 20–25S and is correlated with the presence of the G complexes described above.

Further analysis of these *in vitro* systems exhibiting editing-like activities should allow a distinction between the enzymatic and transesterification mechanisms of editing, and may prove useful in an eventual dissection and reconstitution of editing activities.

11. Evolution of RNA editing in kinetoplastid protozoa

Phylogenetic trees of the kinetoplastid protozoa have been constructed in several laboratories using nuclear small rRNA sequences (106–108). The trees were rooted using the *Euglena* rRNA sequence as an outgroup. As predicted by classical taxonomy, the bodonids/cryptobiids form a sister group to the trypanosomatids. In the trypanosomatids, the earliest diverging species are the African and South American *Trypanosoma*, and the most recently diverging species comprise a monophyletic clade consisting of *Leishmania*, *Crithidia*, *Endotrypanum*, and *Leptomonas*. In the central portion of the tree are several monogenetic (one host) genera such as *Blastocrithidia* and *Herpetomonas* and the digenetic (two hosts) *Phytomonas*. The extent of editing of three cryptogenes—MURF4, ND7, and COIII—was analysed for representative trypanosomatid species (106, 108) (Fig. 14). The most parsimonious interpretation of the results is that pan-editing is a primitive trait and that several times during the evolution of these cells, pan-edited genes were substituted by moderate- or 5′-edited genes (109). In one case, a fully edited COIII gene was apparently substituted for a pan-edited ancestral gene (108).

Two strains of a cryptobiid, *T. borreli*, were also analysed for editing of mitochondrial genes (19, 20). The kDNA in this cell was found to consist of two molecular species, as in the case of the trypanosomatids. A 40–80 kb circular component contains structural genes organized in a different order than in the maxicircle DNA of trypanosomatids. Several of these were shown to be cryptogenes, transcripts of which are edited, and in some cases pan-edited. A pan-edited RPS12 gene and a novel 5′- and 3′-edited CYb gene were identified, and the edited RNAs sequenced. COII and COIII are unedited in *T. borreli*. Putative gRNAs were shown to be encoded as tandemly organized 1 kb units in 270 kb circular molecules (19). These results indicate that gRNA-mediated pan-editing can be traced back to an ancestor of the entire kinetoplastid lineage.

Another apparently primitive trait, at least in the trypanosomatid lineage, is the presence of an extremely large gRNA repertoire with extensive redundancy. In the more recently evolved clade represented by *Leishmania* and *Crithidia*, the gRNA repertoire is much more limited, as discussed previously.

We have speculated, on the basis of the observed culture-induced loss of gRNA genes in the UC strain of *L. tarentolae* and of the distribution of editing patterns in various species, that the evolution of RNA editing in the kinetoplastids involves a replacement of the original pan-edited cryptogenes by 5′-edited or unedited homologues by cDNA copies of partially edited RNAs (108, 109) (Fig. 15). The selective pressure for this retroposition gene replacement could have been the loss of entire gRNA gene families by the loss of the corresponding minicircles. Those cells which had undergone a gene replacement of a pan-edited gene with a partially edited gene would survive the loss of gRNAs. It is possible that all original maxicircle

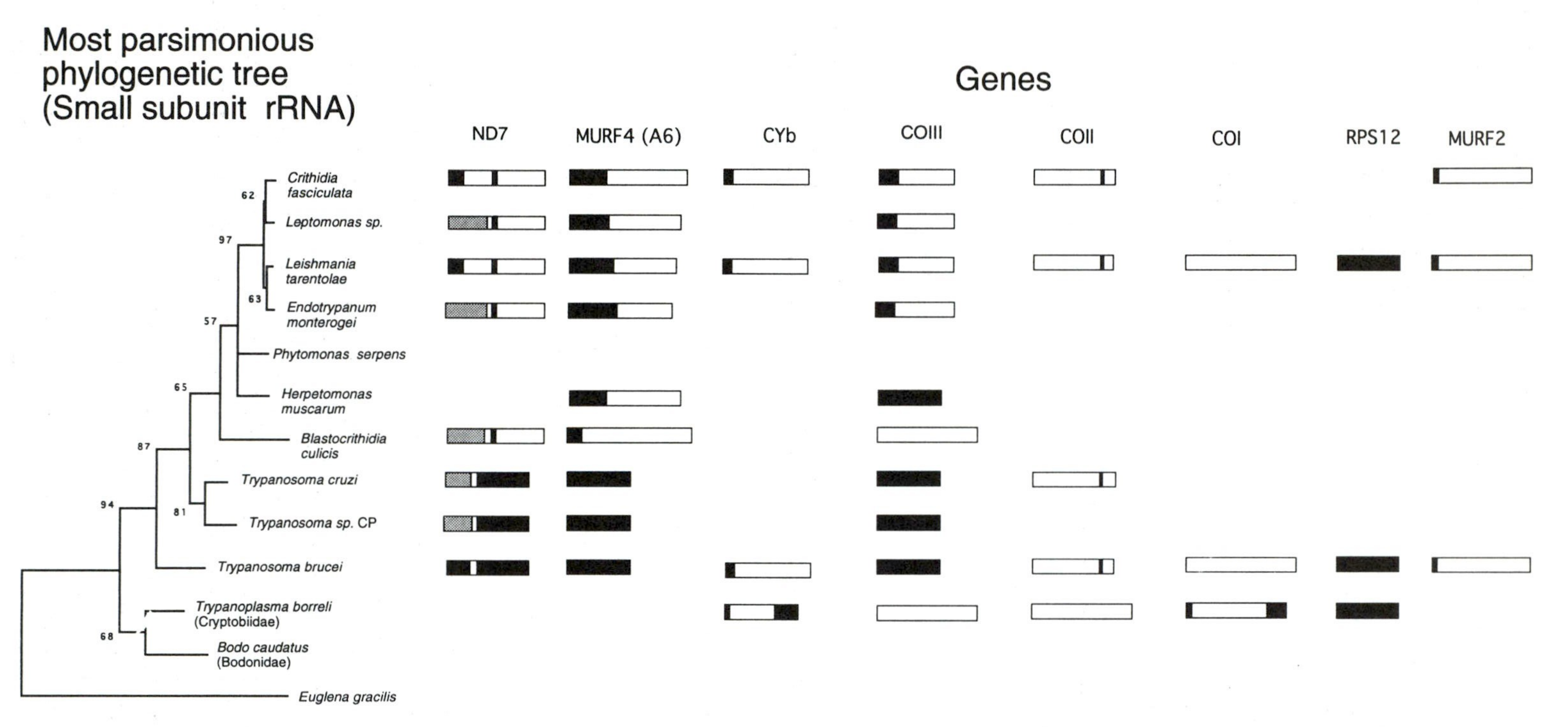

Fig. 14 Phylogeny of kinetoplastid RNA editing. Aligned sequences of 18S rRNAs were used to reconstruct a consensus phylogenetic tree using maximum parsimony. Bootstrap values are indicated for each node. The tree was rooted using the *Euglena gracilis* sequence as an outgroup. A diagram of the extent of editing of eight genes is shown on the right. Dark boxes indicate pan-edited regions, white boxes indicate non-edited regions, and grey boxes indicate a lack of information. The lack of a box indicates an absence of any information on the editing of that gene. All genes are shown 5′ to 3′. 5′-edited sequences are shown at the 5′ end of the RNAs, although there is usually some unedited upstream sequence.

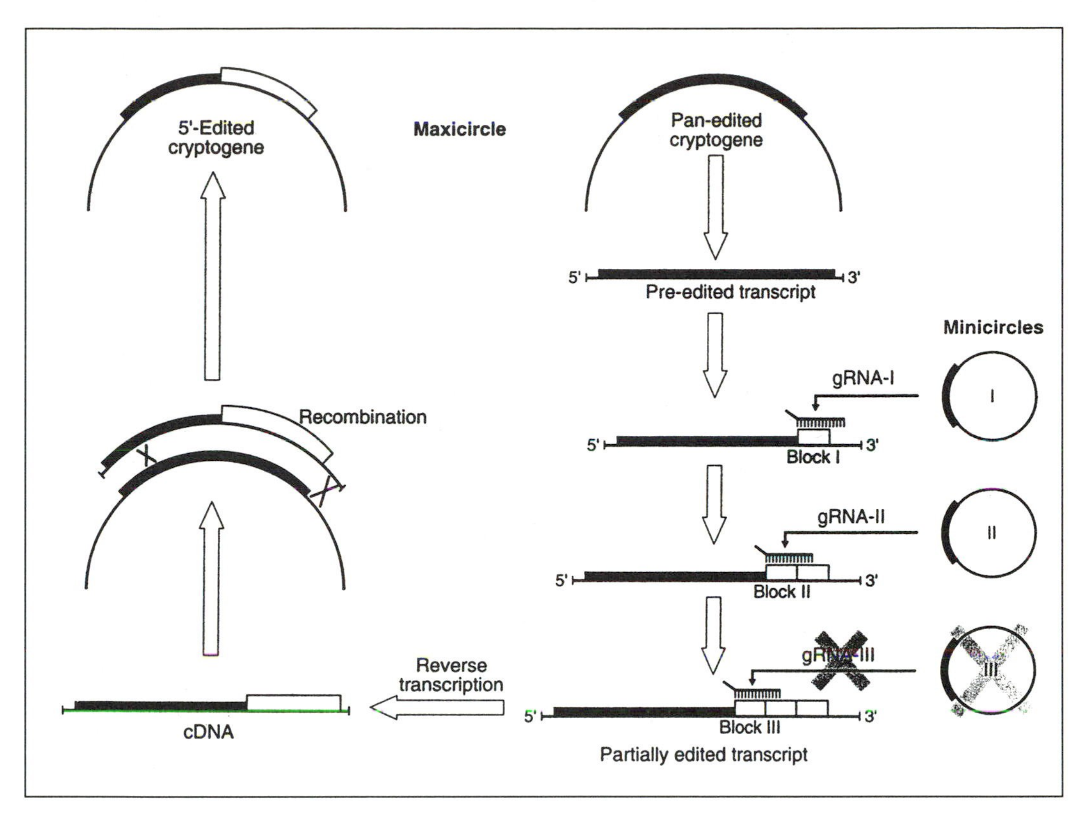

Fig. 15 Evolution of RNA editing in kinetoplastid protozoa. The primary transcript (thick black line) is edited by the first three overlapping gRNAs. Open boxes represent edited sequences. The cDNA for the partially edited transcript replaces the original cryptogene in one of the maxicircles by homologous crossing-over. If the minicircle class encoding one of the three gRNAs is lost, cells lacking the substituted cryptogene could not edit this transcript, and this might be lethal. Cells with a substituted cryptogene would have a selective advantage. (Reprinted from 109 with permission.)

genes were pan-edited GA-rich cryptogene skeletons, and that mature edited RNAs have replaced several of these, giving rise to the presently unedited genes such as ND1, ND4, ND5, and COI.

It is clear that complex gRNA-mediated RNA editing was present in an ancestor of the kinetoplastid flagellate lineage, but the exact time and the mode of origin of this phenomenon remain unclear. Analysis of additional lower eukaryotic cells, and the eubacterial cells that are the likely ancestors of the protomitochondrion endosymbiont, for the presence of the U-insertion/deletion editing should provide information on these issues.

Also, determination of the precise mechanism of this type of editing should provide some constraints for evolutionary speculation. For example, if the mechanism of RNA editing in kinetoplastid mitochondria turns out to involve transesterification, this would suggest that ribozyme-mediated RNA self-splicing or even mRNA splicing shared a common evolutionary history. In this view, introns and gRNAs may both represent relics of the same primitive genetic systems which dealt with the information management of RNA molecules. Modern RNA editing in kinetoplastids would represent a remnant of a primitive mechanism for creating functional mRNA sequences which has been maintained in the mitochondria of these cells. On the other hand, if the mechanism turns out to be cleavage–ligation, then an alternative hypothesis would become viable, in which editing evolved in the mitochondrion of these cells and represents an ancient but derived trait (110-112).

12. Note added on proof

A brief description of significant progress in the field of insertion-deletion RNA editing during the production of this book is provided here.

(i) Putative intermediates in a gRNA-mediated U-deletion/insertion *in vitro* system, using the A6 pre-edited mRNA and a mitochondrial extract from *T. brucei*, were detected by gel analysis (113–115). The intermediates represented the A6 mRNA cleavage and ligation products predicted by the enzyme cascade model for editing at sites 1 and 2; gRNA–mRNA chimeric molecules were also detected, but evidence was presented that these molecules represented byproducts of the editing reaction.

(ii) A stereochemical investigation of an *in vitro* U-insertion activity in *L. tarentolae* mitochondrial extracts also provided results compatible with an enzyme cascade model and incompatible with models involving the transfer of Us from the 3′ end of gRNAs to an editing site (116). Use of a primer extension assay with a pre-edited mRNA mutated in the anchor sequence showed that this activity was independent of both endogenous or exogenous gRNA; this activity was, however, dependent on the secondary structure of the pre-edited mRNA (117). A similar primer extension assay was used to show a U-insertion reaction in site 1 of the ND7 mRNA, which was mediated by the number of guiding nucleotides in the cognate gRNA (118). Blockage of the 3′ end of the gRNA had little effect on the U-insertion activity, a result also compatible with an enzyme cascade model in which the inserted Us are derived from UTP. These results led to a model in which Us are first added to the 3′ end of the 5′ cleavage fragment of the pre-edited mRNA by a TUTase-like activity, and the oligo(U) 3′ overhang not base paired to guiding nucleotides in the gRNA is then removed by a nuclease and the ends are then religated. An imprecise removal of the unpaired Us will lead to the observed gRNA-dependent unguided or misedited U-insertions at this site.

(iii) An analysis of mitochondrial lysates from *L. tarentolae* and *T. brucei* led to the identification of a 20S complex containing RNA ligase activity but lacking gRNA. gRNA was found to be localized within a 10S complex and also within a series of heterodisperse complexes that sedimented from 10S to over 30S. The suggestion was made that these two classes of RNP complexes interact in the editing reaction (119).

(iv) Construction of gRNA libraries from two strains of *L. tarentolae* allowed an analysis of the 3′ ends (120). This analysis showed a remarkable homogeneity of the

3′-uridylylation sites, which could be due either to transcription termination or 3′-end processing.

(v) Minicircle-encoded gRNAs were isolated from *Crithidia fasciculata* (121) and from two strains of *Trypanosoma cruzi* (122), and the genomic organization of the gRNA genes analyzed. In *C. fasciculata*, each 2.5 kb minicircle contains a single gRNA gene in one of the two variable regions situated at a constant distance from the DNA bend. Five gRNAs were identified and all belonged to minor minicircles representing less than 2% of the total minicircle DNA. More than 90% of the kinetoplast DNA was composed of a single minicircle sequence class encoding an unidentified gRNA. These results suggest that the copy number of specific minicircle sequence classes can vary dramatically without an overall effect on the RNA editing system. In *T. cruzi*, each 1.45 kb minicircle encodes four gRNAs situated within the four variable regions. Multiple examples of redundant gRNAs were identified, which encode the same editing information but have different sequences. Extensive sequence polymorphisms, mainly transitions, within the variable regions of homologous minicircles from different strains were shown to account for the known lack of cross hybridization of the kinetoplast DNA from the strains.

(vi) Five gRNAs were identified from *T. borreli* (123) and were shown to be encoded in a 180 kb circular molecule, rather than in minicircles as in the trypanosomatids; structural genes and cryptogenes are encoded in an 80 kb circular molecule, which is the homologue of the maxicircle. The gRNAs are shorter than those from trypanosomatids and possess nonencoded oligo(U) sequences at the 3′ and also at the 5′ ends. The origin and role of the 5′ oligo(U) sequence is unknown.

References

1. Benne, R., Van den Burg, J., Brakenhoff, J., Sloof, P., Van Boom, J., and Tromp, M. (1986) Major transcript of the frameshifted coxII gene from trypanosome mitochondria contains four nucleotides that are not encoded in the DNA. *Cell*, **46**, 819.
2. Shaw, J., Campbell, D., and Simpson, L. (1989) Internal frameshifts within the mitochondrial genes for cytochrome oxidase subunit II and maxicircle unidentified reading frame 3 in *Leishmania tarentolae* are corrected by RNA editing: Evidence for translation of the edited cytochrome oxidase subunit II mRNA. *Proc. Natl Acad. Sci. USA*, **86**, 6220.
3. Shaw, J., Feagin, J. E., Stuart, K., and Simpson, L. (1988) Editing of mitochondrial mRNAs by uridine addition and deletion generates conserved amino acid sequences and AUG initiation codons. *Cell*, **53**, 401.
4. Feagin, J., Jasmer, D., and Stuart, K. (1987) Developmentally regulated addition of nucleotides within apocytochrome b transcripts in *Trypanosoma brucei*. *Cell*, **49**, 337.
5. Feagin, J. E., Shaw, J. M., Simpson, L., and Stuart, K. (1988) Creation of AUG initiation codons by addition of uridines within cytochrome b transcripts of kinetoplastids. *Proc. Natl Acad. Sci. USA*, **85**, 539.
6. Bhat, G. J., Koslowsky, D. J., Feagin, J. E., Smiley, B. M., and Stuart, K. (1990) An extensively edited mitochondrial transcript in kinetoplastids encodes a protein homologous to ATPase subunit 6. *Cell*, **61**, 885.

7. Koslowsky, D. J., Bhat, G. J., Perrollaz, A. M., Feagin, J. E., and Stuart, K. (1990) The MURF3 gene of *T. brucei* contains multiple domains of extensive editing and is homologous to a subunit of NADH dehydrogenase. *Cell*, **62**, 901.
8. Feagin, J. E., Abraham, J., and Stuart, K. (1988) Extensive editing of the cytochrome c oxidase III transcript in *Trypanosoma brucei*. *Cell*, **53**, 413.
9. Van der Spek, H., Speijer, D., Arts, G. -J., Van den Burg, J., Van Steeg, H., Sloof, P., and Benne, R. (1990) RNA editing in transcripts of the mitochondrial genes of the insect trypanosome *Crithidia fasciculata*. *EMBO J.*, **9**, 257.
10. Van der Spek, H., Van den Burg, J., Croiset, A., Van den Broek, M., Sloof, P., and Benne, R. (1988) Transcripts from the frameshifted MURF3 gene from *Crithidia fasciculata* are edited by U insertion at multiple sites. *EMBO J.*, **7**, 2509.
11. Simpson, L. and Shaw, J. (1989) RNA editing and the mitochondrial cryptogenes of kinetoplastid protozoa. *Cell*, **57**, 355.
12. Mahendran, R., Spottswood, M. S., Ghate, A., Ling, M. -I., Jeng, K., and Miller, D. M. (1994) Editing of the mitochondrial small subunit rRNA in *Physarum polycephalum*. *EMBO J.*, **13**, 232.
13. Gott, J. M., Visomirski, L. M., and Hunter, J. M. (1993) Substitutional and insertional RNA editing of the cytochrome *c* oxidase subunit 1 mRNA of *Physarum polycephalum*. *J. Biol. Chem.*, **268**, 25483.
14. Lonergan, K. M. and Gray, M. W. (1993) Editing of transfer RNAs in *Acanthamoeba castellanii* mitochondria. *Science*, **259**, 812.
15. Higuchi, M., Single, F. N., Köhler, M., Sommer, B., Sprengel, R., and Seeburg, P. H. (1993) RNA editing of AMPA receptor subunit GluR-B: A base-paired intron-exon structure determines position and efficiency. *Cell*, **75**, 1361.
16. Bass, B. M. (1993) RNA editing: New uses for old players in the RNA world. In *The RNA world*. Gesteland, R. F., Atkins, J. F. (ed), Cold Spring Harbor Laboratory Press, New York, p. 383.
17. Sogin, M., Gunderson, J., Elwood, H., Alonso, R., and Peattie, D. (1989) Phylogenetic meaning of the kingdom concept: An unusual ribosomal RNA from *Giardia lamblia*. *Science*, **243**, 75.
18. Hajduk, S., Siqueira, A., and Vickerman, K. (1986) Kinetoplast DNA of *Bodo caudatus*: a noncatenated structure. *Mol. Cell. Biol.*, **6**, 4372.
19. Maslov, D. A. and Simpson, L. (1994) RNA editing and mitochondrial genomic organization in the cryptobiid kinetoplastid protozoan, *Trypanoplasma borreli*. *Mol. Cell. Biol.*, **14**, 8174.
20. Lukes, J., Arts, G. J., Van den Burg, J., De Haan, A., Opperdoes, F., Sloof, P., and Benne, R. (1994) Novel pattern of editing regions in mitochondrial transcripts of the cryptobiid *Trypanoplasma borreli*. *EMBO J.*, **13**, 5086.
21. Simpson, L. (1987) The mitochondrial genome of kinetoplastid protozoa: Genomic organization, transcription, replication and evolution. *Ann. Rev. Microbiol.*, **41**, 363.
22. Blum, B., Bakalara, N., and Simpson, L. (1990) A model for RNA editing in kinetoplastid mitochondria: "Guide" RNA molecules transcribed from maxicircle DNA provide the edited information. *Cell*, **60**, 189.
23. Sloof, P., Van den Burg, J., Voogd, A., Benne, R., Agostinelli, M., Borst, P., *et al.* (1985) Further characterization of the extremely small mitochondrial ribosomal RNAs from trypanosomes: a detailed comparison of the 9S and 12S RNAs from *Crithidia fasciculata* and *Trypanosoma brucei* with rRNAs from other organisms. *Nucl. Acids Res.*, **13**, 4171.
24. de la Cruz, V., Simpson, A., Lake, J., and Simpson, L. (1985) Primary sequence and

partial secondary structure of the 12S kinetoplast (mitochondrial) ribosomal RNA from *Leishmania tarentolae*: Conservation of peptidyl-transferase structural elements. *Nucl. Acids Res.*, **13**, 2337.

25. de la Cruz, V., Lake, J. A., Simpson, A. M., and Simpson, L. (1985) A minimal ribosomal RNA: Sequence and secondary structure of the 9S kinetoplast ribosomal RNA from Leishmania tarentolae. *Proc. Natl Acad. Sci. USA*, **82**, 1401.
26. Simpson, L. and Simpson, A. (1978) Kinetoplast RNA from *Leishmania tarentolae*. *Cell*, **14**, 169.
27. Muhich, M., Neckelmann, N., and Simpson, L. (1985) The divergent region of the Leishmania tarentolae kinetoplast maxicircle DNA contains a diverse set of repetitive sequences. *Nucl. Acids Res.*, **13**, 3241.
28. de la Cruz, V., Neckelmann, N., and Simpson, L. (1984) Sequences of six structural genes and several open reading frames in the kinetoplast maxicircle DNA of *Leishmania tarentolae*. *J. Biol. Chem.*, **259**, 15136.
29. Masuda, H., Simpson, L., Rosenblatt, H., and Simpson, A. (1979) Restriction map, partial cloning and localization of 9S and 12S RNA genes on the maxicircle component of the kinetoplast DNA of *Leishmania tarentolae*. *Gene*, **6**, 51.
30. Simpson, L. (1986) Kinetoplast DNA in trypanosomid flagellates. *Int. Rev. Cytol.*, **99**, 119.
31. Simpson, L., Neckelmann, N., de la Cruz, V., Simpson, A., Feagin, J., Jasmer, D., and Stuart, K. (1987) Comparison of the maxicircle (mitochondrial) genomes of *Leishmania tarentolae* and *Trypanosoma brucei* at the level of nucleotide sequence. *J. Biol. Chem., 262*, 6182.
32. Payne, M., Rothwell, V., Jasmer, D., Feagin, J., and Stuart, K. (1985) Identification of mitochondrial genes in *Trypanosoma brucei* and homology to cytochrome c oxidase II in two different reading frames. *Mol. Biochem. Parasitol.*, **15**, 159.
33. Benne, R., DeVries, B., Van den Burg, J., and Klaver, B. (1983) The nucleotide sequence of a segment of Trypanosoma brucei mitochondrial maxi-circle DNA that contains the gene for apocytochrome b and some unusual unassigned reading frames. *Nucl. Acids Res.*, **11**, 6925.
34. de Vries, B., Mulder, E., Brakenhoff, J., Sloof, P., and Benne, R. (1988) The variable region of the *Trypanosoma brucei* kinetoplast maxicircle: sequence and transcript analysis of a repetitive and a non-repetitive fragment. *Mol. Biochem. Parasitol.*, **27**, 71.
35. Sloof, P., Van den Burg, J., Voogd, A., and Benne, R. (1987) The nucleotide sequence of a 3.2 kb segment of mitochondrial maxicircle DNA from *Crithidia fasciculata* containing the gene for cytochrome oxidase subunit III, the N-terminal part of the apocytochrome b gene and a possible frameshift gene; further evidence for the use of unusual initiator triplets in trypanosome mitochondria. *Nucl. Acids Res.*, **15**, 51.
36. Feagin, J., Jasmer, D., Stuart, K. (1985) Apocytochrome b and other mitochondrial DNA sequences are differentially expressed during the life cycle of Trypanosoma brucei. *Nucl. Acids Res.*, **13**, 4577.
37. Souza, A. E., Myler, P. J., and Stuart, K. (1992) Maxicircle CR1 transcripts of Trypanosoma brucei are edited, developmentally regulated, and encode a putative iron-sulfur protein homologous to an NADH dehydrogenase subunit. *Mol. Cell Biol.*, **12**, 2100.
38. Read, L. K., Myler, P. J., and Stuart, K. (1992) Extensive editing of both processed and preprocessed maxicircle CR6 transcripts in Trypanosoma brucei. *J. Biol. Chem.*, **267**, 1123.

39. Maslov, D. A., Sturm, N. R., Niner, B. M., Gruszynski, E. S., Peris, M., and Simpson, L. (1992) An intergenic G-rich region in *Leishmania tarentolae* kinetoplast maxicircle DNA is a pan-edited cryptogene encoding ribosomal protein S12. *Mol. Cell Biol.*, **12**, 56.
40. Thiemann, O. H., Maslov, D. A., and Simpson, L. (1994) Disruption of RNA editing in *Leishmania tarentolae* by the loss of minicircle-encoded guide RNA genes. *EMBO J.*, **13**, 5689.
41. Bhat, G. J., Myler, P. J., and Stuart, K. (1991) The two ATPase 6 mRNAs of *Leishmania tarentolae* differ at their 3′ ends. *Mol. Biochem. Parasitol.*, **48**, 139.
42. Pearson, W. R. (1990) Rapid and sensitive sequence comparison with FASTP and FASTA. *Methods Enzymol.*, **183**, 63.
43. Kanehisa, M. I. (1982) Los Alamos sequence analysis package for nucleic acids and proteins. *Nucl. Acids Res.*, **10**, 183.
44. Gribskov, M., McLachan, A., and Eisenberg, D. (1987) Profile analysis: Detection of distantly related proteins. *Proc. Natl Acad. Sci. USA*, **84**, 4355.
45. Souza, A. E., Shu, H. -H., Read, L. K., Myler, P. J., and Stuart, K. D. (1993) Extensive editing of CR2 maxicircle transcripts of *Trypanosoma brucei* predicts a protein with homology to a subunit of NADH dehydrogenase. *Mol. Cell Biol.*, **13**, 6832.
46. Corell, R. A., Myler, P., and Stuart, K. (1994) *Trypanosoma brucei* mitochondrial CR4 gene encodes an extensively edited mRNA with completely edited sequence only in bloodstream forms. *Mol. Biochem. Parasitol.*, **64**, 65.
47. Sturm, N. R., Maslov, D. A., Blum, B., and Simpson, L. (1992) Generation of unexpected editing patterns in *Leishmania tarentolae* mitochondrial mRNAs: misediting produced by misguiding. *Cell*, **70**, 469.
48. Simpson, A., Neckelmann, N., de la Cruz, V., Muhich, M., and Simpson, L. (1985) Mapping and 5′ end determination of kinetoplast maxicircle gene transcripts from *Leishmania tarentolae*. *Nucl. Acids Res.*, **13**, 5977.
49. Feagin, J. and Stuart, K. (1985) Differential expression of mitochondrial genes between the life cycle stages of Trypanosoma brucei. *Proc. Natl Acad. Sci. USA*, **82**, 3380.
50. Feagin, J. and Stuart, K. (1988) Developmental aspects of uridine addition within mitochondrial transcripts of *Trypanosoma brucei*. *Mol. Cell Biol.*, **8**, 1259.
51. Feagin, J., Jasmer, D., and Stuart, K. (1986) Differential mitochondrial gene expression between slender and stumpy bloodforms of *Trypanosoma brucei*. *Mol. Biochem. Parasitol.*, **20**, 207.
52. Bhat, G. J., Souza, A. E., Feagin, J. E., and Stuart, K. (1992) Transcript-specific developmental regulation of polyadenylation in *Trypanosoma brucei* mitochondria. *Mol. Biochem. Parasitol.*, **52**, 231.
53. Campbell, D. A., Spithill, T. W., Samaras, N., Simpson, A., and Simpson, L. (1989) Sequence of a cDNA for the ND1 gene from *Leishmania major*: potential uridine addition in the polyadenosine tail. *Mol. Biochem. Parasitol.*, **36**, 197.
54. Bakalara, N., Simpson, A. M., and Simpson, L. (1989) The *Leishmania* kinetoplast-mitochondrion contains terminal uridylyltransferase and RNA ligase activities. *J. Biol. Chem.*, **264**, 18679.
55. Adler, B. K., Harris, M. E., Bertrand, K. I., and Hajduk, S. M. (1991) Modification of *Trypanosoma brucei* mitochondrial rRNA by posttranscriptional 3′ polyuridine tail formation. *Mol. Cell Biol.*, **11**, 5878.
56. Simpson, L. (1972) The kinetoplast of the hemoflagellates. *Int. Rev. Cytol.*, **32**, 139.
57. Hill, G. and Cross, G. (1973) Cyanide-resistant respiration and a branched cytochrome chain system in Kinetoplastidae. *Biochim. Biophys. Acta*, **305**, 590.

58. Koslowsky, D. J., Riley, G. R., Feagin, J. E., and Stuart, K. (1992) Guide RNAs for transcripts with developmentally regulated RNA editing are present in both life cycle stages of *Trypanosoma brucei*. *Mol. Cell Biol.*, **12**, 2043.
59. Stuart, K. (1991) RNA editing in trypanosomatid mitochondria. *Annu. Rev. Microbiol.*, **45**, 327.
60. Blum, B. and Simpson, L. (1990) Guide RNAs in kinetoplastid mitochondria have a nonencoded 3′ oligo-(U) tail involved in recognition of the pre-edited region. *Cell*, **62**, 391.
61. Simpson, A. M., Suyama, Y., Dewes, H., Campbell, D., and Simpson, L. (1989) Kinetoplastid mitochondria contain functional tRNAs which are encoded in nuclear DNA and also small minicircle and maxicircle transcripts of unknown function. *Nucl. Acids Res.*, **17**, 5427.
62. Van der Spek, H., Arts, G. -J., Zwaal, R. R., Van den Burg, J., Sloof, P., and Benne, R. (1991) Conserved genes encode guide RNAs in mitochondria of *Crithidia fasciculata*. *EMBO J.*, **10**, 1217.
63. Von Haeseler, A., Blum, B., Simpson, L., Sturm, N., and Waterman, M. S. (1992) Computer methods for locating kinetoplastid cryptogenes. *Nucl. Acids Res.*, **20**, 2717.
64. Sturm, N. R. and Simpson, L. (1990) Kinetoplast DNA minicircles encode guide RNAs for editing of cytochrome oxidase subunit III mRNA. *Cell*, **61**, 879.
65. Sturm, N. R. and Simpson, L. (1991) *Leishmania tarentolae* minicircles of different sequence classes encode single guide RNAs located in the variable region approximately 150 bp from the conserved region. *Nucl. Acids Res.*, **19**, 6277.
66. Pollard, V. W., Rohrer, S. P., Michelotti, E. F., Hancock, K., and Hajduk, S. M. (1990) Organization of minicircle genes for guide RNAs in Trypanosoma brucei. *Cell*, **63**, 783.
67. Riley, G. R., Corell, R. A., and Stuart, K. (1994) Multiple guide RNAs for identical editing of *Trypanosoma brucei* apocytochrome *b* mRNA have an unusual minicircle location and are developmentally regulated. *J. Biol. Chem.*, **269**, 6101.
68. Maslov, D. A., and Simpson, L. (1992) The polarity of editing within a multiple gRNA-mediated domain is due to formation of anchors for upstream gRNAs by downstream editing. *Cell*, **70**, 459.
69. Read, L. K., Corell, R. A., and Stuart, K. (1992) Chimeric and truncated RNAs in *Trypanosoma brucei* suggest transesterifications at non-consecutive sites during RNA editing. *Nucl. Acids Res.*, **20**, 2341.
70. Corell, R. A., Feagin, J. E., Riley, G. R., Strickland, T., Guderian, J. A., Myler, P. J., and Stuart, K. (1993) *Trypanosoma brucei* minicircles encode multiple guide RNAs which can direct editing of extensively overlapping sequences. *Nucl. Acids Res.*, **21**, 4313.
71. Pollard, V. W. and Hajduk, S. M. (1991) *Trypanosoma equiperdum* minicircles encode three distinct primary transcripts which exhibit guide RNA characteristics. *Mol. Cell Biol.*, **11**, 1668.
72. Gajendran, N., Vanhecke, D., Songa, E. B., and Hamers, R. (1992) Kinetoplast minicircle DNA of *Trypanosoma evansi* encode guide RNA genes. *Nucl. Acids Res.*, **20**, 614.
73. Shu, H. -H. and Stuart, K. (1994) Mitochondrial transcripts are processed but are not edited normally in *Trypanosoma equiperdum* (ATCC 30019) which has kDNA sequence deletion and duplication. *Nucl. Acids Res.*, **22**, 1696.
74. Arts, G. J., Van der Spek, H., Speijer, D., Van den Burg, J., Van Steeg, H., Sloof, P., and Benne, R. (1993) Implications of novel guide RNA features for the mechanism of RNA editing in *Crithidia fasciculata*. *EMBO J.*, **12**, 1523.
75. Sturm, N. R. and Simpson, L. (1990) Partially edited mRNAs for cytochrome b and

subunit III of cytochrome oxidase from Leishmania tarentolae mitochondria: RNA editing intermediates. *Cell*, **61**, 871.
76. Sugisaki, H. and Takanami, M. (1993) The 5′-terminal region of the apocytochrome *b* transcript in *Crithidia fasciculata* is successively edited by two guide RNAs in the 3′ to 5′ direction. *J. Biol. Chem.*, **268**, 887.
77. Missel, A. and Goringer, H. U. (1994) *Trypanosoma brucei* mitochondria contain RNA helicase activity. *Nucl. Acids Res.*, **22**, 4050.
78. Lee, S. -T., Tarn, C., and Chang, K. -P. (1993) Characterization of the switch of kinetoplast DNA minicircle dominance during development and reversion of drug resistance in *Leishmania*. *Mol. Biochem. Parasitol.*, **58**, 187.
79. Sloof, P., Arts, G. J., Van den Burg, J., Van den Spek, H., and Benne, R. (1994) RNA editing in mitochondria of cultured trypanosomatids: Translatable mRNAs for NADH-dehydrogenase subunits are missing. *J. Bioenerg. Biomembr.*, **26**, 193.
80. Simpson, A. M., Bakalara, N., and Simpson, L. (1992) A ribonuclease activity is activated by heparin or by digestion with proteinase K in mitochondrial extracts of *Leishmania tarentolae*. *J. Biol. Chem.*, **267**, 6782.
81. Pollard, V. W., Harris, M. E., and Hajduk, S. M. (1992) Native mRNA editing complexes from *Trypanosoma brucei* mitochondria. *EMBO J.*, **11**, 4429.
82. Blum, B., Sturm, N. R., Simpson, A. M., and Simpson, L. (1991) Chimeric gRNA-mRNA molecules with oligo(U) tails covalently linked at sites of RNA editing suggest that U addition occurs by transesterification. *Cell*, **65**, 543.
83. Cech, T. R. (1991) RNA editing: World's smallest introns. *Cell*, **64**, 667.
84. Moerl, M. and Schmelzer, C. (1990) Integration of group II intron bI1 into a foreign RNA by reversal of the self-splicing reaction *in vitro*. *Cell*, **60**, 629.
85. Davies, R., Waring, R., Ray, J., Brown, T., and Scazzocchio, C. (1982) Making ends meet: a model for RNA splicing in fungal mitochondria. *Nature*, **300**, 719.
86. Waring, R. B. and Davies, R. W. (1984) Assessment of a model for intron RNA secondary structure relevant to RNA self-splicing: a review. *Gene*, **28**, 277.
87. Been, M. D. and Cech, T. R. (1986) One binding site determines sequence specificity of *Tetrahymena* pre-rRNA self-splicing, *trans*-splicing, and RNA enzyme activity. *Cell*, **47**, 207.
88. Jacquier, A. and Michel, F. (1987) Multiple exon-binding sites in class II self-splicing introns. *Cell*, **50**, 17.
89. Cech, T. and Bass, B. (1986) Biological catalysis by RNA. *Ann. Rev. Biochem.*, **55**, 599.
90. Klug, J., Wolf, M., and Beato, M. (1991) Creating chimeric molecules by PCR directed homologous DNA recombination. *Nucl. Acids Res.*, **19**, 2793.
91. Abraham, J., Feagin, J., and Stuart, K. (1988) Characterization of cytochrome c oxidase III transcripts that are edited only in the 3′ region. *Cell*, **55**, 267.
92. Maslov, D. A., Thiemann, O., and Simpson, L. (1994) Editing and misediting of transcripts of the kinetoplast maxicircle G5 (ND3) cryptogene in an old laboratory strain of *Leishmania tarentolae*. *Mol. Biochem. Parasitol.*, **68**, 155.
93. Decker, C. J. and Sollner-Webb, B. (1990) RNA editing involves indiscriminate U changes throughout precisely defined editing domains. *Cell*, **61**, 1001.
94. Harris, M., Decker, C., Sollner-Webb, B., and Hajduk, S. (1992) Specific cleavage of pre-edited mRNAs in trypanosome mitochondrial extracts. *Mol. Cell Biol.*, **12**, 2591.
95. Koslowsky, D. J., Bhat, G. J., Read, L. K., and Stuart, K. (1991) Cycles of progressive realignment of gRNA with mRNA in RNA editing. *Cell*, **67**, 537.
96. Harris, M. E., Moore, D. R., and Hajduk, S. M. (1990) Addition of uridines to edited

RNAs in trypanosome mitochondria occurs independently of transcription. *J. Biol. Chem.*, **265**, 11368.
97. Harris, M. E. and Hajduk, S. M. (1992) Kinetoplastid RNA editing: *In vitro* formation of cytochrome b gRNA-mRNA chimeras from synthetic substrate RNAs. *Cell*, **68**, 1091.
98. Koslowsky, D. J., Goringer, H. U., Morales, T. H., and Stuart, K. (1992) *In vitro* guide RNA/mRNA chimaera formation in *Trypanosoma brucei* RNA editing. *Nature*, **356**, 807.
99. Blum, B. and Simpson, L. (1992) Formation of gRNA/mRNA chimeric molecules *in vitro*, the initial step of RNA editing, is dependent on an anchor sequence. *Proc. Natl Acad. Sci. USA*, **89**, 11944.
100. Goringer, H. U., Koslowsky, D. J., Morales, T. H., and Stuart, K. (1994) The formation of mitochondrial ribonucleoprotein complexes involving guide RNA molecules in *Trypanosoma brucei. Proc. Natl Acad. Sci. USA*, **91**, 1776.
101. Read, L. K., Goringer, H. U., and Stuart, K. (1994) Assembly of mitochondrial ribonucleoprotein complexes involves specific guide RNA (gRNA)-binding proteins and gRNA domains but does not require preedited mRNA. *Mol. Cell Biol.*, **14**, 2629.
102. Köller, J., Nörskau, G., Paul, A. S., Stuart, K., and Goringer, H. U. (1994) Different *Trypanosoma brucei* guide RNA molecules associate with an identical complement of mitochondrial proteins *in vitro. Nucl. Acids Res.*, **22**, 1988.
103. Peris, M., Frech, G. C., Simpson, A. M., Bringaud, F., Byrne, E., Bakker, A., and Simpson, L. (1994) Characterization of two classes of ribonucleoprotein complexes possibly involved in RNA editing from *Leishmania tarentolae* mitochondria. *EMBO J.*, **13**, 1664.
104. Frech, G. C., Bakalara, N., Simpson, L., and Simpson, A. M. (1995) *In vitro* RNA editing-like activity in a mitochondrial extract from Leishmania tarentolae. *EMBO J.*, **14**, 178.
105. Seiwert, S. D. and Stuart, K. (1994) RNA editing: transfer of genetic information from gRNA to precursor mRNA *in vitro. Science*, **266**, 114.
106. Landweber, L. F. and Gilbert, W. (1994) Phylogenetic analysis of RNA editing: A primitive genetic phenomenon. *Proc. Natl Acad. Sci. USA*, **91**, 918.
107. Fernandes, A. P., Nelson, K., Beverley, S. M. (1993) Evolution of nuclear ribosomal RNAs in kinetoplastid protozoa: Perspectives on the age and origins of parasitism. *Proc. Natl Acad. Sci. USA*, **90**, 11608.
108. Maslov, D. A., Avila, H. A., Lake, J. A., and Simpson, L. (1994) Evolution of RNA editing in kinetoplastid protozoa. *Nature*, **365**, 345.
09. Simpson, L. and Maslov, D. A. (1994) RNA editing and the evolution of parasites. *Science*, **264**, 1870.
110. Covello, P. S. and Gray, M. W. (1993) On the evolution of RNA editing. *Trends Genet.*, **9**, 265.
111. Simpson, L., Maslov, D. A., and Blum, B. (1993) RNA Editing in *Leishmania* mitochondria. In *RNA Editing - the alteration of protein coding sequences of RNA*. Benne, R. (ed), Ellis Horwood , New York, p. 53.
112. Simpson, L. and Maslov, D. A. (1994) Ancient origin of RNA editing in kinetoplastid protozoa. *Curr. Opin. Genet. Dev.*, **4**, 887.
113. Seiwert, S. D., Heidmann, S. and Stuart, K. (1996) Direct visualization of uridylate deletion *in vitro* suggests a mechanism for kinetoplastid RNA editing. *Cell*, **84**, 1.
114. Kable, M. L., Seiwert, S. D., Heidmann, S. and Stuart, K. (1996) RNA editing: a mechanism for gRNA-specified uridylate insertion into precursor mRNA. *Science*, **273**, 1189.
115. Cruz-Reyes, J. and Sollner-Webb, B. (1996) Trypanosome U-deletional RNA editing involves guide RNA-directed endonuclease cleavage, terminal U exonuclease, and RNA ligase activities. *Proc. Natl Acad. Sci. USA*, **93**, 8901.

116. Frech, G. C. and Simpson, L. (1996) Uridine insertion into preedited mRNA by a mitochondrial extract from *Leishmania tarentolae*: Stereochemical evidence for the enzyme cascade model. *Mol. Cell. Biol.*, **16**, 4584.
117. Connell, G. J., Byrne, E. M. and Simpson, L. (1997) Guide RNA-independent and guide RNA-dependent uridine insertion into cytochrome b mRNA in a mitochondrial extract from *Leishmania tarentolae*. *J. Biol. Chem.*, **272**, 4212.
118. Byrne, E. M., Connell, G. J., and Simpson, L. (1996) Guide RNA-directed uridine insertion RNA editing *in vitro*. *EMBO J.*, **15**, 6758.
119. Peris, M., Simpson, A. M., Grunstein, J., Lilienthal, J. E., Frech, G. C., and Simpson, L. (1997) Native gel analysis of ribonucleoprotein complexes from a *Leishmania tarentolae* mitochondrial extract. *Mol. Biochem. Parasitol.*, in press.
120. Thiemann, O. H. and Simpson, L. (1996) Analysis of the 3′ uridylylation sites of guide RNAs from *Leishmania tarentolae*. *Mol. Biochem. Parasitol.*, **79**, 229.
121. Yasuhira, S. and Simpson, L. (1996) Minicircle-encoded guide RNAs from *Crithidia fasciculata*. *RNA*, **1**, 634.
122. Avila, H. A. and Simpson, L. (1996). Organization and complexity of minicircle-encoded guide RNAs in *Trypanosoma cruzi*. *RNA*, **1**, 939.
123. Yasuhira, S. and Simpson, L. (1996). Guide RNAs and guide RNA genes in the cryptobiid kinetoplastid protozoa, *Trypanoplasma borreli*. *RNA*, **2**, 1153.

Index